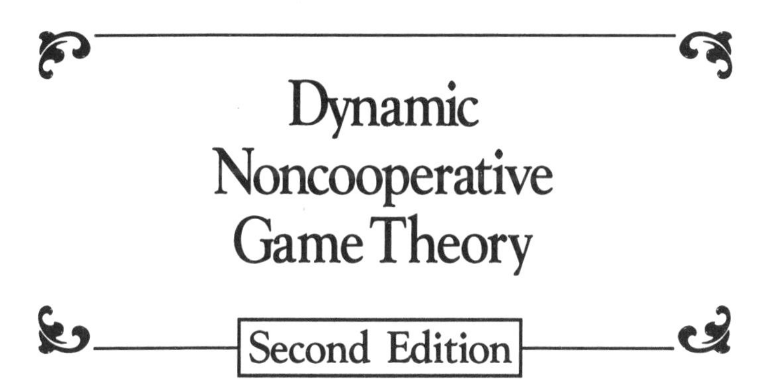

Dynamic Noncooperative Game Theory

Second Edition

SIAM's Classics in Applied Mathematics series consists of books that were previously allowed to go out of print. These books are republished by SIAM as a professional service because they continue to be important resources for mathematical scientists.

Classics in Applied Mathematics

C. C. Lin and L. A. Segel, *Mathematics Applied to Deterministic Problems in the Natural Sciences*

Johan G. F. Belinfante and Bernard Kolman, *A Survey of Lie Groups and Lie Algebras with Applications and Computational Methods*

James M. Ortega, *Numerical Analysis: A Second Course*

Anthony V. Fiacco and Garth P. McCormick, *Nonlinear Programming: Sequential Unconstrained Minimization Techniques*

F. H. Clarke, *Optimization and Nonsmooth Analysis*

George F. Carrier and Carl E. Pearson, *Ordinary Differential Equations*

Leo Breiman, *Probability*

R. Bellman and G. M. Wing, *An Introduction to Invariant Imbedding*

Abraham Berman and Robert J. Plemmons, *Nonnegative Matrices in the Mathematical Sciences*

Olvi L. Mangasarian, *Nonlinear Programming*

*Carl Friedrich Gauss, *Theory of the Combination of Observations Least Subject to Errors: Part One, Part Two, Supplement.* Translated by G. W. Stewart

Richard Bellman, *Introduction to Matrix Analysis*

U. M. Ascher, R. M. M. Mattheij, and R. D. Russell, *Numerical Solution of Boundary Value Problems for Ordinary Differential Equations*

K. E. Brenan, S. L. Campbell, and L. R. Petzold, *Numerical Solution of Initial-Value Problems in Differential-Algebraic Equations*

Charles L. Lawson and Richard J. Hanson, *Solving Least Squares Problems*

J. E. Dennis, Jr. and Robert B. Schnabel, *Numerical Methods for Unconstrained Optimization and Nonlinear Equations*

Richard E. Barlow and Frank Proschan, *Mathematical Theory of Reliability*

Cornelius Lanczos, *Linear Differential Operators*

Richard Bellman, *Introduction to Matrix Analysis, Second Edition*

Beresford N. Parlett, *The Symmetric Eigenvalue Problem*

Richard Haberman, *Mathematical Models: Mechanical Vibrations, Population Dynamics, and Traffic Flow*

Peter W. M. John, *Statistical Design and Analysis of Experiments*

Tamer Başar and Geert Jan Olsder, *Dynamic Noncooperative Game Theory, Second Edition*

*First time in print.

Dynamic Noncooperative Game Theory

Second Edition

Tamer Başar

University of Illinois at Urbana-Champaign
Urbana, Illinois

Geert Jan Olsder

Delft University of Technology
Delft, The Netherlands

Society for Industrial and Applied Mathematics
Philadelphia

This SIAM edition is an unabridged, revised republication of the work first published by Academic Press, New York, in 1982 and revised in 1995.

10 9 8 7 6 5 4 3 2 1

Library of Congress Cataloging-in-Publication Data

Başar, Tamer.
Dynamic noncooperative game theory / Tamer Başar, Geert Jan Olsder. -- 2nd ed.
p. cm. -- (Classics in applied mathematics ; 23)
Revised from the 2nd ed. published in 1995 by Academic Press, New York.
Includes bibliographical references and index.
ISBN 0-89871-429-X (softcover)
1. Differential games. 2. Noncooperative games (Mathematics)
I. Olsder, Geert Jan. II. Title. III. Series.
QA272.B37 1998
519.3--dc21 98-46719

To

Tangül, *Gözen*, and *Elif* (T.B.)

and

Elke, *Rena*, *Theda*, and *Kike* (G.J.O.)

Contents

Part I

Part II

Preface to the Classics Edition

This is the revised second edition of our 1982 book with the same title, which presents a rather comprehensive treatment of static and dynamic noncooperative game theory, with emphasis placed (as in the first and second editions) on the interplay between dynamic information patterns and the structural properties of several different types of equilibria. Whereas the second edition (1995) was a major revision with respect to the original edition, this *Classics* edition only contains some moderate changes with respect to the second one. There has been a number of reasons for the preparation of this edition:

- The second edition was sold out surprisingly fast.
- After some fifty years from its creation, the field of game theory is still very alive and active, as also reinforced by the selection of three game theorists (John Harsanyi, John Nash and Reinhard Selten) to share the 1994 Nobel prize in economics. Quite a few books on game theory have been published during the last ten years or so (though most of them essentially deal with static games only and are at the undergraduate level).
- The recent interest in such fields as biological games, mathematical finance and robust control gives a new impetus to noncooperative game theory.
- The topic of dynamic games has found its way into the curricula of many universities, sometimes as a natural supplement to a graduate level course on optimal control theory, which is actively taught in many engineering, applied mathematics and economics graduate programs.
- At the level of coverage of this book, dynamic game theory is well established by now and has reached a level of maturity, which makes the book a timely addition to SIAM's prestigious *Classics in Applied Mathematics* series.

For a brief description of and the level of the contents of the book, the reader is referred to the Preface of the second edition, which follows. It suffices to mention here the major changes made in this *Classics* edition with respect to the second edition. They are:

- Inclusion of the "Braess paradox" in Chapter 4.
- Inclusion of new material on the relationship between the existence of solutions to Riccati equations on the one hand and the existence of Nash equilibrium solutions to linear-quadratic differential games on the other, in Chapter 6. In the same chapter some new results have been included also on infinite-horizon differential games.

We hope that this revised edition may find its way as its predecessors did.

Tamer Başar
Urbana, June 1998

Geert Jan Olsder
Delft, June 1998

Preface to the Second Edition

This is the second edition of our 1982 book with the same title, which presents an extensive and updated treatment of static and dynamic noncooperative game theory, with emphasis placed again (as in the first edition) on the interplay between dynamic information patterns and the structural properties of several different types of equilibria. There were essentially two reasons for producing this revised edition. One was the favorable reception of the first edition and the other one was the need to include new theoretical developments. Yet another reason was that the topic of dynamic games has found its way into the curricula of many universities. This new edition contains some substantial changes and additions (and also a few deletions), but the flavor and the theme of the original text remain intact.

The first part of the book, part I, which comprises Chapters 2 to 4, covers the material that is generally taught in an advanced undergraduate or first-year graduate course on noncooperative game theory. The coverage includes static finite and infinite games of both the zero-sum and nonzero-sum type and in the latter case both Nash and Stackelberg solution concepts are discussed. Furthermore, this part includes an extensive treatment of the class of dynamic games in which the strategy spaces are finite—the so-called multi-act games. Through an extensive tree formulation, the impact of information patterns on the existence, uniqueness and the nature of different types of noncooperative equilibria of multi-act games is thoroughly investigated. Most of the important concepts of static and dynamic game theory are introduced in these three chapters, and they are supplemented by several illustrative examples. Exposition of the material is quite novel, emphasizing concepts and techniques of multi-person decision making rather than mathematical details. However, mathematical proofs of most of the results are also provided, but without hindering the flow of main ideas. The major changes in this part over the first edition are the inclusion of additional material on: randomized strategies, finite games with repeated decisions and action-dependent information sets (Chapter 2); various refinements on the Nash equilibrium concept, such as trembling-hand, proper and perfect equilibria (Chapter 3); and in the context of static infinite games, stability of Nash equilibria and its relation with numerical schemes, consistent conjectural variations equilibrium, and some new theorems on existence of Nash equilibria (Chapter 4). Some specific types of zero-sum games on the square (Chapter 4) have been left out.

The second part of the book, Part II, which includes Chapters 5 to 8, extends the theory of the first part to infinite dynamic games in both discrete and continuous time (the so-known differential games). Here the emphasis is again on the close interrelation between information patterns and noncooperative equilibria of such multi-person dynamic decision problems. We present a unified treatment of the existing, but scattered, results in the literature on infinite dynamic games as well as some new results on the topic. The treatment is confined to deterministic games and to stochastic games under perfect state information, mainly because inclusion of a complete investigation on stochastic dynamic games under imperfect state information would require presentation of some new techniques and thereby a volume much larger than the present one. Again, some of the major changes in this part over the first edition are the inclusion of new material on: time consistency (Chapters 5-7); viscosity solutions of the Hamilton-Jacobi-Bellman-Isaacs equation (Chapters 5 and 8); affine-quadratic dynamic games and results on infinite-horizon games in discrete and continuous time (Chapters 5 and 6); applications in robust (H^{∞}) controller designs (Chapter 8); incentive theory and relationship with Stackelberg solutions (Chapter 7); and Stackelberg equilibrium in the continuous time (Chapter 7). The material on the dolichobrachistochrone which was in Chapter 8 of the first edition, has been left out. Furthermore, the three appendices (A-C) are expanded versions of the earlier ones, and present the necessary background material on vector spaces, matrix algebra, optimization, probability and stochastic processes and fixed point theorems.

Each chapter (with the exception of the first) is supplemented with a problem section. Each problem section contains standard exercises on the contents of the chapter, which a reader who has carefully followed the text should have no difficulty in solving, as well as exercises which can be solved by making use of the techniques developed in the text, but which require some elaborate thinking on the part of the reader. Following the problem section in each chapter (except the first) is a notes section, which is devoted to historical remarks and sources for further reading on the topics covered in that particular chapter.

A one-semester course on noncooperative game theory, taught using this book, would involve mainly Part I and also an appropriate blend of some of the topics covered in Part II, this latter choice depending to a great extent on the taste of the instructor and the background of the students. Such a course would be suitable for advanced undergraduate or first-year graduate students in engineering, economics, mathematics, operations research and business administration. In order to follow the main flow of Chapters 2 and 3 the student need not have a strong mathematical background—apart from some elementary analysis—but he should be able to think in mathematical terms. However, proofs of some of the theorems in these two chapters, as well as the contents of Chapter 4, require some basic knowledge of real analysis and probability, which is summarized in the three appendices that are included towards the end of the book.

Part II of the book, on the other hand, is intended more for the researcher in the field, to provide him with the state of art in infinite dynamic game theory. However, selected topics from this part could also be used as a part of a graduate course on dynamic optimization, optimal control theory or mathematical economics.

This edition has been prepared by means of LATEX. Toward that end the text of the original edition was scanned; the formulas and figures were added separately. Despite the flexibility of LATEX and an intensive use of electronic mail, the preparation of this new edition has taken us much longer than anticipated, and we are grateful to the publisher for being flexible with respect to deadlines. The preparation would have taken even longer if our secretaries, Francie Bridges and Tatiana Tijanova, respectively, with their expertise of LATEX, had not been around. In the first stage Francie was responsible for the scanning and the retyping of the formulas, and Tatiana for the labeling and the figures, but in later stages they helped wherever necessary, and their assistance in this matter is gratefully acknowledged. The first author would also like to acknowledge his association with the Center for Advanced Study at the University of Illinois during the Fall 1993 semester, which provided him with released time (from teaching) which he spent partly in the writing of the new material in this second edition. The second author would like to thank INRIA Sophia-Antipolis in the person of its director, Pierre Bernhard, for allowing him to work on this revision while he spent a sabbatical there. We both would also like to take this opportunity to thank many of our colleagues and students for their input and suggestions for this second edition. Particular recognition goes to Niek Tholen, of Delft University of Technology, who read through most of the manuscript, caught many misprints and made penetrating comments and suggestions.

We hope that this revised edition may find its way as its predecessor did.

Tamer Başar
Urbana, March 1994

Geert Jan Olsder
Delft, March 1994

Chapter 1

Introduction and Motivation

1.1 Preliminary Remarks

This book is concerned with dynamic noncooperative game theory. In a nutshell, game theory involves multi-person decision making; it is *dynamic* if the order in which the decisions are made is important, and it is *noncooperative* if each person involved pursues his or her[1] own interests which are partly conflicting with others'.

A considerable part of everything that has been written down, whether it is history, literature or a novel, has as its central theme a conflict situation—a collision of interests. Even though the notion of "conflict" is as old as mankind, the scientific approach has started relatively recently, in the years around 1930, with, as a result, a still growing stream of scientific publications. We also see that more and more scientific disciplines devote time and attention to the analysis of conflicting situations. These disciplines include (applied) mathematics, economics, engineering, aeronautics, sociology, politics and mathematical finance.

It is relatively easy to delineate the main ingredients of a conflict situation: an individual has to make a decision and each possible decision leads to a different outcome or result, which are valued differently by that individual. This individual may not be the only one who decides about a particular outcome; a series of decisions by several individuals may be necessary. If all these individuals value the possible outcomes differently, the germs for a conflict situation are there.

The individuals involved, also called *players* or *decision makers*, or simply *persons*, do not always have complete control over the outcome. Sometimes there are uncertainties which influence the outcome in an unpredictable way. Under

[1]Without any preference to sexes, a decision maker, in this book, is most times referred to as a "he". It could equally well be a "she".

Table 1.1: The place of dynamic game theory.

	One player	Many players
Static	Mathematical programming	(Static) game theory
Dynamic	Optimal control theory	Dynamic (and/or differential) game theory

such circumstances, the outcome is (partly) based on data not yet known and not determined by the other players' decisions. Sometimes it is said that such data is under the control of "nature", or "God", and that every outcome is caused by the joint or individual actions of human beings and nature.

The established names of "game theory" (developed from approximately 1930) and "theory of differential games" (developed from approximately 1950, parallel to that of optimal control theory) are somewhat unfortunate. "Game theory", especially, appears to be directly related to parlor games; of course it is, but the notion that it is only related to such games is far too restrictive. The term "differential game" became a generally accepted name for games where differential equations play an important role. Nowadays the term "differential game" is also being used for other classes of games for which the more general term "dynamic game" would be more appropriate.

The applications of "game theory" and the "theory of differential games" mainly deal with economic and political conflict situations, worst-case designs and also modeling of war games. However, it is not only the applications in these fields that are important; equally important is the development of suitable concepts to describe and understand conflict situations. It turns out, for instance, that the role of information—what one player knows relative to others—is very crucial in such problems.

Scientifically, dynamic game theory can be viewed as a child of the parents game theory and optimal control theory.[2] Its character, however, is much more versatile than that of its parents, since it involves a dynamic decision process evolving in (discrete or continuous) time, with more than one decision maker, each with his own cost function and possibly having access to different information. This view is the starting point behind the formulation of "games in extensive form", which started in the 1930s through the pioneering work of Von Neumann, which culminated in his book with Morgenstern (Von Neumann and Morgenstern, 1947), and then made mathematically precise by Kuhn

[2]In almost all analogies there is a deficiency; a deficiency in the present analogy is that the child is as old as one of his parents—optimal control theory. For the relationship between these theories and the theory of mathematical programming, see Table 1.1.

(1953), all within the framework of "finite" games. The general idea in this formulation is that a game evolves according to a road or tree structure, where at every crossing or branching a decision has to be made as how to proceed.

In spite of this original set-up, the evolution of game theory has followed a rather different path. Most research in this field has been, and is being, concentrated on the normal or strategic form of a game. In this form all possible sequences of decisions of each player are set out against each other. For a two-player game this results in a matrix structure. In such a formulation dynamic aspects of a game are completely suppressed, and this is the reason why game theory is classified as basically "static" in Table 1.1. In this framework emphasis has been more on (mathematical) existence questions, rather than on the development of algorithms to obtain solutions.

Independently, control theory gradually evolved from Second World War servomechanisms, where questions of solution techniques and stability were studied. Then followed Bellman's "dynamic programming" (Bellman, 1957) and Pontryagin's "maximum principle" (Pontryagin et al., 1962), which spurred the interest in a new field called optimal control theory. Here the concern has been on obtaining optimal (i.e., minimizing or maximizing) solutions and developing numerical algorithms for one-person single-objective dynamic decision problems. The merging of the two fields, game theory and optimal control theory, which leads to even more concepts and to actual computation schemes, has achieved a level of maturity, which the reader will hopefully agree with after he/she goes through this book.

At this point, at the very beginning of the book, where many concepts have yet to be introduced, it is rather difficult to describe how dynamic game theory evolved in time and what the contributions of relevant references are. We therefore defer such a description until later, to the "notes" section of each chapter (except the present), where relevant historical remarks are included.

1.2 Preview on Noncooperative Games

A clear distinction exists between two-player (or, equivalently, two-person) zero-sum games and the others. In a zero-sum game, as the name implies, the sum of the cost functions of the players is identically zero. Mathematically speaking, if u^i and L^i denote, respectively, the decision variable and the cost function of the ith player (to be written $\mathbf{P}i$), then $\sum_{i=1}^{2} L^i(u^1, u^2) \equiv 0$ in a zero-sum game. If this sum is, instead, equal to a nonzero constant (independent of the decision variables), then we talk about a "constant-sum" game which can, however, easily be transformed to a zero-sum game through a simple translation without altering the essential features of the game. Therefore, constant-sum games can be treated within the framework of zero-sum games, without any loss of generality, which we shall choose to do in this book.

A salient feature of two-person zero-sum games that distinguishes them from other types of games is that they do not allow for any cooperation between the players, since, in a two-person zero-sum game, what one player gains incurs a

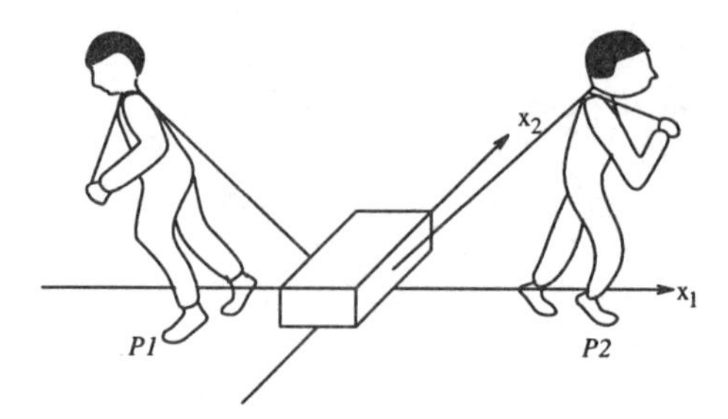

Figure 1.1: The rope-pulling game.

loss to the other player. However, in other games, such as two-player nonzero-sum games (wherein the quantity $\sum_{i=1}^{2} L^i(u^1,u^2)$ is not a constant) or three- or more-player games, the cooperation between two or more players may lead to their mutual advantage.

Example 1.1 A point object (with mass one) can move in a plane which is endowed with the standard (x_1, x_2)-coordinate system. Initially, at $t = 0$, the point mass is at rest at the origin. Two unit forces act on the point mass; one is chosen by **P1**, the other by **P2**. The directions of these forces, measured counter-clockwise with respect to the positive x_1-axis, are determined by the players and are denoted by u^1 and u^2, respectively; they may in general be time-varying. At time $t = 1$, **P1** wants to have the point mass as far in the negative x_1-direction as possible, i.e., he wants to minimize $x_1(1)$, whereas **P2** wants it as far in the positive x_1-direction as possible, i.e., he wants to maximize $x_1(1)$, or equivalently, to minimize $-x_1(1)$. (See Fig. 1.1.) The "solution" to this zero-sum game follows immediately; each player pulls in his own favorite direction, and the point mass remains at the origin—such a solution is known as the *saddle-point* solution.

We now alter the formulation of the game slightly, so that, in the present set-up, **P2** wishes to move the point mass as far in the negative x_2-direction as possible, i.e., he wants to minimize $x_2(1)$. **P1**'s goal is still to minimize $x_1(1)$. This new game is clearly nonzero-sum. The equations of motion for the point mass are

$$\begin{aligned} \ddot{x}_1 &= \cos(u^1) + \cos(u^2), & \dot{x}_1(0) &= x_1(0) = 0; \\ \ddot{x}_2 &= \sin(u^1) + \sin(u^2), & \dot{x}_2(0) &= x_2(0) = 0. \end{aligned}$$

Let us now consider the pair of decisions $\{u^1 \equiv \pi, u^2 \equiv -\pi/2\}$ with the corresponding values of the cost functions being $L^1 = x_1(1) = -\frac{1}{2}$ and $L^2 = x_2(1) = -\frac{1}{2}$. If **P2** sticks to $u^2 \equiv -\pi/2$, the best thing for **P1** to do is to choose $u^1 \equiv \pi$; any other choice of $u^1(t)$ will yield an outcome which is greater than $-\frac{1}{2}$. Analogously, if **P1** sticks to $u^1 \equiv \pi$, **P2** does not have a better choice than $u^2 \equiv -\pi/2$. Hence, the pair $\{u^1 \equiv \pi, u^2 \equiv -\pi/2\}$ exhibits an equilibrium behavior, and this kind of a solution, where one player cannot improve his outcome by altering his decision unilaterally, is called a *Nash equilibrium solution*, or shortly, a *Nash solution.*

If both players choose $u^1 \equiv u^2 \equiv 5\pi/4$, however, then the cost values become $L^1 = L^2 = -\frac{1}{2}\sqrt{2}$, which are obviously better, for both players, than the costs incurred under the Nash solution. But, in this case, the players have to cooperate. If, for instance, **P2** would stick to his choice $u^2 \equiv 5\pi/4$, then **P1** can improve upon his outcome by playing $u^1 \equiv c$, where c is a constant with $\pi \leq c < 5\pi/4$. **P1** is better off, but **P2** is worse off! Therefore, the pair of strategies $\{u^1 \equiv 5\pi/4, u^2 \equiv 5\pi/4\}$ cannot be in equilibrium in a noncooperative mode of decision making, since it requires some kind of faith (or even negotiation), and thereby cooperation, on part of the players. If this is allowable, then the said pair of strategies—known as a *Pareto-optimal solution*—stands out as a reasonable equilibrium solution for the game problem (which is called a *cooperative game*), since it features the property that no other joint decision of the players can improve the performance of at least one of them, without degrading the performance of the other. □

In this book we shall deal only with noncooperative games. The reasons for such a seeming limitation are twofold. Firstly, cooperative games[3] can, in general, be reduced to optimal control problems by determining a single cost function to be optimized by all players, which suppresses the "game" aspects of the problem. Secondly, the size of this book would have increased considerably by inclusion of a complete discussion on cooperative games.

Actions and strategies

Heretofore, we have safely talked about "decisions" made by the players, without being very explicit about what a decision really is. This will be made more precise now in terms of information available to each player. In particular, we shall distinguish between *actions* (also called controls) on the one hand and *strategies* (or, equivalently, decision rules) on the other.

If an individual has to decide about what to do the next day, and the options are fishing and going to work, then a strategy is: "if the weather report early tomorrow morning predicts dry weather, then I will go fishing, otherwise I will go to my office". This is a *strategy* or *decision rule*: what actually will be done depends on quantities not yet known and not controlled by the decision maker; the decision maker cannot influence the course of the events further, once he has fixed his strategy. Any consequence of such a strategy, after the unknown quantities are realized, is called an *action*. In a sense, a constant strategy (such as an irrevocable decision to go fishing without any restrictions or reservations) coincides with the notion of action.

In the example above, the alternative actions are to go fishing and to go to work, and the action to be implemented depends on information (the weather report) which has to be known at the time it is carried out. In general, such information can be of different types. It can, for instance, comprise the previous

[3]What we have in mind here is the class of cooperative games without side payments. If side payments are allowed, we enter the territory of (cooperative) games in characteristic function form, which is an altogether different topic. (See Owen (1968, 1982) and Vorob'ev (1977).)

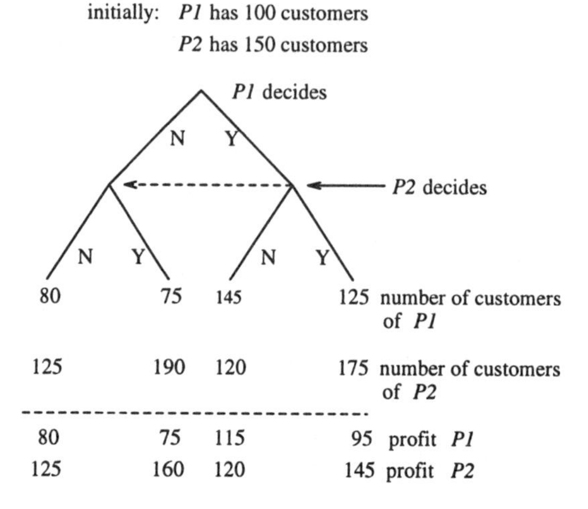

Figure 1.2: One-period advertising game.

actions of all the other players. As an example, consider the following sequence of actions: if he is nice to me, I will be kind to him; if he is cool, I will be cool, etc. The information can also be of a stochastic nature, such as in the fishing example. Then, the actual decision (action) is based on data not yet known and not controlled by other players, but instead determined by "nature". The next example will now elucidate the role of information in a game situation and show how it affects the solution. The information here is deterministic; "nature" does not play a role.

Example 1.2 This example aims at a game-theoretic analysis of the advertising campaigns of two competing firms during two periods of time. Initially, each firm is allotted a certain number of customers; **P1** has 100 and **P2** has 150 customers. Per customer, each firm makes a fixed profit per period, say one dollar. Through advertisement, a firm can increase the number of its customers (some are stolen away from the competitor and others come from outside) and thereby its profit. However, advertising costs money; an advertising campaign for one period costs thirty dollars (for each firm). The figures in this example may not be very realistic; it is, however, only the ratio of the data that matters, not the scale with respect to which it is expressed.

First, we consider the one-period version of this game and assume that the game ends after the first period is over. Suppose that **P1** has to decide first whether he should advertise (Yes) or not (No), and subsequently **P2** makes his choice. The four possible outcomes and the paths that lead to those outcomes are depicted in Fig. 1.2, in the form of a tree diagram. At every branching, a decision has to be made as how to proceed. The objective of each firm is to maximize its profit (for this one-period game). The "best" decisions, in this case, can be found almost by inspection; whatever **P1** does (Y or N), **P2** will always advertise (Y), since that is more profitable to him. **P1**, realizing this, has

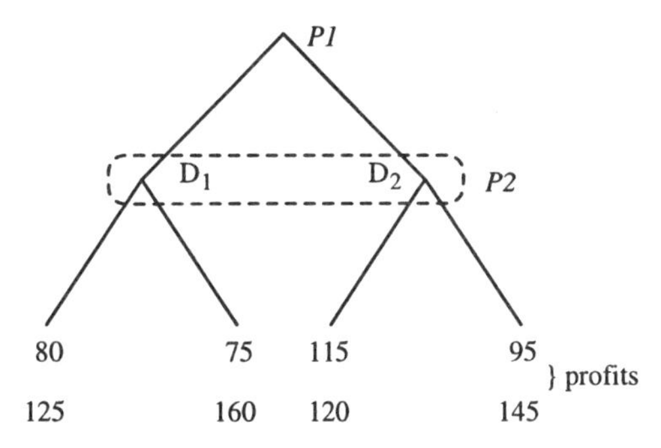

Figure 1.3: **P2** chooses independently of **P1**'s action.

to choose between Y (with profit 95) and N (with profit 75), and will therefore choose Y. Note that, at the point when **P2** makes his decision, he knows **P1**'s choice (action) and therefore **P2**'s choice depends, in principle, on what **P1** has done. **P2**'s best strategy can be described as: "if **P1** chooses N, then I choose Y; if **P1** chooses Y, then I choose Y", which, in this case is a *constant strategy.*

This one-period game will now be modified slightly so that **P1** and **P2** make their decisions simultaneously, or, equivalently, independently of each other. The information structure associated with this new game is depicted in Fig. 1.3; a dashed curve encircling the points D_1 and D_2 has been drawn, which indicates that **P2** cannot distinguish between these two points. In other words, **P2** has to arrive at a decision without knowing what **P1** has actually done. Hence, in this game, the strategy of **P2** has a different domain of definition, and it can easily be verified that the pair $\{Y, Y\}$ provides a Nash solution, in the sense described before in Example 1.1, yielding the same profits as in the previous game.

We now extend the game to two periods, with the objective of the firms being maximization of their respective cumulative profits (over both periods). The complete game is depicted in Fig. 1.4 without the information sets; we shall, in fact, consider three different information structures in the sequel. First, the order of the actions will be taken as **P1** – **P2** – **P1** – **P2**, which means that at the time of his decision each player knows the actions previously taken. Under the second information structure to be considered, the order of the actions is (**P1**, **P2**)-(**P1**, **P2**), which means that during each period the decisions are made independently of each other, but that for the decisions of the second period the (joint) actions of the first period are known. Finally, as a third case, it will be assumed that there is no order in the actions at all; **P1** has to decide on what to do during both periods without any prior knowledge of what **P2** will be doing, and vice versa. We shall be looking for Nash solutions in all three games.

The P1 – P2 – P1 – P2 information structure

The solution is obtained by working backward in time (à la "dynamic programming"). For the decision during the second period, **P2** knows to which point ($D_7 - D_{14}$) the game has proceeded. From each of these points he chooses Y

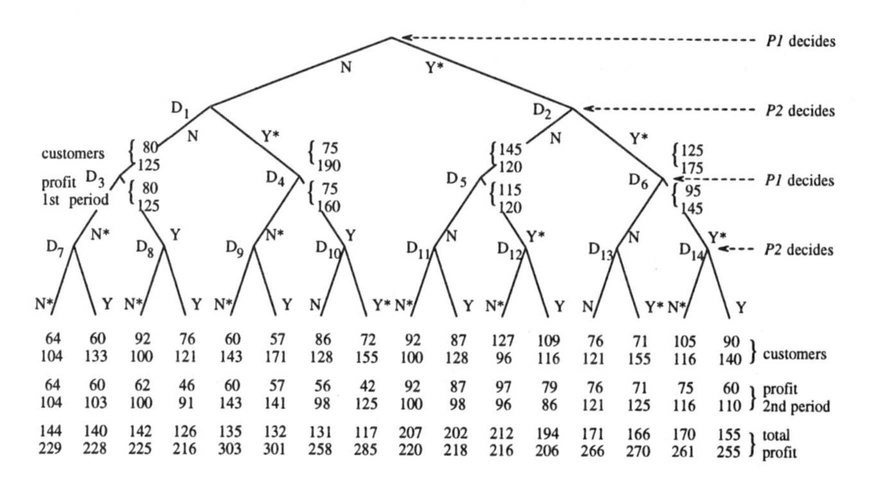

Figure 1.4: The two-period advertising game, without the information sets.

or N, depending on which decision leads to higher profit. These decisions are denoted by an asterisk in Fig. 1.4. At the time of his decision during the second period, **P1** knows that the game has proceeded to one of the points $D_3 - D_6$. He can also guess precisely what **P2** will do after he himself has made his decision (the underlying assumption being that **P2** will behave rationally). Hence, **P1** can determine what will be best for him. At point D_3, for instance, decision N leads to a profit of 144 (for **P1**) and Y leads to 142; therefore, **P1** will choose N if the game would be at D_3. The optimal decisions for **P1** at the second period are indicated by an asterisk also. We continue this way (in retrograde time) until the vertex of the tree is reached; all best actions are designated by an asterisk in Fig. 1.4. As a result, the actual game will evolve along a path passing through the points D_2, D_6, and D_{14}, and the cumulative profits for **P1** and **P2** will be 170 and 261, respectively.

The (P1, P2)-(P1, P2) information structure

During the last period, both players know that the game has evolved to one of the points $D_3 - D_6$, upon which information the players will have to base their actions. This way, we have to solve four "one-period games", one for each of the points $D_3 - D_6$. The reader can easily verify that the optimal (Nash) solutions associated with these games are:

information	action		profit	
(starting point)	**P1**	**P2**	**P1**	**P2**
D_3	N	N	144	229
D_4	N	N	135	303
D_5	Y	N	212	216
D_6	N	Y	166	270

The profit pairs corresponding to these solutions can be attached to the respective points $D_3 - D_6$, after which a one-period game is left to be solved. The game tree is:

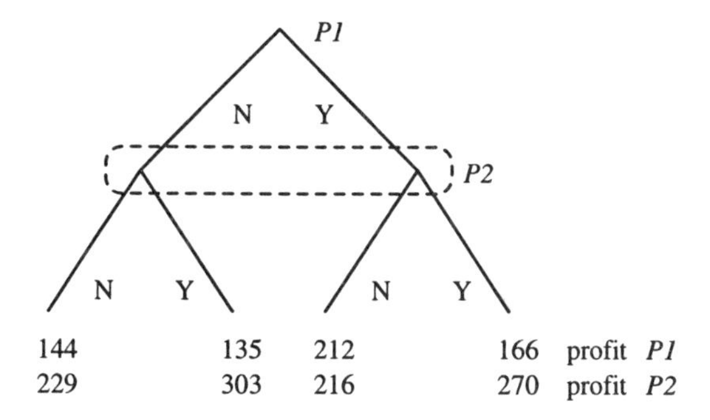

It can be verified that both players should play Y during the first period, since a unilateral deviation from this decision leads to a smaller profit for both players. The realized cumulative profits of **P**1 and **P**2 under this information scheme are therefore 166 and 270, respectively.

The no-information case

Both players have four possible choices: NN, NY, YN and YY, where the first (respectively, second) symbol refers to the first (respectively, second) period. Altogether there are $4 \times 4 = 16$ possible pairs of realizable profits, which can be written in a matrix form:

P1 chooses \ **P2** chooses	NN	NY	YN	YY
NN	144 229	140 228	135 303	132 301
NY	142 225	126 216	131 258	117 285
YN	207 220	202 218	171 266	166 270
YY	212 216	194 206	170 261	155 255

If **P**1 chooses YN and **P**2 chooses YY, then the profits are 166 and 270, respectively. If any one of the players deviates from this decision, he will be worse off. (If, for example, **P**1 deviates, his other profit options are 132, 117 and 155, each one being worse than 166.) Therefore $\{(YN), (YY)\}$ is the Nash solution under the "no-information"-scheme. □

Game problems of this kind will be studied extensively in this book, specifically in Chapter 3 which is devoted to *finite* nonzero-sum games, i.e., games in which each player has only a finite number of possible actions available to him. Otherwise, a game is said to be *infinite*, such as the one treated in Example 1.1. In Chapter 3, we shall also elucidate the reasons why the realized profits are, in general, dependent on the information structure.

The information structure (**P**1, **P**2)-(**P**1, **P**2) in Example 1.2 is called a *feedback* information structure; during each period the players know exactly to which point ("state") the game has evolved and that information is *fed back* into their strategies, which then leads to certain actions. This structure is sometimes referred to as *closed-loop*, though later on in this book a distinction between feedback and closed-loop will be made. The no-information case is referred to as an *open-loop* information structure.

Not every game is well-defined with respect to every possible information structure; even though Example 1.2 was well-defined under both a feedback and an open-loop information structure, this is not always true. To exemplify this situation, we provide, in the sequel, two specific games. The game to be formulated in Example 1.3 does not make much sense under an open-loop information structure, and the one of Example 1.4 can only be played under the open-loop structure.

Example 1.3 "*The lady in the lake.*" A lady is swimming in a circular lake with a maximum speed v_{l}. A man who wishes to intercept the lady, and who has not mastered swimming, is on the side of the lake and can run along the shore with a maximum speed v_{m}. The lady does not want to stay in the lake forever and wants eventually to escape. If, at the moment she reaches the shore, the man cannot intercept her, she "wins" the game since on land she can run faster than the man. Note that open-loop solutions (i.e., solutions corresponding to the open-loop information structure) do not make sense here (at least not for the man). It is reasonable to assume that each player will react immediately to what his/her opponent has done and hence the feedback information structure, whereby the current position of man and lady are fed back, is more appropriate. □

Example 1.4 "*The princess and the monster.*" This game is played in complete darkness. Just like the previous example, this one is also of the pursuit-evasion type. Here, however, the players do not know where their opponent is, unless they bump into each other (or come closer to each other than a given distance ϵ); that moment in fact defines the termination of the game. The velocities of both players are bounded and their positions are confined to a bounded area. The monster wants to bump into the princess as soon as possible, whereas the princess wants to push this moment as far into the future as possible. Since the players cannot react to each other's actions or positions, the information structure is open-loop. □

The solution to Example 1.3 will be given in Chapter 8; for more information on Example 1.4, see Foreman (1977). This latter example also leads, rather naturally, to the concept of "mixed strategies"; the optimal strategies in the "princess and monster" game cannot be deterministic, as were the strategies in Examples 1.1 and 1.2, for instance. For, if the monster would have a deterministic optimal strategy, then the princess would be able to calculate this strategy; if, in addition, she would know the monster's initial position, this would enable her to determine the monster's path and thereby to choose for herself an appropriate strategy so that she can avoid him forever. Therefore, an optimal strategy for the monster (if it exists) should dictate random actions, so that his trajectory cannot be predicted by the princess. Such a strategy is called *mixed*. Equilibria in mixed strategies will be discussed throughout this book.

What is optimal?

In contrast to optimal control problems (one-player games) where optimality has an unambiguous meaning, in multi-person decision making, optimality, in itself, is not a well-defined concept. Heretofore we have considered the Nash equilibrium solution, which is a specific form of "optimality". We also briefly discussed Pareto-optimality. There exist yet other kinds of optimality in nonzero-sum games. Two of them will be introduced now by means of the matrix game encountered in the "no-information"-case of Example 1.2. The equilibrium strategies in the "Nash" sense were YN for **P1** and YY for **P2**. Suppose now that **P2** is a careful and defensive player and wants to protect himself against any irrational behavior on the part of the other player. If **P2** sticks to YY, the worst that can happen to him is that **P1** chooses YY, in which case **P2**'s profit becomes 255 instead of 270. Under such a defensive attitude **P2** might play YN, since then his profit is at least 258. This strategy (or, equivalently, action in this case) provides **P2** with a lower bound for his earnings and the corresponding solution concept is called *minimax*. The player who adopts this solution concept basically solves a zero-sum game, even though the original game might be nonzero-sum.

Yet another solution concept is the one that involves a hierarchy in decision making: one of the players, say **P1**, declares and announces his strategy before the other player chooses his strategy and he (i.e., **P1**) is in a position to enforce this strategy. In the matrix game of the previous paragraph, if **P1** says "I will play YY" and irrevocably sticks to it (by the rules of the game we assume that cheating is not possible), then the best **P2** can do is to choose YN, in which case **P1**'s profit becomes 170 instead of 166 which was obtainable under the Nash solution concept. Such games in which one player (called the *leader*) declares his strategy first and enforces it on the other player (called the *follower*) are referred to as *Stackelberg games*. They will be discussed in Chapter 3 for finite action sets, and in Chapters 4 and 7 for infinite action sets.

Static versus dynamic

As the last topic of this section, we shall attempt to answer the question: "When is a game called dynamic and when is it static?" So far, we have talked rather loosely about these terms; there is, in fact, no uniformly accepted separating line between static games on the one hand and dynamic games on the other. We shall choose to call a game *dynamic* if at least one player is allowed to use a strategy that depends on previous actions. Thus, the games treated in Example 1.2 with the information schemes **P1** – **P2** – **P1** – **P2** and (**P1**, **P2**)-(**P1**, **P2**) are dynamic. The third game of Example 1.2, with the "no-information" scheme, should be called static, but, by an abuse of language, such a game is often also called dynamic. The reason is that the players act more than once and thus time plays a role. A game in which the players act only once and independently of each other is definitely called a *static* game. The game displayed in Fig. 1.3 is clearly static.

1.3 Outline of the Book

The book comprises eight chapters and three appendices. The present (first) chapter serves the purpose of introducing the reader to the contents of the book, and to the conventions and terminology adopted. The next three chapters constitute Part I of the text, and deal with finite static and dynamic games and infinite static games.

Chapter 2 discusses the existence, derivation and properties of saddle-point equilibria in pure, mixed and behavioral strategies for two-person zero-sum finite games. It is in this chapter that the notions of normal and extensive forms of a dynamic game are introduced, and the differences between actions and strategies are delineated. Also treated is the class of two-person zero-sum finite games in which a third (chance) player with a fixed mixed strategy affects the outcome.

Chapter 3 extends the results of Chapter 2 to nonzero-sum finite games under basically two different types of equilibrium solution concepts, viz. the Nash solution and the Stackelberg (hierarchical) solution. The impact of dynamic information on the structure of these equilibria is thoroughly investigated, and in this context the notion of prior and delay commitment modes of play are elucidated.

Chapter 4 deals with static infinite games of both the zero-sum and nonzero-sum type. In this context it discusses the existence, uniqueness and derivation of (pure and mixed) saddle-point, Nash and Stackelberg equilibria, as well as the consistent conjectural variations equilibrium. It provides explicit solutions for some types of games, with applications in microeconomics.

The remaining four chapters constitute Part II of the book, for which Chapter 5 provides a general introduction. It introduces the class of infinite dynamic games to be studied in the remaining chapters, and also gives some background material on optimal control theory. Furthermore, it makes the notions of "representations of strategies on given trajectories" and "time consistency" precise.

The major portion of Chapter 6 deals with the derivation and properties of Nash equilibria with prescribed fixed duration under different types of deterministic information patterns, in both discrete and continuous time. It also presents as a special case saddle-point equilibria in such dynamic games, with important applications in worst-case controller designs (such as H^∞-optimal control).

Chapter 7 discusses the derivation of global and feedback Stackelberg equilibria for the class of dynamic games treated in Chapter 6, and also the relationship with the theory of incentives.

Finally, Chapter 8 deals with the class of zero-sum differential games for which the duration is not fixed *a priori* —the so-called pursuit-evasion games— and under the feedback information pattern. It first presents some necessary and sufficient conditions for the saddle-point solution of such differential games, and then applies these to pursuit-evasion games with specific structures so as to obtain some explicit results.

Each chapter (with the exception of Chapter 1) starts with an introduction section which summarizes its contents, and therefore we have kept the descriptions above rather brief. Following Chapter 8 are the three appendices, two of which (Appendices A and B) provide some background material on sets, vector spaces, matrices, optimization theory and probability theory to the extent to which these notions are utilized in the book. The third appendix, on the other hand, presents some theorems which are used in Chapters 3 and 4. The book ends with a list of references, a table that indicates the page numbers of the Corollaries, Definitions, Examples, Lemmas, Propositions, Remarks and Theorems appearing in the text and an index.

1.4 Conventions, Notation and Terminology

Each chapter of the book is divided into sections, and occasionally sections are divided into subsections. Section 2.1 refers to Section 1 of Chapter 2, and Section 8.5.2 refers to subsection 2 of Section 8.5. The following items appear in this book and they are numbered per chapter, such as Prop. 3.7 referring to the seventh proposition of the third chapter:

Theorem (abbreviated to Thm.)	Corollary	Lemma
Definition (abbreviated to Def.)	Problem	Equation
Figure (abbreviated to Fig.)	Property	Example
Proposition (abbreviated to Prop.)	Remark	

Unlike the numbering of other items, equation numbers appear within parentheses, such as equation (2.5) which refers to the fifth numbered equation in Chapter 2.

References to bibliographical sources (listed alphabetically at the end of the book) are made according to the Harvard system (i.e., by name(s) and date).

The following abbreviations and symbols are adopted in the book, unless stated otherwise in specific contexts:

RHS	right-hand side
LHS	left-hand side
w.p. α	with probability α
LP	linear programming
$=$	equality sign

$\triangleq$	defined by
$\square$	end of proof, remark, example, etc.
$\parallel$	parallel to

$$\operatorname{sgn}(x) = \begin{cases} +1, & \text{if } x > 0 \\ \text{undefined}, & \text{if } x = 0 \\ -1, & \text{if } x < 0 \end{cases}$$

$\partial\Lambda$	boundary of the set Λ
$\mathbf{P}i$	player i
$\mathbf{P}(\mathbf{E})$	pursuer (evader)
L^i	cost function(al) of $\mathbf{P}i$ for a game in extensive form
J^i	cost function(al) of $\mathbf{P}i$ for a game in normal form
N	number of players
$\mathbf{N}$	$\triangleq \{1,\ldots,N\}$ players' set; $i \in \mathbf{N}$
u^i	action (decision, control) variable of $\mathbf{P}i$; $u^i \in U^i$
γ^i	strategy (decision law) of $\mathbf{P}i$; $\gamma^i \in \Gamma^i$
η^i	information available to $\mathbf{P}i$
K	number of stages (levels) in a discrete-time game
$\mathbf{K}$	$\triangleq \{1,\ldots,K\}$; $k \in \mathbf{K}$
$[0,T]$	time interval on which a differential game is defined: $t \in [0,T]$
x	state variable; $x(t)$ in continuous time and x_k in discrete time
$\dot{x}(t)$	time-derivative of $x(t)$
V	value of zero-sum game (in pure strategies)
$\bar{V}$	upper value
$\underline{V}$	lower value
V_{m}	value of a zero-sum game in mixed strategies
$\mathbf{R}$	real line
$\mathbf{R}^n$	n-dimensional Euclidean space
$\lvert x\rvert$	Euclidean norm of a finite-dimensional vector x, i.e., $\{x'x\}^{1/2}$
$\lVert x\rVert$	norm of a vector x in an infinite-dimensional space.

Other symbols or abbreviations used with regard to sets, unions, summation, matrices, optimization, random variables, etc., are introduced in Appendices A and B.

A convention that we have adopted throughout the text (unless stated otherwise) is that in nonzero-sum games all players are cost-minimizers, and in two-person zero-sum games **P1** is the minimizer and **P2** is the maximizer. In two-person matrix games **P1** is the row-selector and **P2** is the column-selector. The word "optimality" is used in the text rather freely for different equilibrium solutions when there is no ambiguity in the context; optimum quantities (such as strategies, controls, etc.) are generally identified by an asterisk (*).

Part I

Chapter 2

Noncooperative Finite Games: Two-Person Zero-Sum

2.1 Introduction

This chapter deals with the class of two-person zero-sum games in which the players have a finite number of alternatives to choose from. There exist two different formulations for such games: the normal (matrix) form and the extensive (tree) form. The former constitutes a suitable representation of a zero-sum game when each player's information is static in nature, since it suppresses all the dynamic aspects of the decision problem. The extensive form, on the other hand, displays explicitly the evolution of the game and the existing information exchanges between the players.

In the first part of the chapter (Sections 2.2 and 2.3), the normal form is introduced together with several related concepts, and then existence and computation of saddle-point equilibria are discussed for both pure and mixed strategies.

In the second part of the chapter (Sections 2.4 and 2.5), extensive form description for zero-sum finite games without chance moves is introduced, and saddle-point equilibria for such games are discussed, also within the class of behavioral strategies. This discussion is first confined to single-act games in which each player is allowed to act only once, and then it is extended to multi-act games.

The third part of the chapter (comprising the remaining three sections) discusses extensions in a number of directions. Section 2.6 is devoted to a brief discussion on the nature of equilibria for zero-sum games which also incorporate chance moves. Section 2.7 deals with two extensions: games with nonstandard information patterns and randomized strategies, and games in extensive form which exhibits cycles. Finally, Section 2.8 deals with games with action-dependent information sets, where the information on which the players can base future moves depends on previous moves.

2.2 Matrix Games

The most elementary type of two-person zero-sum games are matrix games. There are two players, to be referred to as player 1 (**P1**) and player 2 (**P2**), and an $(m \times n)$-dimensional (real-valued) matrix $A = \{a_{ij}\}$. Each entry of this matrix is an outcome of the game, corresponding to a particular pair of decisions made by the players. For **P1**, the alternatives are the m rows of the matrix, while for **P2** the possible choices are the n columns of the same matrix. These alternatives are known as the *strategies* of the players. If **P1** chooses the ith row and **P2** the jth column, then a_{ij} is the outcome of the game, and **P1** pays this amount to **P2**. In case a_{ij} is negative, this should be interpreted as **P2** paying **P1** the positive amount corresponding to this entry (i.e., $-a_{ij}$).

To regard the entries of the matrix as sums of money to be paid by one player to the other is, of course, only a convention. In a more general framework, these outcomes represent utility transfers from one player to the other. Thus, we can view each element of the matrix A (i.e., a possible outcome of the game) as the net change in the utility of **P2** for a particular play of the game, which is equal to minus the net change in the utility of **P1**. Then, regarded as a rational decision maker, **P1** will seek to minimize the outcome of the game, while **P2** will seek to maximize it, by independent decisions.

Assuming that this game is to be played only once, then a reasonable mode of play for **P1** is to secure his losses against any (rational or irrational) behavior of **P2**. Under such an incentive, **P1** is forced to pick that row (i^*) of matrix A, whose largest entry is no bigger than the largest entry of any other row. Hence, if **P1** adopts the i^*th row as his strategy, where i^* satisfies the inequalities

$$\bar{V}(A) \triangleq \max_j a_{i^*j} \leq \max_j a_{ij}, \quad i = 1, \ldots, m, \tag{2.1}$$

then his losses will be no greater than $\bar{V}$ which we call the *loss ceiling* of **P1**, or equivalently, the *security level* for his losses. The strategy "row i^*" that yields this security level will be called a *security strategy* for **P1**.

Adopting a similar mode of play, **P2** will want to secure his gains against any behavior of **P1**, and consequently, he will choose that column (j^*) whose smallest entry is no smaller than the smallest entry of any other column. In mathematical terms, j^* will be determined from the n-tuple of inequalities

$$\underline{V}(A) \triangleq \min_i a_{ij^*} \geq \min_i a_{ij}, \quad j = 1, \ldots, n. \tag{2.2}$$

By deciding on the j^*th column as his strategy, **P2** secures his gains at the level $\underline{V}$ which we call the *gain-floor* of **P2**, or equivalently, the *security level* for his gains. Furthermore, any strategy for **P2** that secures his gain-floor, such as "column j^*", is called a *security strategy* for **P2**.

Theorem 2.1 *In every matrix game* $A = \{a_{ij}\}$,

(i) the security level of each player is unique,

(ii) there exists at least one security strategy for each player,

(iii) the security level of **P1** *(the minimizer) never falls below the security level of* **P2** *(the maximizer), i.e.,*

$$\min_i a_{ij^*} = \underline{V}(A) \leq \bar{V}(A) = \max_j a_{i^*j}, \tag{2.3}$$

where i^* *and* j^* *denote security strategies for* **P1** *and* **P2**, *respectively.*

Proof. The first two parts follow readily from (2.1) and (2.2), since in every matrix game there is only a finite number of alternatives for each player to choose from. To prove part (iii), we first note the obvious set of inequalities

$$\min_i a_{il} \leq a_{kl} \leq \max_j a_{kj}$$

which holds true for all possible k and l. Now, letting $k = i^*$ and $l = j^*$, the desired result, i.e., (2.3), follows. □

The third part of Thm. 2.1 says that, if played by rational players, the outcome of the game will always lie between $\bar{V}$ and $\underline{V}$. It is for this reason that the security levels $\bar{V}$ and $\underline{V}$ are also called, respectively, the *upper value* and *lower value* of the matrix game. Yet another interpretation that could be given to these values is the following.

Consider the matrix game in which players do not make their decisions independently, but there is a predetermined ordering according to which the players act. First **P1** decides on what row he will choose, he transmits this information to **P2** who subsequently chooses a column. In this game, **P2** definitely has an advantage over **P1**, since he knows what his opponent's actual choice of strategy is. Then, it is unquestionable that the best play for **P1** is to choose one of his security strategies (say, row i^*). **P2**'s "optimal" response to this is the choice of a "column j°", where j° is determined from

$$a_{i^*j^\circ} = \max_j a_{i^*j} = \bar{V} = \min_i \max_j a_{ij}, \tag{2.4}$$

with the "min max" operation designating the order of play in this decision process. Thus, we see that the upper value of the matrix game is actually attained in this case. Now, if the roles of the players are interchanged, this time **P1** having a definite advantage over **P2**, then **P2**'s best choice will again be one of his security strategies (say, column j^*), and **P1**'s response to this will be a "row i°", where i° satisfies

$$a_{i^\circ j^*} = \min_i a_{ij^*} = \underline{V} = \max_j \min_i a_{ij}, \tag{2.5}$$

with the "max min" symbol implying that the minimizer acts after the maximizer. Relation (2.5) indicates that the outcome of the game is equal to the lower value $\underline{V}$, when **P1** observes **P2**'s choice.

To illustrate these facets of matrix games, let us now consider the following (3×4) matrix.

P2

P1				
	1	3	3	−2
	0	−1	2	1
	−2	2	0	1

(2.6)

Here **P2** (the maximizer) has a unique security strategy, "column 3" (i.e., $j^* = 3$), securing him a gain-floor of $\underline{V} = 0$. **P1** (the minimizer), on the other hand, has two security strategies, "row 2" and "row 3" (i.e., $i_1^* = 2$, $i_2^* = 3$), yielding him a loss-ceiling of $\bar{V} = \max_j a_{2j} = \max_j a_{3j} = 2$ which is above the security level of **P2**. Now, if **P2** plays first, then he chooses his security strategy "column 3", with **P1**'s unique response being "row 3" ($i^\circ = 3$), resulting in an outcome of $0 = \underline{V}$. If **P1** plays first, he is actually indifferent between his two security strategies. In case he chooses "row 2", then **P2**'s unique response is "column 3" ($j^\circ = 3$),[4] whereas if he chooses "row 3", his opponent's response is "column 2" ($j^\circ = 2$), both pairs of strategies yielding an outcome of $2 = \bar{V}$.

The preceding discussion validates the argument that, when there is a definite order of play, security strategies of the player who acts first make complete sense, and actually they can be considered to be in "equilibrium" with the corresponding response strategies of the other player. By two strategies (of **P1** and **P2**) to be in equilibrium, we roughly mean here that, after the game is over and its outcome is observed, the players should have no ground to regret their past actions. In a matrix game with a fixed order of play, for example, there is no justifiable reason for the player who acts first to regret his security strategy after the game is over. But what about the class of matrix games in which the players arrive at their decisions independently? Do security strategies have any sort of an equilibrium property, in that case?

To shed some light on this question, let us consider the (3×3) matrix game of (2.7) below, with the players acting independently, and the game to be played only once.

P2

P1			
	4	0	−1
	0	−1	3
	1	2	1

(2.7)

Both players have unique security strategies, "row 3" for **P1** and "column 1" for **P2**, with the upper and lower values of the game being $\bar{V} = 2$ and $\underline{V} = 0$, respectively. If both players play their security strategies, then the outcome of the game is 1, which is midway between the security levels of the players. But after the game is over, **P1** might think: "Well I knew that **P2** was going to play his security strategy, it is a pity that I didn't choose row 2 and enjoy an outcome of 0". Thinking along the same lines, **P2** might also regret that he did not play column 2 and achieve an outcome of 2. This, then, indicates that, in this matrix game, the security strategies of the players cannot possibly

[4]Here, and also earlier, it is only coincidental that optimal response strategies are also security strategies.

possess any equilibrium property. On the other hand, if a player chooses a row or column (whichever the case is) different from the one dictated by his security strategy, then he will be taking chances, since there is always a possibility that the outcome of the game might be worse for him than his security level.

For the class of matrix games with equal upper and lower values, however, such a dilemma does not arise. Consider, for example, the (2×2) matrix game

P2

	3	1
P1	−1	1

(2.8)

in which "row 2" and "column 2" are the unique security strategies of **P1** and **P2**, respectively, resulting in the same security level $\bar{V} = \underline{V} = 1$. These security strategies are in equilibrium, since each one is "optimal" against the other one. Furthermore, since the security levels of the players coincide, it does not make any difference, as far as the outcome of the game is concerned, whether the players arrive at their decisions independently or in a predetermined order. The strategy pair {row 2, column 2}, possessing all these favorable features, is clearly the only candidate that could be considered as the equilibrium solution of the matrix game (2.8). Such equilibrium strategies are known as *saddle-point strategies*, and the matrix game, in question, is then said to have a *saddle point* in *pure strategies*. A more precise definition for these terms is given below.

Definition 2.1 *For a given* $(m \times n)$ *matrix game* $A = \{a_{ij}\}$, *let* {*row* i^*, *column* j^*} *be a pair of strategies adopted by the players. Then, if the pair of inequalities*

$$a_{i^*j} \leq a_{i^*j^*} \leq a_{ij^*} \tag{2.9}$$

is satisfied for all $i = 1, \ldots, m$ *and all* $j = 1, \ldots, n$, *the strategies* {*row* i^*, *column* j^*} *are said to constitute a* saddle-point equilibrium *(or simply, they are said to be* saddle-point strategies*), and the matrix game is said to have a* saddle point in pure strategies. *The corresponding outcome* $a_{i^*j^*}$ *of the game is called the* saddle-point value, *or simply the* value, *of the matrix game, and is denoted by* $V(A)$.

Theorem 2.2 *Let* $A = \{a_{ij}\}$ *denote an* $(m \times n)$ *matrix game with* $\bar{V}(A) = \underline{V}(A)$. *Then,*

(i) A has a saddle point in pure strategies,

(ii) an ordered pair of strategies provides a saddle point for A if, and only if, the first of these is a security strategy for **P1** *and the second one is a security strategy for* **P2**,

(iii) $V(A)$ *is uniquely given by* $V(A) = \bar{V}(A) = \underline{V}(A)$.

Proof. Let i^* denote a security strategy for **P1** and j^* denote a security strategy for **P2**, which always exist by Thm. 2.1 (ii). Now, since $\bar{V} = \underline{V}$, we have

$$a_{i^*j} \leq \max_j a_{i^*j} = \bar{V} = \underline{V} = \min_i a_{ij^*} \leq a_{ij^*}$$

for all $i = 1, \ldots, m$ and $j = 1, \ldots, n$; and letting $i = i^*$, $j = j^*$ in this inequality, we obtain $\bar{V} = \underline{V} = a_{i^*j^*}$. Using this result in the same inequality yields

$$a_{i^*j} \leq a_{i^*j^*} \leq a_{ij^*}$$

which is (2.9). This completes the proof of (i) and the sufficiency part of (ii). We now show that the class of saddle-point strategy pairs is no larger than the class of ordered security strategy pairs. Let {row i^*, column j^*} be a saddle-point strategy pair. Then it follows from (2.9) that

$$\max_j a_{i^*j} \leq a_{ij^*} \leq \max_j a_{ij}, \quad i = 1, \ldots, m,$$

thus proving that "row i^*" is a security strategy for **P1**. Analogously, it can be shown that "column j^*" is a security strategy for **P2**. This completes the proof of (ii). Part (iii) then follows readily from (ii). □

We now immediately have the following property of saddle-point solutions, which follows from Thm. 2.2 (ii).

Corollary 2.1 *In a matrix game A, let {row i_1, column j_1} and {row i_2, column j_2} be two saddle-point strategy pairs. Then {row i_1, column j_2}, {row i_2, column j_1} are also in saddle-point equilibrium.*

This feature of the saddle-point strategies, given above in Corollary 2.1, is known as their *ordered interchangeability* property. Hence, in the case of nonunique (multiple) saddle points, each player does not have to know (or guess) the particular saddle-point strategy his opponent will use in the game, since all such strategies are in equilibrium and they yield the same value—indeed a very desirable property that an equilibrium solution is expected to possess.

For the case when the security levels of the players do not coincide, however, no such equilibrium solution can be found within the class of (pure) strategies that we have considered so far. One way of resolving this predicament is, as we have discussed earlier, to assume that one player acts after observing the decision of the other one, in which case the security level of the player who acts first is attained by an equilibrium "strategy pair". Here, of course, the strategy of the player who acts last explicitly depends on the action of the first acting player and hence the game has a "dynamic character". The precise meaning of this will be made clear in Section 2.4, where more details can be found on these aspects of zero-sum games.

Mixed strategies

Yet another approach to obtain an equilibrium solution in matrix games that do not possess a saddle point, and in which players act independently, is to enlarge the strategy spaces, so as to allow the players to base their decisions on the outcome of random events—thus leading to the so-called *mixed* strategies. This is an especially convincing approach when the same matrix game is played over and over again, and the final outcome, sought to be minimized by **P1** and maximized by **P2**, is determined by averaging the outcomes of individual plays.

To introduce the concept of a mixed strategy, let us again consider the matrix game $A = \{a_{ij}\}$ with m rows and n columns. In this game, the "strategy space" of **P1** comprises m elements, since he is allowed to choose one of the m rows of A. If Γ^1 denotes this space of (pure) strategies, then $\Gamma^1 = \{\text{row } 1, \dots, \text{row } m\}$. If we allow mixed strategies for **P1**, however, he will pick probability distributions on Γ^1. That is, an allowable strategy for **P1** is to choose "row 1" with probability (w.p.) y_1, "row 2" w.p. $y_2, \dots$, "row m" w.p. y_m, where $y_1 + y_2 + \cdots + y_m = 1$. The mixed strategy space of **P1**, which we denote by Y, is now comprised of all such probability distributions. Since the probability distributions are discrete, Y simply becomes the space of all nonnegative numbers y_i that add up to 1, which is a simplex. Note that, the m pure strategies of **P1** can also be considered as elements of Y, obtained by setting (i) $y_1 = 1$, $y_i = 0$, $\forall i \neq 1$; (ii) $y_2 = 1$, $y_i = 0$, $\forall i \neq 2$; ...... The mixed strategy space of **P2**, which we denote by Z, can likewise be defined as an n-dimensional simplex. A precise definition for a mixed strategy now follows.[5]

Definition 2.2 *A* mixed strategy *for a player is a probability distribution on the space of his pure strategies. Equivalently, it is a random variable whose values are the player's pure strategies.*

Thus, in matrix games, a mixed strategy for each player can be considered either as an element of a simplex, or as a random variable with a discrete probability distribution. Typical mixed strategies for the players, under the latter convention, would be independent random variables u and v, defined, respectively, by

$$u = \begin{cases} 1 & \text{w.p. } y_1 \\ \dots\dots\dots\dots & \\ m & \text{w.p. } y_m \end{cases}, \qquad \sum_{i=1}^{m} y_i = 1, \quad y_i \geq 0, \tag{2.10a}$$

$$v = \begin{cases} 1 & \text{w.p. } z_1 \\ \dots\dots\dots\dots & \\ n & \text{w.p. } z_n \end{cases}, \qquad \sum_{i=1}^{n} z_i = 1, \quad z_i \geq 0. \tag{2.10b}$$

Now, if the players adopt these random variables as their strategies to be implemented during the course of a repeated game, they will have to base their

[5]This definition of a mixed strategy is valid not only for matrix games but also for other types of finite games to be discussed in this and the following chapter.

actual decisions (as to what specific row or column to choose during each play of the game) on the outcome of a chance mechanism, unless the probability distributions involved happen to be concentrated at one point—the case of pure strategies. If the adopted strategy of **P1** is to pick "row 1" w.p. 1/2 and "row m" w.p. 1/2, for example, then the player in question could actually implement this strategy by tossing a fair coin before each play of the game, playing "row 1" if the actual outcome is "head", and "row m" otherwise. It should be noted, here, that the actual play (action) dictated by **P1**'s strategy becomes known (even to him) only after the outcome of the chance mechanism (tossing of a fair coin, in this case) is observed. Hence, we have a sharp distinction between a player's (*proper mixed*) strategy[6] and its implemented "value" for a particular play, where the latter explicitly depends on the outcome of a chance experiment designed so as to generate the odds dictated by the mixed strategy. Such a dichotomy between strategy and its implemented value does not exist for the class of pure strategies that we have discussed earlier in this section, since if a player adopts the strategy "to play row i", for example, then its implementation is known for sure—he will play "row i".

Given an $(m \times n)$ matrix game A, which is to be played repeatedly and the final outcome to be determined by averaging the scores of the players on individual plays, let us now investigate as to how this final outcome will be related to the (mixed) strategies adopted by the players. Let the independent random variables u and v, defined by (2.10a) and (2.10b), be the strategies adopted by **P1** and **P2**, respectively. Then, as the number of times this matrix game is played gets arbitrarily large, the frequencies with which different rows and columns of the matrix are chosen by **P1** and **P2**, respectively, will converge to their respective probability distributions that characterize the strategies u and v. Hence, the average value of the outcome of the game, corresponding to the strategy pair $\{u, v\}$, will be equal to

$$J(y,z) = \sum_{i=1}^{m} \sum_{j=1}^{n} y_i a_{ij} z_j = y'Az, \tag{2.11}$$

where y and z are the probability distribution vectors defined by

$$y = (y_1, \ldots, y_m)', \quad z = (z_1, \ldots, z_n)'. \tag{2.12}$$

P1 wishes to minimize this quantity, $J(y, z)$, by an appropriate choice of a probability distribution vector $y \in Y$, while **P2** wishes to maximize the same quantity by choosing an appropriate $z \in Z$, where the sets Y and Z are, respectively, the m- and n-dimensional simplices introduced earlier, i.e.,

$$Y = \{y \in \mathbf{R}^m : y \geq 0, \sum_{i=1}^{m} y_i = 1\}, \tag{2.13a}$$

[6]Since, by definition, the concept of mixed strategy also covers pure strategy, we shall sometimes use the term "proper mixed" to emphasize (whenever the case is) that the underlying probability distribution is not one point.

$$Z = \{z \in \mathbf{R}^n : z \geq 0, \sum_{j=1}^{n} z_j = 1\}. \tag{2.13b}$$

The following definitions now follow as obvious extensions to mixed strategies of some of the concepts introduced earlier for matrix games with strategy spaces comprised of only pure strategies.

Definition 2.3 *A vector $y^* \in Y$ is called a* mixed security strategy *for* **P1**, *in the matrix game A, if the following inequality holds for all $y \in Y$:*

$$\bar{V}_{\mathrm{m}}(A) \triangleq \max_{z \in Z} y^{*\prime} A z \leq \max_{z \in Z} y' A z, \qquad y \in Y. \tag{2.14}$$

Here, the quantity $\bar{V}_{\mathrm{m}}$ is known as the average security level *of* **P1** *(or equivalently the* average upper value of the game*). Analogously, a vector $z^* \in Z$ is called a* mixed security strategy *for* **P2**, *in the matrix game A, if the following inequality holds for all $z \in Z$:*

$$\underline{V}_{\mathrm{m}}(A) \triangleq \min_{y \in Y} y' A z^* \geq \min_{y \in Y} y' A z, \qquad z \in Z. \tag{2.15}$$

Here, $\underline{V}_{\mathrm{m}}$ is known as the average security level *of* **P2** *(or equivalently, the* average lower value of the game*).*

Definition 2.4 *A pair of strategies $\{y^*, z^*\}$ is said to constitute a* saddle point *for a matrix game A, in* mixed strategies, *if*

$$y^{*\prime} A z \leq y^{*\prime} A z^* \leq y' A z^*, \qquad \forall y \in Y, z \in Z. \tag{2.16}$$

The quantity $V_{\mathrm{m}}(A) = y^{*\prime} A z^*$ is known as the saddle-point value, *or simply the* value, *of the game, in* mixed strategies.

Remark 2.1 The assumptions of existence of a maximum in (2.14) and a minimum in (2.15) are justifiable since, in each case, the objective functional is linear (thereby continuous) and the simplex over which optimization is performed is closed and bounded (thereby compact). (See Appendix A.) □

As a direct extension of Thm. 2.1, we can now verify the following properties of the security levels and security strategies in matrix games when the players are allowed to use mixed strategies.

Theorem 2.3 *In every matrix game A in which the players are allowed to use mixed strategies, the following properties hold:*

(i) The average security level of each player is unique.

(ii) There exists at least one mixed security strategy for each player.

(iii) The security levels in pure and mixed strategies satisfy the following inequalities:

$$\underline{V}(A) \leq \underline{V}_{\mathrm{m}}(A) \leq \bar{V}_{\mathrm{m}}(A) \leq \bar{V}(A). \tag{2.17}$$

Proof. (i) Uniqueness of $\bar{V}_{\mathrm{m}}$ follows directly from its definition, $\bar{V}_{\mathrm{m}} = \inf_Y \max_Z y'Az$, in view of Remark 2.1. Analogously, the quantity $\underline{V}_{\mathrm{m}} = \sup_Z \min_Y y'Az$ is also unique.

(ii) The quantity, $\max_{z \in Z} y'Az$, on the right-hand side (RHS) of (2.14) is a continuous function of $y \in Y$, a result which can be proven by employing standard methods of analysis. But since Y is a compact set, the minimum of this function is attained on Y, thus proving the desired result for the security strategy of **P1**. Proof of existence of a security strategy for **P2** follows along similar lines.

(iii) The middle inequality follows from the simple reasoning that the maximizer's security level cannot be higher than the minimizer's security level. The other two inequalities follow, since the pure security strategies of the players are included in their mixed strategy spaces. □

As an immediate consequence of part (ii) of Thm. 2.3 above, we obtain the following.

Corollary 2.2 *In a matrix game A, the average upper and lower values in mixed strategies are given, respectively, by*

$$\bar{V}_{\mathrm{m}}(A) = \min_Y \max_Z \; y'Az \tag{2.18a}$$

and

$$\underline{V}_{\mathrm{m}}(A) = \max_Z \min_Y \; y'Az. \tag{2.18b}$$

Now, one of the important results of zero-sum game theory is that these upper and lower values are equal, that is, $\bar{V}_{\mathrm{m}}(A) = \underline{V}_{\mathrm{m}}(A)$. This result is given in Thm. 2.4, below, which is known as the "Minimax Theorem". In its proof, we shall need the following lemma.

Lemma 2.1 *Let A be an arbitrary $(m \times n)$-dimensional matrix. Then, either*

(i) there exists a nonzero vector $y \in \mathbf{R}^m$, $y \geq 0$ such that $A'y \leq 0$,

or

(ii) there exists a nonzero vector $z \in \mathbf{R}^n$, $z \geq 0$ such that $Az \geq 0$.

Proof. Consider in $\mathbf{R}^n$ the unit vectors $e_1, \ldots, e_n$, together with the rows of A, to be indicated by $a_{i\cdot}$, $i = 1, \ldots, m$. Denote the convex hull of these $n + m$ vectors by $\mathcal{C}$. Two possibilities exist: either $0 \in \mathcal{C}$ or $0 \notin \mathcal{C}$.

Assume first that $0 \in \mathcal{C}$. Then there exist nonnegative coefficients y_i ($i = 1, \ldots, m$) and η_j ($j = 1, \ldots, n$), normalized to 1, $\sum_{i=1}^m y_i + \sum_{j=1}^n \eta_j = 1$, such that

$$\sum_{i=1}^m y_i a'_{i\cdot} + \sum_{j=1}^n \eta_j e_j = 0.$$

Componentwise this equation reads $\sum_{i=1}^{m} y_i a_{ij} + \eta_j = 0$, $j = 1, \ldots, n$. Since all the y_i's cannot be zero simultaneously (otherwise all η_j's would also be zero—which is an impossibility due to the normalization), we have

$$\sum_{i=1}^{m} y_i a_{ij} = -\eta_j \leq 0, \quad j = 1, \ldots, n.$$

Now, if we take $y \in \mathbf{R}^m$ as the vector whose components are the y_i's defined above, then the validity of possibility (i) of the lemma follows. Assume next that $0 \notin \mathcal{C}$. Then, there exists a hyperplane passing through the origin (characterized by the equation $z'x = 0$, where $x, z \in \mathbf{R}^n$, and where x is the running variable), such that $\mathcal{C}$ is on one side of it; furthermore, $z'x > 0$ for $x \in \mathcal{C}$. If we choose $x = e_i$ (which belongs to $\mathcal{C}$ by construction), then it follows that $z_i > 0$, and hence the vector z is nonzero. Similarly, if we choose $x = a'_{i\cdot}$, then $z'a'_{i\cdot} > 0$ and hence $Az > 0$, which validates the possibility (ii). □

Theorem 2.4 *(The Minimax Theorem) In any matrix game A, the average security levels of the players in mixed strategies coincide, that is,*

$$\bar{V}_{\mathrm{m}}(A) = \min_{Y} \max_{Z} y'Az = \max_{Z} \min_{Y} y'Az = \underline{V}_{\mathrm{m}}(A). \tag{2.19}$$

Proof. We first show, by applying Lemma 2.1, that for a given matrix game A, at least one of the following two inequalities holds:

$$\bar{V}_{\mathrm{m}}(A) \leq 0, \tag{i}$$

$$\underline{V}_{\mathrm{m}}(A) \geq 0. \tag{ii}$$

Assume that the first alternative of Lemma 2.1 is valid. Then, there exists a vector $y^\circ \in Y$ such that $A'y^\circ \leq 0$, which is equivalent to the statement that the inequality

$$y^{\circ\prime}Az \leq 0$$

holds for all $z \in Z$. This is further equivalent to saying

$$\max_{Z} y^{\circ\prime}Az \leq 0$$

which definitely implies (by also making use of (2.18a))

$$\bar{V}_{\mathrm{m}}(A) = \min_{Y} \max_{Z} y'Az \leq 0.$$

Now, under the second alternative of Lemma 2.1, there exists a vector $z^\circ \in Z$ such that $Az^\circ \geq 0$, or equivalently,

$$y'Az^\circ \geq 0, \qquad \forall y \in Y.$$

This is further equivalent to the inequality

$$\min_{Y} y'Az^\circ \geq 0,$$

which finally implies (ii), by also making use of (2.18b), since

$$\underline{V}_{\rm m}(A) = \max_{Z} \min_{Y} y'Az \geq \min_{Y} y'Az^{\circ}.$$

Thus we see that, under the first alternative of Lemma 2.1, inequality (i) holds; and under the second alternative, inequality (ii) remains valid.

Let us now consider a new $(m \times n)$-dimensional matrix $B = \{b_{ij}\}$ obtained by shifting all entries of A by a constant, c, that is, $b_{ij} = a_{ij} - c$. This results in a shift of the same amount in both $\bar{V}_{\rm m}(A)$ and $\underline{V}_{\rm m}(A)$, that is,

$$\bar{V}_{\rm m}(B) = \bar{V}_{\rm m}(A) - c$$

and

$$\underline{V}_{\rm m}(B) = \underline{V}_{\rm m}(A) - c.$$

Since matrix A was arbitrary in (i) and (ii), we replace A with B, as defined above, in these inequalities, and arrive at the following property with regard to the upper and lower values of a matrix game A, in mixed strategies. For a given matrix game A and an arbitrary constant c, at least one of the following two inequalities holds true:

$$\bar{V}_{\rm m}(A) \leq c, \qquad (iii)$$

$$\underline{V}_{\rm m}(A) \geq c. \qquad (iv)$$

But, for this statement to be valid for arbitrary c, it is necessary that $\bar{V}_{\rm m}(A) = \underline{V}_{\rm m}(A)$. For, otherwise, the only possibility is $\bar{V}_{\rm m}(A) = \underline{V}_{\rm m}(A) + k$, in view of the middle inequality of (2.17), where $k > 0$; and picking $c = \underline{V}_{\rm m}(A) + (1/2)k$ in (iii) and (iv), it readily follows that neither (iii) nor (iv) is satisfied. Thus, this completes the proof of the theorem. □

Corollary 2.3 *Let A denote an $(m \times n)$ matrix game. Then,*

(i) A has a saddle point in mixed strategies,

(ii) a pair of mixed strategies provides a saddle point for A if, and only if, the first of these is a mixed security strategy for **P1**, *and the second one is a mixed security strategy for* **P2**,

(iii) $V_{\rm m}(A)$ is uniquely given by $V_{\rm m}(A) = \bar{V}_{\rm m}(A) = \underline{V}_{\rm m}(A)$,

(iv) in case of multiple saddle points, the mixed saddle-point strategies possess the ordered interchangeability property.

Proof. This corollary to Thm. 2.4 is actually an extension of Thm. 2.2 and Corollary 2.1 to the case of mixed strategies, and its verification is along the same lines as that of Thm. 2.2. It should only be noted that equality of the average upper and lower values of the matrix game is a fact, in the present case, rather than a part of the hypothesis, as in Thm. 2.2. □

We have thus seen that, if the players are allowed to use mixed strategies, matrix games always admit a saddle-point solution which, thereby, manifests itself as the only reasonable equilibrium solution in zero-sum two-person games of that type.

2.3 Computation of Mixed Equilibrium Strategies

We have seen in the previous section that two-person zero-sum matrix games always admit a saddle-point equilibrium in mixed strategies (cf. Thm. 2.4). One important property of mixed saddle-point strategies is that, for each player, the corresponding one also constitutes a mixed security strategy, and conversely, each mixed security strategy is a mixed saddle-point strategy (Corollary 2.3(ii)). This property, then, strongly suggests a possible way of obtaining the mixed saddle-point solution of a matrix game, which is to determine the mixed security strategy(ies) of each player.

To illustrate this approach, let us consider the (2×2) matrix game

P2

P1		
	3	0
	−1	1

(2.20)

which clearly does not admit a pure-strategy saddle point, since $\bar{V} = 1$ and $\underline{V} = 0$. Let the mixed strategies of **P1** and **P2** be denoted by $y = (y_1, y_2)'$ and $z = (z_1, z_2)'$, respectively, with $y_i \geq 0$, $z_i \geq 0$ $(i = 1, 2)$, $y_1 + y_2 = z_1 + z_2 = 1$, and consider first the average security level of **P1**. It should be noted that, while determining the average security level of a player, we can assume, without any loss of generality, that the other player chooses only pure strategies (this follows directly from Defs. 2.3 and 2.4). Hence, in the present case, we can take **P2** to play either $(z_1 = 1, z_2 = 0)$ or $(z_1 = 0, z_2 = 1)$; and under different choices of mixed strategies for **P1**, we can determine the average outcome of the game as shown in Fig. 2.1 by the bold line, which forms the *upper envelope* to the two straight lines drawn. Now, if the mixed strategy $(y_1^* = \frac{2}{5}, y_2^* = \frac{3}{5})$ corresponding to the lowest point of that envelope is adopted by **P1**, then the average outcome will be no greater than $\frac{3}{5}$. For any other mixed strategy of **P1**, however, **P2** can

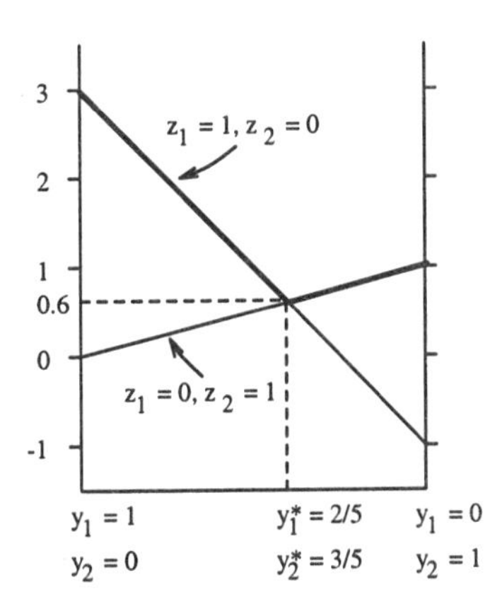

Figure 2.1: Mixed security strategy of **P1** for the matrix game (2.20).

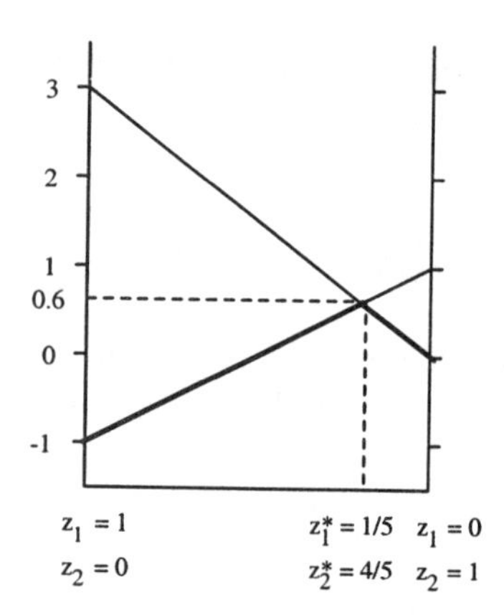

Figure 2.2: Mixed security strategy of **P2** for the matrix game (2.20).

obtain a higher average outcome. (If, for instance, $y_1 = \frac{1}{2}$, $y_2 = \frac{1}{2}$, then **P2** can increase the outcome by playing $z_1 = 1, z_2 = 0$.)

This then implies that the strategy ($y_1 = \frac{2}{5}, y_2 = \frac{3}{5}$) is a mixed security strategy for **P1** (and his only one), and thereby, it is his mixed saddle-point strategy. The mixed saddle-point value can easily be read off from Fig. 2.1 to be $V_{\mathrm{m}} = \frac{3}{5}$.

To determine the mixed saddle-point strategy of **P2**, we now consider his security level, this time assuming that **P1** adopts pure strategies. Then, for different mixed strategies of **P2**, the average outcome of the game can be determined to be the bold line, shown in Fig. 2.2, which forms a *lower envelope* to the two straight lines drawn. Since **P2** is the maximizer, the highest point on this envelope is his average security level, which he can guarantee (on the average) by playing the mixed strategy ($z_1^* = \frac{1}{5}, z_2^* = \frac{4}{5}$) which is also his saddle-point strategy.

The computational technique discussed above is known as the "*graphical solution*" *of matrix games*, since it makes use of a graphical construction directly related to the entries of the matrix. Such an approach is practical not only for (2×2) matrix games but also for general $(2 \times n)$ and $(m \times 2)$ matrix games. Consider, for example, the (2×3) matrix game

P2

P1	1	3	0
	6	2	7

(2.21)

for which **P1**'s average security level in mixed strategies has been determined in Fig. 2.3. Assuming again that **P2** uses only pure strategies ($z_1 = 1, z_2 = z_3 = 0$ or $z_1 = z_3 = 0$, $z_2 = 1$ or $z_1 = z_2 = 0$, $z_3 = 1$), the average outcome of the game for different mixed strategies of **P1** is given by the bold line drawn in Fig. 2.3, which is again the upper envelope (this time formed by three straight lines). The lowest point on this envelope yields the mixed security strategy, and thereby the mixed saddle-point strategy, of **P1**, which is ($y_1^* = \frac{2}{3}, y_2^* = \frac{1}{3}$).

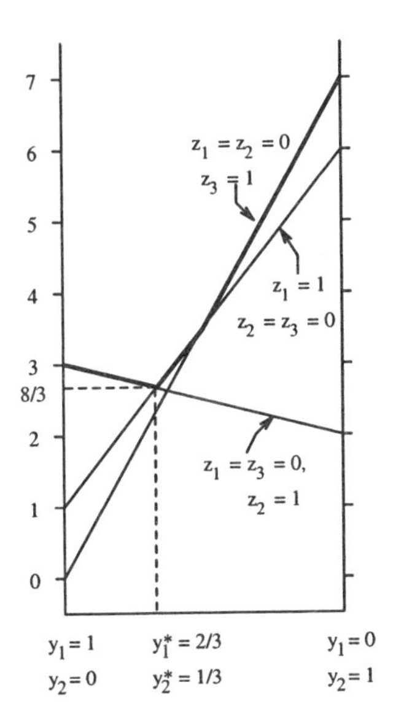

Figure 2.3: Mixed security strategy of **P1** for the matrix game (2.21).

The mixed saddle-point value, in this case, is $V_m = \frac{8}{3}$. To determine the mixed saddle-point strategy of **P2**, we first note that the average security level of **P1** has been determined by the intersection of only two straight lines which correspond to two pure strategies of **P2**, namely "column 1" and "column 2". If **P1** uses his mixed security strategy, and **P2** knows that before he plays, he would of course be indifferent between these two pure strategies, but he would never play the third pure strategy, "column 3", since that will reduce the average outcome of the game. This then implies that, in the actual saddle-point solution, he will "mix" only between the first two pure strategies. Hence, in the process of determining **P2**'s mixed security strategies, we can consider the (2×2) matrix game

	P2	
	1	3
P1	6	2

(2.22)

which is obtained from (2.21) by deleting the last column. Using the graphical approach, **P2**'s mixed security strategy in this matrix game can readily be determined to be $(z_1^* = \frac{1}{6}, z_2^* = \frac{5}{6})$ which is also his mixed saddle-point strategy.

Relation with linear programming

One alternative to the graphical solution when the matrix dimensions are high is to convert the original matrix game into a linear programming (LP) problem and make use of some of the powerful algorithms available for LP. To elucidate the close connection between a two-person zero-sum matrix game and an LP problem, let us start with an $(m \times n)$ matrix game $A = \{a_{ij}\}$ with all entries positive (i.e., $a_{ij} > 0 \; \forall i = 1, \ldots, m; j = 1, \ldots, n$). Then, the average value of this game, in mixed strategies, is given by

$$V_m(A) = \min_Y \max_Z y'Az = \max_Z \min_Y y'Az, \tag{2.23}$$

which is necessarily a positive quantity by our positivity assumption on A. Let us now consider the min-max operation used in (2.23). Here, first a $y \in Y$ is given, and then the resulting expression is maximized over Z; that is, the choice of $z \in Z$ can depend on y. Hence, the middle expression of (2.23) can also be written as

$$\min_{y \in Y} v_1(y),$$

where $v_1(y)$ is a positive function of y, defined by

$$v_1(y) = \max_Z y'Az \geq y'Az \quad \forall z \in Z. \tag{2.24}$$

Since Z is the n-dimensional simplex as defined by (2.13b), the inequality in (2.24) becomes equivalent to the vector inequality

$$A'y \leq 1_n v_1(y),$$

where

$$1_n \triangleq (1, \ldots, 1)' \in \mathbf{R}^n.$$

Further introducing the notation $\tilde{y} = y/v_1(y)$ and recalling the definition of Y from (2.13a), we observe that the optimization problem faced by **P1** in determining his mixed security strategy is

$$\text{minimize } v_1(y) \text{ over } \mathbf{R}^m$$

subject to

$$\begin{aligned} A'\tilde{y} &\leq 1_n, \\ \tilde{y}'1_m &= [v_1(y)]^{-1}, \\ \tilde{y} \geq 0, & \quad\; y = \tilde{y} v_1(y). \end{aligned}$$

This is further equivalent to the maximization problem

$$\max \quad \tilde{y}'1_m \tag{2.25a}$$

subject to

$$A'\tilde{y} \leq 1_n, \tag{2.25b}$$

$$\tilde{y} \geq 0, \tag{2.25c}$$

which is a standard *LP* problem. The solution of this *LP* problem will give the mixed security strategy of **P1**, normalized with the average saddle-point value of the game, $V_m(A)$. Hence, if $y^* \in Y$ is a mixed saddle-point strategy of **P1**, in the matrix game A, then the quantity $\tilde{y}^* = y^*/V_m(A)$ solves the LP problem (2.25a)-(2.25c), whose optimal value is in fact $1/V_m(A)$. Conversely, every solution of the LP problem will correspond to a mixed strategy of **P1** in the same manner.

Now, if we instead consider the right-hand expression of (2.23), and introduce

$$v_2(z) \triangleq \min_Y y'Az \leq y'Az \qquad \forall y \in Y,$$

and

$$\tilde{z} \triangleq z/v_2(z),$$

following similar steps and reasoning leads us to the minimization problem

$$\min \tilde{z}'1_n \tag{2.26a}$$

subject to

$$A\tilde{z} \geq 1_m, \tag{2.26b}$$

$$\tilde{z} \geq 0, \tag{2.26c}$$

which is the "dual" of (2.25a)-(2.25c). We thus arrive at the conclusion that if $z^* \in Z$ is a mixed saddle-point strategy of **P2**, in the matrix game A, then $\tilde{z}^* = z^*/V_m(A)$ solves the LP problem (2.26a)-(2.26c), whose optimal value is again $1/V_m(A)$. Furthermore, the converse statement is also valid.

We have thus shown that, given a matrix game with all positive entries, there exist two LP problems (*duals* of each other) whose solutions yield the saddle-point solution(s) of the matrix game. The positivity of the matrix A, however, is only a convention here, and can easily be dispensed with, as the following lemma shows.

Lemma 2.2 *Let A and B be two $(m \times n)$-dimensional matrices related to each other by the relation*

$$A = B + c1_m 1_n', \tag{2.27}$$

where c is some constant. Then,

(i) every mixed saddle-point strategy pair for matrix game A also constitutes a mixed saddle-point solution for matrix game B, and vice versa,

(ii) $V_m(A) = V_m(B) + c$.

Proof. Let (y^*, z^*) be a saddle-point solution for A, thus satisfying inequalities (2.16). If A is replaced by $B + c1_m 1_n'$ in (2.16), then it is easy to see that (y^*, z^*) also constitutes a saddle-point solution for B, since $y'1_m 1_n' z = 1$ for every $y \in Y$, $z \in Z$. Since the reverse argument also applies, this completes the proof of part (i). Part (ii), on the other hand, follows from the relation
$V_m(A) = y^{*\prime}[B + c1_m 1_n']z^* = y^{*\prime}Bz^* + cy^{*\prime}1_m 1_n' z^* = V_m(B) + c.$ □

Because of property (i) of the above lemma, we call matrix games which satisfy relation (2.27) *strategically equivalent* matrix games. It should now be apparent that, given a matrix game A, we can always find a strategically equivalent matrix game with all entries positive, and thus the transformation of a matrix game into two *LP* problems as discussed prior to Lemma 2.2 is always valid, as far as determination of saddle-point strategies is concerned. This result is summarized below in Thm. 2.5.

Theorem 2.5 *Let B be an $(m \times n)$ matrix game, and A be defined by (2.27) with c chosen to make all its entries positive. Introduce the two* LP *problems*

$$\left.\begin{array}{l} \max y'1_m \\ A'y \leq 1_n \\ y \geq 0 \end{array}\right\} \text{ “primal problem”}, \tag{2.28a}$$

$$\left.\begin{array}{l} \min z'1_n \\ Az \geq 1_m \\ z \geq 0 \end{array}\right\} \text{ “dual problem”}, \tag{2.28b}$$

with their optimal values (if they exist) denoted by V_p and V_d, respectively.

(i) Both **LP** *problems admit a solution, and $V_\text{p} = V_\text{d} = 1/V_\text{m}(A)$.*

(ii) If (y^*, z^*) is a mixed saddle-point solution of the matrix game B, then $y^*/V_\text{m}(A)$ solves (2.28a), and $z^*/V_\text{m}(A)$ solves (2.28b).

(iii) If $\tilde{y}^*$ is a solution of (2.28a), and $\tilde{z}^*$ is a solution of (2.28b), the pair $(\tilde{y}^*/V_\text{p}, \tilde{z}^*/V_\text{d})$ constitutes a mixed saddle-point solution for matrix game B. Furthermore, $V_\text{m}(B) = (1/V_\text{p}) - c$.

Proof. Since A is a positive matrix and strategically equivalent to B, the theorem has already been proven prior to the statement of Lemma 2.2. We should only note that the reason why both *LP* problems admit a solution is that the mixed security strategies of the players always exist in the matrix game A. □

Theorem 2.5 provides one method of making games “essentially” equivalent to two *LP* problems that are duals of each other, thus enabling the use of some powerful algorithms available for *LP* in order to obtain their saddle-point solutions. (The reader is referred to Dantzig (1963), Luenberger (1973), Gonzaga (1992), or Karmarkar (1984) for *LP* algorithms.) There are also other transformation methods available in the literature, which form different kinds of equivalences between matrix games and *LP* problems, but the underlying idea is essentially the same in all these techniques. For one such equivalence the reader is referred to the next chapter, to Corollary 3.2, where an *LP* problem, not equivalent to the one of Thm. 2.5, is obtained as a special case of a more general result on nonzero-sum matrix games. More details on computational techniques for matrix games can be found in Luce and Raiffa (1957) and Singleton and Tyndal (1974).

Dominating strategies

We conclude this section by introducing the concept of "*dominance*" in matrix games, which can sometimes be useful in reducing the dimensions of a matrix game by eliminating some of its rows and/or columns which are known from the very beginning to have no influence on the equilibrium solution. More precisely, given an $(m \times n)$ matrix game $A = \{a_{ij}\}$, we say that "row i" *dominates* "row k" if $a_{ij} \leq a_{kj}$, $j = 1, \ldots, n$, and if, for at least one j, the strict inequality-sign holds. In terms of pure strategies, this means that the choice of the *dominating* strategy, i.e., "row i", is at least as good as the choice of the *dominated* strategy, i.e., "row k". If, in the above set of inequalities, the strict inequality-sign holds for all $j = 1, \ldots, n$, then we say that "row i" *strictly dominates* "row k", in which case, regardless of what **P2** chooses, **P1** does better with "row i" (*strictly dominating* strategy) than with "row k" (*strictly dominated* strategy). It therefore follows that **P1** can always dispense with his strictly dominated strategies and consider only strictly undominated ones, since adoption of a strictly dominated strategy is apt to increase the security level for **P1**. This argument also holds for mixed strategies, and strictly dominated rows are always assigned a probability of zero in an optimal (mixed) strategy for **P1**.

Analogously for **P2**, "column j" of A is said to dominate (respectively, strictly dominate) "column l", if $a_{ij} \geq a_{il}$, $i = 1, \ldots, m$, and if, for at least one i (respectively, for all i), the strict inequality-sign holds. In an optimal (mixed) strategy for **P2**, strictly dominated columns are also assigned a probability of zero. Mathematical verifications of these intuitively obvious assertions, as well as some other properties of strictly dominating strategies, can be found in Vorob'ev (1977).

We have thus seen that, in the computation of (pure or mixed) saddle-point equilibria of matrix games, *strictly* dominated rows and columns can readily be deleted, since they do not contribute to the equilibrium solution—and this could lead to considerable simplifications in the graphical solution or in the solution of the equivalent *LP* problems, since the matrix is now of smaller dimension(s). With (nonstrictly) dominated rows and columns, however, this may not always be so. Even though every saddle-point solution of a matrix game whose dominated rows and columns are deleted is also a saddle-point solution of the original matrix game, there might be other saddle points (of the original game) which are eliminated in this process. As a specific illustration of this possibility, consider the (2×2) matrix game

	P2	
	0	(1)
P1	−1	(1)

which admits two saddle points, as indicated. For **P1**, "row 2" is clearly dominating (but not strictly dominating), and for **P2**, "column 2" is strictly dominating. Thus, by eliminating dominated rows and columns, we end up with the (1×1) game corresponding to the dominating strategies of the players, which trivially has a unique saddle point.

Motivated by this result, we call the saddle-point solutions of the "reduced" game which does not have any dominated rows or columns *dominant saddle-point solutions.* Even though the original game might have other saddle points, it may still be reasonable to confine attention only to dominant saddle points, since any game that possesses a saddle point also possesses (by definition) a dominant saddle point, and furthermore the saddle-point value (in pure or mixed strategies) is unique; in other words, there is nothing essential to gain in seeking saddle-point equilibria which are not dominant.

2.4 Extensive Forms: Single-Act Games

There exist different mathematical descriptions for two-person zero-sum games with finite strategy sets, the matrix form (also known as the *normal form*) being one of these. A normal form does not in general provide the full picture of the underlying decision process, since it only describes the correspondences between different ordered strategy pairs and outcomes. Some important issues like the order of play in the decision process, information available to the players at the time of their decisions, and the evolution of the game in the case of dynamic situations are suppressed in the matrix description of a zero-sum game.

An alternative to the normal form, which explicitly displays the dynamic character of the decision problem, is known as the *extensive form* of a two-person zero-sum game. An extensive form basically involves a tree structure with several nodes and branches, providing an explicit description of the order of play and the information available to each player at the time of his decision(s); the game evolves from the top of the tree to the tip of one of its branches. Two such tree structures are depicted in Fig. 2.4, where in each case **P1** has two alternatives (branches) to choose from, whereas **P2** has three alternatives, and the order of play is such that **P1** acts before **P2** does. The numbers at the end of the lower branches represent the pay-offs to **P2** (or equivalently, losses incurred to **P1**) if the corresponding paths are selected by the players. The uppercase letters L, M, R stand for Left, Middle, and Right branches, respectively.

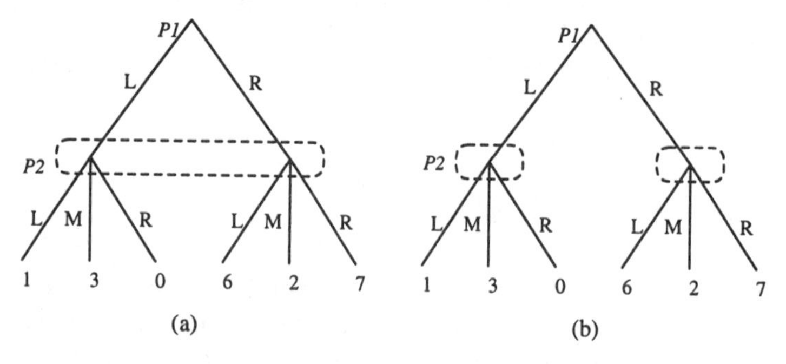

Figure 2.4: Two zero-sum games in extensive form differing only in the information available to **P2**.

It should be noted that the two zero-sum games displayed in Figs 2.4(a) and 2.4(b) are equivalent in every aspect other than the information available to **P2** at the time of his play, which is indicated on the tree diagrams by dotted lines enclosing an area (known as the *information set*) including the relevant nodes. In Fig. 2.4(a), the two possible nodes of **P2** are included in the same dotted area, implying that, even though **P1** acts before **P2** does, **P2** does not have access to his opponent's action. That is to say, at the time of his play, **P2** does not know at what node he really is. This is, of course, equivalent to the case when both players act simultaneously. Hence, simultaneous play can also be represented by a tree structure, provided that the relevant information set is chosen properly. The extensive form of Fig. 2.4(a) can now easily be converted into the normal form, with the equivalent matrix game being the one given by (2.21) in Section 2.3. As it has been discussed there, this matrix game admits a saddle-point solution in mixed strategies, which is ($y_L^* = \frac{2}{3}, y_R^* = \frac{1}{3}; z_L^* = \frac{1}{3}, z_M^* = \frac{2}{3}, z_R^* = 0$), with the mixed saddle-point value being $V_{\rm m} = 8/3$. The extensive form of Fig. 2.4(b), however, admits a different matrix game as its normal form and induces a different behavior on the players. In this case, each node of **P2** is included in a separate information set, thus implying that **P2** has perfect information as to which branch of the tree **P1** has chosen. Hence, if u^1 denotes the actual choice of **P1** and $\gamma^2(\cdot)$ denotes a strategy for **P2**, as a maximizer **P2**'s optimal choice will be

$$\gamma^{2*}(u^1) = \begin{cases} M & \text{if} \quad u^1 = L, \\ R & \text{if} \quad u^1 = R. \end{cases} \tag{2.29a}$$

Not knowing this situation ahead of time, **P1** definitely adopts the strategy

$$\gamma^{1*} \equiv u^{1*} = L \tag{2.29b}$$

with the equilibrium outcome of the game being 3, which is, in fact, the upper value of the matrix game (2.21).

Thus, we have obtained a solution of the zero-sum game of Fig. 2.4(b) by directly making use of the extensive tree formulation. Let us now attempt to obtain the (same) saddle-point solution by transforming the extensive form into an equivalent normal form. To this end, we first delineate the possible strategies of the players. For **P1**, there are two possibilities:

$$\gamma^1 = L \text{ and } \gamma^1 = R.$$

For **P2**, however, since he observes the action of **P1**, there exist $3^2 = 9$ possible strategies, which are $\gamma_1^2(u^1) = u^1$, $\gamma_2^2(u^1) = L$, $\gamma_3^2(u^1) = M$, $\gamma_4^2(u^1) = R$,

$$\gamma_5^2(u^1) = \begin{cases} L & \text{if} \quad u^1 = L, \\ M & \text{if} \quad u^1 = R, \end{cases}$$

$$\gamma_6^2(u^1) = \begin{cases} R & \text{if} \quad u^1 = L, \\ L & \text{if} \quad u^1 = R, \end{cases}$$

$$\gamma_7^2(u^1) = \begin{cases} M & \text{if} \quad u^1 = L, \\ R & \text{if} \quad u^1 = R, \end{cases}$$

$$\gamma_8^2(u^1) = \begin{cases} M & \text{if} \quad u^1 = L, \\ L & \text{if} \quad u^1 = R, \end{cases}$$

$$\gamma_9^2(u^1) = \begin{cases} R & \text{if} \quad u^1 = L, \\ M & \text{if} \quad u^1 = R, \end{cases}$$

where the subscripts $i = 1, \ldots, 9$ denote a particular ordering (labeling) of the possible strategies of **P2**. Hence, the equivalent normal form of the zero-sum game of Fig. 2.4(b) is the 2×9 matrix game

	P2									
	1	1	3	0	1	0	(3*)	(3)	0	L
P1	7	6	2	7	2	6	7	6	2	R
	1	2	3	4	5	6	7	8	9	

which admits two saddle points, as indicated, with the dominant one being $\{L, \text{column } 7\}$. It is this dominant saddle-point solution that corresponds to the one given by (2.29a)-(2.29b), and thus we observe that the derivation outlined earlier, which utilizes directly the extensive tree structure, could cancel out dominated saddle points. This, however, does not lead to any real loss of generality, since, if a matrix game admits a pure-strategy saddle point, it also admits a dominant pure-strategy saddle point, and furthermore, the saddle-point value is unique (regardless of the number of pure-strategy equilibria).

We should now note that the notion of "*(pure) strategy*" introduced above within the context of the zero-sum game of Fig. 2.4(b) is somewhat different from (and, in fact, more general than) the one adopted in Section 2.2. This difference arises mainly because of the dynamic character of the decision problem of Fig. 2.4(b)—**P2** being in a position to know exactly how **P1** acts. A (pure) strategy for **P2**, in this case, is a rule that tells him which one of his alternatives to choose, for each possible action of **P1**. In mathematical language, it is a mapping from the collection of his information sets into the set of his alternatives. We thus see that there is a rather sharp distinction between a strategy and the actual action dictated by that strategy, unless there is only one information set (which is the case in Fig. 2.4(a)). In the case of a single information set, the notions of strategy and action definitely coincide, and that is why we have used these two terms interchangeably in Section 2.2 while discussing matrix games with pure-strategy solutions.

We are now in a position to give precise definitions of some of the concepts introduced above.

Definition 2.5 *An* extensive form *of a two-person zero-sum finite game without chance moves is a finite tree structure with*

(i) a specific vertex *indicating the starting point of the game,*

(ii) *a* pay-off function *assigning a real number to each terminal vertex of the tree, which determines the pay-off (respectively, loss) to* **P2** *(respectively,* **P1***),*

(iii) *a partition of the nodes of the tree into two* player sets *(to be denoted by* $\bar{N}^1$ *and* $\bar{N}^2$ *for* **P1** *and* **P2***, respectively),*

(iv) *a subpartition of each player set* $\bar{N}^i$ *into* information sets $\{\eta_j^i\}$*, such that the same number of immediate branches emanates from every node belonging to the same information set, and no node follows another node in the same information set.*[7]

Remark 2.2 This definition of an extensive form covers more general types of two-person zero-sum games than the ones discussed heretofore since it also allows a player to act more than once in the decision process, with the information sets being different at different levels of play. Such games are known as *multi-act games*, and they will be discussed in Section 2.5. In the remaining parts of this section we shall confine our analysis to single-act games—games in which each player acts only once. □

Remark 2.3 It is possible to extend Def. 2.5 so that it also allows for chance moves by a third party called "nature". Such an extension will be incorporated in Section 2.6. In the present section, and also in Section 2.5, we only consider the class of zero-sum finite games which do not incorporate any chance moves. □

Definition 2.6 *Let* N^i *denote the class of all information sets of* **P***i, with a typical element designated as* η^i*. Let* $U^i_{\eta^i}$ *denote the set of alternatives of* **P***i at the nodes belonging to the information set* η^i*. Define* $U^i = \cup U^i_{\eta^i}$ *where the union is over* $\eta^i \in N^i$*. Then, a* strategy γ^i *for* **P***i is a mapping from* N^i *into* U^i*, assigning one element in* U^i *for each set in* N^i*, and with the further property that* $\gamma^i(\eta^i) \in U^i_{\eta^i}$ *for each* $\eta^i \in N^i$*. The set of all strategies for* **P***i is called his* strategy set *(*space*), and it is denoted by* Γ^i*.*

Remark 2.4 For the starting player, it is convenient to take N^i = a singleton, to eliminate any possible ambiguity. □

Example 2.1 In the extensive form of Fig. 2.4(b), **P2** has two information sets ($N^2 = \{$(node 1), (node 2)$\}$) and three alternatives ($U^2 = \{L,\ M,\ R\}$). Therefore, he has nine possible strategies which have actually been listed earlier during the *normalization phase* of this extensive form. Here, by an abuse of notation, $\eta^2 = \{u^1\}$. □

---

[7]For tree structures in which an information set contains more than one node from any directed path in the tree or for games in which the underlying graph contains cycles (and hence is not a tree anymore), the reader is referred to subsection 2.7.1.

Definition 2.7 *A strategy* $\gamma^i(\cdot)$ *of* **P***i is said to be* constant *if its value does not depend on its argument* η^i.

Saddle-point equilibria of single-act games

The pure-strategy equilibrium solution of single-act two-person zero-sum games in extensive form can be defined in a way analogous to Def. 2.1, with appropriate notational modifications. To this end, let $J(\gamma^1, \gamma^2)$ denote the numerical outcome of a game in extensive form, interpreted as the loss incurred to **P1** when **P1** and **P2** employ the strategies $\gamma^1 \in \Gamma^1$ and $\gamma^2 \in \Gamma^2$, respectively. This loss function in fact defines the correspondences between different ordered strategy pairs and outcomes, and thus describes the equivalent normal form of the zero-sum game. Then, we have the following.

Definition 2.8 *A pair of strategies* $\{\gamma^{1*} \in \Gamma^1, \gamma^{2*} \in \Gamma^2\}$ *is in* saddle-point equilibrium *if the following set of inequalities is satisfied for all* $\gamma^1 \in \Gamma^1$, $\gamma^2 \in \Gamma^2$:

$$J(\gamma^{1*}, \gamma^2) \leq J(\gamma^{1*}, \gamma^{2*}) \leq J(\gamma^1, \gamma^{2^*}). \tag{2.30}$$

The quantity $J(\gamma^{1*}, \gamma^{2*})$ *is known as the* saddle-point value *of the zero-sum game.*

Since a saddle-point equilibrium is defined in terms of the normal form of a game, a direct method for solving two-person single-act games in extensive form would be first to convert them into equivalent normal form and solve for the saddle-point solution of the resulting matrix game, and then to interpret this solution in terms of the original game through the one-to-one correspondences that exist between the rows and columns of the matrix and strategies of the players. This method has already been illustrated in this section in conjunction with the solution of the zero-sum game of Fig. 2.4(b), and as we have observed there, a major disadvantage of such a direct approach is that one has to list down all possible strategies of the players and consider a rather high dimensional matrix game, especially when the second-acting player has several information sets.

An alternative to this approach exists, which makes direct use of the extensive form description of the zero-sum game. This, in fact, is a recursive procedure, and it obtains the solution without necessarily considering all strategy combinations, especially when the second-acting player has more than one information set. The strategies not considered in this procedure are the dominated ones, and this has actually been observed earlier in this section when we employed the technique to arrive at the equilibrium strategy pair (2.29a)-(2.29b) for the zero-sum game of Fig. 2.4(b). Before providing a general outline of the steps involved in this derivation, let us first consider another, somewhat more elaborate, example which is the zero-sum game whose extensive form is depicted in Fig. 2.5.

To determine the saddle-point strategies associated with this zero-sum game, we first note that if $u^1 = R$, then **P2** should choose $\gamma^2(R) = L$, to result in a loss

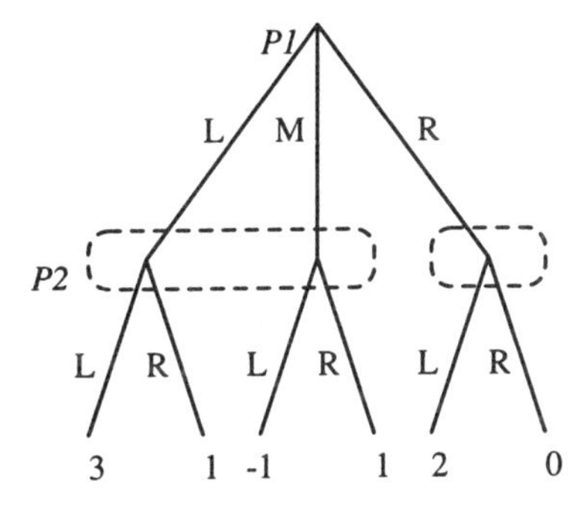

Figure 2.5: An extensive form that admits a saddle-point equilibrium in pure strategies.

of 2 to **P1**. If **P1** picks L or M, however, **P2** cannot differentiate between these two, since they belong to the same information set; and hence now a matrix game has to be solved for that part of the tree. The equivalent normal form is

	P2		
P1	3	1	L
	−1	(1)	M
	L	R	

which clearly admits the unique saddle-point solution $u^1 = M$, $u^2 = R$, yielding a cost of 1 for **P1**. Hence, in the actual play of the game, **P1** will always play $u^1 = \gamma^1 = M$, and **P2**'s optimal response will be $u^2 = R$. However, the saddle-point strategy of **P2** is

$$\gamma^{2*}(\eta^2) = \begin{cases} L & \text{if } u^1 = R, \\ R & \text{otherwise} \end{cases} \tag{2.31}$$

and not the constant strategy $\gamma^2(u^1) = R$, since if **P2** adopts this constant strategy, then **P1** can switch to R to a loss of 0 instead of 1. The strategy for **P1** that is in equilibrium with (2.31) is still

$$\gamma^{1*} = M, \tag{2.32}$$

since N^1 is a singleton. The reader can now easily check that the pair (2.31)-(2.32) satisfies the saddle-point inequalities (2.30).

The preceding analysis now readily suggests a method of obtaining the pure-strategy saddle-point solution of single-act zero-sum games in extensive form, provided that it exists. The steps involved are as follows.

Derivation of pure-strategy saddle-point solutions of single-act games in extensive form

(1) For each information set of the second-acting player, solve the corresponding matrix game (assuming that each such matrix game admits a pure-strategy saddle point). For the case in which an information set is a

singleton, the matrix game involved is a degenerate one with one of the players having a single choice—but such (essentially one-person) games always admit well-defined pure-strategy solutions.

(2) Record the saddle-point value of each of these matrix games. If **P1** is the starting player, then the lowest of these is the saddle-point value of the extensive form; if **P2** is the starting player, then the highest one is.

(3) The saddle-point strategy of the first-acting player is his saddle-point strategy obtained for the matrix game whose saddle-point value corresponds to the saddle-point value of the extensive game.

(4) For the second-acting player, the saddle-point strategy is comprised of his saddle-point solutions in all the matrix games considered, by appropriately identifying them with each of his information sets.

We leave it to the reader to verify that any pair of equilibrium strategies obtained by following the preceding procedure does, indeed, constitute a saddle-point strategy pair that satisfies (2.30). The following conclusions can now readily be drawn.

Proposition 2.1 *A single-act zero-sum two-person finite game in* extensive form *admits a (pure-strategy) saddle-point solution if, and only if, each matrix game corresponding to the information sets of the second-acting player has a saddle point in pure strategies.*

Proposition 2.2 *Every single-act zero-sum two-person finite game in extensive form, in which the information sets of the second-acting player are singletons,*[8] *admits a pure-strategy saddle-point solution.*

If the matrix game corresponding to an information set does not admit a saddle-point solution in pure strategies, then it is clear that the strategy spaces of the players have to be enlarged, in a way similar to the introduction of mixed strategies in Section 2.2 within the context of matrix games. To pave the way for such an enlargement in the case of games in extensive form, let us first consider the single-act two-person zero-sum game whose extensive form is depicted in Fig. 2.6. Here, **P2** is the second-acting player and he has two information sets. If his opponent picks R, then he can observe that choice, and his best response (strategy) in that case is L, yielding an outcome of 1. If **P1** chooses L or M, however, then **P2** is ignorant about that choice, since both of these nodes are included in the same information set of **P2**. The relevant matrix game, in this case, is

P1 \ P2	L	R	
	3	0	L
	−1	1	M

[8]The node corresponding to each such information set is known as the "state" of the game at that level (stage) of play. Games of this type are known as *perfect information games.*

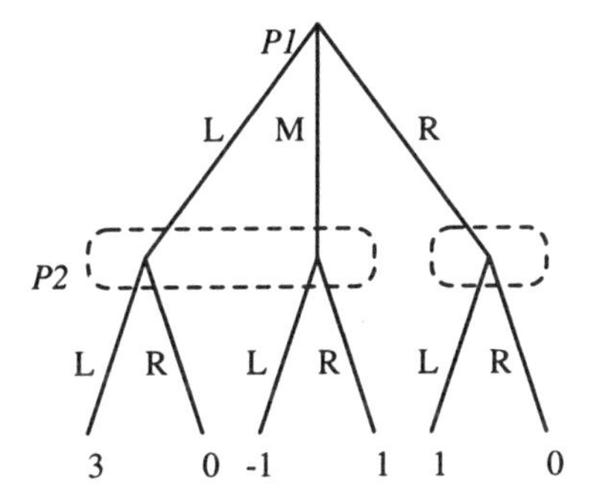

Figure 2.6: A game in extensive form that admits a mixed (behavioral)-strategy saddle-point solution.

which is the matrix game (2.20) considered earlier in Section 2.3, and it is known to admit the mixed saddle-point solution ($y_L^* = \frac{2}{5}, y_M^* = \frac{3}{5}$) for **P1** and ($z_L^* = \frac{1}{5}, z_R^* = \frac{4}{5}$) for **P2**, yielding an average outcome of $\frac{3}{5}$. Since $\frac{3}{5} < 1$, it is clear that **P1** will prefer to stay on that part of the tree and thus stick to the strategy

$$\hat{\gamma}^{1*} = \begin{cases} L & \text{w.p.} \quad 2/5, \\ M & \text{w.p.} \quad 3/5, \\ R & \text{w.p.} \quad 0, \end{cases} \tag{2.33a}$$

which is "mixed" in nature. **P2**'s equilibrium strategy will then be

$$\hat{\gamma}^{2*} = \begin{cases} L \quad \text{w.p.} \quad 1 & \text{if } u^1 = R, \\ \left. \begin{array}{l} L \quad \text{w.p.} \quad 1/5, \\ R \quad \text{w.p.} \quad 4/5 \end{array} \right. & \text{otherwise ,} \end{cases} \tag{2.33b}$$

since he also has to consider the possibility of **P1** playing R (by definition of a strategy). The pair (2.33a)-(2.33b) now yields the average outcome of $\frac{3}{5}$.

In order to declare (2.33a)-(2.33b) as the equilibrium solution pair of the extensive form of Fig. 2.6, we have to specify the spaces in which it possesses such a property. To this end, we first note that **P1** has three possible pure strategies, $\gamma_1^1 = L$, $\gamma_2^1 = M$ and $\gamma_3^1 = R$, and in view of this, (2.33a) is indeed a mixed strategy for **P1** (see Def. 2.2). For **P2**, however, there exist four possible pure strategies, namely

$$\begin{aligned} \gamma_1^2(\eta^2) &= L, \quad \gamma_2^2(\eta^2) = R, \\ \gamma_3^2(\eta^2) &= \begin{cases} L & \text{if } u^1 = R, \\ R & \text{otherwise ,} \end{cases} \end{aligned}$$

and

$$\gamma_4^2(\eta^2) = \begin{cases} R & \text{if } u^1 = R, \\ L & \text{otherwise .} \end{cases}$$

Hence, a mixed strategy for **P2** is a probability law according to which these four strategies will be "mixed" during the play of the game, and (2.33b) provides one such probability distribution which assigns a weight of $\frac{1}{5}$ to the first pure strategy and a weight of $\frac{4}{5}$ to the third one.

We have thus verified that both (2.33a) and (2.33b) are well-defined mixed strategies for **P1** and **P2**, respectively. Now, if $\hat{J}(\hat{\gamma}^1, \hat{\gamma}^2)$ denotes the average outcome of the game when the mixed strategies $\hat{\gamma}^1$ and $\hat{\gamma}^2$ are adopted by **P1** and **P2**, respectively, the reader can easily check that the pair (2.33a)-(2.33b) indeed satisfies the relevant set of saddle-point inequalities

$$\hat{J}(\hat{\gamma}^{1*}, \hat{\gamma}^2) \leq \hat{J}(\hat{\gamma}^{1*}, \hat{\gamma}^{2*}) \leq \hat{J}(\hat{\gamma}^1, \hat{\gamma}^{2*})$$

and thus constitutes a saddle-point solution within the class of mixed strategies.

However, the class of mixed strategies is, in general, an unnecessarily large class for the second-acting player. In the extensive form of Fig. 2.6, for example, since **P2** can tell exactly whether **P1** has played R or not, there is no reason why he should mix between his possible actions in case of $u^1 = R$ and his actions otherwise. That is, **P2** can do equally well by considering probability distributions only on the alternatives belonging to the information set including two nodes, and (2.33b) is, in fact, that kind of a strategy. Such strategies are known as *behavioral strategies*, a precise definition of which is given below.

Definition 2.9[9] *For a given two-person zero-sum finite game in extensive form, let Y_{η^1} denote the set of all probability distributions on $U^1_{\eta^1}$, where the latter is the set of all alternatives of* **P1** *at the nodes belonging to the information set η^1. Analogously, let Z_{η^2} denote the set of all probability distributions on $U^2_{\eta^2}$. Further define $Y = \cup_{N^1} Y_{\eta^1}$, $Z = \cup_{N^2} Z_{\eta^2}$. Then, a* behavioral strategy *$\hat{\gamma}^1$ for* **P1** *is a mapping from the class of all his information sets (N^1) into Y, assigning one element in Y for each set in N^1, such that $\hat{\gamma}^1(\eta^1) \in Y_{\eta^1}$ for each $\eta^1 \in N^1$. A typical behavioral strategy $\hat{\gamma}^2$ for* **P2** *is defined, analogously, as a restricted mapping from N^2 into Z. The set of all behavioral strategies for* **P*i*** *is called his* behavioral strategy set, *and it is denoted by $\hat{\Gamma}^i$.*

Remark 2.5 Every behavioral strategy is a mixed strategy, but every mixed strategy is not necessarily a behavioral strategy, unless the player has a single information set. □

For games in extensive form, the concept of a saddle-point equilibrium in behavioral strategies can now be introduced as in Def. 2.8, but with some slight obvious modifications. If $\hat{J}(\hat{\gamma}_1, \hat{\gamma}_2)$ denotes the average outcome of the game resulting from the strategy pair $\{\hat{\gamma}^1 \in \hat{\Gamma}^1, \hat{\gamma}^2 \in \hat{\Gamma}^2\}$, then we have the following.

Definition 2.10 *A pair of strategies $\{\hat{\gamma}^{1^*} \in \hat{\Gamma}^1, \hat{\gamma}^{2^*} \in \hat{\Gamma}^2\}$ is said to constitute a* saddle point in behavioral strategies *for a zero-sum game in extensive form, if the set of inequalities*

$$\hat{J}(\hat{\gamma}^{1*}, \hat{\gamma}^2) \leq \hat{J}(\hat{\gamma}^{1*}, \hat{\gamma}^{2*}) \leq \hat{J}(\hat{\gamma}^1, \hat{\gamma}^{2*}) \tag{2.34}$$

[9]This definition is valid not only for single-act games, but also for multi-act games which are treated in section 2.5.

is satisfied for all $\hat{\gamma}^1 \in \hat{\Gamma}^1$, $\hat{\gamma}^2 \in \hat{\Gamma}^2$. *The quantity* $\hat{J}(\hat{\gamma}^{1*}, \hat{\gamma}^{2*})$ *is known as the saddle-point value of the game in behavioral strategies.*

For a given single-act zero-sum game in extensive form, derivation of a behavioral-strategy saddle-point solution basically follows the four steps outlined earlier for pure-strategy saddle points, with the only difference being that this time mixed equilibrium solutions of the matrix games are also allowed in this process. The reader should note that this routine extension has already been illustrated within the context of the extensive form of Fig. 2.6.

Since every matrix game admits a saddle point in mixed strategies (cf. Corollary 2.3), and further, since the above derivation involves only a finite number of comparisons, it readily follows that every single-act zero-sum game in extensive form admits a saddle point in behavioral (or, equivalently, mixed) strategies. We thus conclude this section with a precise statement of this property, and by recording another immediate feature of the saddle-point solution in extensive games.

Corollary 2.4 *In every single-act two-person zero-sum finite game,*

(i) there exists at least one saddle point in behavioral strategies,

(ii) if there exist more than one saddle-point solution in behavioral strategies, then they possess the ordered interchangeability property, and

(iii) every saddle point in behavioral strategies is also a saddle point in the larger class of mixed strategies.

We should note that a finite single-act game in extensive form might also admit a mixed-strategy saddle point (which is not behavioral) in addition to a behavioral saddle point. The corresponding average saddle-point values, however, will all be the same, because of the interchangeability property, and therefore we may confine our attention only to behavioral strategies, particularly in view of Corollary 2.4(i).

2.5 Extensive Games: Multi-Act Games

Zero-sum games in which at least one player is allowed to act more than once and with possibly different information sets at each level of play, are known as *multi-act zero-sum games.*[10] In the study of extensive forms of such games, and in accordance with Def. 2.5, we consider the case when the number of alternatives available to a player at each information set is finite. This leads to a finite tree structure, incorporating possibly different information sets for each

[10] Such games are also referred to as "multi-stage zero-sum games" in the literature. They may be considered also as "dynamic games", since the information sets of a player do, in general, have a dynamic character, providing information concerning the past actions of the players. In this respect, the single-act games of the previous section may also be referred to as "dynamic games" if the second-acting player has at least two information sets.

player at each level of play. A (pure) strategy of a player is again defined as a restricted mapping from the collection of his information sets to the finite set of all his alternatives, exactly as in Def. 2.6, and the concept of a saddle-point solution then follows the lines of Def. 2.8. This general framework, however, suppresses the dynamic nature of the decision process and is not constructive as far as the saddle-point solution is concerned. An alternative to this set-up exists, which brings out the dynamic nature of the problem, for an important class of multi-act zero-sum games known as "feedback games", a precise definition of which is given below.

Definition 2.11 *A multi-act two-person zero-sum game in extensive form is called a* two-person zero-sum feedback game in extensive form, *if*

(i) at the time of his action, each player has perfect information concerning the current level of play, i.e., no information set contains nodes of the tree belonging to different levels of play,

(ii) information sets of the first-acting player at every level of play are singletons, and the information sets of the second-acting player at every level of play are such that none of them include nodes corresponding to branches emanating from two or more different information sets of the other player, i.e., each player knows the state of the game at every level of play.

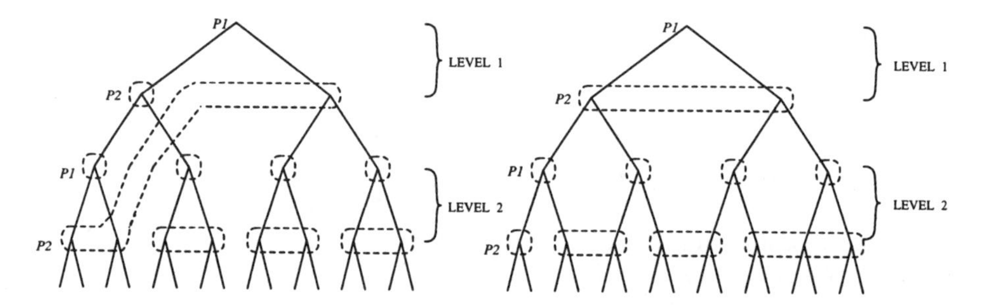

Figure 2.7: Two multi-act zero-sum games which are not feedback games.

Figure 2.7 displays two multi-act zero-sum games in extensive form, which are not feedback games, since the first one violates (i) above, and the second one violates (ii). The one displayed in Fig. 2.8, however, is a feedback game.

Now, let the number of levels of play in a zero-sum feedback game be K. Then, a typical strategy γ^i of $\mathbf{P}i$ in such a game can be viewed as composed of K components $(\gamma_1^i, \ldots, \gamma_K^i)$, where γ_j^i stands for the corresponding strategy of $\mathbf{P}i$ at his jth level of action. Moreover, because of the nature of a feedback game, γ_j^i can be taken, without any loss of generality, to have as its domain only those information sets of $\mathbf{P}i$ which pertain to the jth level of play. Let us denote the collection of all such strategies for $\mathbf{P}i$ at level j by Γ_j^i. Then, we can rewrite the saddle-point inequality (2.30), for a feedback game in extensive

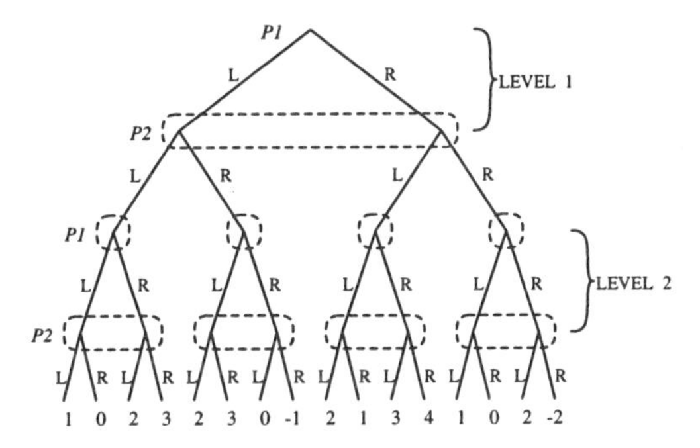

Figure 2.8: A zero-sum feedback game in extensive form.

form, as

$$\begin{aligned} J(\gamma_1^{1*},\dots,\gamma_K^{1*};\gamma_1^2,\dots,\gamma_K^2) &\le J(\gamma_1^{1*},\dots,\gamma_K^{1*};\gamma_1^{2*},\dots,\gamma_K^{2*}) \\ &\triangleq J^* \le J(\gamma_1^1,\dots,\gamma_K^1;\gamma_1^{2*},\dots,\gamma_K^{2*}) \end{aligned} \tag{2.35}$$

which is to be satisfied for all $\gamma_j^i \in \Gamma_j^i$, $i = 1,2$; $j = 1,\dots,K$. Such a decomposition of the strategies of the players now allows a recursive procedure for the determination of the saddle-point solution $\{\gamma^{1*},\gamma^{2*}\}$ of the feedback game, if we impose some further restrictions on the pair $\{\gamma^{1*},\gamma^{2*}\}$. Namely, let the pair $\{\gamma^{1*},\gamma^{2*}\}$ also satisfy (recursively) the following set of K pairs of inequalities for all $\gamma_j^i \in \Gamma_j^i$, $i = 1,2$; $j = 1,\dots,K$:

$$\left.\begin{aligned} &J(\gamma_1^1,\dots,\gamma_{K-1}^1,\gamma_K^{1*};\gamma_1^2,\dots,\gamma_K^2) \le J(\gamma_1^1,\dots,\gamma_{K-1}^1,\gamma_K^{1*};\gamma_1^2,\dots,\\ &\qquad\qquad \gamma_{K-1}^2,\gamma_K^{2*}) \le J(\gamma_1^1,\dots,\gamma_K^1;\gamma_1^2,\dots,\gamma_{K-1}^2,\gamma_K^{2*}),\\ &J(\gamma_1^1,\dots,\gamma_{K-2}^1,\gamma_{K-1}^{1*},\gamma_K^{1*};\gamma_1^2,\dots,\gamma_{K-1}^2,\gamma_K^{2*})\\ &\qquad \le J(\gamma_1^1,\dots,\gamma_{K-2}^1,\gamma_{K-1}^{1*},\gamma_K^{1*};\gamma_1^2,\dots,\gamma_{K-2}^2,\gamma_{K-1}^{2*},\gamma_K^{2*})\\ &\qquad \le J(\gamma_1^1,\dots,\gamma_{K-1}^1,\gamma_K^{1*};\gamma_1^2,\dots,\gamma_{K-2}^2,\gamma_{K-1}^{2*},\gamma_K^{2*}),\\ &\cdots\cdots\cdots\cdots\cdots\cdots\cdots\cdots\\ &\cdots\cdots\cdots\cdots\cdots\cdots\cdots\cdots\\ &J(\gamma_1^{1*},\dots,\gamma_K^{1*};\gamma_1^2,\gamma_2^{2*},\dots,\gamma_K^{2*}) \le J(\gamma_1^{1*},\dots,\gamma_K^{1*};\gamma_1^{2*},\dots,\gamma_K^{2*})\\ &\qquad \le J(\gamma_1^1,\gamma_2^{1*},\dots,\gamma_K^{1*};\gamma_1^{2*},\dots,\gamma_K^{2*}). \end{aligned}\right\} \tag{2.36}$$

Definition 2.12 *For a two-person zero-sum feedback game in extensive form with K levels of play, let $\{\gamma^{1*},\gamma^{2*}\}$ be a pair of strategies satisfying (2.35) and (2.36) for all $\gamma_j^i \in \Gamma_j^i$, $i = 1,2$; $j = 1,\dots,K$. Then, $\{\gamma^{1*},\gamma^{2*}\}$ is said to constitute a pair of (pure)* feedback saddle-point strategies *for the feedback game.*

Proposition 2.3 *Every pair $\{\gamma^{1*},\gamma^{2*}\}$ that satisfies the set of inequalities (2.36) also satisfies the pair of saddle-point inequalities (2.35). Hence, the requirement for satisfaction of (2.35) in Def. 2.12 is redundant, and it can be deleted without any loss of generality.*

Proof. Replace $\gamma_1^1, \ldots, \gamma_{K-1}^1$ on the LHS inequalities of the set (2.36) by $\gamma_1^{1*}, \ldots, \gamma_{K-1}^{1*}$, respectively. Then, the LHS inequalities of (2.36) read

$$J(\gamma_1^{1*}, \ldots, \gamma_K^{1*}; \gamma_1^2, \ldots, \gamma_K^2) \le J(\gamma_1^{1*}, \ldots, \gamma_K^{1*}; \gamma_1^2, \ldots, \gamma_{K-1}^2, \gamma_K^{2*}),$$

$$J(\gamma_1^{1*}, \ldots, \gamma_K^{1*}; \gamma_1^2, \ldots, \gamma_{K-1}^2, \gamma_K^{2*}) \le J(\gamma_1^{1*}, \ldots, \gamma_K^{1*}; \gamma_1^2, \ldots, \gamma_K^2, \gamma_{K-1}^{2*}, \gamma_K^{2*}),$$

$$\cdots\cdots\cdots\cdots\cdots\cdots\cdots\cdots$$

$$\cdots\cdots\cdots\cdots\cdots\cdots\cdots\cdots$$

$$J(\gamma_1^{1*}, \ldots, \gamma_K^{1*}; \gamma_1^2, \gamma_2^{2*}, \ldots, \gamma_K^{2*}) \le J(\gamma_1^{1*}, \ldots, \gamma_K^{1*}; \gamma_1^{2*}, \ldots, \gamma_K^{2*})$$

for all $\gamma_j^2 \in \Gamma_j^2$, $j = 1, \ldots, K$. But this set of inequalities reduces to the single inequality

$$J(\gamma_1^{1*}, \ldots, \gamma_K^{1*}; \gamma_1^2, \ldots, \gamma_K^2) \le J(\gamma_1^{1*}, \ldots, \gamma_K^{1*}; \gamma_1^{2*}, \ldots, \gamma_K^{2*}),$$

since they characterize the recursive dynamic programming conditions for a one-person maximization problem (Bellman (1957)). This then establishes the validity of the LHS inequality of (2.35) under satisfaction of (2.36). Validity of the RHS inequality of (2.35) can likewise be verified by considering the RHS inequalities of (2.36). □

The appealing feature of a feedback saddle-point solution is that it can be computed recursively, by solving a number of static games at each level of play. Details of this recursive procedure readily follow from the set of inequalities (2.36) and the specific structure of the information sets of the game, as it is described below.

A recursive procedure to determine the (pure) feedback saddle-point strategies of a feedback game

(1) Starting at the last level of play, K, solve each single-act game corresponding to the information sets of the first-acting player at that level K. The resulting (pure) saddle-point strategies are the saddle-point strategies $\{\gamma_K^{1*}, \gamma_K^{2*}\}$ sought for the feedback game at level K. Record the value of each single-act game corresponding to each of the information sets of the first-acting player, and assign these values to the corresponding nodes (states) of the extensive form at level K.

(2) Cross out the Kth level of play, and consider the resulting $(K-1)$ level feedback game. Now solve the last level single-act games of that extensive form, with the resulting saddle-point strategies denoted by $\{\gamma_{K-1}^{1*}, \gamma_{K-1}^{2*}\}$, and repeat the remaining deeds of step 1 with K replaced by $K-1$.

....

....

(K) Cross out the second level of play, and consider the resulting single-act game in extensive form. Denote its saddle-point strategies by $\{\gamma_1^{1*}, \gamma_1^{2*}\}$, and its value by V. Then, the original feedback game in extensive form admits a saddle-point solution $\{(\gamma_1^{1*},\ldots,\gamma_K^{1*}),(\gamma_1^{2*},\ldots,\gamma_K^{2*})\}$ and has value $J^* = V$.

Example 2.2 As an illustration of the above procedure, let us consider the 2-level feedback game whose extensive form is depicted in Fig. 2.8. At the last (second) level, there are four single-act games to be solved, which are all equivalent to matrix games. From left to right, we have the normal forms

(1)	0
2	3

2	3
(0)	−1

(2)	1
3	4

(1)	0
2	−2

with the encircled entries indicating the location of the saddle-point solution in each case. The saddle-point strategies γ_2^{1*} and γ_2^{2*} are thus given by

$$\gamma_2^{1*}(\eta_2^1) = \begin{cases} R & \text{if } u_1^1 = L, u_1^2 = R, \\ L & \text{otherwise,} \end{cases} \quad \gamma_2^{2*}(\eta_2^2) = L, \tag{2.37a}$$

where u_1^i denotes the choice of $\mathbf{P}i$ at level 1. It is noteworthy that the saddle-point strategy of **P2** is a constant at this level of play. Now, crossing out the second level of play, but retaining the encircled quantities in the four matrix games as costs attached to the four nodes, we end up with the equivalent single-act game

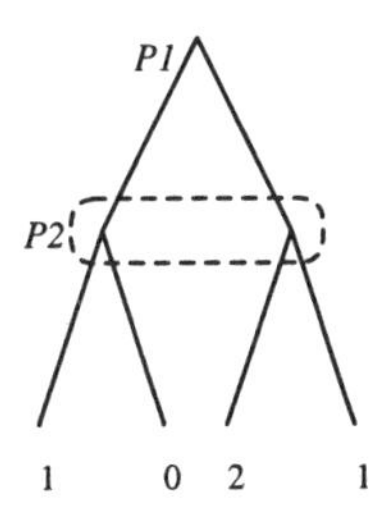

whose normal form is

(1)	0
2	1

admitting the unique saddle-point solution

$$\gamma_1^{1*} = L, \quad \gamma_1^{2*} = L, \tag{2.37b}$$

and with the value being $V = 1$, which is also the value of the original 2-level feedback game. The feedback saddle-point strategies of this feedback game are now given by (2.37a)-(2.37b), with the actual play dictated by them being

$$u_1^1 = u_2^1 = u_1^2 = u_2^2 = L. \tag{2.38}$$

□

If a feedback game in extensive form does not admit a feedback saddle point in pure strategies, then it could still admit a saddle point in accordance with inequalities (2.35). If that is not possible either, then one has to extend the strategy spaces of the players so as to include also behavioral strategies. Definition of a behavioral strategy in multi-act games is precisely the one given by Def. 2.9 and the relevant saddle-point inequality is still (2.34), of course under the right kind of interpretation. Now, in a zero-sum feedback game with K levels of play, a typical behavioral strategy $\hat{\gamma}^i$ of $\mathbf{P}i$ can also be viewed as composed of K components $(\hat{\gamma}_1^i, \ldots, \hat{\gamma}_K^i)$, where $\hat{\gamma}_j^i$ stands for the corresponding behavioral strategy of $\mathbf{P}i$ at his jth level of play—a feature that follows readily from Defs. 2.9 and 2.11. By the same token, $\hat{\gamma}_j^i$ can be taken, without any loss of generality, to have as its domain only those information sets of $\mathbf{P}i$, which pertain to the jth level of play. Let us now denote the collection of all such behavioral strategies for $\mathbf{P}i$ at level j by $\hat{\Gamma}_j^i$. Then, since the saddle-point inequality (2.34) corresponding to a K-level feedback game can equivalently be written as

$$\begin{aligned}\hat{J}(\hat{\gamma}_1^{1*}, \ldots, \hat{\gamma}_K^{1*}; \hat{\gamma}_1^{2}, \ldots, \hat{\gamma}_K^{2}) \le \hat{J}(\hat{\gamma}_1^{1*}, \ldots, \hat{\gamma}_K^{1*}; \hat{\gamma}_1^{2*}, \ldots, \hat{\gamma}_K^{2*}) \triangleq \hat{J}^* \\ \le \hat{J}(\hat{\gamma}_1^{1}, \ldots, \hat{\gamma}_K^{1}; \hat{\gamma}_1^{2*}, \ldots, \hat{\gamma}_K^{2*}), \forall \hat{\gamma}_j^i \in \hat{\Gamma}_j^i, i = 1, 2; j = 1, \ldots, K,\end{aligned} \tag{2.39}$$

it follows that a recursive procedure can be devised to determine the behavioral saddle-point strategies if we further require satisfaction of a set of inequalities analogous to (2.36). Such a recursive procedure basically follows the K steps outlined in this section within the context of pure feedback saddle-point solutions, with the only difference being that now also behavioral strategies are allowed in the equilibrium solutions of the single-act games involved. Since this is a routine extension, we do not provide details of this recursive procedure here, but only illustrate it in the sequel in Example 2.3.

It is worth noting at this point that since every single-act two-person zero-sum game admits a saddle-point solution in behavioral strategies (cf. Corollary 2.4), every feedback game will also admit a saddle-point solution in behavioral strategies. More precisely, we have the following proposition.

Proposition 2.4 *Every two-person zero-sum feedback game, which has an extensive form comprised of a finite number of branches, admits a saddle point in behavioral strategies.*

Example 2.3 As an illustration of the derivation of behavioral saddle point(s) in feedback games, let us reconsider the game of Fig. 2.8 with two modifications: The outcomes of the game corresponding to the paths $u_1^1 = R$, $u_1^2 = L$, $u_2^1 = L$, $u_2^2 = L$ and $u_1^1 = R$, $u_1^2 = L$, $u_2^1 = L$, $u_2^2 = R$, respectively, are now taken as 0 and -1, respectively. Then, the four single-act games to be solved at the last (second) level are equivalent to the matrix games (respectively, from left to right) which all admit pure-strategy solutions, as indicated.

(1)	0
2	3

2	3
(0)	−1

(0)	−1
3	4

(1)	0
2	−2

The saddle-point strategies at this level are, in fact, as given by (2.37a).

Now, crossing out the second level of play, but retaining the encircled quantities in the four matrix games as costs attached to the four nodes, we end up with the equivalent single act game

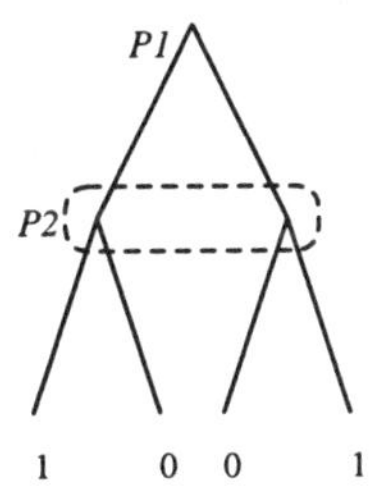

whose normal form

1	0
0	1

admits a mixed saddle-point solution:

$$\hat{\gamma}_1^{1*} = \begin{cases} L & \text{w.p. } 1/2, \\ R & \text{w.p. } 1/2, \end{cases} \quad \hat{\gamma}_1^{2*} = \begin{cases} L & \text{w.p. } 1/2, \\ R & \text{w.p. } 1/2. \end{cases} \tag{2.40}$$

Furthermore, the average saddle-point value is $V_{\mathrm{m}} = 1/2$. Since, within the class of behavioral strategies at level 2, the pair (2.37a) can be written as

$$\hat{\gamma}_2^{1*} = \begin{cases} \left.\begin{matrix} R & \text{w.p. } 1 \\ L & \text{w.p. } 0 \end{matrix}\right\} & \text{if } u_1^1 = L, u_1^2 = R, \\ \left.\begin{matrix} L & \text{w.p. } 1 \\ R & \text{w.p. } 0 \end{matrix}\right\} & \text{otherwise ,} \end{cases} \qquad \hat{\gamma}_2^{2*} = \begin{cases} L & \text{w.p. } 1, \\ R & \text{w.p. } 0, \end{cases} \tag{2.41}$$

(2.40) and (2.41) now constitute a set of behavioral saddle-point strategies for the feedback game under consideration, leading to a behavioral saddle-point value of $\hat{J}^* = 1/2$. □

For multi-act games which are not feedback games, neither Prop. 2.4 nor the recursive procedures discussed in this section are, in general, applicable, and there is no general procedure that would aid in the derivation of pure or behavioral saddle-point solutions, even if they exist. If we allow for mixed-strategy solutions, however, it directly follows from Corollary 2.3 that an equilibrium solution exists, since every multi-act zero-sum finite game in extensive form has an equivalent normal form which is basically a matrix game with a finite number of rows and columns. More precisely, we have the following.

Proposition 2.5 *Every two-person zero-sum multi-act finite game in extensive form admits a saddle-point solution in mixed strategies.*

The following example now illustrates derivation of mixed-strategy saddle-point equilibria in multi-act games, which basically involves solution of an appropriate matrix game.

Example 2.4 Let us reconsider the extensive form of Fig. 2.8, but with a single information set for each player at each level of play, as depicted in Fig. 2.9.

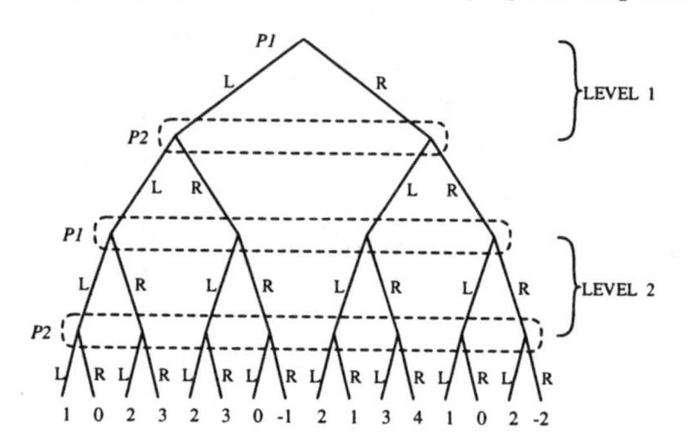

Figure 2.9: A zero-sum multi-act game that does not admit a behavioral saddle-point solution.

Each player has four possible ordered choices (LL, LR, RL, RR), and thus the equivalent normal form of the matrix game is

P1 \ P2	LL	LR	RL	RR
LL	1	0	2	3
LR	2	3	0	−1
RL	2	1	1	0
RR	3	4	2	−2

which admits the unique mixed saddle-point solution

$$\hat{\gamma}^{1*} = \begin{cases} LL & \text{w.p.} \quad 3/5, \\ RR & \text{w.p.} \quad 2/5, \end{cases} \tag{2.42a}$$

$$\hat{\gamma}^{2*} = \begin{cases} LL & \text{w.p.} \quad 4/5, \\ RR & \text{w.p.} \quad 1/5 \end{cases} \tag{2.42b}$$

with the saddle-point value in mixed strategies being $V_m = 7/5$. (The reader can check this result by utilizing Thm. 2.5.) □

Even though every zero-sum multi-act game in extensive form admits a saddle point in mixed strategies, such a result is no longer valid within the class of behavioral strategies, unless the mixed saddle-point strategies happen to be behavioral strategies. A precise version of this property is given below in Prop. 2.6.

Proposition 2.6 *Every saddle-point equilibrium of a zero-sum multi-act game in behavioral strategies also constitutes a saddle-point equilibrium in the larger class of mixed strategies for both players.*

Proof. Assume, to the contrary, that the behavioral saddle-point value ($\hat{J}^*$) is not equal to the mixed saddle-point value (J_m), where the latter always exists by Prop. 2.5, and consider the case $\hat{J}^* < J_m$. This implies that **P1** does better with his behavioral saddle-point strategy than with his mixed saddle-point strategy. But this is not possible, since the set of mixed strategies is a much larger class. Hence, we can only have $\hat{J}^* > J_m$. Now, repeating the same argument on this strict inequality from **P2**'s point of view, it follows that only the equality $\hat{J}^* = J_m$ can hold. □

Proposition 2.6 now suggests a method for derivation of behavioral saddle-point strategies of zero-sum multi-act games in extensive form. First, determine all mixed saddle-point strategies of the equivalent normal form, and then investigate whether there exists, in that solution set, a pair of behavioral strategies. If we apply this method on the extensive form of Example 2.4, we first note that it admits a *unique* saddle-point solution in mixed strategies, which is the one given by (2.42a)-(2.42b). Now, since every behavioral strategy for **P1** is in the form

$$\hat{\gamma}^1 = \begin{cases} LL & \text{w.p.} \quad y_1 y_2, \\ LR & \text{w.p.} \quad y_1(1-y_2), \\ RL & \text{w.p.} \quad y_2(1-y_1), \\ RR & \text{w.p.} \quad (1-y_1)(1-y_2) \end{cases}$$

for some $0 \le y_1 \le 1$, $0 \le y_2 \le 1$,[11] and every behavioral strategy for **P2** is given by

$$\hat{\gamma}^2 = \begin{cases} LL & \text{w.p.} \quad z_1 z_2, \\ LR & \text{w.p.} \quad z_1(1-z_2), \\ RL & \text{w.p.} \quad z_2(1-z_1), \\ RR & \text{w.p.} \quad (1-z_1)(1-z_2) \end{cases}$$

for some $0 \le z_1 \le 1$, $0 \le z_2 \le 1$, it follows (by picking $y_1 = 1$, $y_2 = 3/5$) that (2.42a) is indeed a behavioral strategy for **P1**; however, (2.42b) is not a behavioral strategy for **P2**.

Hence the conclusion is that the extensive form of Fig. 2.9 does not admit a saddle point in behavioral strategies, from which we deduce the following corollary to Prop. 2.6.

Corollary 2.5 *A zero-sum multi-act game does not necessarily admit a saddle-point equilibrium in behavioral strategies, unless it is a feedback game.*

We now introduce a special class of zero-sum multi-act games known as *open-loop games*, in which the players do not acquire any dynamic information throughout the decision process, and they only know the level of play that corresponds to their action. More precisely, see below.

[11]Note that y_1 and y_2 (and also z_1 and z_2, in the sequel) are picked independently, and hence do not necessarily add up to 1.

Definition 2.13 *A multi-act game is said to be an* open-loop game, *if, at each level of play, each player has a single information set.*

The multi-act game whose extensive form is depicted in Fig. 2.9 is an open-loop game. Such games can be viewed as one extreme class of multi-act games in which both players are completely ignorant about the evolution of the decision process. Feedback games, on the other hand, constitute another extreme case (with regard to the structure of the information sets) in which both players have full knowledge of their past actions. Hence, the open-loop and feedback versions of the same multi-act game could admit different saddle-point solutions in pure, behavioral or mixed strategies. A comparison of Examples 2.2 and 2.4, in fact, provides a validation of the possibility of such a situation: the extra information embodied in the feedback formulation actually helps **P1** in that case. It is of course possible to devise examples in which extra information in the above sense instead makes **P2** better off or does not change the value of the game at all. A precise condition for the latter property to hold is given below in Prop. 2.7 whose proof is in the same spirit as that of Prop. 2.6 and is thus left as an exercise for the reader.

Proposition 2.7 *A zero-sum open-loop game in extensive form admits a (pure-strategy) saddle-point solution if, and only if, its feedback version admits constant saddle-point strategies at each level of play.*[12]

Prior and delayed commitment models

Since there is no extra information available to the players throughout the duration of an open-loop game, the players can decide on their actions at the very beginning, and then there is no real incentive for them to renege during the actual play of the game. Such games in which decisions are made at the outset, with no incentive to deviate from them later, are known as "*prior commitment*" games. Mixed saddle-point strategies can then be considered to constitute a reasonable equilibrium solution within such a framework. Feedback games, on the other hand, are of "*delayed commitment*" type, since each player could wait until he finds out at what information set he really is, and only then announces his action. This then implies that mixed strategies are not the "right type" of strategies to be considered for such games, but behavioral strategies are. Since feedback games always admit a saddle point in behavioral strategies (cf. Prop. 2.3), equilibrium is again well defined for this class of games.

There are also other classes of multi-act games of the delayed commitment type, which might admit a saddle point in behavioral strategies. One such class of multi-act zero-sum games are those in which players recall their past actions but are ignorant about the past actions of their opponent. As an illustrative example, let us reconsider the extensive form of Fig. 2.9, but under the present set-up. The information sets of the players would then look as displayed in

[12] It is implicit here that these constant strategies might differ from one level to another and that they might not be feedback saddle-point strategies, i.e., they might not satisfy (2.36).

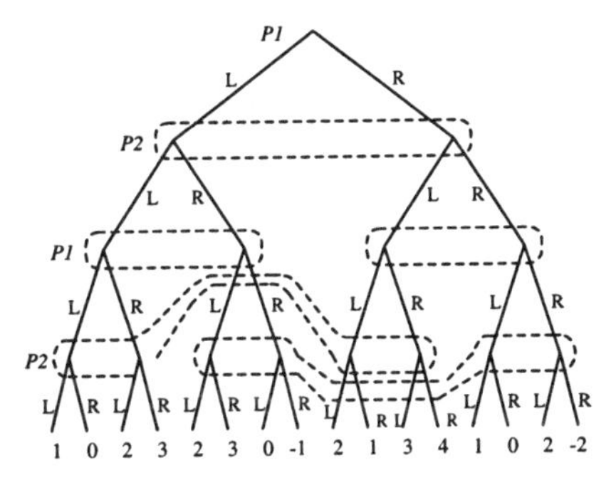

Figure 2.10: A multi-act game of the delayed commitment type that admits a behavioral saddle-point solution.

Fig. 2.10. Now, permissible behavioral strategies of the players in this game are of the form

$$\hat{\gamma}_1^1 = \begin{cases} L & \text{w.p.} \quad y_1, \\ R & \text{w.p.} \quad 1 - y_1, \end{cases} \tag{2.43a}$$

$$\hat{\gamma}_2^1(u_1^1) = \begin{cases} \left.\begin{matrix} L & \text{w.p.} & y_2 \\ R & \text{w.p.} & 1-y_2 \end{matrix}\right\} & \text{if } u_1^1 = L, \\ \left.\begin{matrix} L & \text{w.p.} & y_3 \\ R & \text{w.p.} & 1-y_3 \end{matrix}\right\} & \text{if } u_1^1 = R, \end{cases} \tag{2.43b}$$

$$\hat{\gamma}_1^2 = \begin{cases} L & \text{w.p.} \quad z_1, \\ R & \text{w.p.} \quad 1 - z_1, \end{cases} \tag{2.43c}$$

$$\hat{\gamma}_2^2(u_1^2) = \begin{cases} \left.\begin{matrix} L & \text{w.p.} & z_2 \\ R & \text{w.p.} & 1-z_2 \end{matrix}\right\} & \text{if } u_1^2 = L, \\ \left.\begin{matrix} L & \text{w.p.} & z_3 \\ R & \text{w.p.} & 1-z_3 \end{matrix}\right\} & \text{if } u_1^2 = R, \end{cases} \tag{2.43d}$$

where $0 \le y_i < 1$, $0 \le z_i \le 1$, $i = 1, 2, 3$, and the average outcome of the game corresponding to these strategies can be determined to be

$$\begin{aligned} \hat{J}(\hat{\gamma}^1, \hat{\gamma}^2) &= z_1 z_2 (2y_1 y_2 - 2y_1 y_3 + 2y_3 - 1) \\ &\quad + z_1 (5y_1 y_3 - 2y_1 - 7y_1 y_2 - 5y_3 + 6) \\ &\quad + z_3 (z_1 - 1)(2y_1 y_2 + 3y_1 + 3y_3 - 3y_1 y_3 - 4) \\ &\quad + 4y_1 y_2 + y_1 + 2y_3 - y_1 y_3 - 2 \\ &\triangleq F(y_1, y_2, y_3; z_1, z_2, z_3). \end{aligned}$$

The reader can now check that the values $(y_1^* = 1, y_2^* = 3/5, y_3^* = 0; z_1^* = 4/5, z_2^* = 1, z_3^* = 0)$ satisfy the inequalities

$$\begin{aligned} F(y_1^*, y_2^*, y_3^*; z_1, z_2, z_3) &\le F(y_1^*, y_2^*, y_3^*; z_1^*, z_2^*, z_3^*) \\ &\le F(y_1, y_2, y_3; z_1^*, z_2^*, z_3^*) \end{aligned} \tag{2.44}$$

for all $0 \le y_i \le 1$, $0 \le z_i \le 1$, $i = 1, 2, 3$; and hence, when they are used in (2.43a)-(2.43d); the resulting behavioral strategies constitute a saddle-point solution for the multi-act game under consideration, with the average value being 7/5. We thus observe that a modified version of the extensive form of Fig. 2.9, in which the players recall their past actions, does admit a saddle-point solution in behavioral strategies.

Derivation of the behavioral saddle-point solutions of multi-act games of the delayed commitment type involve, in general, solution of a pair of saddle-point inequalities of the type (2.44). For the present case (i.e., for the specific example of Fig. 2.10) it so happens that F admits a saddle-point solution which, in turn, yields the behavioral strategies sought; however, in general it is not guaranteed that it will admit a well-defined solution. The kernel F will always be a continuous function of its arguments which belong to a convex compact set (in fact a simplex), but this is not sufficient for existence of a saddle point for F (see Section 4.3). The implication, therefore, is that there will exist multi-act games of the delayed commitment type which do not admit a saddle-point solution in pure or behavioral strategies.

Randomized strategies

As observed above, and also in Example 2.4, not every multi-act finite game would admit a behavioral saddle-point solution, but whenever it exists, one method for obtaining the behavioral saddle-point policies would be first to construct the average cost of the game on the set of behavioral strategies (such as the kernel F defined prior to (2.44)), and then find the saddle point of that kernel in pure strategies. The (pure) strategies in this case are in fact the probability weights attached to different actions at each information set, which belong to closed and bounded subsets of finite dimensional spaces. Let us denote these subsets for **P**1 and **P**2 by $\hat{Y}$ and $\hat{Z}$, respectively. For the game of Example 2.4, for instance, both $\hat{Y}$ and $\hat{Z}$ are positive unit squares, whereas for the multi-act game of Fig. 2.10, they are positive unit cubes. In all cases, the kernel F will be continuous on $\hat{Y} \times \hat{Z}$, and $\hat{Y}$ and $\hat{Z}$ will further be convex. We shall study the derivation of saddle-point equilibria of such continuous-kernel games in Chapter 4, where we will see that F does not necessarily admit a pure-strategy saddle point on $\hat{Y} \times \hat{Z}$ unless it is convex-concave. Clearly, neither F defined earlier in conjunction with the multi-act game of Fig. 2.10, nor the one that could be constructed for the game of Example 2.4, is convex-concave,[13] which is one of the reasons why these two games do not admit behavioral saddle points. The theory of Chapter 4, and particularly of Section 4.3, will tell us, however, that such continuous-kernel games defined on closed-bounded convex sets will admit mixed saddle-point equilibria, where *mixed strategies* in this case are probability distributions on $\hat{Y}$ and $\hat{Z}$. For present application, since the elements of $\hat{Y}$ and $\hat{Z}$ already correspond to behavioral strategies for the original finite game

[13]We leave it as an exercise for the reader to construct the kernel F for the game of Example 2.4, and show that it is not convex-concave (i.e., convex over $\hat{Y}$ and concave over $\hat{Z}$).

(themselves being special types of mixed strategies), and to avoid confusion, the further mixed strategies defined on $\hat{Y}$ and $\hat{Z}$ are often called *randomized strategies* (for the original finite game). Hence, the message here is that every finite multi-act game admits a saddle point in randomized strategies—a full verification of which is postponed until Chapter 4. Some further discussion on randomized strategies, in this chapter, can be found in subsection 2.7.1.

2.6 Zero-Sum Games with Chance Moves

In this section, we briefly discuss the class of two-person zero-sum finite games which also incorporate chance moves. In such decision problems, the final outcome is determined not only by the decisions (actions) of the two players, but also by the outcome of a chance mechanism whose odds between different alternatives are known *a priori* by both players. One can also view this situation as a three-player game wherein the third player, commonly known as "nature", has a fixed mixed strategy.

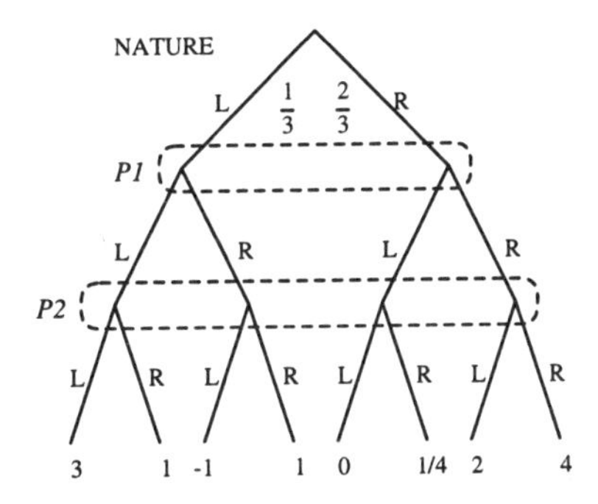

Figure 2.11: A zero-sum single-act game in extensive form that incorporates a chance move.

As an illustration of such a zero-sum game, let us consider the tree structure of Fig. 2.11, where N (nature) picks L w.p. 1/3 and R w.p. 2/3. Not knowing the realized outcome of the chance mechanism, the players each decide on whether to play L or R. Let us now assume that such a game is played a sufficiently large number of times and the final outcome is determined as the arithmetic mean of the outcomes of the individual plays. In each of these plays, the matrix of possible outcomes is either

P2

P1	L	R	
	3	1	L
	−1	(1)	R

(2.45a)

or

P2

	L	R	
P1	0	($\frac{1}{4}$)	L
	2	4	R,

(2.45b)

the former occurring w.p. 1/3 and the latter w.p. 2/3. Each of these matrix games admits a pure-strategy saddle-point solution as indicated. If N's choice were known to both players, then **P2** would always play R, and **P1** would play R if N's choice is L and play L otherwise. However, since N's choice is not known *a priori*, and since the equilibrium strategies to be adopted by the players cannot change from one individual play to another, the matrix game of real interest is the one whose entries correspond to the possible average outcomes of a sufficiently large number of such games. This matrix game can readily be obtained from (2.45a) and (2.45b) to be

P2

	L	R	
P1	(1)	1/2	L
	1	3	R

(2.46a)

which admits, as indicated, the unique saddle-point solution

$$\gamma^{1*} = \gamma^{2*} = L \tag{2.46b}$$

with an average outcome of 1. This solution is, in fact, the only reasonable equilibrium solution of the zero-sum single-act game of Fig. 2.11, since it also uniquely satisfies the pair of saddle-point inequalities

$$\bar{J}(\gamma^{1*}, \gamma^2) \leq \bar{J}(\gamma^{1*}, \gamma^{2*}) \leq \bar{J}(\gamma^1, \gamma^{2*}) \tag{2.47}$$

for all $\gamma^1 \in \Gamma^1$, $\gamma^2 \in \Gamma^2$, where

$$\Gamma^1 \equiv \Gamma^2 = \{L, R\}$$

and $\bar{J}(\gamma^1, \gamma^2)$ denotes the expected (average) outcome of the game corresponding to a pair of permissible strategies $\{\gamma^1, \gamma^2\}$ with the expectation operation taken with respect to the probability weights assigned *a priori* to different choices of N. For this problem, the correspondences between $\bar{J}$ and different choices of (γ^1, γ^2) are in fact displayed as the entries of matrix (2.46a).

It is noteworthy that **P2**'s equilibrium strategy is L for the extensive form of Fig. 2.11 (as discussed above), whereas if N's choice were known to both players (Fig. 2.12), he would have chosen the constant strategy R, as shown earlier. Such a feature might, at first sight, seem to be counter-intuitive—after all, when his information sets can distinguish between different choices of N, **P2** disregards N's choice and adopts a constant strategy (R), whereas in the case when this information is not available to him, he plays quite a different one. Such an argument is misleading, however, because of the following reason:

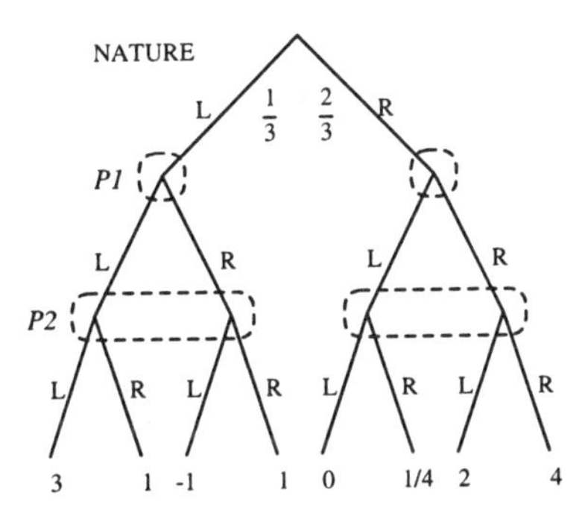

Figure 2.12: A modified version of the zero-sum game of Fig. 2.11, with both players having access to nature's actual choice.

In the tree structure of Fig. 2.12, **P**1's equilibrium strategy is not a constant but is a variant of N's choice; whereas in Fig. 2.11 he is confined to constant strategies, and hence the corresponding zero-sum game is quite a different one. Consequently there is no reason for **P**2's equilibrium strategies to be the same in both games, even if they are constants.

Remark 2.6 It should be noted that the saddle-point solution of the zero-sum game of Fig. 2.12 also satisfies a pair of inequalities similar to (2.47), but this time each player has four permissible strategies, i.e., Γ^i is comprised of four elements. It directly follows from (2.45a)-(2.45b) that the average saddle-point value, in this case, is $(1/3)(1) + (2/3)(1/4) = 1/2$, which is less than the average equilibrium value of the zero-sum game of Fig. 2.11. Hence, the increase in information with regard to the action of N helps **P**1 to reduce his average losses but works against **P**2. □

The zero-sum games of both Fig. 2.11 and Fig. 2.12 are in extensive form, but they do not fit the framework of Def. 2.5 since the possibility of chance moves was excluded there. We now provide below a more general definition of a zero-sum game in extensive form that also accounts for chance moves.

Definition 2.14 *An* extensive form *of a two-person zero-sum finite game that also incorporates chance moves is a tree structure with*

(i) a specific vertex *indicating the starting point of the game,*

(ii) a pay-off function *assigning a real number to each terminal vertex of the tree, which determines the pay-off (respectively, loss) to* **P**2 *(respectively,* **P**1*) for each possible set of actions of the players together with the possible choices of nature,*

(iii) a partition of the nodes of the tree into three player sets (to be denoted by $\bar{N}^1$ *and* $\bar{N}^2$ *for* **P**1 *and* **P**2*, respectively, and by* $\bar{N}^0$ *for nature),*

(iv) a probability distribution, defined at each node of $\bar{N}^0$, *among the immediate branches (alternatives) emanating from this node,*

(v) a subpartition of each player set $\bar{N}^i$ *into* information sets $\{\eta_j^i\}$ *such that the same number of immediate branches emanates from every node belonging to the same information set, and no node follows another node in the same information set.*

This definition of a zero-sum game in extensive form clearly covers both single-act and multi-act games, and nature's chance moves can also be at intermediate levels of play. For such a game, let us denote the strategy sets of **P1** and **P2** again as Γ^1 and Γ^2, respectively. Then, for each pair $\{\gamma^1 \in \Gamma^1, \gamma^2 \in \Gamma^2\}$ the pay-off function, $J(\gamma^1, \gamma^2)$, is a random variable, and hence the real quantity of interest is its expected value which we denote as $\bar{J}(\gamma^1, \gamma^2)$. The saddle-point equilibrium can now be introduced precisely in the same fashion as in Def. 2.8, with only J replaced by $\bar{J}$. Since this formulation converts the extensive form into an equivalent normal form, and since both Γ^1 and Γ^2 are finite sets, it follows that the saddle-point equilibrium solution of such a game problem can be obtained by basically solving a matrix game whose entries correspond to the possible values of $\bar{J}(\gamma^1, \gamma^2)$. This has actually been the procedure followed earlier in this section in solving the single-act game of Fig. 2.11, and it readily extends also to multi-act games.

When a pure-strategy saddle point does not exist, it is possible to extend the strategy sets so as to include behavioral or mixed strategies and to seek equilibria in these larger classes of strategies. Proposition 2.5 clearly also holds for the class of zero-sum games covered by Def. 2.14, and so does Prop. 2.6. Existence of a saddle point in behavioral strategies is again not guaranteed, and even if such a saddle point exists there is no systematic way to obtain it, unless the players have access to nature's actions and the game possesses a feedback structure (as in Def. 2.11). For games of this specific structure, one can use an appropriate extension of the recursive derivation developed in Section 2.5 for feedback games, which we do not further discuss here. Before concluding we should mention that the notions of prior and delayed commitment introduced in the previous section also fit within the framework of games with chance moves, and so does the concept of randomized strategies, with the extensions being conceptually straightforward.

2.7 Two Extensions

In this section we discuss two possible extensions of the extensive form description of finite zero-sum games—one where the restriction of (cf. Def. 2.5 and Def. 2.14) "no nodes following another node in the same information set" is dispensed with (see subsection 2.7.1), and another one where the tree structure, viewed as a *directed graph*, is allowed to have cycles (see subsection 2.7.2).

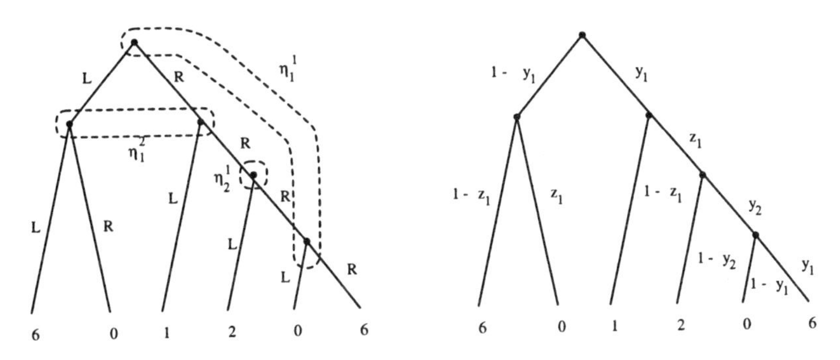

Figure 2.13: A zero-sum game in extensive form in which the tree enters an information set twice.

2.7.1 Games with repeated decisions

We start this subsection with a specific example.

Example 2.5 Consider the zero-sum game in extensive form shown in Fig. 2.13.
P1 has two information sets, one of which (η_1^1) contains two nodes from one path in the tree; **P2** has a single information set. If the probability of going right from η_1^1 is given by y_1, and going left by $1 - y_1$, and if y_2 and z_1 are the corresponding probabilities for η_2^1 and η_1^2, respectively, then the behavioral normal form is given by the kernel[14]

$$6(1-y_1)(1-z_1) + y_1(1-z_1) + 2y_1z_1(1-y_2) + 6y_1z_1y_2y_1. \tag{2.48}$$

First minimizing this expression with respect to $y_2 \in [0,1]$ leads to the kernel

$$K(y_1, z_1) \triangleq \begin{cases} 6(1-y_1)(1-z_1) + y_1(1-z_1) + 6(y_1)^2 z_1 & \text{if} \quad y_1 < 1/3, \\ 6(1-y_1)(1-z_1) + y_1(1-z_1) + 2y_1z_1 & \text{if} \quad y_1 \geq 1/3, \end{cases}$$

with a corresponding minimizing y_2 being

$$y_2 = \begin{cases} 0 & \text{if} \quad y_1 \geq 1/3, \\ 1 & \text{if} \quad y_1 < 1/3. \end{cases}$$

Now, the function K above is to be minimized by **P1** and maximized by **P2**. Toward this end consider the max-min problem: $\max_{z_1} \min_{y_1} K(y_1, z_1)$, defined on the unit square; see Fig. 2.14 for a sketch of the function K. Some calculus leads to the result that this max-min is uniquely achieved for $z_1 = \beta \triangleq (170 + 40\sqrt{6})/386 = 0.694$. This behavioral strategy is optimal for **P2**. To obtain

[14]When there are two (or more) nodes following each other in the same information set, a behavioral strategy is defined as one where the odds for different alternatives are the same at each node (since a player cannot differentiate between two nodes belonging to the same information set), but the random mechanism that generates the actions (moves) according to these odds is independent from one node to another. Hence, moves from different nodes belonging to the same information set can be considered as independent random variables with identical probability distributions.

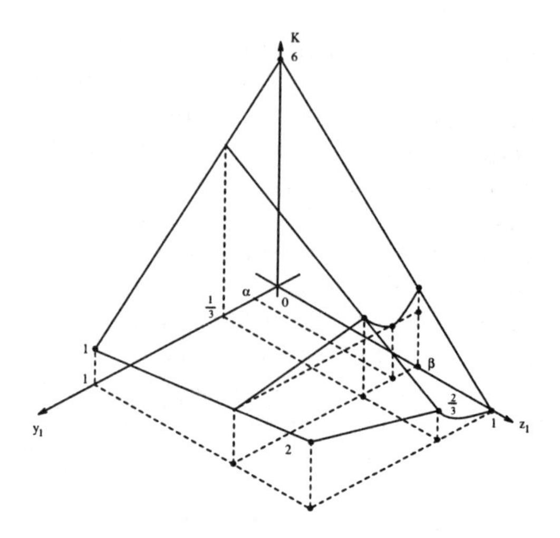

Figure 2.14: The function K.

the optimal strategy of **P1**, consider the minimization problem $\min_{y_1} K(y_1, \beta)$, where the minimum is attained at only two points: $y_1 = \alpha \triangleq 5(1-\beta)/12\beta = .184$ and $y_1 = 1$. Hence it follows that **P1** should use two behavioral strategies, viz. $y_{(1)} \triangleq (y_1 = \alpha, y_2 = 1)$ and $y_{(2)} \triangleq (y_1 = 1, y_2 = 0)$, in an appropriate mix. To find this right "mix", consider the 2×2

	P2		
P1	$6 - 5\alpha$	$6\alpha^2$	$y_{(1)}$
	1	2	$y_{(2)}$
	$z_1 = 0$	$z_1 = 1$	

The solution to this game is given by the following: choose $z_1 = 1$ with probability β and $z_1 = 0$ with probability $1 - \beta$; choose $y_{(1)}$ with probability $-1/(6\alpha^2 + 5\alpha - 7) = 0.17$ and $y_{(2)}$ with probability 0.83. Thus it has been established that **P1** mixes between two behavioral strategies, the result of which is a *randomized strategy*. □

The preceding example has shown features which have also surfaced in the discussion on randomized strategies in Section 2.5. In both cases the starting point was a behavioral normal form which more generally has the form

$$\sum_{k=1}^{K} a_k y_1^{i_1(k)} y_2^{i_2(k)} \cdots y_m^{i_m(k)} z_1^{j_1(k)} z_2^{j_2(k)} \cdots z_n^{j_n(k)},$$

where the exponents $i_l(k)$ and $j_l(k)$ are nonnegative integers. This form is sometimes called a *multinomial*. It is to be minimized by **P1** by choosing a vector $(y_1, \ldots, y_m)$ on $[0,1]^m$ and to be maximized by **P2** by choosing $(z_1, \ldots, z_n)$ on $[0,1]^n$. It is fairly easy to see that an arbitrary game in extensive form, of the form described in this subsection, and without chance moves, has a behavioral normal form which is a multinomial. The converse is also true: every multinomial is the behavioral normal form of an appropriately defined game in extensive form, as shown by Alpern (1988). The theory of Chapter 4 will tell us that the zero-sum game just described in the behavioral normal form does indeed have a saddle point in mixed strategies. This means that the optimal solution of the underlying game in extensive form exists in the class of randomized strategies. In Alpern (1988), an even stronger result can be found: for the saddle-point solution of finite games in randomized strategies, each player needs to average over only a finite number of behavioral strategies.

Counterparts of these results do exist for games in extensive form which also include chance moves. Since the basic features are the same as above, and no new difficulties emerge, we will not treat them separately here.

2.7.2 Extensive forms with cycles

In this subsection we consider finite zero-sum dynamic games without chance moves, in which the underlying (directed) graph is no longer a finite tree; it will contain cycles, though the number of nodes remains finite. Moreover, it is assumed that the players have perfect information, i.e., each information set consists of a single node. If the players prefer to do so, such games can go on forever by moving and staying within one (or more) cycle(s). Thus infinite play is possible. It is of course possible to 'unfold' the underlying graph with cycles so as to obtain a tree whose nodes are paths of the original game. However, the resulting tree will be infinite and no useful general results for such trees are known.

Games with terminal pay-off

We now first discuss the case when the game has only a terminal pay-off.

Definition 2.15 *A* two-person zero-sum deterministic graph game with terminal pay-off *consists of a finite directed graph of which the nodes are partitioned into three sets:*

(i) a specific vertex *indicating the starting point of the game,*

(ii) a pay-off function *assigning a real number to each terminal vertex of the graph, which determines the pay-off (respectively, loss) to* **P2** *(respectively,* **P1***) for each possible set of actions of the players,*

(iii) a partition of the nonterminal nodes of the graph into two sets (to be denoted by $\bar{N}^1$ and $\bar{N}^2$ for **P1** *and* **P2***, respectively).*

If the players choose their actions in such a way that the game does not end in a finite number of steps (i.e., a terminal node is never reached), the pay-off will, by definition, be zero. A pure strategy *for* **P***i in the current context is a rule that assigns to any path that ends in $\bar{N}^i$ one of its successor nodes deterministically.*

A stationary strategy *is a pure strategy that only depends on the final node of the path.*

It is possible to devise an algorithm that yields a pair of stationary strategies which form a saddle-point solution for the game. To introduce such an algorithm, we assign a value V to each node x of the graph, with $V(x)$ denoting the value of the game if node x is chosen as the starting point of the game. The set of nodes for which the value has already been calculated while running the algorithm is denoted by H; note that H will be increasing as long as the algorithm is running, until the entire graph has been covered by H. Let us also introduce the notation $\sigma(x)$ to denote the set of all immediate successor nodes of node x. Then, the algorithm proceeds as follows.

1. For all terminal nodes, the function $V(x)$ equals the pay-off if the game terminates at x; the set of all terminal nodes is denoted by H.

2. Consider all $x \notin H$ for which $\sigma(x) \cap \bar{N}^1 \in H$. The value for such an x is defined as $V(x) = \min_{\sigma(x)} V(\sigma(x))$. The argument(s) for which this minimum is achieved is then part of the saddle-point strategy for **P1**. Increase H with new nodes for which a value has been obtained. Repeat this as long as new nodes $x \in \bar{N}^1$ are found such that $\sigma(x) \in H$.

3. Consider all $x \notin H$ for which $\sigma(x) \cap \bar{N}^2 \in H$. The value for such an x is defined as $V(x) = \max_{\sigma(x)} V(\sigma(x))$. The argument(s) for which this maximum is achieved is then part of the saddle-point strategy for **P2**. Increase H with new nodes for which a value has been obtained. Repeat this as long as new nodes $x \in \bar{N}^2$ can be found such that $\sigma(x) \in H$.

4. Repeat steps 2 and 3 until no new nodes can be found.

5. Consider those $x \notin H$ for which $\sigma(x) \cap \bar{N}^1 \cap H \neq \emptyset$. Define node y to be that node (or those nodes) of the set $\sigma(x) \cap H$ which has (have) the smallest value, i.e., $V(y) = \min_{z \in \sigma(x) \cap H} V(z)$. If $V(y) < 0$, then increase H by node x and define $V(x) = V(y)$. The arc between x and y then constitutes part of the saddle-point strategy for **P1**. (If $V(y) \geq 0$, then H is not increased: **P1** prefers nontermination with value 0 over a positive pay-off.) Repeat this as long as H can be increased.

6. Consider those $x \notin H$ for which $\sigma(x) \cap \bar{N}^2 \cap H \neq \emptyset$. Define node y to be that node (or those nodes) of the set $\sigma(x) \cap H$ which has (have) the largest value, i.e., $V(y) = \max_{z \in \sigma(x) \cap H} V(z)$. If $V(y) > 0$, then increase H by node x and define $V(x) = V(y)$. The arc between x and y then constitutes part of the saddle-point strategy for **P2**. (If $V(y) \leq 0$, then H is not increased: **P2** prefers nontermination with value 0 to a negative pay-off.) Repeat this as long as H can be increased.

7. Repeat steps 5 and 6 until no new nodes can be found.

8. Repeat steps 2-7 until no new nodes can be found.

9. The remaining nodes x get assigned the value 0. Any arc departing from such an x to another node not belonging to H is part of the saddle-point strategies. (The players do not have any desire to terminate the game once they reach any one of these remaining nodes.)

Intuitively it should be clear that this algorithm indeed yields the saddle-point strategies for the game (the algorithm is in the spirit of dynamic programming and is identical to the recursive procedures introduced earlier in the context of feedback games, but here one also takes into account the possibility of nontermination). A formal proof is given in Washburn (1990).

Games with local pay-offs

So far we have discussed only games with terminal pay-offs. Other pay-off structures are also possible, such as assigning a 'local' pay-off to each arc of the underlying graph. For games in extensive form in which the underlying graph is a finite tree, this is not a real generalization, since the local pay-offs assigned to arcs can be moved to the terminal nodes (and added together so as to obtain the 'total' pay-off at each terminal node). For extensive games with cycles in which pay-offs are assigned to arcs, this is not possible and other means are necessary to study saddle-point solutions (if they exist). For such games it brings in no loss of generality to assume that the underlying directed graph does not have terminal nodes (games with some terminal nodes can be modeled with a loop at those nodes: a terminal pay-off can then be assigned to the corresponding loop). Thus all plays are infinite. The total pay-off associated with a play of the game is defined to be the long term average of the local pay-offs obtained during the play. If necessary one can consider *Césaro limits* and/or take the *lim inf* or *lim sup*. Note that if a 'terminal' node is reached (if such a node existed in the original game) to which a loop has been added, then the total pay-off converges to the local pay-off corresponding to that loop.

For a theory of zero-sum games on graphs without terminal nodes, but with perfect information and with local pay-offs, the reader is referred to Ehrenfeucht and Mycielski (1979) and Alpern (1991). The theory and results are built around an auxiliary game defined on the same graph and with the same rules except for the fact that it ends as soon as a node is repeated. This auxiliary game has therefore a finite number of moves.

2.8 Action-Dependent Information Sets

The classes of zero-sum dynamic games we have seen heretofore all carried one common feature, which is that the information sets were fixed *a priori*. There are other important classes of games, however, where this side condition does not hold, which means that the information sets change with the actions of the players. In this section we discuss the theory behind such games, within the context of two specific examples—a class of *duels*, and a *searchlight game*.

2.8.1 Duels

Consider a duel in which two men (**P1** and **P2**) walk towards each other, taking one step at a time, at time instants $t = 0, 1, \ldots, N$. If nothing intervenes, then they will meet at $t = N$. Each player has a pistol with exactly one bullet in it, and he may fire the pistol at any time $t \in \{0, 1, \ldots, N\}$. If one of them fires and hits the other, the duel is immediately over and the one who has fired successfully is declared the winner. If neither one fires successfully or if both fire simultaneously and successfully, the duel becomes a stand-off. The probability of hitting the opponent after the pistol has been fired is a function of the distance between the two men. If a player fires at $t \in \{0, 1, \ldots, N\}$, then the probability of hitting the opponent is t/N.

To complete the formulation of the game, we have to specify whether the game is 'silent' or 'noisy'. In the silent duel a player does not know whether his opponent has already fired, unless, of course, he is hit. In a noisy duel a player knows when his opponent has fired and if he has not yet fired himself and is not hit he will wait until $t = N$ to fire his own pistol, because at $t = N$ he is certain to hit his opponent.

Suppose that $\mathbf{P}i$ fires at t^i. Then, if $t^1 < t^2$, the probability that **P2** is hit is t^1/N, and the probability that **P1** will be hit at $t^2 > t^1$ is $(1 - t^1/N)t^2/N$ for the silent duel and is $1 - t^1/N$ for the noisy duel (in which case **P2** fires at $t^2 = N$). *Mutatis mutandis* similar results are obtained if $t^1 > t^2$. With these probabilities one can construct a matrix A of size $(N+1) \times (N+1)$, the $(t^1 + 1, t^2 + 1)$-th element of which, with $t^1 < t^2$ and $t^i, t^j \in \{0, 1, \ldots, N\}$, is $t^1/N - (1 - t^1/N)t^2/N$ for the silent case and $(N - 2t^1)/N$ for the noisy case. The $(t^1 + 1, t^2 + 1)$-th element with $t^1 > t^2$ is $t^2/N - (1 - t^2/N)t^1/N$ for the silent case and $1 - 2t^2/N$ for the noisy case. The diagonal elements of A equal 0 in both cases. Finding the optimal strategies for the duel now boils down to obtaining a saddle-point solution for the matrix game characterized by matrix A in which the minimizer (**P1**) chooses the row vector and the maximizer (**P2**) the column vector. We now note that for the extensive-form description of the noisy game, the extent of a player's knowledge on the evaluation of the game depends explicitly on his own and/or his opponent's actions. Hence, the noisy duel is a game with action-dependent information sets.

2.8.2 A searchlight game

In this game, two players, **P1** and **P2**, move in a certain closed domain and are not aware of each other's position unless one of them flashes a searchlight which illuminates an area of known shape around him. By flashing his searchlight a player thus discloses his own position to the other player, regardless of their relative positions. Termination of the game occurs only if a player flashes his searchlight and the other player finds himself trapped in the area illuminated. Each player's objective is to catch the other player in his searchlight before he himself is caught. Therefore flashing has two competing consequences: in order for a player to win, he must flash; if, however, during such a flash the other player is not caught, then the flashing player is in a more vulnerable position because he has disclosed his location to the other player.

We will provide here some explicit results for the case when the players are confined to n points on the circumference of a circle, which makes the underlying decision process a finite game. At each time step (the time is considered here to be discrete) a player can either move to one of the two adjacent nodes or stay where he is. Even if the players occupy the same position, they will not be aware of this, unless of course one of them (or both) flashes (flash). Flashing illuminates the position of the player who flashes, as well as the two adjacent positions.

In the specific situation when only one of the players (say **P1**) has a searchlight at his disposal, the game becomes a survival game: **P1** wants to catch the other player (**P2**), whereas **P2** wants to avoid capture. A practical motivation for this game might be the following. **P2** is a smuggler who wants to steer his boat to the shore and **P1** is a police boat. For obvious reasons **P2** wants to avoid **P1**, and therefore he performs his practices only during the night, and that is exactly the reason why the police boat will use a searchlight. Of course in this case the search domain will not be the circumference of a circle, but rather a two-dimensional plane.

The basic game

The basic game to be formulated below is a building block for the more general game described above. Initially both players know each other's position on the circle (but not thereafter if mixed strategies are involved); **P1** is at position p_1 and **P2** is at position p_2, $p_i \in \{1, 2, \ldots, n\}$. Player **P1** can flash only once during the time instants $t = 1, 2, \ldots, T$, where T is the final time which is fixed and known to both players. **P2** does not have any flashes at his disposal. Once the time proceeds the players do not acquire any new information on each other's new position (unless **P1** flashes).

The optimal strategies of this basic game can be found by solving the matrix game

$$\min_{y_{p_2} \in S_{3^T}} \max_{z_{p_1} \in S_{d_T}} y'_{p_2} A_{p_1, p_2, T} z_{p_1},$$

where the notation is as follows. At each instant of time, **P2** can choose from three options (move to one of the two adjacent nodes or stay where he is) and therefore he has 3^T pure strategies. Each component of y_{p_2} denotes the probability with respect to which one of the 3^T pure strategies will be chosen. The set S_{3^T} is the simplex to which the 3^T-vector y_{p_2} belongs. The components of z_{p_1}, on the other hand, indicate **P1**'s pure strategies, determined by when and where to flash. After one time step **P1** can flash at three different positions, after two time steps at five different positions, etc. Hence the number of pure strategies equals $3 + 5 + \cdots + (2T + 1) = T^2 + 2T$, provided that $T \leq [n/2]$. (For $T \geq [n/2]$, all points on the circle can be reached by **P1**.) Vector z_{p_1} has $d_T = T^2 + 2T$ components and S_{d_T} denotes the simplex from which z_{p_1} must be chosen. The size of $A_{p_1,p_2,T}$ is $3^T \times d_T$ and the elements of this matrix are either 1 (corresponding to the cases when for the particular strategies capture takes place) or 0 (if no capture takes place). The first index of A refers to **P1**'s initial position, the second index to **P2**'s initial position, and the third index is the T introduced earlier.

If the saddle-point value of this matrix game is denoted by $J_{p_1,p_2,T}$, then it should be intuitively clear (and this has been rigorously proven in Olsder and Papavassilopoulos (1988a)) that

$$J_{p_1,p_2,1} \leq J_{p_1,p_2,2} \leq \cdots \leq J_{p_1,p_2,[n/2]} = J_{p_1,p_2,[n/2]+1} = \cdots,$$

for all $p_1, p_2 \in \{1, 2, \ldots, n\}$. Hence, in the solution of the game, we can restrict the time horizon to $[n/2]$ steps, without any loss of generality. The idea behind the proof of this result is that in $[n/2]$ steps each player can reach any point on the circle with an arbitrary underlying probability distribution.

Now, consider the following generalization of the basic game. At $t = 0$ **P2** still knows where **P1** is (at node p_1), but **P1** only knows the probability distribution of **P2**'s position. The probability for **P2** to be at node p_2 at $t = 0$ is v_{p_2}. The probability vector $v = (v_1, \ldots, v_n)'$ is assumed to be known to both players. Of course **P2** knows his own position at $t = 0$. Such a generalization arises if one considers the following game: two players move on the circumference of a circle and do not know each other's initial position (other than in a probabilistic way). Each player has exactly one flash at his disposal. Suppose that **P1** flashes first and that he does not capture **P2** in this flash. Then **P2** knows where **P1** is at the time of the flash, but **P1** only knows that **P2** is not in the area just illuminated. If **P1** does not have any other information about **P2**'s position, then he may assume a uniform distribution on the nodes which were not illuminated at the time of the flash. This is the starting point for this generalization (with all v_i's corresponding to nonilluminated nodes being equal and adding up to one).

This generalization can again be formulated as a matrix game:

$$\min_{y_{p_2} \in S_{3^T}, p_2 = 1, \ldots, n} \max_{z_{p_1} \in S_{d_T}} \sum_{i=1}^{n} v_i y_i' A_{p_1,p_2,T} z_{p_1}.$$

This minimax problem is not in the standard form treated earlier in Section 2.3, but it can also be solved by means of LP, viz.,

$$\max \sum_{j=1}^{3^T} (\tilde{y}_1)_j$$

subject to

$$(\tilde{y}_1', \ldots, \tilde{y}_n') \begin{pmatrix} v_1 A_{p_1,1,T} \\ \vdots \\ v_n A_{p_n,n,T} \end{pmatrix} \leq (1, 1, \ldots, 1);$$

$$\textstyle\sum_{j=1}^{3^T} (\tilde{y}_1)_j = \sum_{j=1}^{3^T} (\tilde{y}_i)_j, \; i = 2, \ldots, n;$$

$$(\tilde{y}_i)_j \geq 0, \; i = 1, \ldots, n; \; j = 1, \ldots, 3^T.$$

The saddle-point value of this game when $v_i = 0$ for $i = 1, 2, 3$ and $v_i = 1/(n-3)$ for $i = 4, \ldots, n$ will be indicated by $\tilde{J}_T$, tacitly assuming that $n > 3$.

Both players have one flash each

We will now solve the problem where each player has exactly one flash at his disposal and the players do not know each other's initial position, other than that they are given by the uniform distribution on the n nodes with no correlation between the two initial positions.

If both players flash at $t = 1$, then the probability with which **P2** captures **P1** is $3/n$ (since three positions are illuminated) and so is the probability for **P1** to capture **P2**. Thus the pay-off is $(3/n) - (3/n) = 0$. Let us now consider the situation where **P2** flashes at $t = 1$ and **P1** does not. Then the pay-off for **P2** is $(3/n) - (1 - 3/n)\tilde{J}_{T-1}$. In other words, the pay-off equals the probability that **P2** captures **P1** minus the probability that **P2** does not capture **P1**, times the probability that **P1** captures **P2** during the remaining $T-1$ period game. Similarly, if **P2** flashes at $t = 2$ and **P1** does not flash at $t = 1$ or $t = 2$, then the pay-off for **P2** is $(3/n) - (1 - 3/n)\tilde{J}_{T-2}$, etc. Define $c_k = (3/n) - (1 - 3/n)\tilde{J}_{T-k}$, and consider the matrix

$$M \triangleq \begin{pmatrix} 0 & c_1 & c_1 & \cdots & \cdots & c_1 \\ -c_1 & 0 & c_2 & \cdots & \cdots & c_2 \\ -c_1 & -c_2 & 0 & \cdots & \cdots & c_3 \\ \vdots & \vdots & & \ddots & & \vdots \\ \vdots & \vdots & & & 0 & c_{T-1} \\ -c_1 & -c_2 & -c_3 & \cdots & -c_{T-1} & 0 \end{pmatrix}.$$

Let $s = (s_1 \ldots, s_T)'$ and $r = (r_1, \ldots, r_T)'$ represent probability vectors according to which **P1** and **P2** choose to flash at times $t = 1, \ldots, T$. Then the players face the following game:

$$\min_s \max_r \; r' M s.$$

The solution(s) to this latter matrix game provides the solution to our problem. Since the matrix M is skew-symmetric, it turns out that the value of the game is 0 (see Problem 9 in Section 2.9). Some further analysis regarding the signs and monotonicity properties of the elements c_k shows that the optimal strategies as when to flash are pure strategies. This is no longer true, however, if the illuminated areas for both players have different sizes. See Olsder and Papavassilopoulos (1988a) for further details and extensions.

In conclusion, for the searchlight game as described above to be solved, one must solve a hierarchy of matrix games. The first matrix game determines the coefficients c_k, which are subsequently used as elements in the second matrix game defined by means of the matrix M.

2.9 Problems

1. Obtain the security strategies and the security levels of the players in the following matrix game. Does it admit a (pure) saddle-point solution?

P2

	−2	1	−1	1
	2	3	−1	2
P1	1	2	3	4
	−1	1	0	1

2. Obtain the mixed security strategies and the average security levels of the players in each of the following matrix games.

P2

	4	0
P1	0	2
	3	1

P2

	1	4
P1	3	−1

3. Verify that the quantity $\sup_Z \min_Y \; y'Az$ is unique.

4. Prove that the quantity $\max_Z y'Az$ is a continuous function of $y \in Y$.

5. Determine graphically the mixed saddle-point solutions of each of the following matrix games.

P2

	1	3	−1	2
P1	−3	−2	2	1
	0	2	−2	1

P2

	2	1	0	−1
P1	−1	3	1	4

6. Convert the following matrix games into linear programming problems and thereby numerically evaluate their saddle-point solutions.

P2

	3	1
P1	2	2
	1	3

P2

	0	1	2	3
	1	0	1	2
P1	0	1	0	1
	−1	0	1	0

7. The space of all real $(m \times n)$ matrices $A = \{a_{ij}\}$ can be considered to be isomorphic to the mn-dimensional Euclidean space R^{mn}. Let S_{mn} denote the set of all real $(m \times n)$ matrices which, considered as the pay-off matrix of a zero-sum game, admit pure strategy saddle-point equilibria. Then S_{mn} is isomorphic to a subset of R^{mn}, to be denoted by $S^{m,n}$.

   (i) Is $S^{m,n}$ a subspace?

   (ii) Is $S^{m,n}$ closed, convex?

   (iii) Describe $S^{2,3}$ explicitly.

8. Let $T^i(A)$ denote the set of all mixed saddle-point strategies of $\mathbf{P}i$ in a matrix game A. Show that $T^i(A)$ is nonempty, convex, closed and bounded.

9. A matrix game A, where A is a skew-symmetric matrix, is called a *symmetric* game. For such a matrix game prove, using the notation of Problem 8, that $T^1(A) = T^2(A)$, and $V_{\mathrm{m}}(A) = 0$.

10. Obtain the pure or behavioral saddle-point solutions of the following single-act games in extensive form.

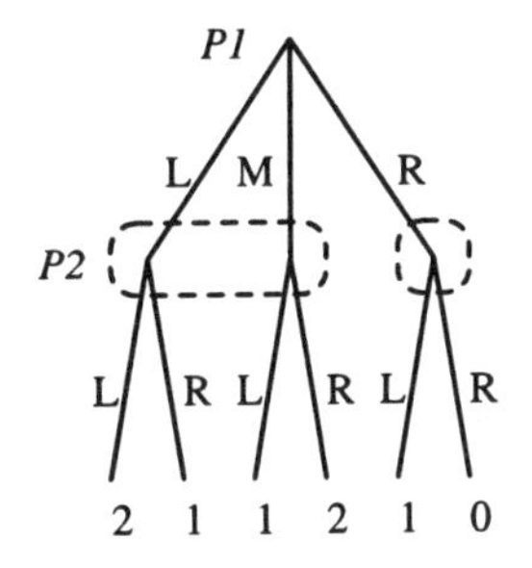

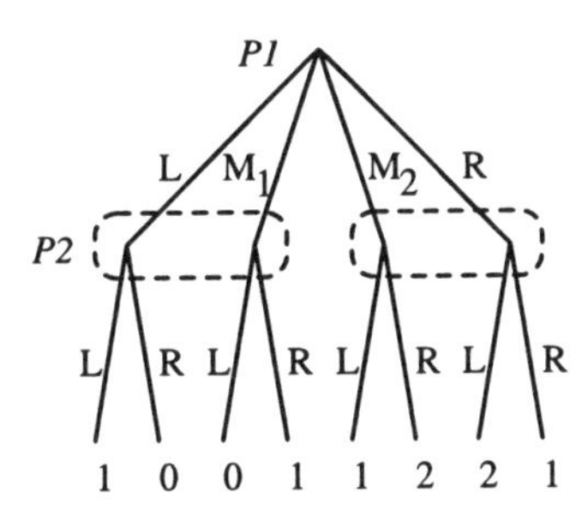

11. Obtain a feedback saddle-point solution for the feedback game in extensive form depicted in Fig. 2.15. What is the value of the game? What is the actual play dictated by the feedback saddle-point solution? Show that these constant strategies also constitute a saddle-point solution for the feedback game.

12. Show that the feedback game in extensive form depicted in Fig. 2.16 does not admit a pure-strategy feedback saddle-point solution. Obtain its behavioral saddle-point solution and the behavioral saddle-point value. Compare the latter with the value of the game of Problem 11. What is the actual play dictated by the feedback saddle-point solution?

13. What is the open-loop version of the feedback game of Fig. 2.16? Obtain its saddle-point solution and value.

14. Prove that if a feedback game in extensive form admits a feedback saddle-point solution with the value J_{f}, and its open-loop version (if it exists)

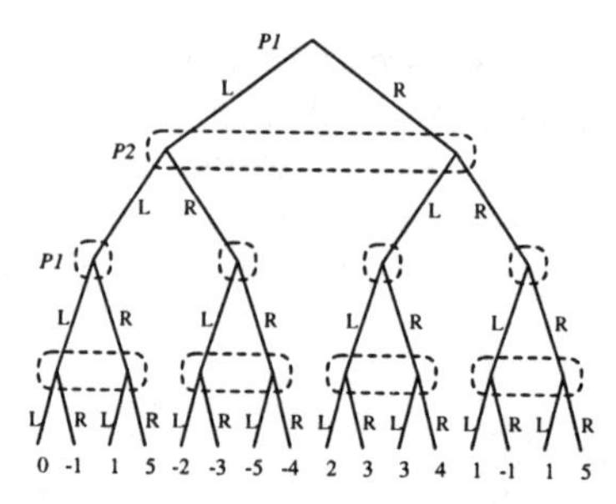

Figure 2.15: Feedback game of Problem 11.

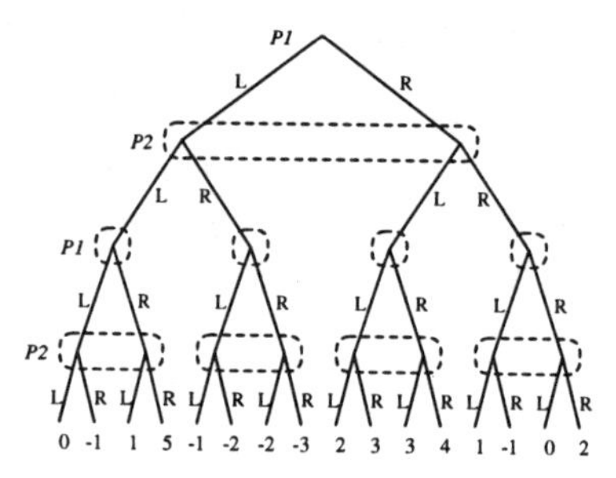

Figure 2.16: Feedback game of Problem 12.

admits a unique saddle-point solution with the value J_o, then $J_f = J_o$. Furthermore, show that the open-loop saddle-point solution is the actual play dictated by the feedback saddle-point solution. (**Hint**: make use of Prop. 2.7 and the interchangeability property of saddle-point strategies.)

15. Verify the following conjecture: If a feedback game in extensive form admits a behavioral saddle-point solution which actually dictates a random choice at least at one level of play, then its open-loop version (if it exists) cannot admit a pure-strategy saddle-point equilibrium.

16. Investigate whether the following multi-act game of the delayed commitment type admits a behavioral saddle-point solution or not.

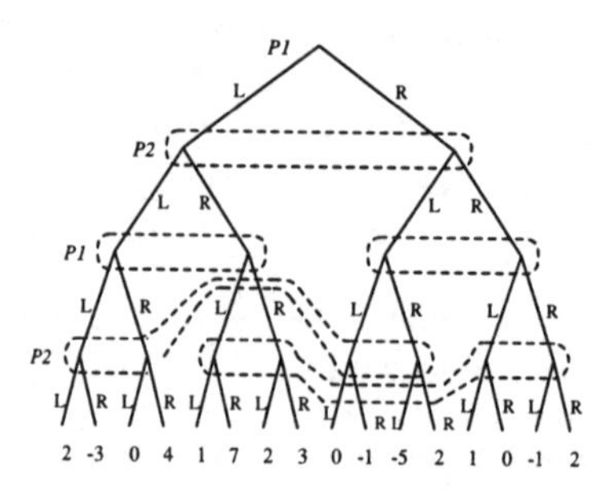

17. Determine the pure or behavioral saddle-point solution of the game in extensive form depicted in Fig. 2.17.

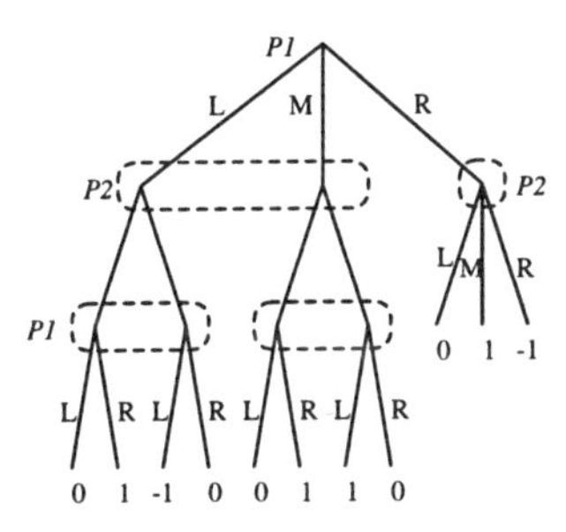

Figure 2.17: The two-person dynamic game of Problem 17.

18. Determine the pure or behavioral saddle-point solution of the game in extensive form depicted in Fig. 2.18, which also incorporates a chance move.

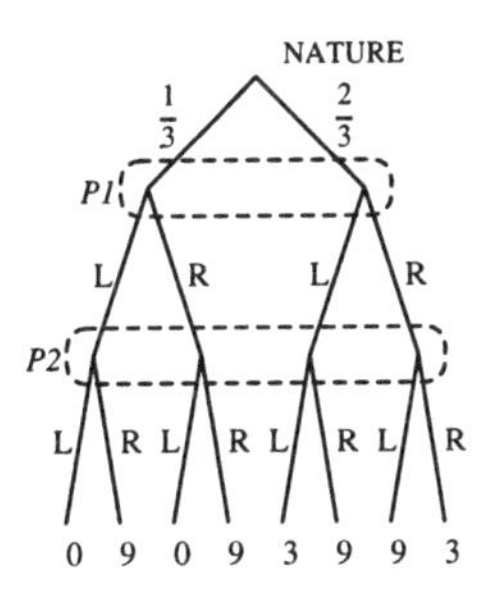

Figure 2.18: The game of Problem 18.

19. Determine the solutions to the previous problem in the cases in which (i) **P1** has access to nature's choice; (ii) **P2** has access to nature's choice; and (iii) both players have access to nature's choice. What is the value of this extra information to individual players in each case?

20. Construct an example of a matrix with only real eigenvalues, which, when considered as a zero-sum game, has a value in mixed or pure strategies that is smaller than the minimum eigenvalue.

21. A matrix game is said to be *completely mixed* if it admits a mixed saddle-point solution which assigns positive probability weight to all pure strategies of the players. Show that in a completely mixed game it does not matter whether **P1** is minimizing and **P2** is maximizing or the other way around, i.e., **P1** is maximizing and **P2** is minimizing. In both cases the saddle-point value and the strategies are the same. (For further results on completely mixed games the reader is referred to Bapat and Raghavan (1997).)

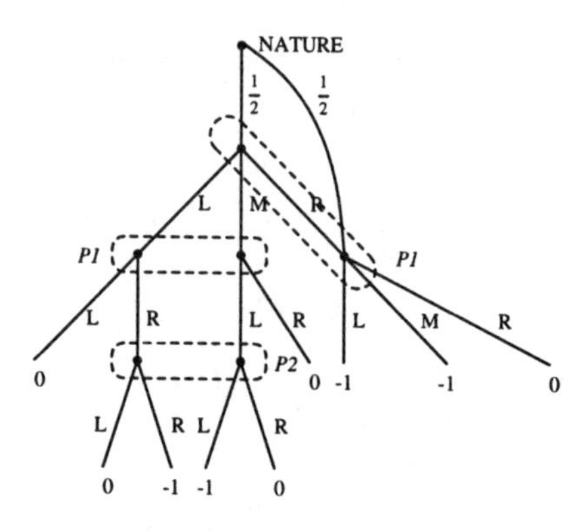

Figure 2.19: The game of Problem 18.

22. Obtain for **P1** his minimum guaranteed (average) costs in mixed, behavioral and randomized strategies, for the game with incomplete information, described by the tree of Fig. 2.19.

    (Answers: In mixed strategies, **P1** can guarantee an average (minimum) outcome of $-3/4$, in behavioral strategies his security level is $-25/64$, and in randomized strategies it is $-9/16$.)

23. 23. Consider the following matrix game.

	P2	
P1	$11+x$	7
	5	$9+y$

If the parameters x and y would be known to both players, the matrix game is a standard one. Parameter x is known to **P1** only, however, and y is only known to **P2**. What both players do know is that both parameters are stochastic, and independently and uniformly distributed on the interval $[0, T]$, with $T = 1$. Show that the optimal solutions are

$$u^{1*} = \begin{cases} \text{first row, if } 0 \leq x < c^1, \\ \text{second row, if } c^1 < x \leq 1, \end{cases}$$

$$u^{2*} = \begin{cases} \text{first column, if } 0 \leq y < c^2, \\ \text{second column, if } c^2 < y \leq 1, \end{cases}$$

where the real constants c^i are (uniquely) determined by $0 < c^i < 1$ and

$$2c^2(8 + c^1) + (c^2)^2 - 5 = 0, \; 2c^1(8 + c^2) + (c^1)^2 - 8 - 2c^2 = 0.$$

For $x = c^1$ and/or $y = c^2$ the solution is not unique. Investigate the solution for $\lim T \downarrow 0$ and compare the limit behavior with the solution of the original matrix game with $x = y = 0$.

2.10 Notes

Section 2.2. The theory of finite zero-sum games dates back to Borel in the early 1920s whose work on the subject was later translated into English (Borel, 1953). Borel introduced the notion of a conflicting decision situation that involves more than one decision maker, and the concepts of pure and mixed strategies, but he did not really develop a complete theory of zero-sum games. He even conjectured that the minimax theorem was false. It was Von Neumann who first came up with a proof of the minimax theorem and laid down the foundations of game theory as we know of today (Von Neumann, 1928, 1937). His pioneering book with Morgenstern definitely culminates his research on the theory of games (Von Neumann and Morgenstern, 1947). A full account of the historical development in finite zero-sum games, as well as possible applications in social sciences, is given in Luce and Raiffa (1957). The reader is also referred to McKinsey (1952) and the two edited volumes by Kuhn and Tucker (1950, 1953). The original proof of the minimax theorem given by Von Neumann is nonelementary and rather complicated. Since then, the theorem has been proven in several different ways, some being simpler and more illuminating than others. The proof given here seems to be the simplest one available in the literature.

Section 2.3. Several illustrative examples of the graphical solution of zero-sum matrix games can be found in the nontechnical book by Williams (1954). Graphical solution and the linear programming approach are not the only two applicable in the computation of mixed saddle-point solutions of matrix games; there is also the method of Brown and Von Neumann (1950) which relates the solution of symmetric games (see Problem 9 in Section 2.9 for a definition) to the solution of a particular differential equation. Furthermore, there is the iterative solution method of Brown (1951) by fictitious play (see also (Robinson, 1951)). A good account of these two numerical procedures, as well as another numerical technique of Von Neumann (1954) can be found in Luce and Raiffa (1957, Appendix 6). For a more recent development in this area, see Lakshmivarahan and Narendra (1981).

Sections 2.4, 2.5 and 2.6. Description of single- and multi-act games in extensive form again goes back to the original work of Von Neumann and Morgenstern (1947), which was improved upon considerably by Kuhn (1953). For a critical discussion of prior and delayed commitment approaches to the solution of zero-sum games in extensive form the reader is referred to Aumann and Maschler (1972).

Section 2.7. The theory presented in this section is based on Alpern (1988, 1991), Isbell (1957), and Washburn (1990). Washburn (1990) also discusses computational complexity of the solutions of zero-sum games on finite graphs with cycles and terminal pay-off. Extensions to discounting are also given in that reference, but extensions to games with random moves lead to complications in the solution algorithm in an essential way. Alpern (1991), on (not necessarily zero-sum) games with perfect information on finite graphs with cycles and with local pay-offs, gives a constructive way to find optimal stationary strategies. See also Ehrenfeucht and Mycielski (1979) for earlier work on this topic. The former reference also discusses m-automated strategies; such strategies can be played by an automaton with m internal states and where inputs and outputs are coded to the nodes of the graph.

Section 2.8. The two duels presented, the silent and the noisy one, are discretized versions (discretization with respect to time) of duels played in continuous time as given in Karlin (1959). For further discussion on extensions of game of timing see the Notes section of Chapter 4, with reference to Section 4.7. The theory of the searchlight game as presented here has been taken from Olsder and Papavassilopoulos (1988a, 1988b). Such games with a finite state space belong to the realm of Markov chain games; for a survey see Parthasarathy and Stern (1977), and Raghavan and Filar (1991). For other types of search games, see Gal (1980).

Section 2.9. Problem 20: For more results on the relation between games and their spectra, see Weil (1968). Problem 22 has been taken from Isbell (1957).

Chapter 3

Noncooperative Finite Games: N-Person Nonzero-Sum

3.1 Introduction

This chapter develops a general theory for static and dynamic nonzero-sum finite games under two different types of noncooperative solution concepts—those named after J. Nash and H. von Stackelberg. The analysis starts in Section 3.2 with bimatrix games which basically constitute a normal form description of two-person nonzero-sum finite games, and in this context the Nash equilibrium solution (in both pure and mixed strategies) is introduced and its properties and features are thoroughly investigated. The section concludes with some computational aspects of mixed-strategy Nash equilibria in bimatrix games.

Section 3.3 extends the analysis of Section 3.2 to N-person games in normal form and includes a proof for an important theorem (Thm. 3.2) which states that every N-person finite game in normal form admits Nash equilibria in mixed strategies. In Section 3.4, the computational aspects of mixed-strategy Nash equilibria are briefly discussed, and in particular the close connection between mixed-strategy Nash equilibria in bimatrix games and the solution of a nonlinear programming problem is elucidated.

Section 3.5 deals with properties and derivation of Nash equilibria (in pure, behavioral and mixed strategies) for dynamic N-person games in extensive form. The analysis is first confined to single-act games and then extended to multi-act games. In both contexts, the concept of "informational inferiority" between two finite games in extensive form is introduced, which readily leads to "informational nonuniqueness" of Nash equilibria in dynamic games. To eliminate this feature, the additional concept of "delayed commitment" type equilibrium strategy is introduced, and such equilibria of special types of single-act and multi-act finite games are obtained through recursive procedures. Other refinement schemes discussed in this section are the "perfectness" and "properness" of Nash equilibria.

Section 3.6 is devoted to an extensive discussion of hierarchical equilibria in N-person finite games. First, the two-person games are treated, and in this context both the (global) Stackelberg and feedback Stackelberg solution concepts are introduced, also within the class of mixed and behavioral strategies. Several examples are included to illustrate these concepts and the derivation of the Stackelberg solution. The analysis is then extended to many-player games.

The final section of the chapter discusses extension of the results of the earlier sections to N-person finite games in extensive form which also incorporate chance moves.

3.2 Bimatrix Games

Before developing a general theory of nonzero-sum noncooperative games, it is most appropriate first to investigate equilibria of bimatrix games which in fact constitute the most elementary type of decision problems within the class of nonzero-sum games. In spite of their simple structure, bimatrix games still carry most of the salient features and intricacies of noncooperative decision making.

A bimatrix game can be considered as a natural extension of the matrix game of Section 2.2, to cover situations in which the outcome of a decision process does not necessarily dictate the verdict that what one player gains the other one has to lose. Accordingly, a bimatrix game is comprised of two $(m \times n)$-dimensional matrices, $A = \{a_{ij}\}$ and $B = \{b_{ij}\}$, with each pair of entries (a_{ij}, b_{ij}) denoting the outcome of the game corresponding to a particular pair of decisions made by the players. As in Section 2.2, we call the alternatives available to the players *strategies*. If **P1** adopts the strategy "row i" and **P2** adopts the strategy "column j", then a_{ij} (respectively, b_{ij}) denotes the loss incurred to **P1** (respectively, **P2**). Being a rational decision maker, each player will strive for an outcome which provides him with the lowest possible loss.

Stipulating that there exists no cooperation between the players and that they make their decisions independently, let us now investigate which pair(s) of strategies to declare as equilibrium pair(s) of a given bimatrix game; in other words, how do we define a noncooperative equilibrium solution in bimatrix games? To recall the basic property of a saddle-point equilibrium solution in zero-sum games would definitely help in this investigation: "A pair of strategies is in saddle-point equilibrium if there is no incentive for any unilateral deviation by any one of the players" (Section 2.2). This is, in fact, a most natural property that an equilibrium solution is expected to possess, and surely it finds relevance also in nonzero-sum games. Hence, we have the following definition.

Definition 3.1 *A pair of strategies* $\{row\ i^*,\ column\ j^*\}$ *is said to constitute a* noncooperative (Nash) equilibrium solution *to a bimatrix game* $(A = \{a_{ij}\}, B =$

$\{b_{ij}\}$) *if the following pair of inequalities is satisfied for all* $i = 1, \ldots, m$ *and all* $j = 1, \ldots, n$:

$$a_{i^*j^*} \leq a_{ij^*}, \tag{3.1a}$$
$$b_{i^*j^*} \leq b_{i^*j}. \tag{3.1b}$$

Furthermore, the pair $(a_{i^*j^*}, b_{i^*j^*})$ *is known as a* noncooperative (Nash) equilibrium outcome *of the bimatrix game.*

Example 3.1 Consider the following (2×2) bimatrix game

$$\begin{array}{c} \quad\mathbf{P2} \\ A = \begin{array}{|c|c|}\hline \boxed{1} & 0 \\ \hline 2 & \textcircled{-1} \\ \hline\end{array}\ \mathbf{P1} \end{array} \qquad \begin{array}{c} \quad\mathbf{P2} \\ B = \begin{array}{|c|c|}\hline \boxed{2} & 3 \\ \hline 1 & \textcircled{0} \\ \hline\end{array}\ \mathbf{P1}. \end{array}$$

It admits two Nash equilibria, as indicated: {row 1, column 1} and {row 2, column 2}. The corresponding equilibrium outcomes are $(1, 2)$ and $(-1, 0)$. □

We now readily observe from the preceding example that a bimatrix game can admit more than one Nash equilibrium solution, with the equilibrium outcomes being different in each case. This then raises the question of whether it would be possible to order different Nash equilibrium solution pairs among themselves so as to declare only one of them as the most favorable equilibrium solution. This, however, is not completely possible, since a total ordering does not exist between pairs of numbers; but, notions of "betterness" and "admissibility" can be introduced through a partial ordering.

Definition 3.2 *A pair of strategies* {*row* i_1, *column* j_1} *is said to be* better *than another pair of strategies* {*row* i_2 *column* j_2} *if* $a_{i_1j_1} \leq a_{i_2j_2}$, $b_{i_1j_1} \leq b_{i_2j_2}$ *and if at least one of these inequalities is strict.*

Definition 3.3 *A Nash equilibrium strategy pair is said to be* admissible *if there exists no better Nash equilibrium strategy pair.*

In Example 3.1, out of a total of two Nash equilibrium solutions only one of them is admissible ({row 2, column 2}), since it provides uniformly lower costs for both players. This pair of strategies can therefore be declared as the most "reasonable" noncooperative equilibrium solution of the bimatrix game. In the case when a bimatrix game admits more than one admissible Nash equilibrium, however, such a clean choice cannot be made. Consider, for example, the (2×2) bimatrix game

$$\begin{array}{c} \quad\mathbf{P2} \\ A = \begin{array}{|c|c|}\hline \boxed{-2} & 1 \\ \hline -1 & \textcircled{-1} \\ \hline\end{array}\ \mathbf{P1}, \end{array} \qquad \begin{array}{c} \quad\mathbf{P2} \\ B = \begin{array}{|c|c|}\hline \boxed{-1} & 1 \\ \hline 2 & \textcircled{-2} \\ \hline\end{array}\ \mathbf{P1} \end{array} \tag{3.2}$$

which admits two admissible Nash equilibrium solutions: {row 1, column 1}, {row 2, column 2}, with the equilibrium outcomes being $(-2,-1)$ and $(-1, -2)$, respectively. These Nash equilibria are definitely not interchangeable, and hence a possibility for the outcome of the game to be a nonequilibrium one in an actual play emerges. In effect, since there is no cooperation between the players by the nature of the problem, **P1** might stick to one equilibrium strategy (say, row l) and **P2** might adopt the other one (say, column 2), thus yielding an outcome of $(1, 1)$ which is unfavorable to both players. This is indeed one of the dilemmas of noncooperative nonzero-sum decision making (if a given problem does not admit a unique admissible equilibrium), but there is really no remedy for it unless one allows some kind of communication between the players, at least to ensure that a nonequilibrium outcome will not be realized. In fact, a possibility of collusion saves only that aspect of the difficulty in the above example, since the "best" choice still remains indeterminate. One way of resolving this indeterminacy is to allow a repetition and to choose one or the other Nash equilibrium solution at consecutive time points. The average outcome is then clearly (-3/2,-3/2), assuming that the players make their choices dependent on each other. However, this favorable average outcome is not an equilibrium outcome in mixed strategies, as we shall see in the last part of this section. The solution proposed above is acceptable in the "real" decision situation, known as the *battle of the sexes*, that the bimatrix game (3.2) represents. The story goes as follows.

A couple has to make a choice out of two alternatives for an evening's entertainment. The alternatives are a basketball game which is preferred by the husband (**P1**) and a musical comedy which is preferred by the wife (**P2**). But they would rather go together, instead of going separately to their top choices. This then implies, also in view of the bimatrix game (3.2), that they will eventually decide to go together either to the game or to the comedy. It should be noted that there is a flavor of cooperation in this example, but the solution is not unique in any sense. If repetition is possible, however, then they can go one evening to the game, and another evening to the comedy, thus resolving the problem in a "fair" way. Such a solution, however, does not carry any game-theoretic meaning as we shall see in the last part of this section. The message here is that if a bimatrix game admits more than one admissible Nash equilibrium solution, then a valid approach would be to go back to the actual decision process that the bimatrix game represents, and see whether a version of the bimatrix game that also accounts for repetition admits a solution which is favored by both parties. This, of course, is all valid provided that some communication is allowed between the players, in which case the "favorable" solution thus obtained will not be a noncooperative equilibrium solution.

Thus, we have seen that if a bimatrix game admits more than one admissible Nash equilibrium solution, then the equilibrium outcome of the game becomes rather ill-defined—mainly due to the fact that multiple Nash equilibria are in general not interchangeable. This ambiguity disappears, of course, whenever the equilibrium strategies are interchangeable, which necessarily requires the corresponding outcomes to be the same. Since zero-sum matrix games are special types of bimatrix games, in which case the equilibrium solutions are known to be interchangeable (Corollary 2.1), it follows that there exists some nonempty class of bimatrix games whose equilibrium solutions possess such a property. This class is, in fact, larger than the set of all $(m \times n)$ zero-sum games, as it is shown below.

Definition 3.4 *Two $(m \times n)$ bimatrix games (A, B) and (C, D) are said to be* strategically equivalent *if there exist positive constants α_1, α_2, and scalars β_1, β_2, such that*

$$a_{ij} = \alpha_1 c_{ij} + \beta_1, \tag{3.3a}$$
$$b_{ij} = \alpha_2 d_{ij} + \beta_2, \tag{3.3b}$$

for all $i = 1, \ldots, m$; $j = 1, \ldots, n$.

Remark 3.1 The reader can easily verify that the pair (3.3a)-(3.3b) is indeed an equivalence relation since it is symmetric, reflexive and transitive. □

Proposition 3.1 *All strategically equivalent bimatrix games have the same Nash equilibria.*

Proof. Let (A, B) be a bimatrix game with a Nash equilibrium solution {row i^*, column j^*}. Then, by definition, inequalities (3.1a)-(3.1b) are satisfied. Now let (C, D) be any other bimatrix game that is strategically equivalent to (A, B). Then there exist $\alpha_1 > 0$, $\alpha_2 > 0$, β_1, β_2 such that (3.3a)-(3.3b) are satisfied for all possible i and j. Using these relations in (3.1a) and (3.1b), we obtain, in view of the positivity of α_1 and α_2,

$$c_{i^*j^*} \leq c_{ij^*}$$
$$d_{i^*j^*} \leq d_{i^*j},$$

for all $i = 1, \ldots, m$ and all $j = 1, \ldots, n$, which proves the proposition since (A, B) was an arbitrary bimatrix game. □

Proposition 3.2 *Multiple Nash equilibria of a bimatrix game (A, B) are interchangeable if (A, B) is strategically equivalent to $(A, -A)$.*

Proof. The result readily follows from Corollary 2.1 since, by hypothesis, the bimatrix game (A, B) is strategically equivalent to a zero-sum game A. □

The undesirable features of the Nash equilibrium solutions of bimatrix games, that we have witnessed, might at first sight lead to questioning appropriateness and suitability of the Nash equilibrium solution concept for such *noncooperative* multi-person decision problems. Such a verdict, however, is not fair, since the undesirable features detected so far are due to the noncooperative nature of the decision problem under consideration, rather than to the weakness of the Nash equilibrium solution concept. To supplement our argument, let us consider one extreme case (the so-called "team" problem) that involves two players, identical goals (i.e., $A = B$), and noncooperative decision making. Note that, since possible outcomes are single numbers in this case, a total ordering of outcomes is possible, which then makes the admissible Nash equilibrium solution the only possible equilibrium solution for this class of decision problems. Moreover, the admissible Nash equilibrium outcome will be unique and be equal to the smallest entry of the matrix that characterizes the game. Note, however, that even for such a reasonable class of decision problems, the players might end up at a nonequilibrium point in case of multiple equilibria; this being mainly due to the noncooperative nature of the decision process which requires the players to make their decisions independently. To illustrate such a possibility, let us consider the identical goal game (team problem) with the cost matrices

$$A = B = \overset{\mathbf{P2}}{\begin{array}{|c|c|} \hline 0 & 1 \\ \hline 1 & 0 \\ \hline \end{array}}\ \mathbf{P1}, \tag{3.4}$$

which both parties try to minimize, by appropriate choices of strategies. There are clearly two equilibria, {row 1, column 1} and {row 2, column 2}, both yielding the minimum cost of 0 for both players. These equilibria, however, are not interchangeable, thus leaving the players at a very difficult position if there is no communication between them. (Perhaps tossing a fair coin and thereby securing an average cost of $\frac{1}{2}$ would be preferred by both sides to any pure strategy which could easily lead to an unfortunate cost of 1.) This then indicates that even the well-established team-equilibrium solution could be frowned upon in case of multiple equilibria when the decision mode is noncooperative.

One can actually make things look even worse by replacing matrix (3.4) with a nonsymmetric one

$$A = B = \overset{\mathbf{P2}}{\begin{array}{|c|c|} \hline 0 & 2 \\ \hline 1 & 0 \\ \hline \end{array}}\ \mathbf{P1}, \tag{3.5}$$

which admits the same two equilibria as in the previous matrix game. But what are the players actually going to play in this case? To us, there seems to be only one logical play for each player: **P1** would play "row 2", hoping that **P2** would play his second equilibrium strategy, but at any rate securing a loss ceiling of 1. **P2**, reasoning along similar lines, would rather play "column 1". Consequently, the outcome of their joint decision is 1 which is definitely not an equilibrium outcome. But, in an actual play of this matrix game, this outcome is more likely to occur than the equilibrium outcome. One might counter-argue at this point and say that **P2** could very well figure out **P1**'s way of thinking and pick "column 2" instead, thus enjoying the low equilibrium outcome. But what if **P1** thinks the same way? There is clearly no end to this so-called *second-guessing iteration* procedure if adopted by both players as a method that guides them to the "optimum" strategy.

One might wonder at this point as to why, while the multiple Nash equilibria of bimatrix games (including identical-goal games) possess all these undesirable features, the saddle-point equilibrium solutions of zero-sum (matrix) games are quite robust—in spite of the fact that both are noncooperative decision problems. The answer to this lies in the nature of zero-sum games: they are completely antagonistic, and the noncooperative equilibrium solution fits well within this framework. In other nonzero-sum games, however, the antagonistic nature is rather suppressed or is completely absent, and consequently the "noncooperative decision making" framework does not totally suit such problems. If some cooperation is allowed, or if there is a hierarchy in decision making, then equilibrium solutions of nonzero-sum (matrix) games could possess more desirable features. We shall, in fact, observe this later in Section 3.6 where we discuss equilibria in nonzero-sum games under the latter mode of decision making.

The minimax solution

One important property of the saddle-point strategies in zero-sum games was that they were also the security strategies of the players (see Thm. 2.2). In bimatrix games, one can still introduce security strategies for the players in the same way as it was done in Section 2.2, and with the properties cited in Thm. 2.1 being valid with the exception of the third one. It should of course be clear that, in a bimatrix game (A, B) the security strategies of **P1** involve only the entries of matrix A, while the security strategies of **P2** involve only those of matrix B. In Example 3.1, considered earlier, the unique security strategy of **P1** is "row 1", and the unique security strategy of **P2** is "column 1". Thus, considered as a pair, the security strategies of the players correspond to one of the Nash equilibrium solutions, but not to the admissible one, in this case. One can come up with examples of bimatrix games in which the security strategies do not correspond to any one of the equilibrium solutions, or examples of bimatrix games in which they coincide with an admissible Nash solution. As an illustration of the latter possibility, consider the following bimatrix game:

$$A = \overset{\textbf{P2}}{\begin{array}{|c|c|}\hline \circled{8} & 0 \\ \hline 30 & 2 \\ \hline\end{array}}\ \textbf{P1}, \qquad B = \overset{\textbf{P2}}{\begin{array}{|c|c|}\hline \circled{8} & 30 \\ \hline 0 & 2 \\ \hline\end{array}}\ \textbf{P1}. \tag{3.6}$$

This admits a unique pair of Nash equilibrium strategies, as indicated, which are also the security strategies of the players. Moreover, the equilibrium outcome is determined by the security levels of the players.

The preceding bimatrix game is known in the literature as the *prisoners' dilemma*. It characterizes a situation in which two criminals, suspected of having committed a serious crime, are detained before a trial. Since there is no direct evidence against them, their conviction depends on whether they confess or not. If the prisoners both confess, then they will be sentenced to 8 years. If neither one confesses, then they will be convicted of a lesser crime and sentenced to

2 years. If only one of them confesses and puts the blame on the other one, then he is set free according to the laws of the country and the other one is sentenced to 30 years. The unique Nash equilibrium in this case dictates that they both should confess. Note, however, that there is in fact a better solution for both criminals, which is that they both should refuse to confess; but implementation of such a solution would require a cooperation of some kind (and also trust), which is clearly out of the question in the present context. Besides, this second solution (which is not Nash) is extremely unstable, since a player will find it to his advantage to unilaterally deviate from this position at the last minute.

Even if the security strategies might not be in noncooperative equilibrium, they can still be employed in an actual game, especially in cases when there exist two or more noninterchangeable Nash equilibria or when a player is not completely sure of the cost matrix, or even the rationality of the other player. This reasoning leads to the following second solution concept in bimatrix games.

Definition 3.5 *A pair of strategies {row $\hat{i}$, column $\hat{j}$} is known as a pair of* minimax strategies *for the players in a bimatrix game (A, B) if the former is a security strategy for* **P1** *in the matrix game A, and the latter is a security strategy for* **P2** *in the matrix game B. The corresponding security levels of the players are known as the* minimax values *of the bimatrix game.*

Remark 3.2 It should be noted that minimax values of a bimatrix game are definitely not lower (in an ordered way) than the pair of values of any Nash equilibrium outcome. Even if the unique Nash equilibrium strategies correspond to the minimax strategies, the minimax values could be higher than the values of the Nash equilibrium outcome, mainly because the minimax strategy of a player might not constitute optimal response to the minimax strategy of the other player. □

Mixed strategies

In our discussion of bimatrix games in this section, we have so far encountered only cases in which a given bimatrix game admits a unique or a multiple of Nash equilibria. There are other cases, however, in which Nash equilibrium strategies might not exist, as illustrated in the following example. In such a case we simply say that a Nash equilibrium does not exist in pure strategies.

Example 3.2 Consider the (2×2) bimatrix game

$$A = \overset{\textbf{P2}}{\begin{array}{|c|c|}\hline 1 & 0 \\ \hline 2 & -1 \\ \hline \end{array}}\ \textbf{P1}, \qquad B = \overset{\textbf{P2}}{\begin{array}{|c|c|}\hline 3 & 2 \\ \hline 0 & 1 \\ \hline \end{array}}\ \textbf{P1}.$$

It clearly does not admit a Nash equilibrium solution in pure strategies. The minimax strategies, however, exist (as they always do, by Thm. 2.1 (ii)) and are given as {row 1, column 2}. □

Paralleling our analysis in the case of nonexistence of a saddle point in zero-sum games (Section 2.2), we now enlarge the class of strategies so as to include all mixed strategies, defined as the set of all probability distributions on the set of pure strategies of each player (see Def. 2.2). This extension is in fact sufficient to ensure the existence of a Nash equilibrium solution (in mixed strategies, in a bimatrix game)—a result which we state below in Thm. 3.1, after making the concept of a Nash equilibrium in this extended space precise. Using the same notation as in Defs. 2.3-2.5, wherever appropriate, we first have the following definition.

Definition 3.6 *A pair* $\{y^* \in Y, z^* \in Z\}$ *is said to constitute a* noncooperative (Nash) equilibrium solution *to a bimatrix game* (A, B) *in* mixed strategies, *if the following inequalities are satisfied for all* $y \in Y$ *and* $z \in Z$*:*

$$y^{*\prime}Az^* \leq y'Az^*, \quad y \in Y, \tag{3.7a}$$

$$y^{*\prime}Bz^* \leq y^{*\prime}Bz, \quad z \in Z. \tag{3.7b}$$

Here, the pair $(y^{*\prime}Az^*, y^{*\prime}Bz^*)$ *is known as a* noncooperative (Nash) equilibrium outcome *of the bimatrix game in mixed strategies.*

Theorem 3.1 *Every bimatrix game has at least one Nash equilibrium solution* in mixed strategies.

Proof. We postpone the proof of this important result until the next section where it is proven in a more general context (Thm. 3.2). □

Computation of mixed Nash equilibrium strategies of bimatrix games is more involved than the computation of mixed saddle-point solutions, and in the literature there are very few generally applicable methods in that context. One of these methods converts the original game problem into a nonlinear programming problem, and it will be discussed in Section 3.4. What we intend to include here is an illustration of the computation of mixed Nash equilibrium strategies in (2×2) bimatrix games by directly making use of inequalities (3.7a)-(3.7b). Specifically, consider again the bimatrix game of Example 3.2 which is known not to admit a pure-strategy Nash equilibrium. In this case, since every $y \in Y$ can be written as $y = (y_1, (1 - y_1))'$, with $0 \leq y_1 \leq 1$, and similarly every $z \in Z$ as $z = (z_1, (1 - z_1))'$, with $0 \leq z_1 \leq 1$, we first obtain the following equivalent inequalities for (3.7a) and (3.7b), respectively, for the bimatrix game under consideration:

$$-2y_1^* z_1^* + y_1^* \leq -2y_1 z_1^* + y_1, \quad 0 \leq y_1 \leq 1,$$

$$2y_1^* z_1^* - z_1^* \leq 2y_1^* z_1 - z_1, \quad 0 \leq z_1 \leq 1.$$

Here $(y_1^*, (1-y_1^*))'$ and $(z_1^*, (1-z_1^*))'$ denote the mixed Nash equilibrium strategies of **P1** and **P2**, respectively, which have yet to be computed so that the preceding pair of inequalities is satisfied. A way of obtaining this solution would be to find that value of z_1^* (if it exists in the interval $[0, 1]$) which would make the RHS of the first inequality independent of y_1, and also the value of y_1^* that would make the RHS of the second inequality independent of z_1. The solution then readily turns out to be $(y_1^* = \frac{1}{2}, z_1^* = \frac{1}{2})$ which can easily be checked to be the unique solution set of the two inequalities. Consequently the bimatrix game of Example 3.2 admits a unique Nash equilibrium solution in mixed strategies, which is $\{y^* = (\frac{1}{2}, \frac{1}{2})', z^* = (\frac{1}{2}, \frac{1}{2})'\}$.

Remark 3.3 In the above solution to Example 3.2 we observe an interesting, and rather counter-intuitive, feature of the Nash equilibrium solution in mixed strategies. The solution obtained, i.e., $\{y^*, z^*\}$, has the property that the inequalities (3.7a) and (3.7b) in fact become independent of $y \in Y$ and $z \in Z$. In other words, direction of the inequality loses its significance. This leads to the important conclusion that if the players were instead seeking to maximize their average costs, then we would have obtained the same mixed equilibrium solution. This implies that, while computing his mixed Nash equilibrium strategy, each player pays attention only to the average cost function of his co-player, rather than optimizing his own average cost function. Hence, the nature of the optimization (i.e., minimization or maximization) becomes irrelevant in this case. The same feature can also be observed in mixed saddle-point solutions, yielding the conclusion that a zero-sum matrix game A could have the same mixed saddle-point solution as the zero-sum game with matrix $-A$. (See Problem 21 in section 2.9.) □

The following proposition now makes the conclusions of the preceding remark precise.

Proposition 3.3 *Let* $\mathring{Y}$ *and* $\mathring{Z}$ *denote the sets of inner points (interiors) of* Y *and* Z*, respectively. If a bimatrix game* (A, B) *admits a mixed-strategy Nash equilibrium solution* $\{y^* \in \mathring{Y}, z^* \in \mathring{Z}\}$*,*[15] *then this also serves as a mixed-strategy Nash equilibrium solution for the bimatrix game* $(-A, -B)$*.*

Proof. By hypothesis, inequalities (3.7a) and (3.7b) are satisfied by a pair $\{y^* \in \mathring{Y}, z^* \in \mathring{Z}\}$. This implies that $\min_Y y'Az^*$ is attained by an inner point of Y, and hence, because of linearity of $y'Az^*$ in y, this expression has to be independent of y. Similarly, $y^{*\prime}Bz$ is independent of z. Consequently, (3.7a)-(3.7b) can be written as

$$\begin{aligned} y^{*\prime}Az^* &= y'Az^*, \quad \forall y \in Y, \\ y^{*\prime}Bz^* &= y^{*\prime}Bz, \quad \forall z \in Z, \end{aligned}$$

which further leads to the inequality-pair

$$\begin{aligned} y^{*\prime}(-A)z^* &\leq y'(-A)z^*, \quad \forall y \in Y, \\ y^{*\prime}(-B)z^* &\leq y^{*\prime}(-B)z, \quad \forall z \in Z, \end{aligned}$$

which verifies the assertion of the proposition. □

[15] Such a solution is also known as a completely mixed Nash equilibrium solution or as an inner mixed-strategy Nash equilibrium solution.

As a special case of this proposition, we now immediately have the following result for zero-sum matrix games.

Corollary 3.1 *Any inner mixed saddle point of a zero-sum matrix game A also constitutes an inner mixed saddle point for the zero-sum matrix game $-A$.*

Even though we have introduced the concept of a mixed noncooperative equilibrium solution for bimatrix games which do not admit a Nash equilibrium in pure strategies, it is quite possible for a bimatrix game that admits (pure) Nash equilibria to admit a mixed Nash equilibrium as well. To verify this possibility, let us again consider the bimatrix game of the "battle of the sexes", given by (3.2). We have already seen that it admits two noninterchangeable Nash equilibria with outcomes $(-2,-1)$ and $(-1,-2)$. Let us now investigate whether it admits a third equilibrium, this time in mixed strategies.

For this bimatrix game, letting $y = (y_1, (1-y_1))' \in Y$ and $z = (z_1, (1-z_1))' \in Z$, we first rearrange the inequalities (3.7a)-(3.7b) and write them in the equivalent form

$$\begin{aligned} -5y_1^* z_1^* + 2y_1^* &\leq -5y_1 z_1^* + 2y_1, \quad 0 \leq y_1 \leq 1, \\ -5y_1^* z_1^* + 3z_1^* &\leq -5y_1^* z_1 + 3z_1, \quad 0 \leq z_1 \leq 1. \end{aligned}$$

Now it readily follows that these inequalities admit the unique inner point solution ($y_1^* = \frac{3}{5}, z_1^* = \frac{2}{5}$) dictating the mixed Nash equilibrium solution $\{y^* = (\frac{3}{5}, \frac{2}{5})', z^* = (\frac{2}{5}, \frac{3}{5})'\}$, and an average outcome of $(-\frac{1}{5}, -\frac{1}{5})$. Of course, the above inequalities also admit the solutions ($y_1^* = 1, z_1^* = 1$) and ($y_1^* = 0, z_1^* = 0$) which correspond to the two pure Nash equilibria which we have discussed earlier at some length. Our interest here lies in the nature of the mixed equilibrium solution. Interpreted within the context of the "battle of the sexes", this mixed solution dictates that the husband chooses to go to the game with probability $\frac{3}{5}$ and to the musical comedy with probability $\frac{2}{5}$. The wife, on the other hand, decides on the game with probability $\frac{2}{5}$ and on the comedy with probability $\frac{3}{5}$. Since, by the very nature of the noncooperative equilibrium solution concept, these decisions have to be made independently, there is a probability of 13/25 with which the husband and wife will have to spend the evening separately. In other words, even if repetition is allowed, the couple will be together for less than half the time, as dictated by the unique mixed Nash equilibrium solution; and this so in spite of the fact that the husband and wife explicitly indicate in their preference ranking that they would rather be together in the evening. The main reason for such a dichotomy between the actual decision process and the solution dictated by the theory is that the noncooperative equilibrium solution concept requires decisions to be made independently, whereas in the bimatrix game of the "battle of the sexes" our way of thinking for a reasonable equilibrium solution leads us to a cooperative solution which inevitably asks for dependent decisions. (See Problem 4, Section 3.8, for more variations on the "battle of the sexes".)

The conclusion we draw here is that bimatrix games with pure-strategy Nash equilibria could also admit mixed-strategy Nash solutions which, depending on the relative magnitudes of the entries of the two matrices, could yield better or worse (and of course also noncomparable) average equilibrium outcomes.

3.3 N-Person Games in Normal Form

The class of N-person nonzero-sum finite static games in normal form models a decision making process similar in nature to that modeled by bimatrix games, but this time with $N(> 2)$ interacting decision makers (players). Decisions are again made independently and out of a finite set of alternatives for each player. Since there exist more than two players, a matrix formulation on the plane is not possible for such games, thus making the display of possible outcomes and visualization of equilibrium strategies rather difficult. However, a precise formulation is still possible, as it is provided below together with the notation to be used in describing N-person finite static games in normal form.

Formulation of an N-person finite static game in normal form

(1) There are N players to be denoted by $\mathbf{P1}, \mathbf{P2}, \ldots, \mathbf{P}N$. Let us further denote the index set $\{1, 2, \ldots, N\}$ by $\mathbf{N}$.

(2) There is a finite number of alternatives for each player to choose from. Let m_i denote the number of alternatives available to $\mathbf{P}i$, and further denote the index set $\{1, 2, \ldots, m_i\}$ by $\mathbf{M}_i$, with a typical element of $\mathbf{M}_i$ designated as n_i.

(3) If $\mathbf{P}j$ chooses a strategy $n_j \in \mathbf{M}_j$, and this so for all $j \in \mathbf{N}$, then the loss incurred to $\mathbf{P}i$ is a single number $a^i_{n_1,n_2,\ldots,n_N}$. The ordered N-tuple of all these numbers (over $i \in \mathbf{N}$), i.e., $(a^1_{n_1,\ldots,n_N}, a^2_{n_1,\ldots,n_N}, a^N_{n_1,\ldots,n_N})$, constitutes the corresponding unique outcome of the game.

(4) Players make their decisions independently and each one unilaterally seeks the minimum possible loss, of course by also taking into account the possible rational choices of the other players.

The noncooperative equilibrium solution concept within the context of this N-person game can be introduced as follows as a direct extension of Def. 3.1.

Definition 3.7 *An N-tuple of strategies $\{n_1^*, n_2^*, \ldots, n_N^*\}$, with $n_i^* \in \mathbf{M}_i$, $i \in \mathbf{N}$, is said to constitute a* noncooperative (Nash) equilibrium solution *for an N-person nonzero-sum static finite game in normal form, as formulated above, if the following N inequalities are satisfied for all $n_i \in \mathbf{M}_i$, $i \in \mathbf{N}$:*

$$\left.\begin{aligned} a^{1*} &\triangleq a^1_{n_1^*,n_2^*,\ldots,n_N^*} \le a^1_{n_1,n_2^*,\ldots,n_N^*}, \\ a^{2*} &\triangleq a^2_{n_1^*,n_2^*,\ldots,n_N^*} \le a^2_{n_1^*,n_2,n_3^*,\ldots,n_N^*}, \\ &\cdots\cdots\cdots\cdots\cdots \\ a^{N*} &\triangleq a^N_{n_1^*,n_2^*,\ldots,n_N^*} \le a^N_{n_1^*,\ldots,n_{N-1}^*,n_N}. \end{aligned}\right\} \tag{3.8}$$

Here, the N-tuple $(a^{1*}, a^{2*}, \ldots, a^{N*})$ is known as a noncooperative (Nash) equilibrium outcome *of the N-person game in normal form.*

There is actually no simple method to determine the Nash equilibrium solutions of N-person finite games in normal form. One basically has to check exhaustively all possible combinations of N-tuples of strategies, to see which ones provide a Nash equilibrium. This enumeration, though straightforward, could at times be rather strenuous, especially when N and/or m_i, $i \in \mathbf{N}$, are large. However, given an N-tuple of strategies asserted to be in Nash equilibrium, it is relatively simpler to verify their equilibrium property, since one then has to check only unilateral deviations from the given equilibrium solution. To get a flavor of the enumeration process and the method of verification of a Nash equilibrium solution, let us now consider the following example.

Example 3.3 Consider a 3-person game in which each player has two alternatives to choose from. That is, $N = 3$ and $m_1 = m_2 = m_3 = 2$. To complete the description of the game, the $2^3 = 8$ possible outcomes are given as

$$
\begin{aligned}
(a^1_{1,1,1} a^2_{1,1,1} a^3_{1,1,1}) &= (1, -1, 0), \\
(a^1_{1,2,1} a^2_{1,2,1} a^3_{1,2,1}) &= (2, 1, 1), \\
(a^1_{2,1,1} a^2_{2,1,1} a^3_{2,1,1}) &= (2, 0, 1), \\
(a^1_{2,2,1} a^2_{2,2,1} a^3_{2,2,1}) &= (0, -2, 1), \\
(a^1_{1,1,2} a^2_{1,1,2} a^3_{1,1,2}) &= (1, 1, 1), \\
(a^1_{1,2,2} a^2_{1,2,2} a^3_{1,2,2}) &= (0, 1, 0), \\
(a^1_{2,1,2} a^2_{2,1,2} a^3_{2,1,2}) &= (0, 1, 2), \\
(a^1_{2,2,2} a^2_{2,2,2} a^3_{2,2,2}) &= (-1, 2, 0).
\end{aligned}
$$

These possible outcomes can actually be displayed in the form of two (2×2) matrices

$$
n_3 = 1 \quad \overset{\mathbf{P2}}{\begin{array}{|c|c|} \hline {}^* \ (1,\ -1,\ 0) & (2,\ 1,\ 1) \\ \hline (2,0,1) & (0,-2,1) \\ \hline \end{array}} \ \mathbf{P1}, \tag{3.9a}
$$

$$
n_3 = 2 \quad \overset{\mathbf{P2}}{\begin{array}{|c|c|} \hline (1,1,1) & (0,1,0) \\ \hline (0,1,2) & (-1,2,0) \\ \hline \end{array}} \ \mathbf{P1}, \tag{3.9b}
$$

where the entries of the former matrix correspond to the possible outcomes if **P3**'s strategy is fixed at $n_3 = 1$, and the latter matrix provides possible outcomes if his strategy is fixed at $n_3 = 2$. We now assert that the "starred" entry is a Nash equilibrium outcome for this game. A verification of this assertion would involve three separate checks, concerning unilateral deviation of each player. If **P1** deviates from this asserted equilibrium strategy $n_1 = 1$, then his loss becomes 2 which is not favorable. If **P2** deviates from $n_2 = 1$, his loss becomes 1 which is not favorable either. Finally, if **P3** deviates from $n_3 = 1$, his loss becomes 1 which is higher than his asserted equilibrium loss 0. Consequently, the first entry of the first matrix, i.e., $(1, -1, 0)$, indeed provides a Nash equilibrium outcome, with the corresponding equilibrium strategies being $\{n_1^* = 1, n_2^* = 1, n_3^* = 1\}$. The reader can now check by enumeration that this is actually the only Nash equilibrium solution of this 3-person game. □

When the Nash equilibrium solution of an N-person nonzero-sum game is not unique, we can again introduce the concept of "*admissible Nash equilibrium solution*" as direct extensions of Defs. 3.2 and 3.3 to the N-person case. It is of course possible that a given N-person game will admit more than one admissible Nash equilibrium solution which are also not interchangeable. This naturally leads to an ill-defined equilibrium outcome; but we will not elaborate on these aspects of the equilibrium solution here since they were extensively discussed in the previous section within the context of bimatrix games, and that discussion can readily be carried over to fit the present framework. Furthermore, Prop. 3.1 has a natural version in the present framework, which we quote below without proof, after introducing an extension of Def. 3.4. Proposition 3.2, on the other hand, has no direct counterpart in an N-person game with $N > 2$.

Definition 3.8 *Two nonzero-sum finite static games in normal form are said to be* strategically equivalent *if the following three conditions are satisfied:*

(i) The two games have the same number of players (say N),

(ii) each player has the same number of alternatives in both games,

(iii) if $\{(a^1_{n_1,\dots,n_N}, \dots, a^N_{n_1,\dots,n_N}), n_i \in \mathbf{M}_i, i \in \mathbf{N}\}$ is the set of possible outcomes in one game, and $\{(b^1_{n_1,\dots,n_N}, \dots, b^N_{n_1,\dots,n_N}), n_i \in \mathbf{M}_i, i \in \mathbf{N}\}$ is the set of possible outcomes in the other, then there exist positive constants α_i, $i \in \mathbf{N}$, and scalars β_i, $i \in \mathbf{N}$, such that

$$a^i_{n_1,\dots,n_N} = \alpha_i b^i_{n_1,\dots,n_N} + \beta_i, \quad i \in \mathbf{N}, \tag{3.10}$$

for all $n_j \in \mathbf{M}_j$, $j \in \mathbf{N}$.

Proposition 3.4 *All strategically equivalent nonzero-sum finite static games in normal form have the same set of Nash equilibria.*

Minimax strategies introduced earlier within the context of bimatrix games (Def. 3.5) find applications also in N-person games, especially when there exist more than one admissible Nash equilibrium which are not interchangeable. In the present context their definition involves a natural extension of Def. 3.5, which we do not provide here.

Example 3.4 In the 3-person game of Example 3.3, the minimax strategies of **P1**, **P2** and **P3** are $\hat{n}_1 = 1$ or 2, $\hat{n}_2 = 1$ and $\hat{n}_3 = 1$, respectively. It should be noted that for **P1** any strategy is minimax. The security levels of the players, however, are unique (as they should be) and are 2, 1 and 1, respectively. Note that these values are higher than the unique Nash equilibrium outcome $(1, -1, 0)$, in accordance with the statement of Remark 3.2 altered to fit the present context. □

Mixed-strategy Nash equilibria

Mixed noncooperative equilibrium solutions of an N-person finite static game in normal form can be introduced by extending Def. 3.6 to N players, i.e., by replacing inequalities (3.7a) and (3.7b) by an N-tuple of similar inequalities. To this end, let us first introduce the notation Y^i to denote the mixed strategy space of **P**i, and further denote a typical element of this space by y^i, and its kth component by y^i_k. Then, we have the following definition.

Definition 3.9 *An N-tuple $\{y^{i*} \in Y^i; i \in \mathbf{N}\}$ is said to constitute a* mixed-strategy noncooperative (Nash) equilibrium solution *for an N-person finite static game in normal form if the following N inequalities are satisfied for all $y^j \in Y^i$, $j \in \mathbf{N}$:*

$$\left.\begin{aligned}
\bar{J}^{1*} &\triangleq \sum_{\mathbf{M}_1} \cdots \sum_{\mathbf{M}_N} y^{1*}_{n_1} y^{2*}_{n_2} \cdots y^{N*}_{n_N} a^1_{n_1,\ldots,n_N} \\
&\quad \leq \sum_{\mathbf{M}_1} \cdots \sum_{\mathbf{M}_N} y^{1}_{n_1} y^{2*}_{n_2} \cdots y^{N*}_{n_N} a^1_{n_1,\ldots,n_N}, \\
\bar{J}^{2*} &\triangleq \sum_{\mathbf{M}_1} \cdots \sum_{\mathbf{M}_N} y^{1*}_{n_1} y^{2*}_{n_2} \cdots y^{N*}_{n_N} a^2_{n_1,\ldots,n_N} \\
&\quad \leq \sum_{\mathbf{M}_1} \cdots \sum_{\mathbf{M}_N} y^{1*}_{n_1} y^{2}_{n_2} y^{3*}_{n_3} \cdots y^{N*}_{n_N} a^2_{n_1,\ldots,n_N}, \\
&\cdots\cdots\cdots\cdots\cdots\cdots \\
&\cdots\cdots\cdots\cdots\cdots\cdots \\
\bar{J}^{N*} &\triangleq \sum_{\mathbf{M}_1} \cdots \sum_{\mathbf{M}_N} y^{1*}_{n_1} y^{2*}_{n_2} \cdots y^{N*}_{n_N} a^N_{n_1,\ldots,n_N} \\
&\quad \leq \sum_{\mathbf{M}_1} \cdots \sum_{\mathbf{M}_N} y^{1*}_{n_1} \cdots y^{N-1*}_{n_{N-1}} y^{N}_{n_N} a^N_{n_1,\ldots,n_N}.
\end{aligned}\right\} \tag{3.11}$$

Here, the N-tuple $(\bar{J}^{1*}, \ldots, \bar{J}^{N*})$ is known as a noncooperative (Nash) equilibrium outcome *of the N-person game in* mixed strategies.

One of the important results of static game theory is that every N-person game of the type discussed in this section admits a mixed-strategy Nash equilibrium solution (note that pure strategies are also included in the class of mixed strategies). We now state and prove this result which is also an extension of Thm. 3.1.

Theorem 3.2 *Every N-person static finite game in normal form admits a noncooperative (Nash) equilibrium solution in mixed strategies.*

Proof. Let $\{y^i \in Y^i; i \in \mathbf{N}\}$ denote an N-tuple of mixed strategies for the N-person game, and introduce

$$\psi^i_{n_i}(y^1,\ldots,y^N) \triangleq \sum_{\mathbf{M}_1}\cdots\sum_{\mathbf{M}_N} y^1_{n_1}y^2_{n_2}\cdots y^N_{n_N}a^i_{n_1,\ldots,n_N}$$
$$-\sum_{\mathbf{M}_1}\cdots\sum_{\mathbf{M}_{i-1}}\sum_{\mathbf{M}_{i+1}}\cdots\sum_{\mathbf{M}_N} y^1_{n_1}\cdots y^{i-1}_{n_{i-1}}y^{i+1}_{n_{i+1}}\cdots y^N_{n_N}a^i_{n_1,\ldots,n_N},$$

for each $n_i \in \mathbf{M}_i$, and $i \in \mathbf{N}$. Note that this function is continuous on the product set $\Pi_N Y^i$. Now, let $c^i_{n_i}$ be related to $\psi^i_{n_i}$ by

$$c^i_{n_i} \triangleq \max\{\psi^i_{n_i}, 0\}$$

for each $n_i \in \mathbf{M}_i$ and $i \in \mathbf{N}$. It readily follows that $c^i_{n_i}(y^1,\ldots,y^N)$ is also a continuous function on $\Pi_N Y^i$. Then, introducing the transformation

$$\bar{y}^i_{n_i} = \frac{y^i_{n_i} + c^i_{n_i}}{1+\sum_{j\in\mathbf{M}_i} c^i_j}, \qquad n_i \in \mathbf{M}_i, i \in \mathbf{N}, \tag{3.12}$$

and denoting it by T:

$$(\bar{y}^1,\ldots,\bar{y}^N) = T(y^1,\ldots,y^N), \tag{3.13}$$

we observe that T is a continuous mapping of $\Pi_N Y^i$ into itself. Since each Y^i is a simplex of appropriate dimension (see Section 2.2), the product set $\Pi_N Y^i$ becomes closed, bounded and convex. Hence, by *Brouwer's fixed point theorem* (see Appendix C), the mapping T has at least one fixed point. We now prove that every fixed point of this mapping is necessarily a mixed-strategy Nash equilibrium solution of the game, and conversely that every mixed-strategy Nash equilibrium solution is a fixed point of T, thereby concluding the proof of the theorem.

We first verify the latter assertion: If $\{y^{i*}; i \in \mathbf{N}\}$ is a mixed-strategy equilibrium solution, then, by definition (i.e., from inequalities (3.11)), the function $\psi^i_{n_i}(y^{1*},\ldots,y^{N*})$ is nonpositive for every $n_i \in \mathbf{M}_i$ and $i \in \mathbf{N}$, and thus $c^i_{n_i} = 0$ for all $n_i \in \mathbf{M}_i$, $i \in \mathbf{N}$. This readily yields the conclusion

$$T(y^{1*},\ldots,y^{N*}) = (y^{1*},\ldots,y^{N*}),$$

which is that $\{y^{i*}; i \in \mathbf{N}\}$ is a fixed point of T.

For verification of the former assertion, suppose that $\{y^i; i \in N\}$ is a fixed point of T, but not a mixed equilibrium solution. Then, for some $i \in \mathbf{N}$ (say $i = 1$) there exists a $\tilde{y}^1 \in Y^1$ such that

$$\sum_{\mathbf{M}_1}\cdots\sum_{\mathbf{M}_N} y^1_{n_1}y^2_{n_2}\cdots y^N_{n_N}a^1_{n_1,\ldots,n_N} > \sum_{\mathbf{M}_1}\cdots\sum_{\mathbf{M}_N} \tilde{y}^1_{n_1}y^2_{n_2}\cdots y^N_{n_N}a^1_{n_1,\ldots,n_N}. \tag{3.14}$$

Now let $\tilde{n}_1$ denote an index for which the quantity

$$\sum_{\mathbf{M}_1}\cdots\sum_{\mathbf{M}_N} y^2_{n_2}\cdots y^N_{n_N}a^1_{n_1,\ldots,n_N} \tag{3.15}$$

attains its minimum value over $n_1 \in \mathbf{M}_1$. Then, since $\tilde{y}^1$ is a mixed strategy, the RHS of (3.14) can be bounded below by (3.15) with $n_1 = \tilde{n}_1$, thus yielding the strict inequality

$$\sum_{\mathbf{M}_1} \cdots \sum_{\mathbf{M}_N} y^1_{n_1} \cdots y^N_{n_N} a^1_{n_1,\ldots,n_N} > \sum_{\mathbf{M}_2} \cdots \sum_{\mathbf{M}_N} y^2_{n_2} \cdots y^N_{n_N} a^1_{\tilde{n}_1,\ldots,n_N}$$

which can further be seen to be equivalent to

$$\psi^1_{\tilde{n}_1}(y^1, y^2, \ldots, y^N) > 0. \tag{3.16}$$

This inequality, when used in the definition of $c^1_{n_1}$, implies that $c^1_{\tilde{n}_1} > 0$. But since $c^1_{n_1}$ is nonnegative for all $n_1 \in \mathbf{M}_1$, the summation term $\sum_{\mathbf{M}_1} c^1_{n_1}$ becomes positive.

Now, again referring back to inequality (3.14), this time we let $\hat{n}_1$ denote an index for which (3.15) attains its maximum value over $n_1 \in \mathbf{M}_1$. Then, going through an argument similar to the one used in the preceding paragraph, we bound the LHS of (3.14) from above by (3.15) with $n_1 = \hat{n}_1$, and arrive at the strict inequality

$$\psi^1_{\hat{n}_1}(\tilde{y}^1, y^2, \ldots, y^N) < 0.$$

This then implies that $c^1_{\hat{n}_1} = 0$, which, when used in (3.12) with $i = 1$ and $n_1 = \hat{n}_1$, yields the conclusion

$$\bar{y}^1_{\hat{n}_1} < y^1_{\hat{n}_1}$$

since $\sum_{\mathbf{M}_1} c^1_{n_1} > 0$. But this contradicts the hypothesis that $\{y^i; i \in \mathbf{N}\}$ is a fixed point of T. □

The proof of Thm. 3.2, as given above, is not constructive. For a constructive proof, which also leads to a numerical scheme to obtain the equilibrium solution, the reader is referred to Scarf (1967). But, in general, it is not possible to obtain the equilibrium solution explicitly. If a given game admits an inner mixed-strategy equilibrium solution, however, then a possibility of obtaining the corresponding strategies in explicit form emerges, as the following proposition (which could also be considered as an extension of Remark 3.3 and Prop. 3.3) indicates.

Proposition 3.5 *Let $\mathring{Y}^i$ denote the interior of Y^i with $i \in \mathbf{N}$. Then, any inner mixed Nash equilibrium solution $\{y^{i*} \in \mathring{Y}^i; i \in \mathbf{N}\}$ of an N-person finite static game in normal form satisfies the set of equations*

$$\left.\begin{array}{c}
\sum\limits_{\mathbf{M}_2} \cdots \sum\limits_{\mathbf{M}_N} y^{2*}_{n_2} \cdots y^{N*}_{n_N} (a^1_{n_1,\ldots,n_N} - a^1_{1,n_2,\ldots,n_N}) = 0, \\
n_1 \in \mathbf{M}_1,\ n_1 \neq 1, \\
\sum\limits_{\mathbf{M}_1} \sum\limits_{\mathbf{M}_3} \cdots \sum\limits_{\mathbf{M}_N} y^{1*}_{n_1} y^{3*}_{n_3} \cdots y^{N*}_{n_N} (a^2_{n_1,\ldots,n_N} - a^2_{n_1,1,n_3,\ldots,n_N}) = 0, \\
n_2 \in \mathbf{M}_2,\ n_2 \neq 1, \\
\cdots\cdots\cdots\cdots\cdots\cdots\cdots\cdots \\
\sum\limits_{\mathbf{M}_1} \cdots \sum\limits_{\mathbf{M}_{N-1}} y^{1*}_{n_1} \cdots y^{N-1*}_{n_{N-1}} (a^N_{n_1,\ldots,n_N} - a^N_{n_1,\ldots,n_{N-1},1}) = 0, \\
n_N \in \mathbf{M}_N,\ n_N \neq 1.
\end{array}\right\} \tag{3.17}$$

Furthermore, these mixed strategies also provide a mixed-strategy Nash equilibrium solution for the N-person game structured in a similar way but with some (or all) of the players maximizing their average costs instead of minimizing.

Proof. Since $\{y^{i*} \in \mathring{Y}^i; i \in \mathbf{N}\}$ is an equilibrium solution, y^{i*} minimizes the quantity

$$\sum_{\mathbf{M}_1} \cdots \sum_{\mathbf{M}_N} y^{1*}_{n_1} \cdots y^{i-1*}_{n_{i-1}} y^i_{n_i} y^{i+1*}_{n_{i+1}} \cdots y^{N*}_{n_N} a^i_{n_1,\ldots,n_N}$$

over $\mathring{Y}^i$, and with $i \in \mathbf{N}$. Let us first take $i = 1$, and rewrite the preceding expression (in view of the relation $\sum_{\mathbf{M}_1} y^1_{n_1} = 1$) as

$$\sum_{\mathbf{M}_2} \cdots \sum_{\mathbf{M}_N} \left[\left(1 - \sum_{\substack{n_1 \in \mathbf{M}_1 \\ n_1 \neq 1}} y^1_{n_1} \right) y^{2*}_{n_2}, \ldots, y^{N*}_{n_N} a^1_{1,n_2,\ldots,n_N} \right.$$

$$\left. + \sum_{\substack{n_1 \in \mathbf{M}_1 \\ n_1 \neq 1}} y^1_{n_1} y^{2*}_{n_2} \cdots y^{N*}_{n_N} a^1_{n_1,n_2,\ldots,n_N} \right].$$

Now, since the minimizing y^{1*} is an inner point of Y^1 by hypothesis, and since the preceding expression is linear in $\{y^1_{n_1}; n_1 \in \mathbf{M}_1, n_1 \neq 1\}$, it readily follows that the coefficient of $y^1_{n_1}$ has to vanish for each $n_1 \in \mathbf{M}_1$, $n_1 \neq 1$. This condition then yields the first set of equations of (3.17). The remaining ones can be verified analogously, by taking $i = 2, 3, \ldots, N$. Finally, the last part of Prop. 3.5 follows, as in Prop. 3.3, from the property that at the equilibrium point $\{y^{i*} \in \mathring{Y}^i, i \in \mathbf{N}, i \neq j\}$ $\mathbf{P}j$'s average cost is actually independent of y^j (which has just been proven), and thus it becomes immaterial whether he minimizes or maximizes his average cost. □

Since the hypothesis of Prop. 3.5 requires the mixed-strategy Nash equilibrium solution to be an inner point of the product set $Y^1 \times \cdots \times Y^N$, the set of equations (3.17) is satisfied under quite restrictive conditions, and even more so under the conjecture of Problem 5 (Section 3.8), since then it is required that the number of alternatives of each player be the same. Nevertheless, the proposition still provides a characterization of the equilibrium solution, and enables direct computation of the corresponding strategies by merely solving coupled algebraic equations, under conditions which are not totally void. As an illustration of this point we now provide the following example.

Example 3.5 Consider a 3-person game in which each player has two alternatives to choose from and in which the possible outcomes are given by

$n_3 = 1$ (rows: **P1**; columns: **P2**)

(1,−1,0)	(0,1,0)
(2,0,0)	(0,0,1)

P1,

$n_3 = 2$ (rows: **P1**; columns: **P2**)

(1,0,1)	(0,0,0)
(0,3,0)	(−1,2,0)

P1.

Since $\mathbf{M}_i$ consists of two elements for each $i = 1, 2, 3$, the set (3.17) involves only three equations for this game, which can be written as

$$\left.\begin{aligned} &1 - y_2^2 - 2y_2^3 + y_2^2 y_2^3 = 0, \\ &2 - 2y_2^1 - 2y_2^3 + y_2^1 y_2^3 = 0, \\ &y_2^1 + y_2^2 - 1 = 0. \end{aligned}\right\} \tag{3.18}$$

This coupled set of equations admits a unique solution with the property $0 < y_2^1, y_2^2, y_2^3 < 1$, which is

$$y_2^{1*} = \sqrt{3} - 1; \quad y_2^{2*} = 2 - \sqrt{3}; \quad y_2^{3*} = 1 - \frac{1}{3}\sqrt{3}.$$

Hence, this 3-person game admits a unique inner Nash equilibrium solution in mixed strategies, which is

$$y^{1*} = (2 - \sqrt{3}, \sqrt{3} - 1)', \quad y^{2*} = (\sqrt{3} - 1, 2 - \sqrt{3})', \quad y^{3*} = \left(\frac{1}{3}\sqrt{3}, 1 - \frac{1}{3}\sqrt{3}\right)'.$$

It should be noted that this game admits also two pure-strategy Nash equilibria with outcomes $(1, -1, 0)$ and $(-1, 2, 0)$. □

3.4 Computation of Mixed-Strategy Nash Equilibria in Bimatrix Games

We have discussed in Section 3.3 a possible approach toward obtaining mixed-strategy Nash equilibria of N-person games in normal form when these equilibria have the property of being inner solutions. In other words, when a mixed-strategy equilibrium solution assigns positive probabilities to all possible pure-strategy choices of a player, and this so for all players, then the corresponding equilibrium strategies can be determined by solving a set of algebraic equations (cf. Prop. 3.5).

For the special case of bimatrix games, this requirement of Prop. 3.5 says that the pair of mixed Nash equilibrium strategies y^* and z^* should be elements of $\mathring{Y}$ and $\mathring{Z}$, the interiors of Y and Z, respectively, which is a rather restrictive condition. If this condition is not satisfied, then one natural approach would be to set some components of the strategies y and z equal to zero and obtain algebraic equations for the other components, in the hope that the solutions of these algebraic equations will be nonnegative. If such a solution does not exist, then one can set some other components equal to zero and look for a nonnegative solution of the algebraic equation obtained from the other components, and so on. Since there is only a finite number of possibilities, it is obvious that such an approach will eventually yield mixed-strategy Nash equilibria of bimatrix games, also in view of Thm. 3.2. But, then, the natural question that comes to mind is whether this search can be done in a systematic way. Indeed it can, and this has been established by Lemke and Howson (1964) in their pioneering work. In that paper, the authors' intention has actually been to give an algebraic proof of the existence of mixed equilibrium solutions in bimatrix games (i.e., the result of Thm. 3.1), but, as a byproduct, they also obtained an efficient scheme for computing mixed Nash equilibria, which was thereafter referred to as the "*Lemke-Howson*" *algorithm.* The reader should consult the original work (Lemke and Howson, 1964) and the expository article by Shapley (1974) for the essentials of this algorithm and its application in bimatrix games.

Relation to nonlinear programming

Yet another general method for the solution of a bimatrix game is to transform it into a nonlinear (in fact, a bilinear) programming problem, and to utilize the numerical techniques developed for solutions of nonlinear programming problems. In the sequel, we establish this equivalence between bimatrix games and a specific class of bilinear programming problems with linear constraints (see Prop. 3.6); we refer the reader to Luenberger (1973) for algorithms on the numerical solution of nonlinear programming problems.

Proposition 3.6 *A pair* $\{y^*, z^*\}$ *constitutes a mixed-strategy Nash equilibrium solution to a bimatrix game* (A, B) *if, and only if, there exists a pair* (p^*, q^*) *such that* $\{y^*, z^*, p^*, q^*\}$ *is a solution of the following bilinear programming problem:*

$$\min_{y,z,p,q} \; [y'Az + y'Bz + p + q] \tag{3.19}$$

subject to

$$\left.\begin{array}{ll} Az \geq -pl_m, & B'y \geq -ql_n, \\ y \geq 0, z \geq 0, & y'l_m = 1, z'l_n = 1. \end{array}\right\} \tag{3.20}$$

Proof. The constraints evidently imply that

$$y'Az + y'Bz + p + q \geq 0, \tag{i}$$

which shows that the optimal value of the objective function is nonnegative. If $\{y^*, z^*\}$ is an equilibrium pair, then the quadruple

$$y^*, z^*, p^* = y^{*\prime}Az^*, \quad q^* = y^{*\prime}Bz^* \tag{ii}$$

is feasible and the corresponding value of the objective function is zero. Hence (ii) is an optimal solution to the bilinear programming problem.

Conversely, let $\{\bar{y}, \bar{z}, \bar{p}, \bar{q}\}$ be a solution to the bilinear programming problem. The fact that feasible solutions to this problem exist has already been established by (ii). Specifically, by Thm. 3.2, the bimatrix game (A, B) has an equilibrium pair (y^*, z^*) and hence (ii) is a feasible solution. Furthermore, (ii) also yields the conclusion that the minimum value of the objective function is nonpositive, which, in view of (i), implies

$$\bar{y}'A\bar{z} + \bar{y}'B\bar{z} + \bar{p} + \bar{q} = 0.$$

For any $y \geq 0$, $z \geq 0$ such that $y'l_m = 1$ and $z'l_n = 1$, we have

$$y'A\bar{z} \geq -\bar{p}, \quad \bar{y}'Bz \geq -\bar{q},$$

and hence,

$$y'A\bar{z} + \bar{y}'Bz \geq -\bar{p} - \bar{q} = \bar{y}'A\bar{z} + \bar{y}'B\bar{z}.$$

In particular, we have $\bar{y}'A\bar{z} \geq -\bar{p}$, $\bar{y}'B\bar{z} \geq -\bar{q}$, $\bar{y}'A\bar{z} + \bar{y}'B\bar{z} = p + q$. Thus $\bar{y}'A\bar{z} = -p$, $\bar{y}'B\bar{z} = -q$, and therefore,

$$y'A\bar{z} \geq \bar{y}'A\bar{z}, \qquad \bar{y}'Bz \geq \bar{y}'B\bar{z}.$$

This last set of inequalities verifies that $\{\bar{y}, \bar{z}\}$ is indeed a mixed-strategy Nash equilibrium solution of the bimatrix game (A, B). □

For the special case of zero-sum matrix games, the following corollary now readily follows from Prop. 3.6, and establishes an equivalence between two-person zero-sum matrix games and LP problems.

Corollary 3.2 *A pair $\{y^*, z^*\}$ constitutes a mixed-strategy saddle-point solution for a two-person zero-sum matrix game A if, and only if, there exists a pair (p^*, q^*) such that $\{y^*, z^*, p^*, q^*\}$ is a solution of the LP problem*

$$\min_{y,z,p,q} [p + q] \tag{3.21}$$

subject to

$$\left.\begin{array}{ll} Az \geq -p l_m, & A'y \leq q l_n, \\ y \geq 0, z \geq 0, & y'l_m = 1, z'l_n = 1. \end{array}\right\} \tag{3.22}$$

Remark 3.4 Note that Prop. 3.6 directly extends to N-person finite games in normal form. Furthermore, it is noteworthy that the LP problem of Corollary 3.2 is structurally different from the LP problem considered in Section 2.3 (Thm. 2.5), which also provided a means of obtaining mixed-strategy saddle-point solutions for matrix games. □

3.5 Nash Equilibria of N-Person Games in Extensive Form

This section is devoted to noncooperative (Nash) equilibria of N-person finite games in extensive form, which do not incorporate chance moves. Extensive tree formulation for finite games without chance moves has already been introduced in Chapter 2 within the context of zero-sum games (cf. Def. 2.5), and such a formulation is equally valid for finite nonzero-sum games, with certain appropriate modifications. Hence first, as a direct extension of Def. 2.5, we have the following definition which covers both single-act and multi-act games.

Definition 3.10 *An* extensive form *of an N-person nonzero-sum finite game without chance moves is a tree structure with*

(i) a specific vertex indicating the starting point of the game,

(ii) N cost functions, *each one assigning a real number to each terminal vertex of the tree, where the ith cost function determines the loss to be incurred to* $\mathbf{Pi}$,

(iii) a partition of the nodes of the tree into N player sets,

(iv) a subpartition of each player set into information sets $\{\eta_j^i\}$, *such that the same number of branches emanates from every node belonging to the same information set and no node follows another node in the same information set.*

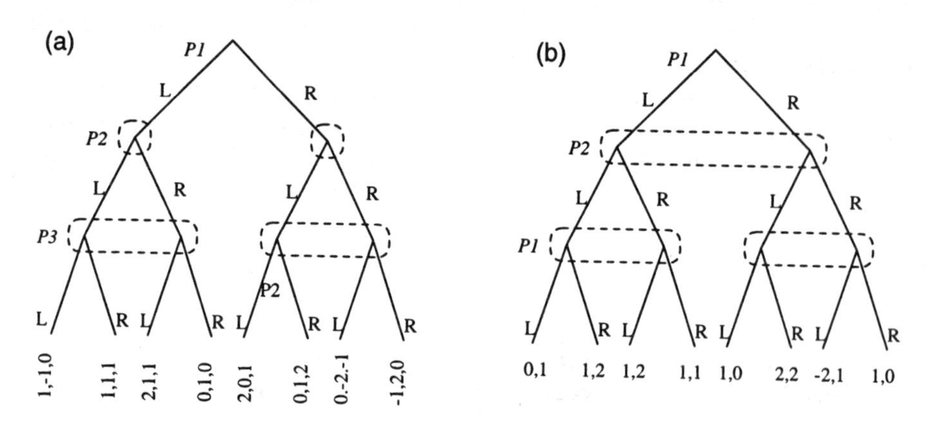

Figure 3.1: Two typical nonzero-sum finite games in extensive form.

Two typical nonzero-sum finite games in extensive form are depicted in Fig. 3.1. The first one represents a 3-player single-act nonzero-sum finite game in extensive form in which the information sets of the players are such that both **P**2 and **P**3 have access to the action of **P**1. The second extensive form of Fig. 3.1, on the other hand, represents a 2-player multi-act nonzero-sum finite game in which **P**1 acts twice and **P**2 only once. In both extensive forms, the set of alternatives for each player is the same at all information sets and it consists of two elements. The outcome corresponding to each possible path is denoted by an ordered N-tuple of numbers $(a^1, \ldots, a^N)$, where N stands for the number of players and a^i stands for the corresponding cost to $\mathbf{P}i$.

Pure-strategy Nash equilibria

We now introduce the concept of a noncooperative (Nash) equilibrium solution for N-person nonzero-sum finite games in extensive form, which covers both single-act and multi-act games. To this end, let us first recall the definition of a strategy from Chapter 2 (cf. Def. 2.6).

Definition 3.11 *Let* N^i *denote the class of all information sets of* $\mathbf{P}i$, *with a typical element designated as* η^i. *Let* $U^i_{\eta^i}$ *denote the set of alternatives of* $\mathbf{P}i$

at the nodes belonging to the information set η^i. *Define* $U^i = \cup\, U^i_{\eta^i}$, *where the union is over* $\eta^i \in \mathbf{N}^i$. *Then, a* strategy γ^i *for* $\mathbf{P}i$ *is a mapping from* N^i *into* U^i, *assigning one element in* U^i *for each set in* N^i, *and with the further property that* $\gamma^i(\eta^i) \in U^i_{\eta^i}$ *for each* $\eta^i \in \mathbf{N}^i$. *The set of all strategies of* $\mathbf{P}i$ *is called his* strategy set *(*space*), and it is denoted by* Γ^i.

Let $J^i(\gamma^1,\ldots,\gamma^N)$ denote the loss incurred to $\mathbf{P}i$ when the strategies $\gamma^1 \in \Gamma^1,\ldots,\gamma^N \in \Gamma^N$ are adopted by the players. Then, the noncooperative (Nash) equilibrium solution concept for such games can be introduced as follows, as a direct extension of Def. 3.7.

Definition 3.12 *An* N*-tuple of strategies* $\{\gamma^{1*},\gamma^{2*},\ldots,\gamma^{N*}\}$ *with* $\gamma^{i*} \in \Gamma^i$, $i \in \mathbf{N}$, *is said to constitute a* noncooperative (Nash) equilibrium solution *for an* N*-person nonzero-sum finite game in extensive form, if the following* N *inequalities are satisfied for all* $\gamma^i \in \Gamma^i$, $i \in \mathbf{N}$:

$$\left.\begin{array}{l} J^{1*} \triangleq J^1(\gamma^{1*},\gamma^{2*},\ldots,\gamma^{N*}) \le J^1(\gamma^1,\gamma^{2*},\ldots,\gamma^{N*}), \\ J^{2*} \triangleq J^2(\gamma^{1*},\gamma^{2*},\gamma^{3*},\ldots,\gamma^{N*}) \le J^2(\gamma^{1*},\gamma^2,\gamma^{3*},\ldots,\gamma^{N*}), \\ \cdots\cdots\cdots\cdots\cdots\cdots\cdots\cdots \\ \cdots\cdots\cdots\cdots\cdots\cdots\cdots\cdots \\ J^{N*} \triangleq J^N(\gamma^{1*},,\ldots,\gamma^{N-1*},\gamma^{N*}) \le J^N(\gamma^{1*},\ldots,\gamma^{N-1*},\gamma^N). \end{array}\right\} \quad (3.23)$$

The N*-tuple of quantities* $\{J^{1*},\ldots,J^{N*}\}$ *is known as* a Nash equilibrium outcome *of the nonzero-sum finite game in extensive form.*

We emphasize the word **a** in the last sentence of the preceding definition, since the Nash equilibrium solution could possibly be nonunique with the corresponding set of Nash values being different. This then leads to a partial ordering in the set of all Nash equilibrium solutions, as in the case of Defs. 3.2 and 3.3 whose extensions to the N-person case are along similar lines and are therefore omitted.

Nash equilibria in mixed and behavioral strategies

The concepts of noncooperative equilibria in mixed and behavioral strategies for nonzero-sum finite games in extensive form can be introduced as straightforward (natural) extensions of the corresponding ones presented in Chapter 2 within the context of saddle-point equilibria. We should recall that a mixed strategy for a player ($\mathbf{P}i$) is a probability distribution on the set of all his pure strategies, i.e., on Γ^i. A behavioral strategy, on the other hand, is an appropriate mapping whose domain of definition is the class of all the information sets of the player. By denoting the behavioral strategy set of $\mathbf{P}i$ by $\hat{\Gamma}^i$, and the average loss incurred to $\mathbf{P}i$ as a result of adoption of the behavioral strategy N-tuple $\{\hat{\gamma}^1 \in \hat{\Gamma}^1,\ldots,\hat{\gamma}^N \in \hat{\Gamma}^N\}$ by $\hat{J}^i(\hat{\gamma}^1,\ldots,\hat{\gamma}^N)$, the definition of a Nash equilibrium solution in behavioral strategies may be obtained directly from Def. 3.12 by replacing γ^i, Γ^i and J^i by $\hat{\gamma}^i$, $\hat{\Gamma}^i$ and $\hat{J}^i$, respectively.

We now discuss, in the four subsections to follow, properties and derivation of these different types of Nash equilibria for finite games in extensive form—first for single-act games, and then for multi-act games.

3.5.1 Single-act games: Pure-strategy Nash equilibria

In single-act games, each player acts only once, and the order in which the players act could be a variant of their strategies. If the order is fixed *a priori*, and further if the single-act game is static in nature (i.e., if each player has a single information set), then there is no basic difference between extensive and normal form descriptions, and consequently no apparent advantage of one form over the other. Since such static games in normal form have already been extensively studied in Sections 3.2 and 3.3, our concern here will be primarily on dynamic single-act games, that is, games in which at least one of the players has access to some nontrivial information concerning the action of some other player. Figure 3.1(a), for example, displays one such game wherein both **P2** and **P3** have access to **P1**'s action, but they are ignorant about each other's actions. Yet another single-act game with dynamic information is the 2-person nonzero-sum game depicted in Fig. 3.2, wherein **P2**'s information sets can differentiate between whether **P1** has played L or not, but not between his possible actions M and R.

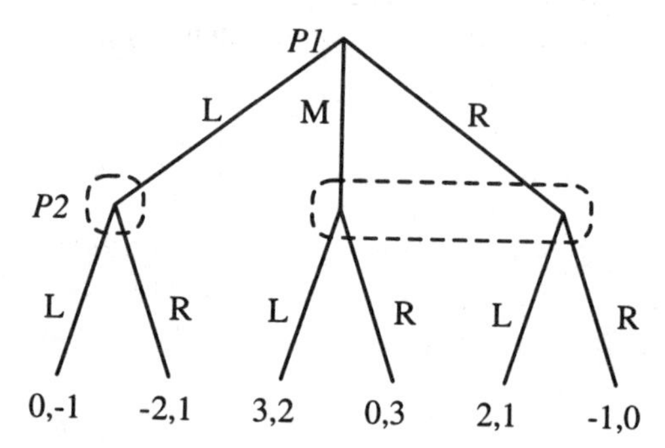

Figure 3.2: A 2-person nonzero-sum single-act game with dynamic information.

One method of obtaining Nash equilibria of a dynamic single-act finite game in extensive form is to transform it into an equivalent normal form and to make use of the theory of Sections 3.2 and 3.3. This direct method has already been discussed earlier in Section 2.4 within the context of similarly structured zero-sum games, and it has the basic drawback that one has to take into consideration all possible strategy combinations. As in the case of dynamic zero-sum games, an alternative to this direct method exists for some special types of nonzero-sum games, which makes explicit use of the extensive form description, and obtains a subclass of all Nash equilibria through a systematic (recursive) procedure. But in order to aid in understanding the essentials and the limitations of such a recursive procedure, it will be instructive first to consider the following specific example. This simple example, in fact, displays most of the intricacies of Nash equilibria in dynamic finite games.

Example 3.6 Consider the 2-person nonzero-sum single-act game whose extensive form is displayed in Fig. 3.2. The following is an intuitively appealing recursive procedure that would generate a Nash equilibrium solution for this game.

At the first of his two information sets (counting from left), **P2** can tell precisely whether **P1** has played L or not. Hence, in case $u^1 = L$, the unique decision of **P2** would be $u^2 = L$, yielding him a cost of -1 which is in fact the lowest possible level **P2** can hope to attain. Now, if $u^1 = M$ or R, however, **P2**'s information set does not differentiate between these two actions of **P1**, thus forcing him to play a static nonzero-sum game with **P1**. The corresponding bimatrix game (under the usual convention of A denoting the loss matrix of **P1** and B denoting that of **P2**) is

$$A = \begin{array}{c|c|c|} & \multicolumn{2}{c}{\mathbf{P2}} \\ M & 3 & 0 \\ R & 2 & \textcircled{-1} \\ & L & R \end{array}\ \mathbf{P1}, \qquad B = \begin{array}{c|c|c|} & \multicolumn{2}{c}{\mathbf{P2}} \\ M & 2 & 3 \\ R & 1 & \textcircled{0} \\ & L & R \end{array}\ \mathbf{P1}. \tag{3.24}$$

This bimatrix game admits (as indicated) a unique Nash equilibrium $\{R, R\}$ with an equilibrium cost pair of $(-1, 0)$. This then readily suggests

$$\gamma^{2*}(\eta^2) = \begin{cases} L & \text{if } u^1 = L, \\ R & \text{otherwise} \end{cases} \tag{3.25a}$$

as a candidate equilibrium strategy for **P2**.

P1, on the other hand, has two possible strategies, $u^1 = L$ and $u^1 = R$, since the third possibility $u^1 = M$ is ruled out by the preceding argument. Now, the former of these leads (under (3.25a)) to a cost of $J^1 = 0$, while the latter leads to $J^1 = -1$. Hence, he would clearly decide on playing R, i.e.,

$$\gamma^{1*}(\eta^1) = R \tag{3.25b}$$

is a candidate equilibrium strategy for **P1**. The cost pair corresponding to (3.25a)-(3.25b) is

$$J^{1*} = -1, \quad J^{2*} = 0. \tag{3.25c}$$

The strategy pair (3.25a)-(3.25b) is thus what an intuitively appealing (and rather straightforward) recursive procedure would provide us with, as a reasonable noncooperative equilibrium solution for this extensive form. It can, in fact, be directly verified by referring to inequalities (3.23) that the pair (3.25a)-(3.25b) is indeed in Nash equilibrium. To see this, let us first fix **P2**'s strategy at (3.25a), and observe that (3.25b) is then the unique cost-minimizing decision for **P1**. Now, fixing **P1**'s strategy at (3.25b), it is clear that **P2**'s unique cost-minimizing decision is $u^2 = R$ which is indeed implied by (3.25a) under (3.25b).

The recursive scheme adopted surely resulted in a unique strategy pair which is also in Nash equilibrium. But is this the only Nash equilibrium solution that the extensive form of Fig. 3.2 admits? The reply is, in fact, no! To obtain the additional Nash equilibrium solution(s), we first transform the extensive form of Fig. 3.2 into an equivalent normal form. To this end, let us first note that $\Gamma^1 = \{L, M, R\}$ and $\Gamma^2 = \{\gamma_1^2, \gamma_2^2, \gamma_3^2, \gamma_4^2\}$, where $\gamma_1^2(\eta^2) = L$, $\gamma_2^2(\eta^2) = R$,

$$\gamma_3^2(\eta^2) = \begin{cases} L & \text{if } u^1 = L, \\ R & \text{otherwise}, \end{cases} \qquad \gamma_4^2(\eta^2) = \begin{cases} R & \text{if } u^1 = L, \\ L & \text{otherwise}. \end{cases}$$

Hence, the equivalent normal form is the 3×4 bimatrix game

$A =$ (rows: P1, columns: P2)

	1	2	3	4
L	0	−2	0	−2
M	3	0	0	3
R	2	−1	(−1)	2

$B =$ (rows: P1, columns: P2)

	1	2	3	4
L	−1	1	−1	1
M	2	3	3	2
R	1	0	(0)	1

which admits two pure-strategy Nash equilibria, as indicated. The encircled one is $\{R; \gamma_3^2\}$ which corresponds to the strategy pair (3.25a)-(3.25b) obtained through the recursive procedure. The other Nash equilibrium solution is the constant strategy pair

$$\gamma^2(\eta^2) = L, \qquad (3.26a)$$

$$\gamma^1(\eta^1) = L, \qquad (3.26b)$$

with a corresponding cost pair of

$$J^1 = 0, \quad J^2 = -1. \qquad (3.26c)$$

A comparison of (3.25c) and (3.26c) clearly indicates that both of these equilibria are admissible.

It is noteworthy that, since (3.26a) is a constant strategy, the pair (3.26a)-(3.26b) also constitutes a Nash equilibrium solution for the static single-act game obtained from the one of Fig. 3.2 by replacing the dynamic information of **P2** with static information (i.e., by allowing him a single information set). It is, in fact, the unique Nash equilibrium solution of this static game which admits the normal form (bimatrix) description

$A =$ (rows: P1, columns: P2)

	L	R
L	0	−2
M	3	0
R	2	−1

$B =$ (rows: P1, columns: P2)

	L	R
L	−1	1
M	2	3
R	1	0

The actual play (R, R) dictated by (3.25a)-(3.25b), however, does not possess such a property as a constant strategy pair. □

The preceding example has displayed certain important features of Nash equilibria of a single-act game, which we now list below. We will shortly see, in this section, that these features are, in fact, valid on a broader scale for both single-act and multi-act nonzero-sum games.

Features of the single-act game of Example 3.6

(i) The single-act game admitted multiple (two) Nash equilibria.

(ii) The (unique) Nash equilibrium solution of the static version of the single-act game (obtained by replacing the information sets of the second-acting player by a single information set) also constituted a Nash equilibrium solution for the original single-act game with the dynamic information.

(iii) A recursive procedure that involves solution of only bimatrix games (which could also be degenerate games) at each information set yielded only one of the two Nash equilibria. The actual play dictated by this equilibrium solution did not constitute an equilibrium solution as constant strategies.

We now show that appropriate (and more general) versions of these features are retained in a more general framework for N-person single-act games. To this end, we first introduce some terminology and also a partial ordering of extensive forms in terms of their information sets.

Definition 3.13 *A single-act N-person finite game in extensive form (say,* I*) is the* static version *of a dynamic single-act N-person finite game in extensive form (say,* II*), if* I *can be obtained from* II *by replacing the information sets of each player with a single information set encompassing all the nodes pertaining to that player.*[16] *The equivalent normal form of the static game* I *is called the* static normal form *of* II.

Definition 3.14 *Let* I *and* II *be two single-act N-person games in extensive form, and further let Γ^i_{I} and Γ^i_{II} denote the strategy sets of* $\mathbf{P}i$ *($i \in \mathbf{N}$) in* I *and* II*, respectively. Then,* I *is said to be* informationally inferior *to* II *if $\Gamma^i_{\mathrm{I}} \subseteq \Gamma^i_{\mathrm{II}}$ for all $i \in \mathbf{N}$, with strict inclusion for at least one i.*

Remark 3.5 In Fig. 3.3, the single-act game I is informationally inferior to II and III. The latter two, however, do not admit any such comparison. □

Proposition 3.7 *Let* I *be an N-person single-act game that is informationally inferior to some other single-act N-person game, say* II*. Then,*

(i) any Nash equilibrium solution of I *also constitutes a Nash equilibrium solution for* II*,*

---

[16] It should be noted that a single-act finite game admits a static version only if the order in which the players act is fixed *a priori* and the possible actions of each player are the same at all of his information sets.

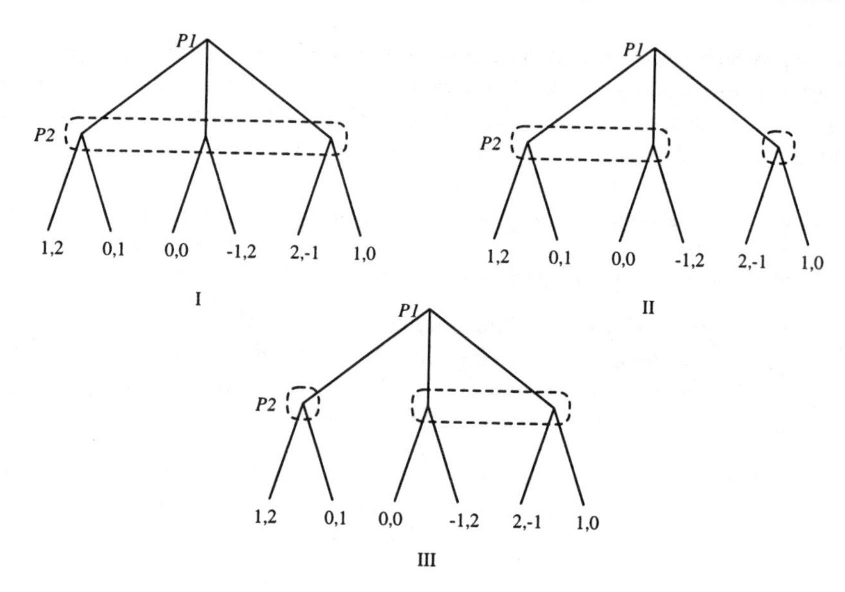

Figure 3.3: Extensive forms displaying informational inferiority.

(ii) if $\{\gamma^1, \ldots, \gamma^N\}$ *is a Nash equilibrium solution of* II *so that* $\gamma^i \in \Gamma^i_{\mathrm{I}}$ *for all* $i \in \mathbf{N}$, *then it also constitutes a Nash equilibrium solution for* I.

Proof. (i) If $\{\gamma^{1*} \in \Gamma^1_{\mathrm{I}}, \ldots, \gamma^{N*} \in \Gamma^N_{\mathrm{I}}\}$ constitutes a Nash equilibrium solution for I, then inequalities (3.23) are satisfied for all $\gamma^i \in \Gamma^i_{\mathrm{I}}$, $i \in \mathbf{N}$. But, since $\Gamma^i_{\mathrm{I}} \subseteq \Gamma^i_{\mathrm{II}}$, $i \in \mathbf{N}$, we clearly also have $\gamma^i \in \Gamma^i_{\mathrm{II}}$, $i \in \mathbf{N}$. Now assume, to the contrary, that $\{\gamma^{1*}, \ldots, \gamma^{N*}\}$ is not a Nash equilibrium solution of II. Then, this implies that there exists at least one i (say, $i = N$, without any loss of generality) for which the corresponding inequality of (3.23) is not satisfied for all $\gamma^i \in \Gamma^i_{\mathrm{II}}$. In particular, there exists a $\tilde{\gamma}^N \in \Gamma^N_{\mathrm{II}}$ such that

$$J^{N*} > J^N(\gamma^{1*}; \ldots; \gamma^{N-1*}; \tilde{\gamma}^N). \qquad (i)$$

Now, the N-tuple of strategies $\{\gamma^{1*}, \ldots, \gamma^{N-1*}, \tilde{\gamma}^N\}$ leads to a unique path of action, and consequently to a unique outcome, in the single-act game II. Let us denote the information set of $\mathbf{P}N$, which is actually traversed by this path, by $\tilde{\eta}^N_{\mathrm{II}}$, and the specific element (node) of $\tilde{\eta}^N_{\mathrm{II}}$ intercepted by $\tilde{n}^N$. Let us further denote the information set of $\mathbf{P}N$ in game I, which includes the node $\tilde{n}^N$, by $\tilde{\eta}^N_{\mathrm{I}}$. Then, there exists at least one element in Γ^N_{I} (say, $\bar{\gamma}^N$) with the property $\bar{\gamma}^N(\tilde{\eta}^N_{\mathrm{I}}) = \tilde{\gamma}^N(\tilde{\eta}^N_{\mathrm{II}})$. If this strategy replaces $\tilde{\gamma}^N$ on the RHS of inequality (i), the value of J^N clearly does not change, and hence we equivalently have

$$J^{N*} > J^N(\gamma^{1*}; \ldots; \gamma^{N-1*}; \bar{\gamma}^N).$$

But this inequality contradicts the initial hypothesis that the N-tuple of policies $\{\gamma^{1*}; \ldots; \gamma^{N*}\}$ was in Nash equilibrium for the game I. This then completes the proof of part (i).

(ii) Part (ii) of the proposition can be proven analogously. □

Remark 3.6 Proposition 3.7 now verifies a general version of feature (ii) of the single-act game of Example 3.6 for N-person single-act finite games in extensive form. A counterpart of feature (i) also readily follows from this proposition, which is that such games with dynamic information will in general admit multiple Nash equilibria. A more definite statement cannot be made, since there would always be exceptional cases when the Nash equilibrium solution of a single-act dynamic game is unique and is attained in constant strategies which definitely also constitute a Nash equilibrium solution for its static version, assuming that it exists. It is almost impossible to single out all these special games, but we can comfortably say that they are "rarely" met, and existence of multiple Nash equilibria is a rule in dynamic single-act games rather than an exception. Since this sort of nonuniqueness emerges mainly because of the dynamic nature of the information sets of the players, we call it *informational nonuniqueness.* □

Now, to elaborate on the third (and the last) feature listed earlier, we have to impose some further structure on the "relative nestedness" of the information sets of the players, which is already implicit in the extensive form of Fig. 3.2. It is, in fact, possible to develop a systematic recursive procedure, as an extension of the one adopted in the solution of Example 3.6, if the order in which the players act is fixed *a priori* and the extensive form is in "ladder-nested" form— a notion that we introduce below.

Definition 3.15 *In an extensive form of a single-act nonzero-sum finite game with a fixed order of play, a player* **P***i is said to be a* precedent *of another player* **P***j if the former is situated closer to the vertex of the tree than the latter. The extensive form is said to be* nested *if each player has access to the information acquired by all his precedents. If, furthermore, the only difference (if any) between the information available to a player* (**P***i*) *and his closest (immediate) precedent (say* **P***i* $-$ 1*) involves only the actions of* **P***i* $-$ 1*, and only at those nodes corresponding to the branches of the tree emanating from singleton information sets of* **P***i* $-$ 1*, and this so for all players, the extensive form is said to be* ladder-nested.[17] *A single-act nonzero-sum finite game is said to be nested (respectively,* ladder-nested*) if it admits an extensive form that is nested (respectively, ladder-nested).*

Remark 3.7 The single-act extensive forms of Figs. 3.1(a) and 3.2 are both ladder-nested. If the extensive form of Fig. 3.1(a) is modified so that both nodes of **P**2 are included in the same information set, then it is only nested, but not ladder-nested, since **P**3 can differentiate between different actions of **P**1 but **P**2 cannot. Finally, if the extensive form of Fig. 3.1(a) is modified so that this time **P**3 has a single information set (see Fig. 3.4(a)), then the resulting extensive form becomes non-nested, since then even though **P**2 is a precedent of **P**3 he actually knows more than **P**3 does. The single-act game that this

[17]Note that in 2-person single-act games the concepts of "nestedness" and "ladder-nestedness" coincide, and every extensive form is, by definition, ladder-nested.

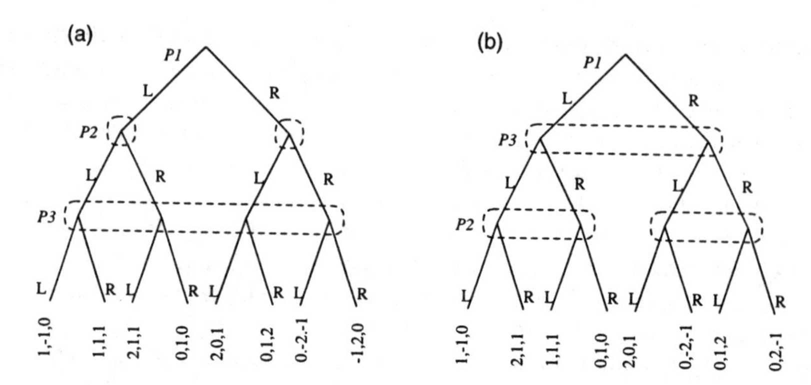

Figure 3.4: Two extensive forms of the same nested single-act game.

extensive form describes is, however, nested since it also admits the extensive form description depicted in Fig. 3.4(b). □

One advantage of dealing with ladder-nested extensive forms is that they can recursively be decomposed into simpler tree structures which are basically static in nature. This enables one to obtain a class of Nash equilibria of such games recursively, by solving static games at each step of the recursive procedure. Before providing the details of this recursive procedure, let us introduce some terminology.

Definition 3.16 *For a given single-act dynamic game in nested extensive form (say,* I*), let* η *denote a singleton information set of* **P***i's immediate follower (say* **P***j); consider the part of the tree structure of* I*, which is cut off at* η*, has* η *as its vertex and has as immediate branches only those that enter into that information set of* **P***j. Then, this tree structure is called a* sub-extensive form *of* I*. (Here, we adopt the convention that the starting vertex of the original extensive form is the singleton information set of the first-acting player.)*

Remark 3.8 The single-act game depicted in Fig. 3.2 admits two sub-extensive forms which are as follows:

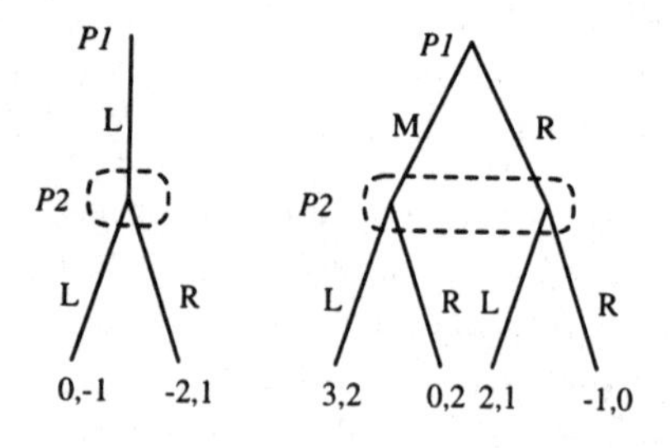

The extensive form of Fig. 3.1(a), on the other hand, admits a total of four sub-extensive forms which we do not display here. It should be noted that each sub-extensive form is itself an extensive form describing a simpler game. The first one displayed above describes a degenerate 2-player game in which **P1** has only one alternative. The second one again describes a 2-player game in which the players each have two alternatives. Both of these sub-extensive forms will be called static since the first one is basically a one-player game and the second one describes a static 2-player game. A precise definition follows. □

Definition 3.17 *A sub-extensive form of a nested extensive form of a single-act game is* static *if every player appearing in this tree structure has a single information set.*

We are now in a position to extend the recursive procedure adopted in Example 3.6 to obtain (some of) the Nash equilibrium solutions of N-person single-act games in ladder-nested extensive form.

A recursive procedure to determine pure-strategy Nash equilibria of N-person single-act games in ladder-nested extensive form

(1) For each fixed information set of the last-acting player (say, **P**N), single out the players who have precisely the same information as **P**N, determine the static sub-extensive form that includes all these players and their single information sets, and solve the static game that this sub-extensive form describes. Assuming that each of these static games admits a unique pure-strategy Nash equilibrium solution, record the equilibrium strategies and the corresponding Nash outcomes by identifying them with the players and their information sets. (If any one of the static games admits more than one Nash equilibrium solution, then this procedure as well as the following steps are repeated for each of these multiple equilibria.)

(2) Replace each static sub-extensive form considered at step 1 with the immediate branch emanating from its vertex that corresponds to the Nash strategy of the starting player of that sub-extensive form; furthermore, attach the corresponding N-tuple of Nash equilibrium values to the end of this branch.

(3) For the remaining game in extensive form, repeat steps 1 and 2 until the extensive form left is a tree structure comprising only a vertex and some immediate branches with an N-tuple of numbers attached to each of them.

(4) The branch of this final tree structure which corresponds to a minimum loss for the starting player is his Nash equilibrium strategy, and the corresponding N-tuple of numbers at the end of this branch determines a Nash outcome for the original game. The corresponding Nash equilibrium strategies of the other players in the original single-act game can then be determined from the solutions of the static sub-extensive forms, by appropriately identifying them with the players and their information sets.

Remark 3.9 Because of the ladder-nestedness property of the single-act game, it can readily be shown that every solution obtained through the foregoing recursive procedure is indeed a Nash equilibrium solution of the original single-act game. □

A most natural question to raise now is whether the preceding recursive procedure generates all the Nash equilibrium solutions of a given N-person single-act game in ladder-nested extensive form. We have already seen in Example 3.6 that this is, in fact, not so, since out of the two Nash equilibrium solutions, only one of them was generated by this recursive procedure. Then, the question may be rephrased as: precisely what subclass of Nash equilibria is generated by the preceding scheme? In other words, what common feature (if any) can we attribute to the Nash equilibria that can be derived recursively? To shed some light on this matter, it will be instructive to refer again to Example 3.6 and to take a closer look at the two Nash equilibria obtained:

EXAMPLE 3.6 (continued) For the single-act game of Fig. 3.2, the recursive procedure generated the solution (3.25a)-(3.25b) with Nash cost pair $(-1, 0)$. When compared with the other Nash cost pair which is $(0, -1)$, it is clear that the former is more advantageous for **P1**. Hence, **P2** would rather prefer to play the constant strategy (3.26a) and thereby attain a favorable cost level of $J^2 = -1$. But, since he acts later, the only way he can ensure such a cost level is by announcing his constant strategy ahead of time and by irrevocably sticking to it. If he has the means of doing this, and further if **P1** has strong reasons to believe in such an attitude on the part of **P2**, then $(0, -1)$ will clearly be the only cost pair to be realized. Hence, the constant-strategy Nash solution in this case corresponds to a "*prior commitment*" mode of play on the part of the last acting player—a notion which has been introduced earlier in Section 2.5 within the context of multi-act zero-sum games, and which also has connections with the Stackelberg mode of play to be discussed in the next section of this chapter.

Now, yet another mode of play for **P2** (the second-acting player) would be to wait until he observes at what information set he is at the time of his play, and only then decide on his action. This is known as the "*delayed commitment*" type of attitude, which has also been introduced earlier in Section 2.5, within the context of zero-sum games and in conjunction with behavioral strategies. For the single-act game of Example 3.6, a Nash equilibrium solution of the delayed commitment type is the one given by (3.25a)-(3.25b), which has been obtained using the recursive procedure. □

This discussion now leads us to the following definition.

Definition 3.18 *A Nash equilibrium solution of a ladder-nested single-act finite N-person game is of the* delayed commitment *type if it can be obtained through the recursive procedure outlined earlier, i.e., by solving only static single-act games.*

Remark 3.10 As a side remark, it is noteworthy that in the single-act game of Example 3.6 additional information for **P2** concerning the action of **P1** could be detrimental. For, if his information set was a single one (i.e., the static version), then the game would admit the unique Nash solution (3.26a)-(3.26b) with **P2**'s Nash cost being $J^2 = -1$. In the set-up of the extensive form of Fig. 3.2, however, he could end up with a higher Nash cost of $J^2 = 0$. Hence, in nonzero-sum games, additional information could be detrimental for the player who receives it. □

We now illustrate, giving two examples, the steps involved in the outlined recursive procedure to obtain admissible Nash equilibrium strategies of the delayed commitment type.

Example 3.7 Consider the 3-person single-act game whose extensive form (of the ladder-nested type) is depicted in Fig. 3.1(a). Here, both **P2** and **P3** have access to the action of **P1** (and to no more information), and hence they are faced with a bimatrix game at each of their information sets. Specifically, if $u^1 = L$, then the corresponding bimatrix game is

$$A = \begin{array}{c} \\ L \\ R \\ \\ \end{array} \begin{array}{c} \textbf{P3} \\ \begin{array}{|c|c|} \hline (-1) & 1 \\ \hline 1 & 1 \\ \hline \end{array} \\ L \quad R \end{array} \ \textbf{P2}, \qquad B = \begin{array}{c} \\ L \\ R \\ \\ \end{array} \begin{array}{c} \textbf{P3} \\ \begin{array}{|c|c|} \hline (0) & 1 \\ \hline 1 & 0 \\ \hline \end{array} \\ L \quad R \end{array} \ \textbf{P2},$$

where the entries of matrix A denote the possible losses to **P2** and the entries of matrix B denote the possible losses to **P3**. This bimatrix game admits a unique admissible Nash equilibrium solution which is $\{u^{2*} = L, u^{3*} = L\}$, with the corresponding Nash values being -1 and 0 for **P2** and **P3**, respectively. If $u^1 = R$, on the other hand, the bimatrix game of interest is

$$A = \begin{array}{c} \\ L \\ R \\ \\ \end{array} \begin{array}{c} \textbf{P3} \\ \begin{array}{|c|c|} \hline 0 & 1 \\ \hline (-2) & 2 \\ \hline \end{array} \\ L \quad R \end{array} \ \textbf{P2}, \qquad B = \begin{array}{c} \\ L \\ R \\ \\ \end{array} \begin{array}{c} \textbf{P3} \\ \begin{array}{|c|c|} \hline 1 & 2 \\ \hline (-1) & 0 \\ \hline \end{array} \\ L \quad R \end{array} \ \textbf{P2}.$$

This game admits a unique Nash solution which is $\{u^{2*} = R, u^{3*} = L\}$ with a Nash cost pair of $(-2, -1)$. Hence, regardless of what **P1** plays, the delayed commitment type admissible Nash equilibrium strategies of **P2** and **P3** in this ladder-nested game are unique, and they are given as

$$\gamma^{2*}(\eta^2) = u^1, \quad \gamma^{3*}(\eta^3) = L. \tag{i}$$

(Note that **P3**'s equilibrium strategy is a constant.) Now, if **P1** picks $u^1 = L$, his loss under (i) will be 1, but otherwise his loss is 0. Hence he also has a unique equilibrium strategy

$$\gamma^{1*} = R. \tag{ii}$$

Strategies (i) and (ii) now constitute the unique delayed commitment type admissible Nash equilibrium solution of the single-act game of Fig. 3.1(a), which can also be directly verified by again referring to the set of inequalities (3.23). The corresponding Nash equilibrium outcome is $(0, -2, -1)$. □

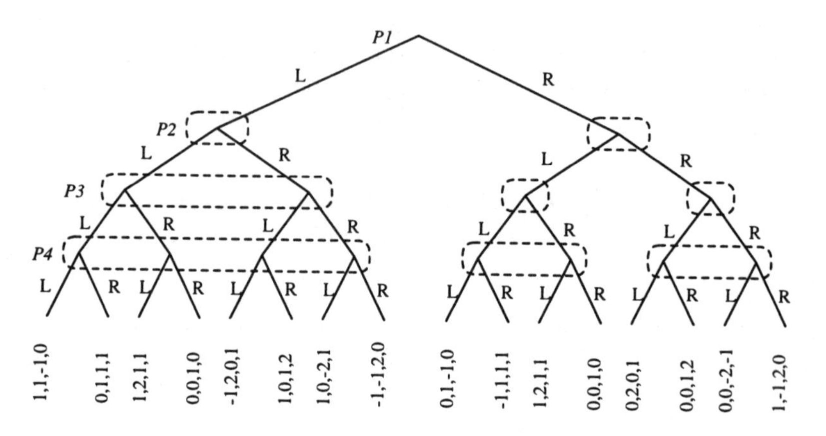

Figure 3.5: A 4-person single-act ladder-nested game in extensive form.

Example 3.8 Consider the 4-person single-act game whose extensive form (of the ladder-nested type) is depicted in Fig. 3.5. The last-acting player is **P4** and he has three information sets. The first one (counting from left) tells him whether **P1** has actually played L or not, but nothing concerning the actions of the other players. The same is true for **P2** and **P3**. The corresponding three-player static game admits an equivalent normal form which is the one considered in Example 3.3, with only $\mathbf{P}j$ replaced by $\mathbf{P}j+1$ in the present context. Furthermore, we now also have the costs of an additional player (**P1**) showing up at the end of the branches (i.e., in the present context we have a quadruple cost instead of a triple). It is already known that this game admits a unique Nash equilibrium solution which, under the present convention, is

$$u^{2*} = L, \; u^{3*} = L, \; u^{4*} = L,$$

and this is so if **P1** has picked $u^1 = L$, with the corresponding cost quadruple being $(1, 1, -1, 0)$.

The second information set of **P4** tells him whether **P1** has picked R and **P2** has picked L; and **P3** also has access to this information. Hence, we now have a bimatrix game under consideration, which is essentially the first bimatrix game of Example 3.6, admitting the unique solution

$$u^{3*} = L, \quad u^{4*} = L,$$

and this is so if $u^1 = R$, $u^2 = L$, with the corresponding cost quadruple being $(0, 1, -1, 0)$.

The third information set of **P4** provides him with the information $u^1 = R$, $u^2 = R$. **P3** has access to the same information, and thus the equivalent normal form is the second bimatrix game of Example 3.7, which admits the unique Nash solution

$$u^{3*} = R, u^{4*} = L,$$

with the corresponding cost quadruple being $(0, 0, -2, 1)$.

Using all these, we can now write down the unique Nash equilibrium strategies of **P3** and **P4**, which are given as

$$\gamma^{3*}(\eta^3) = \begin{cases} R & \text{if } u^1 = R, u^2 = R, \\ L & \text{otherwise} , \end{cases} \tag{i}$$

$$\gamma^{4*}(\eta^4) = L. \tag{ii}$$

Deletion of all the branches of the extensive tree already used now leads to the simpler tree structure depicted below:

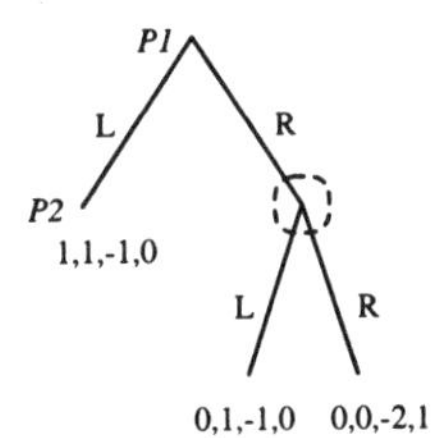

In this game, **P2** acts only if $u^1 = R$, in which case he definitely picks $u^2 = R$, since this yields him a lower cost (0) as compared with 1 which he would obtain otherwise. This also completely determines the delayed commitment type Nash equilibrium strategy of **P2** in the game of Fig. 3.5, which is unique and is given by

$$\gamma^{2*}(\eta^2) = u^1. \tag{iii}$$

Then, the final equivalent form that the game takes is

from which it is clear that the optimal unique strategy for **P1** is

$$\gamma^{1*} = R. \tag{iv}$$

The strategies (i)-(iv) constitute the unique delayed commitment type Nash equilibrium solution of the extensive form of Fig. 3.8 (which is clearly also admissible), and the reader is also encouraged to check this result by verifying satisfaction of inequalities (3.23). The corresponding unique Nash equilibrium outcome is $(0, 0, -2, 1)$. □

Nested extensive forms which are not ladder-nested cannot be decomposed into static sub-extensive forms, and hence the recursive derivation does not directly apply to such games. However, some nested extensive forms still admit a recursive decomposition into simpler sub-extensive forms, some of which will be dynamic in nature. Hence, an extended version of the recursive procedure applies for nested games, which involves solution of simpler dynamic sub-extensive forms. In this context, the delayed commitment mode of play also makes sense as to be introduced in the sequel.

Definition 3.19 *A nested extensive (or sub-extensive) form of a single-act game is said to be* undecomposable *if it does not admit any simpler sub-extensive form. It is said to be* dynamic, *if at least one of the players has more than one information set.*

Definition 3.20 *For an N-person single-act dynamic game in nested undecomposable extensive form* (**I**) *let* **J** *denote the set of games which are informationally inferior to* **I**. *Let* $\{\gamma^{1*},\ldots,\gamma^{N*}\}$ *be an N-tuple of strategies in Nash equilibrium for* **I**, *and further let* j^* *denote the number of games in* **J** *to which this N-tuple provides an equilibrium solution. Then,* $\{\gamma^{1*},\ldots,\gamma^{N*}\}$ *is of* delayed commitment *type if there exists no other N-tuple of Nash equilibrium strategies of* **I** *which constitutes an equilibrium solution to a smaller (than* j^**) number of games in* **J**.[18]

The following example now provides an illustration of these notions as well as a derivation of delayed commitment type Nash equilibria in nested games.

Example 3.9 Consider the 3-person single-act game whose extensive form is displayed in Fig. 3.6. In this game **P3** has access to **P1**'s action, but **P2** does not. Hence, the game is nested but not ladder-nested. Furthermore, it is both dynamic and undecomposable, with the latter feature eliminating the possibility for a recursive derivation of any of its Nash equilibria. Then, the only method is to bring the single-act game into equivalent normal form and to obtain the Nash equilibria of this normal form. Toward this end, let us first note that **P1** and **P2** each have two possible strategies: $\{\gamma_1^1 = L, \gamma_2^1 = R; \gamma_1^2 = L, \gamma_2^2 = R\}$. **P3**, on the other hand, has four possible strategies:

$$\gamma_1^3(\eta^3) = L, \quad \gamma_2^3(\eta^3) = R, \quad \gamma_3^3(\eta^3) = u^1,$$

and

$$\gamma_4^3(\eta^3) = \begin{cases} L & \text{if } u^1 = R, \\ R & \text{otherwise}. \end{cases}$$

[18]The reader should verify that every single-act game that admits Nash equilibria in pure strategies has at least one equilibrium solution of the delayed commitment type, i.e., the set of such equilibria is not empty unless the game does not admit any pure-strategy Nash equilibrium solution.

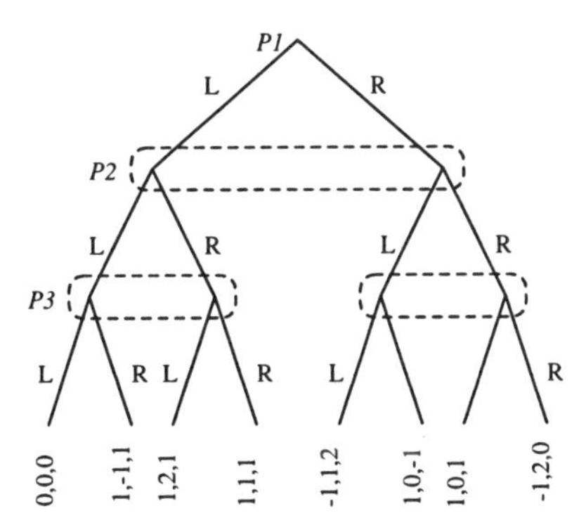

Figure 3.6: A nested single-act game in extensive form which is dynamic and undecomposable.

Using the terminology of Section 3.3, we now display the normal form as two (2×4) matrices, one for each strategy of **P1**:

$n_1 = 1$: (rows: **P2**, columns: **P3**)

	1	2	3	4
1	$(0, 0, 0)$	$(1, -1, 1)$	$(0,0,0)^*$	$(1, -1, 1)$
2	$(1, 2, 1)$	$(1, 1, 1)$	$(1, 2, 1)$	$(1, 1, 1)$

$n_1 = 2$: (rows: **P2**, columns: **P3**)

	1	2	3	4
1	$(-1, 1, 2)$	$(1,0,-1)^*$	$(1, 0, -1)$	$(-1, 1, 2)$
2	$(1, 0, 1)$	$(-1, 2, 0)$	$(-1, 2, 0)$	$(1, 0, 1)$

Here, the integers $1, 2, \ldots$ identify particular strategies of the players in accordance with the subscripts attached to γ^i's. This normal form admits two Nash equilibria, as indicated, which correspond, in the original game, to the strategy triplets

$$\{\gamma^{1*} = L, \gamma^{2*} = L, \gamma^{3*}(\eta^3) = u^1\} \qquad (i)$$

and

$$\{\gamma^{1*} = R, \gamma^{2*} = L, \gamma^{3*}(\eta^3) = R\}. \qquad (ii)$$

There exists only one game which is informationally inferior to the single-act game under consideration, which is the one obtained by allowing a single information set to **P3** (i.e., the static version). Since the second Nash solution dictates a constant action for **P3**, it clearly also constitutes a Nash solution for this informationally inferior static game (cf. Prop. 3.7). Hence, for the second set of Nash strategies, $j^* = 1$, using the terminology of Def. 3.20. The first set of Nash strategies, on the other hand, does not constitute a Nash equilibrium solution to the informationally inferior game, and hence in this case $j^* = 0$. This then yields the conclusion that the nested single-act game of Fig. 3.6 admits a unique delayed commitment type Nash equilibria, given by (i) and with an outcome of $(0, 0, 0)$. □

For more general types of single-act finite games in nested extensive form, a recursive procedure could be used to simplify the derivation by decomposing the original game into static and dynamic undecomposable sub-extensive forms. Essential steps of such a recursive procedure are given below. Nash equilibria obtained through this recursive procedure will be called the delayed commitment type.

A recursive procedure to determine delayed commitment type pure-strategy Nash equilibria of N-person single-act games in nested extensive form

(1) Each information set of the last-acting player is included in either a static sub-extensive form or a dynamic undecomposable sub-extensive form. Single out these extensive forms and obtain their delayed commitment type Nash equilibria. (Note that in static games every Nash equilibrium is, by definition, of the delayed commitment type.) Assuming that each of these games admits a unique pure-strategy Nash equilibrium solution of the delayed commitment type, record the equilibrium strategies and the corresponding Nash outcomes by identifying them with the players and their information sets. (If any one of these games admits more than one Nash equilibrium solution of the delayed commitment type, then this implies that the original game admits more than one such equilibrium, and this procedure as well as the following steps will have to be repeated for each of these multiple equilibria.)

(2) Replace each sub-extensive form considered at step 1 with the immediate branch emanating from its vertex that corresponds to the delayed commitment Nash strategy of the starting player of that sub-extensive form; furthermore, attach the corresponding N-tuple of Nash equilibrium values to the end of this branch.

(3) For the remaining game in nested extensive form, repeat steps 1 and 2 until the extensive form left is a tree structure comprised of only a vertex and some immediate branches with an N-tuple of numbers attached to each of them.

(4) The branch of this final tree structure which corresponds to a minimum loss for the starting player is his Nash equilibrium strategy of the delayed commitment type, and the corresponding N-tuple of numbers at the end of this branch determines a Nash outcome for the original game. The corresponding delayed commitment type Nash equilibrium strategies of the other players in the original single-act game can then be captured from the solution(s) of the sub-extensive forms considered, by appropriately identifying them with the players and their information sets.

Example 3.10 To illustrate the steps of the foregoing recursive procedure, let us consider the 3-person single-act game of Fig. 3.7, which is in nested extensive

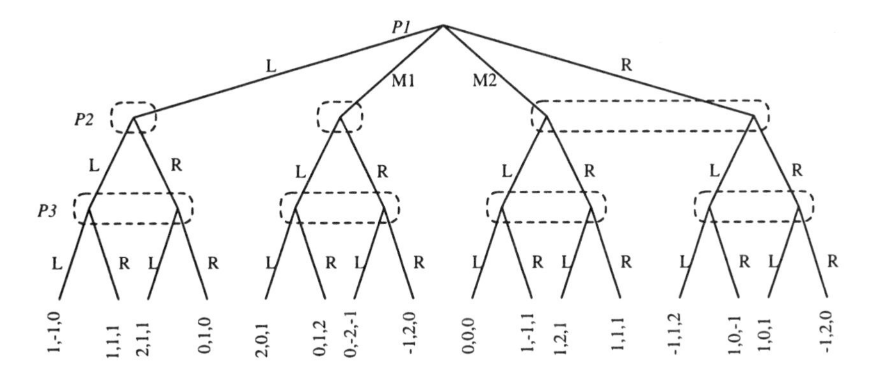

Figure 3.7: The single-act game of Example 3.10.

form. At step 1 of the recursive procedure, we have two 2-person static sub-extensive forms and one 3-person dynamic undecomposable sub-extensive form. Counting from the left, the first static sub-extensive form admits the unique Nash equilibrium solution (from Example 3.7):

$$\{u^{2*} = L, u^{3*} = L\}. \qquad (i)$$

The second static sub-extensive form also admits a unique solution which is (again from Example 3.7):

$$\{u^{2*} = R, u^{3*} = L\}. \qquad (ii)$$

It should be noted that (i) corresponds to the information sets $\eta^2 = \eta^3 = \{u^1 = L\}$ and (ii) corresponds to $\eta^2 = \eta^3 = \{u^1 = M1\}$. Now, the dynamic undecomposable sub-extensive form is in fact the single-act game of Example 3.9, which is already known to admit the unique delayed-commitment type Nash equilibrium solution

$$\gamma^{1*} = M2, \quad \gamma^{2*} = L, \quad \gamma^{3*}(\eta^3) = \begin{cases} L & \text{if } u^1 = M2, \\ R & \text{if } u^1 = R. \end{cases}$$

At step 2, therefore, the extensive form of Fig. 3.7 becomes

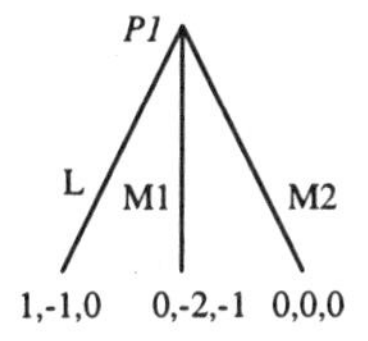

and it readily follows from this tree structure that **P1** has two Nash equilibrium strategies:

$$\gamma^{1*} = M1 \text{ and } \gamma^{1*} = M2.$$

Consequently, the original single-act game admits two Nash equilibria of the delayed commitment type:

$$\gamma^{1*} = M1, \quad \gamma^{2*}(\eta^2) = \begin{cases} R & \text{if } u^1 = M1, \\ L & \text{otherwise}, \end{cases}$$

$$\gamma^{3*}(\eta^3) = \begin{cases} R & \text{if } u^1 = R, \\ L & \text{otherwise}, \end{cases}$$

and

$$\begin{aligned} \gamma^{1*} = M2, \quad \gamma^{2*}(\eta^2) &= \begin{cases} R & \text{if } u^1 = M1, \\ L & \text{otherwise}, \end{cases} \\ \gamma^{3*}(\eta^2) &= \begin{cases} R & \text{if } u^1 = R, \\ L & \text{otherwise}, \end{cases} \end{aligned}$$

with equilibrium outcomes being $(0, -2, -1)$ and $(0, 0, 0)$, respectively. □

Remark 3.11 As we have discussed earlier in this section, single-act nonzero-sum finite dynamic games admit, in general, a multiple of Nash equilibria, mainly due to the fact that Nash equilibria of informationally inferior games also provide Nash solutions to the original game (cf. Prop. 3.7) the so-called informational nonuniqueness of Nash equilibria. The delayed commitment mode of play introduced for nested and ladder-nested games, however, eliminates this informational nonuniqueness, and therefore strengthens the Nash equilibrium solution concept for dynamic single-act games. □

3.5.2 Single-act games: Nash equilibria in behavioral and mixed strategies

If an N-person single-act game in extensive form does not admit a Nash equilibrium solution (in pure strategies), then an appropriate approach is to investigate Nash equilibria within the enlarged class of behavioral strategies, as has been done in Section 2.5 for zero-sum games. If the nonzero-sum single-act finite game under consideration is of the "ladder-nested" type, then it is relatively simpler to obtain Nash equilibria in behavioral strategies (which also fit well within a delayed commitment framework), since one still follows the recursive procedure developed in this section for pure strategies, but this time by also considering the mixed-strategy Nash equilibria of the static sub-extensive forms. Then, these mixed strategies, obtained for each equivalent normal form, will have to be appropriately concatenated and written in the form of a behavioral strategy. The following example now illustrates the steps in the derivation of behavioral Nash equilibria.

Example 3.11 Consider the 3-person single-act game depicted in Fig. 3.8. This is of the ladder-nested type, and both **P2** and **P3** have access to the

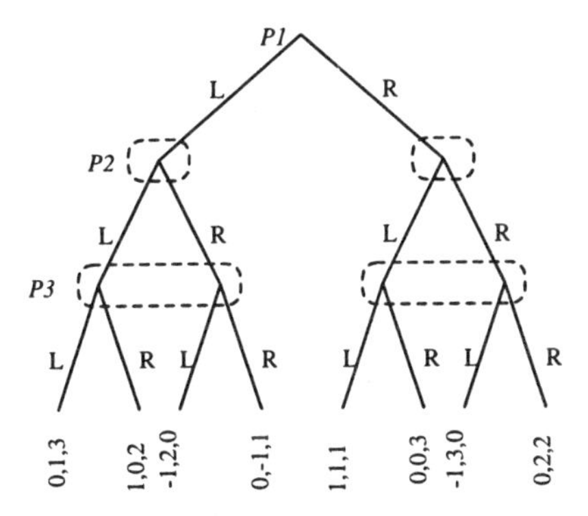

Figure 3.8: The 3-person single-act game of Example 3.11.

action of **P1**, but to no other information. Hence, we first have to consider two bimatrix games corresponding to the possible actions L and R of **P1**. If $u^1 = L$, then the bimatrix game is the one treated in Example 3.2, which admits the unique mixed-strategy Nash equilibrium

$$u^{2*} = \begin{cases} L & \text{w.p. } 1/2, \\ R & \text{w.p. } 1/2, \end{cases} \qquad u^{3*} = \begin{cases} L & \text{w.p. } 1/2, \\ R & \text{w.p. } 1/2, \end{cases}$$

with the corresponding average cost triple being $(0, 1/2, 3/2)$. If $u^1 = R$, on the other hand, the bimatrix game is

$$A = \begin{matrix} & \mathbf{P3} & \\ L & \begin{array}{|c|c|} \hline \textcircled{1} & 0 \\ \hline 3 & 2 \\ \hline \end{array} & \mathbf{P2}, \\ R & & \\ & L \quad R & \end{matrix} \qquad B = \begin{matrix} & \mathbf{P3} & \\ L & \begin{array}{|c|c|} \hline \textcircled{1} & 3 \\ \hline 0 & 2 \\ \hline \end{array} & \mathbf{P2}, \\ R & & \\ & L \quad R & \end{matrix}$$

which admits the unique Nash solution (in pure strategies)

$$u^{2*} = L, \qquad u^{3*} = L,$$

with the corresponding cost triple being $(1, 1, 1)$. Comparing this outcome with the previous one, it is obvious that the unique equilibrium strategy of **P1** is

$$u^{1*} = \hat{\gamma}^{1*} = L.$$

Appropriate concatenation now yields the following unique behavioral Nash equilibrium strategies for **P2** and **P3**, respectively:

$$\hat{\gamma}^{2*}(\eta^2) = \left\{ \begin{array}{lll} L & \text{w.p.} & 1/2 \\ R & \text{w.p.} & 1/2 \\ L & \text{w.p.} & 1 \\ R & \text{w.p.} & 0 \end{array} \right. \begin{array}{l} \left.\vphantom{\begin{array}{l} L \\ R \end{array}}\right\} \text{if } u^1 = L, \\ \left.\vphantom{\begin{array}{l} L \\ R \end{array}}\right\} \text{if } u^1 = R, \end{array}$$

$$\hat{\gamma}^{3*}(\eta^3) = \left\{ \begin{array}{lll} L & \text{w.p.} & 1/2 \\ R & \text{w.p.} & 1/2 \\ L & \text{w.p.} & 1 \\ R & \text{w.p.} & 0 \end{array} \right. \begin{array}{l} \left.\vphantom{\begin{array}{l} L \\ R \end{array}}\right\} \text{if } u^1 = L, \\ \left.\vphantom{\begin{array}{l} L \\ R \end{array}}\right\} \text{if } u^1 = R. \end{array}$$

The unique average Nash equilibrium outcome in behavioral strategies is $(0, 1/2, 3/2)$. □

Since the recursive procedure involves solution of static games in normal form at every step of the derivation, and since every N-person nonzero-sum finite game in normal form admits a mixed strategy Nash equilibrium solution (cf. Thm. 3.2), it follows that a Nash equilibrium solution in behavioral strategies always exists for single-act games in ladder-nested extensive form. This result is summarized below in Prop. 3.8, which is the counterpart of Corollary 2.4 in the present framework.

Proposition 3.8 *In every single-act N-person game which has a ladder-nested extensive form comprising a finite number of branches, there exists at least one Nash equilibrium solution in behavioral strategies.*

Remark 3.12 It should be noted that, as opposed to Corollary 2.4, Prop. 3.8 does not claim an ordered interchangeability property in case of multiple Nash equilibria, for reasons which have extensively been discussed in Sections 3.2 and 3.3. Furthermore, the result might fail to hold if the single-act game is not ladder-nested, in which case one has to consider the even larger class of mixed strategies. □

Remark 3.13 Since the players have only a finite number of possible pure strategies, it readily follows from Thm. 3.2 that there always exists a Nash equilibrium in mixed strategies in single-act finite games of the general type. However, computation of these mixed equilibrium strategies is quite a nontrivial problem, since a recursive procedure cannot be developed for finite single-act games which are not ladder-nested. □

3.5.3 Multi-act games: Pure-strategy Nash equilibria

In the discussion of the Nash-equilibria of multi-act nonzero-sum finite games in extensive form, we shall follow rather closely the analysis of Section 2.5 which was devoted to similarly structured zero-sum games, since the results to be obtained for multi-act nonzero-sum games will be direct extensions of their counterparts in the case of zero-sum games. Not all conclusions drawn in Section 2.5, however, are valid in the present context and these important differences between the properties of saddle-point and Nash equilibrium solutions will also be elucidated in the sequel.

In order to be able to develop a systematic (recursive) procedure to obtain Nash equilibria of multi-act nonzero-sum finite games, we will have to impose (as in Section 2.5) some restrictions on the nature of the informational coupling among different levels of play. To this end, we confine ourselves to multi-act games whose Nash equilibria can be obtained by solving a sequence of single-act games and by appropriate concatenation of the equilibrium strategies determined at each level of play. This reasoning, then, brings us to the so-called "feedback games", a precise definition of which is given below (as a counterpart of Def. 2.11).

Definition 3.21 *A multi-act N-person nonzero-sum game in extensive form with a fixed order of play is called an* N-person nonzero-sum feedback game in extensive form, *if*

(i) at the time of his act, each player has perfect information concerning the current level of play, i.e., no information set contains nodes of the tree belonging to different levels of play,

(ii) information sets of the first-acting player at every level of play are singletons, and the information sets of the other players at every level of play are such that none of them includes nodes corresponding to branches emanating from two or more different information sets of the first-acting player, i.e., each player knows the state of the game at every level of play.

If, furthermore,

(iii) the single-act games corresponding to the information sets of the first-acting player at each level of play are of the ladder-nested (respectively, nested) type (cf. Def. 3.15), then the multi-act game is called an N-person nonzero-sum feedback game in ladder-nested *(respectively,* nested*)* extensive form.

Remark 3.14 It should be noted that, in the case of two-person multi-act games, every feedback game is, by definition, also in ladder-nested extensive form; hence, the zero-sum feedback game introduced earlier by Def. 2.11 is also ladder-nested. □

Now, paralleling the analysis of Section 2.5, let the number of levels of play in an N-person nonzero-sum feedback game in extensive form be K, and consider a typical strategy γ^i of $\mathbf{P}i$ in such a game to be composed of K components $(\gamma_1^i, \ldots, \gamma_K^i)$. Here, γ_j^i stands for the corresponding strategy of $\mathbf{P}i$ at his jth level of act, and it can be taken to have as its domain the class of only those information sets of $\mathbf{P}i$ which pertain to the jth level of play. Let us denote the collection of all such strategies for $\mathbf{P}i$ at level j by Γ_j^i. Then, for such a game, the set of inequalities (3.23) can be written as (as a counterpart of (2.35))

$$\left.\begin{array}{l} J^{1*} \leq J^1(\gamma_1^1, \ldots, \gamma_K^1; \gamma^{2*}; \ldots; \gamma^{N*}), \\ \quad \cdots\cdots\cdots\cdots\cdots\cdots \\ J^{N*} \leq J^N(\gamma^{1*}; \ldots; \gamma^{N-1*}; \gamma_1^N, \ldots, \gamma_K^N), \end{array}\right\} \tag{3.27}$$

which are to be satisfied for all $\gamma_j^i \in \Gamma_j^i$, $i = 1, \ldots, N$; $j = 1, \ldots, K$. On any N-tuple of Nash equilibrium strategies $\{\gamma^{1*}, \ldots, \gamma^{N*}\}$ that satisfies (3.27), let us impose the further restriction that it also satisfies (recursively) the following K N-tuple inequalities for all $\gamma_j^i \in \Gamma_j^i$, $i = 1, \ldots, N$; $j = 1, \ldots, K$:

$$
\text{level } K \left\{
\begin{aligned}
& J^1(\gamma_1^1, \ldots, \gamma_{K-1}^1, \gamma_K^{1*}; \gamma_1^2, \ldots, \gamma_{K-1}^2, \gamma_K^{2*}; \ldots; \gamma_1^N, \ldots, \\
& \qquad \gamma_{K-1}^N, \gamma_K^{N*}) \\
& \quad \le J^1(\gamma^1; \gamma_1^2, \ldots, \gamma_{K-1}^2, \gamma_K^{2*}; \ldots; \gamma_1^N, \ldots, \gamma_{K-1}^N, \gamma_K^{N*}) \\
& J^2(\gamma_1^1, \ldots, \gamma_{K-1}^1, \gamma_K^{1*}; \gamma_1^2, \ldots, \gamma_{K-1}^2, \gamma_K^{2*}; \ldots; \gamma_1^N, \ldots, \\
& \qquad \gamma_{K-1}^N, \gamma_K^{N*}) \\
& \quad \le J^2(\gamma_1^1, \ldots, \gamma_{K-1}^1, \gamma_K^{1*}; \gamma^2; \ldots; \gamma_1^N, \ldots, \gamma_{K-1}^N, \gamma_K^{N*}) \\
& \cdots\cdots\cdots\cdots\cdots\cdots\cdots\cdots \\
& J^N(\gamma_1^1, \ldots, \gamma_{K-1}^1, \gamma_K^{1*}; \gamma_1^2, \ldots, \gamma_{K-1}^2, \gamma_K^{2*}; \ldots; \gamma_1^N, \ldots, \\
& \qquad \gamma_{K-1}^N, \gamma_K^{N*}) \\
& \quad \le J^N(\gamma_1^1, \ldots, \gamma_{K-1}^1, \gamma_K^{1*}; \gamma_1^2, \ldots, \gamma_{K-1}^2, \gamma_K^{2*}; \ldots; \gamma^N)
\end{aligned}
\right.
$$

$$
\text{level } K-1 \left\{
\begin{aligned}
& J^1(\gamma_1^1, \ldots, \gamma_{K-2}^1, \gamma_{K-1}^{1*}, \gamma_K^{1*}, \gamma_1^2, \ldots, \gamma_{K-1}^{2*}, \gamma_K^{2*}; \ldots; \\
& \qquad \gamma_1^N, \ldots, \gamma_{K-2}^N, \gamma_{K-1}^{N*}, \gamma_K^{N*}) \\
& \quad \le J^1(\gamma_1^1, \ldots, \gamma_{K-2}^1, \gamma_{K-1}^1, \gamma_K^{1*}; \gamma_1^2, \ldots, \gamma_{K-2}^2, \gamma_{K-1}^{2*}, \gamma_K^{2*}; \ldots; \\
& \qquad \gamma_1^N, \ldots, \gamma_{K-2}^N, \gamma_{K-1}^{N*}, \gamma_K^{N*}) \\
& J^2(\gamma_1^1, \ldots, \gamma_{K-2}^1, \gamma_{K-1}^{1*}, \gamma_K^{1*}; \gamma_1^2, \ldots, \gamma_{K-2}^2, \gamma_{K-1}^{2*}, \gamma_K^{2*}; \ldots; \\
& \qquad \gamma_1^N, \ldots, \gamma_{K-2}^N, \gamma_{K-1}^{N*}, \gamma_K^{N*}) \\
& \quad \le J^2(\gamma_1^1, \ldots, \gamma_{K-2}^1, \gamma_{K-1}^{1*}, \gamma_K^{1*}; \gamma_1^2, \ldots, \gamma_{K-2}^2, \gamma_{K-1}^2, \gamma_K^{2*}; \ldots; \\
& \qquad \gamma_1^N, \ldots, \gamma_{K-2}^N, \gamma_{K-1}^{N*}, \gamma_K^{N*}) \\
& \cdots\cdots\cdots\cdots\cdots\cdots\cdots\cdots \\
& J^N(\gamma_1^1, \ldots, \gamma_{K-2}^1, \gamma_{K-1}^{1*}, \gamma_K^{1*}, \gamma_1^2, \ldots, \gamma_{K-2}^2, \gamma_{K-1}^{2*}, \gamma_K^{2*}; \ldots; \\
& \qquad \gamma_1^N, \ldots, \gamma_{K-2}^N, \gamma_{K-1}^{N*}, \gamma_K^{N*}) \\
& \quad \le J^N(\gamma_1^1, \ldots, \gamma_{K-2}^1, \gamma_{K-1}^{1*}, \gamma_K^{1*}; \gamma_1^2, \ldots, \gamma_{K-2}^2, \gamma_{K-1}^{2*}, \gamma_K^{2*}; \ldots; \\
& \qquad \gamma_1^N, \ldots, \gamma_{K-2}^N, \gamma_{K-1}^N, \gamma_K^{N*})
\end{aligned}
\right.
$$

$$\cdots\cdots\cdots\cdots\cdots\cdots\cdots\cdots\cdots\cdots$$

$$\cdots\cdots\cdots\cdots\cdots\cdots\cdots\cdots\cdots\cdots$$

$$\tag{3.28}$$

$$
\text{level } 1 \left\{
\begin{aligned}
& J^1(\gamma^{1*}; \gamma^{2*}; \cdots; \gamma^{N*}) \le J^1(\gamma_1^1, \gamma_2^{1*}, \ldots, \gamma_K^{1*}; \gamma^{2*}; \ldots; \gamma^{N*}) \\
& J^2(\gamma^{1*}; \gamma^{2*}; \cdots; \gamma^{N*}) \le J^2(\gamma^{1*}; \gamma_1^2, \gamma_2^{2*}, \ldots, \gamma_K^{2*}; \gamma^{2*}; \ldots; \gamma^{N*}) \\
& \cdots\cdots\cdots\cdots\cdots\cdots\cdots\cdots \\
& J^N(\gamma^{1*}; \gamma^{2*} \ldots; \gamma^{N*}) \le J^N(\gamma^{1*}; \gamma^{2*}; \ldots; \gamma_1^N, \gamma_2^{N*}, \ldots, \gamma_K^{N*}).
\end{aligned}
\right.
$$

Definition 3.22 *For an N-person nonzero-sum feedback game in extensive form with K levels of play, let $\{\gamma^{1*}, \ldots, \gamma^{N*}\}$ be an N-tuple of strategies satisfying (3.27) and (3.28) for all $\gamma_j^i \in \Gamma_j^i$, $i = 1, \ldots, N$; $j = 1, \ldots, K$. Then, $\{\gamma^{1*}, \ldots, \gamma^{N*}\}$ constitutes a (pure)* feedback Nash equilibrium *for the feedback game.*

Proposition 3.9 *Every N-tuple $\{\gamma^{1*}, \ldots, \gamma^{N*}\}$ that satisfies the set of inequalities (3.28) also satisfies the set of N inequalities (3.27). Hence, the requirement for satisfaction of (3.27) in Def. 3.22 is redundant, and it can be dispensed with without any loss of generality.*

Proof. This basically follows from the lines of argument used in the proof of Prop. 2.3, with two inequalities at each level of play replaced by N inequalities in the present case. □

The set of inequalities (3.28) now readily suggests a recursive procedure for derivation of feedback Nash equilibria in feedback nonzero-sum games. Starting at level K, we have to solve a single-act game for each information set of the first-acting player at each level of play, and then appropriately concatenate all the equilibrium strategies (with restricted domains) thus obtained. If the single-act games encountered in this recursive procedure admit more than one (pure-strategy) Nash equilibrium, the remaining analysis will have to be repeated for each one of these equilibria so as to determine the complete set of feedback Nash equilibria the multi-act game admits.

For nested multi-act feedback games one might wish to restrict the class of Nash equilibria of interest even further, so that only those obtained by concatenation of delayed commitment type equilibria of single-act games are considered. The motivations behind such a further restriction are: (i) the delayed commitment mode of play conforms well with the notion of feedback Nash equilibria and it eliminates informational nonuniqueness, and (ii) Nash equilibria at every level of play can then be obtained by utilizing the recursive technique developed earlier in this section for single-act games (which is computationally attractive, especially for ladder-nested games). A precise definition now follows.

Definition 3.23 *For a nonzero-sum feedback game in extensive form, a feedback Nash equilibrium solution is of the* delayed commitment *type if its appropriate restriction at each level of play constitutes a delayed commitment type Nash equilibrium (cf. Defs. 3.18, 3.20) for each corresponding single-act game to be encountered in the recursive derivation.*

Heretofore, the discussion on multi-act feedback games has been confined only to feedback Nash equilibria (of the delayed commitment type or otherwise). However, such games could very well admit other types of Nash equilibria which satisfy inequalities (3.23) but which do not fit the framework of Def. 3.22. The main reason for this phenomenon is the "informational nonuniqueness" feature of Nash equilibria in dynamic games, which has been discussed earlier in this section within the context of single-act games. In the present context, a natural extension of Prop. 3.7 is valid, which we provide below after introducing some terminology and also the concept of informational inferiority in nonzero-sum feedback games.

Definition 3.24 *Let* I *be an N-person multi-act nonzero-sum game with a fixed order of play, which satisfies the first requirement of Def. 3.21. An N-person* *open-loop game in extensive form (say,* II*) (cf. Def. 3.13) is said to be the* static (open-loop) version *of* I *if it can be obtained from* I *by replacing, at each level of play, the information sets of each player with a single information set. The equivalent normal form of the open-loop version of* I *is the* static normal form *of* I.

Definition 3.25 *Let* I *and* II *be two N-person multi-act nonzero-sum games with fixed orders of play, which satisfy the first requirement of Def. 3.21. Further let Γ_{I}^i and Γ_{II}^i denote the strategy sets of* **P***i in* I *and* II*, respectively. Then* I *is* informationally inferior *to* II *if $\Gamma_{\mathrm{I}}^i \subseteq \Gamma_{\mathrm{II}}^i$ for all $i \in \mathbf{N}$, with strict inclusion for at least one i.*

Proposition 3.10 *Let* I *and* II *be two N-person multi-act nonzero-sum games as introduced in Def. 3.25, so that* I *is informationally inferior to* II*. Then,*

(i) any Nash equilibrium solution for I *is also a Nash equilibrium solution for* II*,*

(ii) if $\{\gamma^1, \ldots, \gamma^N\}$ is a Nash equilibrium solution for II *so that $\gamma^i \in \Gamma_{\mathrm{I}}^i$ for all $i = 1, \ldots, N$, then it is also a Nash equilibrium solution for* I.

Proof. The proof is analogous to the proof of Prop. 3.7, and is therefore omitted. □

Remark 3.15 It should now be apparent why informationally nonunique Nash equilibria exist in nonzero-sum multi-act feedback games. Every nontrivial multi-act feedback game in which the players have at least two alternatives at each level of play admits several informationally inferior multi-act games. These different games in general admit different Nash equilibria which are, moreover, not interchangeable. Since every one of these also constitutes a Nash equilibrium for the original multi-act game (cf. Prop. 3.10), existence of a plethora of informationally nonunique equilibria readily follows. The further restrictions of Defs. 3.22 and 3.23, imposed on the concept of noncooperative equilibrium, clearly eliminate this informational nonuniqueness, and they stand out as providing one possible criterion according to which a further selection can be made.

This "informational nonuniqueness" property of noncooperative equilibrium in dynamic games is, of course, also featured by zero-sum games, since they are special types of nonzero-sum games, in which case the Nash equilibrium solution coincides with the saddle-point solution. However, since all saddle-point strategies have the ordered interchangeability property and further since the saddle-point value is unique, nonuniqueness of equilibrium strategies does not create a problem in zero-sum dynamic games—the main reason why we have not included a discussion on informational nonuniqueness in Chapter 2. It should be noted, however, that for two-person zero-sum multi-act feedback games, Def. 3.23 is redundant and Def. 3.22 becomes equivalent to Def. 2.12, which clearly dictates a delayed commitment mode of play. □

The following example now serves to illustrate both the recursive procedure and the informational nonuniqueness of Nash equilibria in feedback nonzero-sum games.

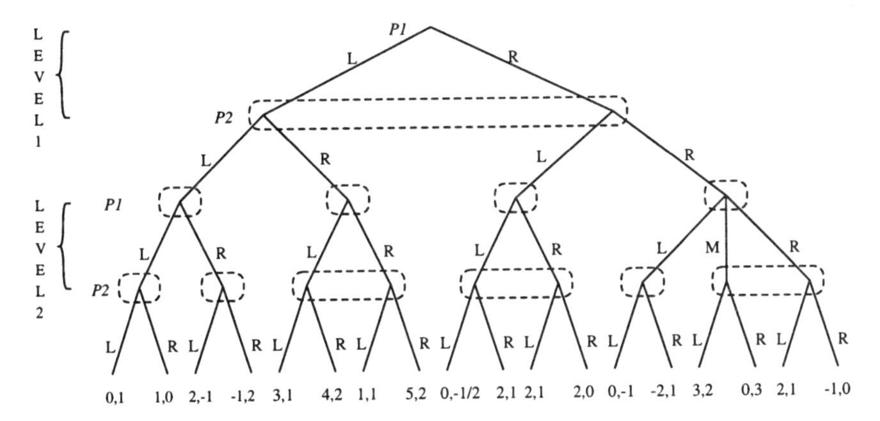

Figure 3.9: The multi-act feedback nonzero-sum game of Example 3.12.

Example 3.12 Consider the multi-act feedback game whose extensive form is depicted in Fig. 3.9. Since this is a 2-person game (of the feedback type) it is definitely also ladder-nested (or equivalently, nested), and hence derivation of its delayed commitment type feedback Nash equilibria involves solutions of only static sub-extensive forms. At the second (and the last) levels of play, there are four single-act games, one for each singleton information set of **P**1. The first of these (counting from left) admits the unique Nash equilibrium solution of the delayed commitment type

$$\gamma_2^{1*}(\eta_2^1) = L, \quad \gamma_2^{2*}(\eta_2^2) = \begin{cases} R & \text{if } u_2^1 = L, \\ L & \text{otherwise} \end{cases}, \tag{i}$$

with the Nash equilibrium outcome being $(1, 0)$. The second one admits the unique Nash equilibrium solution

$$\gamma_2^{1*}(\eta_2^1) = R, \quad \gamma_2^{2*}(\eta_2^2) = L, \tag{ii}$$

with an outcome of $(1, 1)$. The third single-act game again has a unique Nash equilibrium, given by

$$\gamma_2^{1*}(\eta_2^1) = L, \quad \gamma_2^{2*}(\eta_2^2) = L, \tag{iii}$$

with an outcome of $(0, -1/2)$. The last one is the single-act game of Example 3.6, which admits the unique delayed commitment type Nash equilibrium solution

$$\gamma_2^{1*}(\eta_2^1) = R, \quad \gamma_2^{2*}(\eta_2^2) = \begin{cases} L & \text{if } u_2^1 = L, \\ R & \text{otherwise} \end{cases}, \tag{iv}$$

with the Nash outcome being $(-1, 0)$. When (i)-(iv) are collected together, we obtain the delayed commitment type Nash equilibrium solution of the feedback game at level 2 to be uniquely given as

$$\left.\begin{aligned}\gamma_2^{1*}(\eta_2^1) &= u_1^2, \\ \gamma_2^{2*}(\eta_2^2) &= \begin{cases} R & \text{if } \begin{cases} u_2^1 = L, u_1^1 = L, u_1^2 = L, \\ u_1^1 = R, u_1^2 = R, u_2^1 = M \text{ or } R, \end{cases} \\ L & \text{otherwise}. \end{cases}\end{aligned}\right\} \quad (3.29a)$$

Now, crossing out the second level of play, we end up with the single-act game

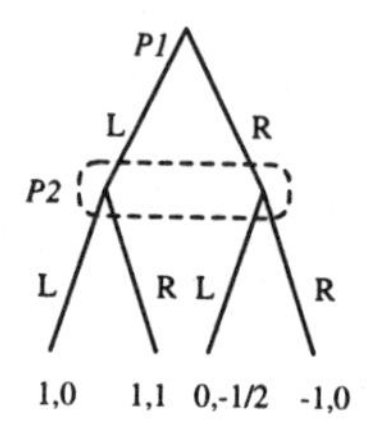

whose equivalent normal form is the bimatrix game

$$A = \begin{array}{c} \\ L \\ R \\ \end{array} \overset{\textbf{P2}}{\begin{array}{|c|c|} \hline 1 & 1 \\ \hline \textcircled{0} & -1 \\ \hline \end{array}} \textbf{ P1} \qquad \begin{array}{c} \\ L \\ R \\ \end{array} \overset{\textbf{P2}}{\begin{array}{|c|c|} \hline 0 & 1 \\ \hline \textcircled{-\frac{1}{2}} & 0 \\ \hline \end{array}} \textbf{ P1}$$

$$\qquad\quad L \quad R \qquad\qquad\qquad\qquad L \quad R$$

which admits, as indicated, the unique Nash equilibrium solution

$$\gamma_1^{1*}(\eta_1^1) = R, \qquad \gamma_1^{2*}(\eta_1^2) = L \tag{3.29b}$$

with the corresponding outcome being

$$(0, -1/2). \tag{3.29c}$$

In conclusion, the feedback game of Fig. 3.9 admits a unique feedback Nash equilibrium solution of the delayed commitment type, which is given by (3.29a)-(3.29b), with the corresponding Nash outcome being (3.29c).

This recursively obtained Nash equilibrium solution, although it is unique as a delayed commitment type feedback Nash equilibrium, is clearly not unique as a feedback Nash equilibrium solution, since it is already known from Example 3.6 that the fourth single-act game encountered at the first step of the recursive derivation admits another Nash equilibrium solution which is not of the delayed commitment type. This equilibrium solution is the constant strategy pair given by (3.26a)-(3.26b), that is,

$$\gamma_2^{1*}(\eta_2^1) = L, \qquad \gamma_2^{2*}(\eta_2^2) = L$$

which has to replace (iv) in the recursive derivation. The Nash outcome corresponding to this pair of strategies is $(0, -1)$. The strategy pair (3.29a) will now be replaced by

$$\begin{aligned}\gamma_2^{1*}(\eta_2^1) &= \begin{cases} R & \text{if } u_1^1 = L, u_1^2 = R, \\ L & \text{otherwise}\ , \end{cases} \\ \gamma_2^{2*}(\eta_2^2) &= \begin{cases} R & \text{if } u_2^1 = L, u_1^1 = L, u_1^2 = L, \\ L & \text{otherwise}\ , \end{cases}\end{aligned} \tag{3.30a}$$

which also provides a Nash equilibrium solution at level 2. Now, crossing out the second level of play, this time we have the single-act game

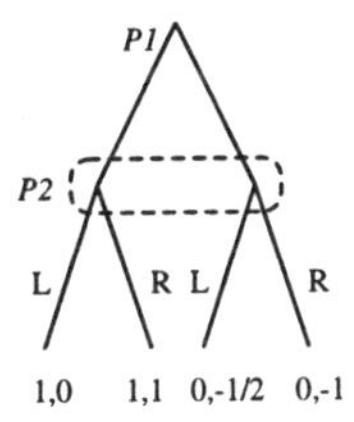

whose equivalent normal form is

$$A = \begin{array}{c} \\ L \\ R \\ \\ \end{array}\overset{\mathbf{P2}}{\begin{array}{|c|c|}\hline 1 & 1 \\ \hline 0 & \textcircled{0} \\ \hline \multicolumn{1}{c}{L} & \multicolumn{1}{c}{R} \end{array}}\ \mathbf{P1}, \qquad B = \begin{array}{c} \\ L \\ R \\ \\ \end{array}\overset{\mathbf{P2}}{\begin{array}{|c|c|}\hline 0 & 1 \\ \hline -1/2 & \textcircled{-1} \\ \hline \multicolumn{1}{c}{L} & \multicolumn{1}{c}{R} \end{array}}\ \mathbf{P1},$$

which admits, as indicated, the unique Nash equilibrium solution

$$\gamma_1^{1*}(\eta_1^1) = R, \quad \gamma_1^{2*}(\eta_1^2) = R \tag{3.30b}$$

with the corresponding outcome being

$$(0, -1). \tag{3.30c}$$

We therefore arrive at the conclusion that (3.30a)-(3.30b) provides another feedback Nash equilibrium solution for the multi-act game of Fig. 3.9, which is, however, not of the delayed commitment type. It is noteworthy that the unique feedback Nash solution of the delayed commitment type is in this case inadmissible, which follows immediately by a comparison of (3.29c) and (3.30c).

To determine the other (non-feedback type) Nash equilibria of this multi-act game, the only possible approach is to bring it into equivalent normal form and to investigate Nash equilibria of the resulting bimatrix game. This is, however, quite a strenuous exercise, since the bimatrix game is of rather high dimensions. Specifically, at level 1 each player has two possible strategies, while at level 2 **P1** has $2^3 \times 3 = 24$ and **P2** has $2^6 = 64$ possible strategies. This then implies that the bimatrix game is of dimensions 48×128. Two Nash equilibria of this bimatrix game are the ones given by (3.29a)-(3.29b) and (3.30a)-(3.30b), but there are possibly others which in fact correspond to Nash equilibria of multi-act games which are informationally inferior to the one of Fig. 3.9; one of these games is the open-loop game which is the static version of the original feedback game. □

3.5.4 Multi-act games: Behavioral and mixed equilibrium strategies

Behavioral equilibrium strategies for an N-person feedback game in ladder-nested extensive form can be derived by essentially following the recursive procedure developed in Section 3.5.3 for pure-strategy feedback equilibria, but this time by also considering the behavioral-strategy Nash equilibria of the single-act games to be encountered in the derivation. The justification of applicability of such a recursive procedure is analogous to the one of behavioral saddle-point equilibria of feedback games discussed in Section 2.5. Specifically, if K denotes the number of levels of play in the N-person feedback game of the ladder-nested type, then a typical behavioral strategy $\hat{\gamma}^i$ of $\mathbf{P}i$ in such a game can be decomposed into K components $\hat{\gamma}_1^i, \ldots, \hat{\gamma}_K^i$, where $\hat{\gamma}_j^i \in \hat{\Gamma}_j^i$ is the corresponding behavioral strategy of $\mathbf{P}i$ at his jth level of play. Furthermore, $\hat{\gamma}_j^i$ has, as its domain, only the class of those information sets of $\mathbf{P}i$ which pertain to the jth level of play. Then, the set of inequalities that determines the behavioral Nash equilibrium solution can be written as

$$\left.\begin{array}{l} \hat{J}^{1*} \leq \hat{J}^1(\hat{\gamma}_1^1, \ldots, \hat{\gamma}_K^1; \hat{\gamma}^{2*}; \ldots; \hat{\gamma}^{N*}) \\ \cdots\cdots\cdots\cdots\cdots\cdots \\ \hat{J}^{N*} \leq \hat{J}^N(\hat{\gamma}^{1*}; \ldots; \hat{\gamma}^{N-1*}; \hat{\gamma}_1^N, \ldots, \hat{\gamma}_K^N) \end{array}\right\} \quad (3.31)$$

which have to be satisfied for all $\hat{\gamma}_j^i \in \hat{\Gamma}_j^i$, $j = 1, \ldots, K$; $i \in \mathbf{N}$. Under the delayed commitment mode of play, we impose the further restriction that they satisfy the K N-tuple inequalities similar to (3.28), with only γ_j^i replaced by $\hat{\gamma}_j^i$ and J^i by $\hat{J}^i$. A counterpart of Prop. 3.9 also holds true in the present framework and this readily leads to recursive derivation of behavioral-strategy Nash equilibria in feedback games of the ladder-nested type. This procedure always leads to a Nash equilibrium solution in behavioral strategies as the following proposition states.

Proposition 3.11 *Every finite N-person nonzero-sum feedback game in ladder-nested extensive form admits a Nash equilibrium solution in behavioral strategies which can be determined recursively.*

Proof. This result is an immediate consequence of Prop. 3.8, since every feedback game in ladder-nested extensive form can be decomposed recursively into ladder-nested single-act games. □

For feedback games which are not of the ladder-nested type, and also for other types of (non-feedback) multi-act games, Prop. 3.11 is not in general valid, and the Nash equilibrium in behavioral strategies (even if it exists) cannot be obtained recursively. Then, the only approach would be to bring the extensive form into an equivalent normal form, obtain all mixed-strategy Nash equilibria of this normal form, and to seek whether any one of these equilibria would constitute a Nash equilibrium in behavioral strategies. This argument is now supported by the following two propositions which are direct counterparts of Props 2.5 and 2.6, respectively.

Proposition 3.12 *Every N-person nonzero-sum multi-act finite game in extensive form admits a Nash equilibrium solution in mixed strategies.*

Proof. The result readily follows from Thm. 3.2, since every such game can be transformed into an equivalent normal form in which each player has a finite number of strategies. □

Proposition 3.13 *Every Nash equilibrium of a finite N-person nonzero-sum multi-act game in behavioral strategies also constitutes a Nash equilibrium in the larger class of mixed strategies.*

Proof. Let $\{\hat{\gamma}^{1*} \in \hat{\Gamma}^1, \ldots, \hat{\gamma}^{N*} \in \hat{\Gamma}^N\}$ denote an N-tuple of behavioral strategies stipulated to be in Nash equilibrium, and let $\bar{\Gamma}^i$ denote the mixed-strategy set of $\mathbf{P}i$, $i \in \mathbf{N}$. Since $\hat{\Gamma}^i \subset \bar{\Gamma}^i$, we clearly have $\hat{\gamma}^{i*} \in \bar{\Gamma}^i$, for every $i \in \mathbf{N}$. Assume, to the contrary, that $\{\hat{\gamma}^{1*}, \ldots, \hat{\gamma}^{N*}\}$ is not a mixed-strategy Nash equilibrium solution; then this implies that there exists at least one i (say, $i = N$, without any loss of generality) for which the corresponding inequality of (3.31)[19] is not satisfied for all $\gamma^i \in \bar{\Gamma}^i$. In particular, there exists a $\bar{\gamma}^N \in \bar{\Gamma}^N$ such that

$$\hat{J}^{N*} > \hat{J}^N(\hat{\gamma}^{1*}; \ldots; \hat{\gamma}^{N-1*}; \bar{\gamma}^N) \triangleq F^*. \qquad (i)$$

Now, abiding by our standard convention, let Γ^N denote the pure-strategy set of $\mathbf{P}N$, and consider the quantity

$$F(\gamma^N) \triangleq \hat{J}^N(\hat{\gamma}^{1*}; \ldots; \hat{\gamma}^{N-1*}; \gamma^N)$$

defined for each $\gamma^N \in \Gamma^N$. (This is well defined since $\Gamma^N \subset \hat{\Gamma}^N$.) The infimum of this quantity over Γ^N is definitely achieved (say, by $\gamma^{N*} \in \Gamma^N$), since Γ^N is a finite set. Furthermore, since $\bar{\Gamma}^N$ is comprised of all probability distributions on Γ^N,

$$\inf_{\bar{\Gamma}^N} F(\gamma^N) = \inf_{\Gamma^N} F(\gamma^N) = F(\gamma^{N*}).$$

We therefore have

$$F^* = F(\gamma^{N*}), \qquad (ii)$$

and also the inequality

$$\hat{J}^{N*} > F(\gamma^{N*})$$

in view of (i). But this is impossible since $\hat{J}^{N*} = \inf_{\hat{\Gamma}^N} F(\gamma^N)$ and $\Gamma^N \subset \hat{\Gamma}^N$, thus completing the proof of the proposition. □

[19] Here, by an abuse of notation, we take $\hat{J}^i$ to denote the average loss to $\mathbf{P}i$ under also mixed-strategy N-tuples.

3.5.5 Other refinements on Nash equilibria

As we have seen heretofore in this chapter, existence of multiple Nash equilibria for a given dynamic game is more a rule rather than an exception, with the multiplicity arising because of the informational richness of the underlying decision problem as well as the structure of the players' cost matrices. As a means of shrinking the set of Nash equilibria in a rational way, we have introduced heretofore the notions of "delayed commitment", "informational inferiority", "feedback games" and "admissibility". There are, however, several other refinement schemes which have been introduced in the literature, some of which we discuss below.

To motivate the discussion, let us start with a two-player matrix game (A, B) where the players have identical cost matrices.

$$A = B = \begin{array}{c} \textbf{P2} \\ \begin{array}{c|c|c|} & \hline L & 0 & 1 \\ \hline R & 1 & 1 \\ \hline & L & R \end{array} \end{array} \ \textbf{P1}. \tag{3.32}$$

The game admits two pure-strategy Nash equilibria: (L, L) and (R, R). Note, however, that if we perturb the entries of the two matrices slightly, and independently:

$$A + \Delta A = \overset{\textbf{P2}}{\begin{array}{|c|c|} \hline \epsilon_{11}^1 & 1+\epsilon_{12}^1 \\ \hline 1+\epsilon_{21}^1 & 1+\epsilon_{22}^1 \\ \hline \end{array}} \ \textbf{P1}, \quad B + \Delta B = \overset{\textbf{P2}}{\begin{array}{|c|c|} \hline \epsilon_{11}^2 & 1+\epsilon_{12}^2 \\ \hline 1+\epsilon_{21}^2 & 1+\epsilon_{22}^2 \\ \hline \end{array}} \ \textbf{P1},$$

where ϵ_{ij}^k, i, j, $k = 1, 2$, are infinitesimally small (positive or negative) numbers, then (L, L) will still retain its equilibrium property (as long as $|\epsilon_{ij}^k| < 1/2$), but (R, R) will not. More precisely, there will exist infinitely many perturbed versions of the original game for which (R, R) will not constitute a Nash equilibrium. Hence, in addition to admissibility, (L, L) can be singled out in this case as the Nash solution that is *robust* to infinitesimal perturbations in the entries of the cost matrices.

Can such perturbations be induced naturally by some behavioral assumptions imposed on the players? The answer is yes, as we discuss next. Consider the scenario where a player who intends to play a particular pure strategy (out of a set of n possible alternatives) errs and plays with some small probability one of the other $n - 1$ alternatives. In the matrix game (3.32), for example, if both players err with equal (independent) probability $\epsilon > 0$, the resulting matrix game is (A_ϵ, B_ϵ), where

$$A_\epsilon = B_\epsilon = \begin{array}{c} \textbf{P2} \\ \begin{array}{c|c|c|} & \hline L & \epsilon(2-\epsilon) & 1-\epsilon+\epsilon^2 \\ \hline R & 1-\epsilon+\epsilon^2 & 1-\epsilon^2 \\ \hline & L & R \end{array} \end{array} \ \textbf{P1}.$$

Note that for all $\epsilon \in (0, 1/2)$ this matrix game admits the unique Nash equilibrium (L, L), with a cost pair of $(\epsilon(2-\epsilon), \epsilon(2-\epsilon))$, which converges to $(0, 0)$

as $\epsilon \downarrow 0$, thus recovering one of the Nash cost pairs of the original game. A Nash equilibrium solution that can be recovered this way is known as a *perfect equilibrium*, which was first introduced in precise terms by Selten (1975), in the context of N-player games in extensive form. Given a game of perfect recall, denoted $\mathcal{G}$, the idea is to generate a sequence of games, $\mathcal{G}_1, \mathcal{G}_2, \ldots, \mathcal{G}_k, \ldots$, a limiting equilibrium solution of which (in behavioral strategies, and as $k \to \infty$) is an equilibrium solution of $\mathcal{G}$. If $\mathcal{G}_k$ is obtained from $\mathcal{G}$ by forcing the players at each information set to choose every possible alternative with positive probability (albeit small, for those alternatives that are not optimal), then the equilibrium solution(s) of $\mathcal{G}$ that are recovered as a result of the limiting procedure above is (are) called *perfect equilibrium (equilibria)*.[20] Selten (1975) has shown that every finite game in extensive form with perfect recall (and as a special case in normal form) admits at least one perfect equilibrium, thus making this refinement scheme a legitimate one.

The procedure discussed above, which amounts to "completely" perturbing a game with multiple equilibria, is one way of obtaining perfect equilibria; yet another one, as introduced by Myerson (1978), is to restrict the players to use completely mixed strategies (with some lower positive bound on the probabilities) at each information set. Again referring back to the matrix game (A, B) of (3.32), let the players' mixed strategies be restricted to the class

$$\hat{\gamma}^1 = \begin{cases} L & \text{w.p.} \quad y \\ R & \text{w.p.} \quad 1-y \end{cases} ; \quad \hat{\gamma}^2 = \begin{cases} L & \text{w.p.} \quad z, \\ R & \text{w.p.} \quad 1-z, \end{cases}$$

where $\epsilon \le y \le 1-\epsilon$, $\epsilon \le z \le 1-\epsilon$, for some (sufficiently small) positive ϵ. Over this class of strategies, the average cost functions of the players will be

$$\hat{J}^1 = \hat{J}^2 \quad = \quad -yz + 1,$$

which admits (assuming that $0 \le \epsilon < \frac{1}{2}$) a unique Nash equilibrium:

$$\hat{\gamma}_\epsilon^{1*} = \hat{\gamma}_\epsilon^{2*} = \begin{cases} L & \text{w.p.} \quad 1-\epsilon \\ R & \text{w.p.} \quad \epsilon \end{cases} ; \quad \hat{J}_\epsilon^{1*} = \hat{J}_\epsilon^{2*} = 1 - (1-\epsilon)^2.$$

Such a solution is called an *ϵ-perfect equilibrium* (Myerson, 1978), which in the limit as $\epsilon \downarrow 0$ clearly yields the perfect Nash equilibrium obtained earlier. Myerson in fact proves, for N-person games in normal form, that every perfect equilibrium can be obtained as the limit of an appropriate ϵ-perfect equilibrium, with the converse statement also being true. More precisely, using the notation of Section 3.3, we have the following.

Proposition 3.14 *For an N-person finite game in normal form, a mixed-strategy Nash equilibrium $\{y^{i*} \in Y^i, i \in \mathbf{N}\}$ is a* perfect equilibrium *if, and only if, there exist some sequences $\{\epsilon_k\}_{k=1}^\infty$, $\{y^i_{\epsilon_k} \in Y^i, i \in \mathbf{N}\}_{k=1}^\infty$ such that*

[20] This is also called "trembling hand equilibrium", as the process of erring at each information set is reminiscent of a "trembling hand" making unintended choices with small probability. Here, as $k \to \infty$, this probability of unintended plays converges to zero.

i) $\epsilon_k > 0$ *and* $\lim_{k\to\infty} \epsilon_k = 0$

ii) $\{y^i_{\epsilon_k}, i \in \mathbf{N}\}$ *is an* ϵ_k*-perfect equilibrium*

iii) $\lim_{k\to\infty} y^i_{\epsilon_k} = y^{i*}$, $i \in \mathbf{N}$.

Furthermore, a perfect equilibrium necessarily exists, and every perfect equilibrium is a Nash equilibrium.

Even though the notion of perfect equilibrium provides a refinement of the notion of Nash equilibrium with some appealing properties, it also carries some undesirable features as the following example of an identical cost matrix game (due to Myerson (1978)) exhibits:

P2

$A = B =$				
L	0	1	10	
M	1	1	8	**P1**.
R	10	8	8	
	L	M	R	

(3.33)

Note that this is a matrix game derived from (3.32) by adding a completely dominated row and a completely dominated column. It now has three Nash equilibria: (L, L), (M, M), (R, R), the first two of which are perfect equilibria, while the last one is not.[21] Hence, inclusion of completely dominated rows and columns could create additional perfect equilibria not present in the original game—a feature that is clearly not desirable. To remove this shortcoming of perfect equilibrium, Myerson (1978) has introduced the notion of *proper equilibrium*, which corresponds to a particular construction of the sequence of strategies used in Prop. 3.14. *Proper equilibrium* is defined as in Prop. 3.14, with only the ϵ_k*-perfect equilibrium* in ii) replaced by the notion of ϵ_k*-proper equilibrium* to be introduced next. Toward this end, let $\bar{J}^i(j; y_\epsilon)$ denote the average cost to $\mathbf{P}i$ when he uses his jth strategy (such as jth column or row of the matrix) in the game and all the other players use their mixed strategies y^k_ϵ, $k \in \mathbf{N}$, $k \neq i$. Furthermore, let $y^{i,j}_\epsilon$ be the probability attached to his jth strategy under the mixed strategy y^i_ϵ. Then, the N-tuple $\{y^i_\epsilon, i \in \mathbf{N}\}$ is said to be in ϵ*-proper equilibrium* if the strict inequality

$$\bar{J}^i(j; y_\epsilon) \quad > \quad \bar{J}^i(k; y_\epsilon)$$

implies that $y^{i,j}_\epsilon \le \epsilon\, y^{i,k}_\epsilon$, this being so for every j, $k \in \mathbf{M}_i$, and every $i \in \mathbf{N}$. In other words, an ϵ-proper equilibrium is one in which every player is giving his better responses much more probability weight than his worse responses (by a factor $1/\epsilon$), regardless of whether those "better" responses are "best" or not. Myerson (1978) proves that such an equilibrium necessarily exists as follows.

Proposition 3.15 *Every finite N-player game in normal form admits at least one proper equilibrium. Furthermore, every proper equilibrium is a perfect equilibrium (but not vice versa).*

---

[21]The reader is encouraged to verify this conclusion.

Remark 3.16 The reader should verify that in the matrix game (3.33) there is only one proper equilibrium, which is (L, L), the perfect equilibrium of (3.32). □

Another undesirable feature of a perfect equilibrium is that it is very much dependent on whether the game is in extensive or normal form (whereas the Nash equilibrium property is form-independent). As it has been first observed by Selten (1975), and further elaborated on by van Damme (1984), a perfect equilibrium of the extensive form of a game need not be perfect in the normal form, and conversely a perfect equilibrium of the normal form need not be perfect in the extensive form. To remove this undesirable feature, van Damme (1984) has introduced the concept of *quasi-perfect* equilibria for games in extensive form, and has shown that a proper equilibrium of a normal form game induces a quasi-perfect equilibrium in every extensive form game having this normal form. *Quasi-perfect equilibrium* is defined as a behavioral strategy combination which prescribes at every information set a choice that is optimal against mistakes ("trembling hands") of the other players; its difference from perfect equilibrium is that here in the construction of perturbed matrices each player ascribes "trembling hand" behavior to all other players (with positive probability), but not to himself.

Other types of refinement have also been proposed in the literature, such as *sequential equilibria* (Kreps and Wilson, 1982), and *strategic equilibria* (Kohlberg and Mertens, 1986), which we do not further discuss here. None of these, however, are uniformly powerful, in the sense of shrinking the set of Nash equilibria to the smallest possible set. We will revisit the topic of "refinement on Nash equilibria" later in Chapters 5 and 6, in the context of infinite dynamic games and with emphasis placed on the issue of time consistency.

3.6 The Stackelberg Equilibrium Solution

The Nash equilibrium solution concept that we have heretofore studied in this chapter provides a reasonable noncooperative equilibrium solution for nonzero-sum games when the roles of the players are symmetric, that is to say, when no single player dominates the decision process. However, there are yet other types of noncooperative decision problems wherein one of the players has the ability to enforce his strategy on the other player(s), and for such decision problems one has to introduce a hierarchical equilibrium solution concept. Following the original work of H. von Stackelberg (1934), the player who holds the powerful position in such a decision problem is called the *leader*, and the other players who react (rationally) to the leader's decision (strategy) are called the *followers*. There are, of course, cases of multi-levels of hierarchy in decision making, with many leaders and followers; but for purposes of brevity and clarity in exposition we will first confine our discussion here to hierarchical decision problems which incorporate two players (decision makers)—one leader and one follower.

To set the stage to introduce the hierarchical (Stackelberg) equilibrium solution concept, let us first consider the bimatrix game (A, B) displayed (under our standard convention) as

$$A = \begin{array}{c} \\ L \\ M \\ R \\ \\ \end{array} \overset{\textbf{P2}}{\begin{array}{|c|c|c|} \hline 0^{S_1} & 2 & 3/2^{S_2} \\ \hline 1 & 1^{N} & 3 \\ \hline -1 & 2 & 2 \\ \hline \multicolumn{1}{c}{L} & \multicolumn{1}{c}{M} & \multicolumn{1}{c}{R} \end{array}} \ \textbf{P1}, \qquad B = \begin{array}{c} \\ L \\ M \\ R \\ \\ \end{array} \overset{\textbf{P2}}{\begin{array}{|c|c|c|} \hline -1^{S_1} & 1 & -2/3^{S_2} \\ \hline 2 & 0^{N} & 1 \\ \hline 0 & 1 & -1/2 \\ \hline \multicolumn{1}{c}{L} & \multicolumn{1}{c}{M} & \multicolumn{1}{c}{R} \end{array}} \ \textbf{P1}. \tag{3.34}$$

This bimatrix game clearly admits a unique Nash equilibrium solution in pure strategies, which is $\{M, M\}$, with the corresponding outcome being $(1, 0)$. Let us now stipulate that the roles of the players are not symmetric and **P1** can enforce his strategy on **P2**. Then, before he announces his strategy, **P1** has to take into account possible responses of **P2** (the follower), and in view of this, he has to decide on the strategy that is most favorable to him. For the decision problem whose possible cost pairs are given as entries of A and B, above, let us now work out the reasoning that **P1** (the leader) will have to go through. If **P1** chooses L, then **P2** has a unique response (that minimizes his cost) which is L, thereby yielding a cost of 0 to **P1**. If the leader chooses M, **P2**'s response is again unique (which is M), with the corresponding cost incurred to **P1** being 1. Finally, if he picks R, **P2**'s unique response is also R, and the cost to **P1** is 2. Since the lowest of these costs is the first one, it readily follows that L is the most reasonable choice for the leader (**P1**) in this hierarchical decision problem. We then say that L is the Stackelberg strategy of the leader (**P1**) in this game, and the pair $\{L, L\}$ is the Stackelberg solution with **P1** as the leader. Furthermore, the cost pair $(0, -1)$ is the Stackelberg (equilibrium) outcome of the game with **P1** as the leader. It should be noted that this outcome is actually more favorable for both players than the unique Nash outcome—this latter feature, however, is not a rule in such games. If, for example, **P2** is the leader in the bimatrix game (3.34), and **P1** the follower, then the unique Stackelberg solution is $\{L, R\}$ with the corresponding outcome being $(3/2, -2/3)$ which is clearly not favorable for **P1** (the follower) when compared with his unique Nash cost. For **P2** (the leader), however, the Stackelberg outcome is again better than his Nash outcome.

The Stackelberg equilibrium solution concept introduced above within the context of the bimatrix game (3.34) is applicable to all two-person finite games in normal form, but provided that they exhibit one feature which was inherent to the bimatrix game (3.34) and was used implicitly in the derivation: *the follower's response to every strategy of the leader should be unique.* If this requirement is not satisfied, then there is ambiguity in the possible responses of the follower and thereby in the possible attainable cost levels of the leader. As an explicit example to demonstrate such a decision situation, consider the bimatrix game

$$A = \begin{array}{c} \\ L \\ R \\ \\ \end{array} \overset{\textbf{P2}}{\begin{array}{|c|c|c|} \hline 0 & 1 & 3 \\ \hline 2 & 2 & -1 \\ \hline \multicolumn{1}{c}{L} & \multicolumn{1}{c}{M} & \multicolumn{1}{c}{R} \end{array}} \ \textbf{P1}, \qquad B = \begin{array}{c} \\ L \\ R \\ \\ \end{array} \overset{\textbf{P2}}{\begin{array}{|c|c|c|} \hline 0 & 0 & 1 \\ \hline -1 & 0 & -1 \\ \hline \multicolumn{1}{c}{L} & \multicolumn{1}{c}{M} & \multicolumn{1}{c}{R} \end{array}} \ \textbf{P1} \tag{3.35}$$

and with **P1** acting as the leader. Here, if **P1** chooses (and announces) L, **P2** has two optimal responses L and M, whereas if **P1** picks R, **P2** again has two optimal responses, L and R. Since this multiplicity of optimal responses for the follower results in a multiplicity of cost levels for the leader for each one of his strategies, the Stackelberg solution concept introduced earlier cannot directly be applied here. However, this ambiguity in the attainable cost levels of the leader can be resolved if we stipulate that the leader's attitude is toward securing his possible losses against the choices of the follower within the class of his optimal responses, rather than toward taking risks. Then, under such a mode of play, **P1**'s secured cost level corresponding to his strategy L would be 1, and the one corresponding to R would be 2. Hence, we declare $\gamma^{1*} = L$ as the unique Stackelberg strategy of **P1** in the bimatrix game of (3.35), when he acts as the leader. The corresponding Stackelberg cost for **P1** (the leader) is $J^{1*} = 1$. It should be noted that, in the actual play of the game, **P1** could actually end up with a lower cost level, depending on whether the follower chooses his optimal response $\gamma^2 = L$ or the optimal response $\gamma^2 = M$. Consequently, the outcome of the game could be either $(1,0)$ or $(0,0)$, and hence we cannot talk about a unique Stackelberg equilibrium outcome of the bimatrix game (3.35) with **P1** acting as the leader. Before concluding our discussion on this example, we finally note that the admissible Nash outcome of the bimatrix game (3.35) is $(-1,-1)$ which is more favorable for both players than the possible Stackelberg outcomes given above.

We now provide a precise definition for the Stackelberg solution concept introduced above within the context of two bimatrix games, so as to encompass all two-person finite games of the single-act and multi-act type which do not incorporate any chance moves. For such a game, let Γ^1 and Γ^2 again denote the pure-strategy spaces of **P1** and **P2**, respectively, and $J^i(\gamma^1, \gamma^2)$ denote the cost incurred to **P**i corresponding to a strategy pair $\{\gamma^1 \in \Gamma^1, \gamma^2 \in \Gamma^2\}$. Then, we have the following definitions.

Definition 3.26 *In a two-person finite game, the set $R^2(\gamma^1) \subset \Gamma^2$ defined for each $\gamma^1 \in \Gamma^1$ by*

$$R^2(\gamma^1) = \{\xi \in \Gamma^2 : J^2(\gamma^1, \xi) \leq J^2(\gamma^1, \gamma^2), \quad \forall \gamma^2 \in \Gamma^2\} \tag{3.36}$$

is the optimal response (rational reaction) set *of* **P2** *to the strategy $\gamma^1 \in \Gamma^1$ of* **P1**.

Definition 3.27 *In a two-person finite game with* **P1** *as the leader, a strategy $\gamma^{1*} \in \Gamma^1$ is called a* Stackelberg equilibrium strategy *for the leader, if*

$$\max_{\gamma^2 \in R^2(\gamma^{1*})} J^1(\gamma^{1*}, \gamma^2) = \min_{\gamma^1 \in \Gamma^1} \max_{\gamma^2 \in R^2(\gamma^1)} J^1(\gamma^1, \gamma^2) \triangleq J^{1*}. \tag{3.37}$$

The quantity J^{1*} is the Stackelberg cost *of the leader. If, instead,* **P2** *is the leader, the same definition applies with only the superscripts 1 and 2 interchanged.*

Theorem 3.3 *Every two-person finite game admits a Stackelberg strategy for the leader.*

Proof. Since Γ^1 and Γ^2 are finite sets, and $R^2(\gamma^1)$ is a subset of Γ^2 for each $\gamma^1 \in \Gamma^1$, the result readily follows from (3.37). □

Remark 3.17 The Stackelberg strategy for the leader does not necessarily have to be unique. But nonuniqueness of the equilibrium strategy does not create any problem here (as it did in the case of Nash equilibria), since the Stackelberg cost for the leader is unique. □

Remark 3.18 If $R^2(\gamma^1)$ is a singleton for each $\gamma^1 \in \Gamma^1$, then there exists a mapping $T^2 : \Gamma^1 \to \Gamma^2$ such that $\gamma^2 \in R^2(\gamma^1)$ implies $\gamma^2 = T^2\gamma^1$. This corresponds to the case in which the optimal response of the follower (given by T^2) is unique for every strategy of the leader, and it leads to the following simplified version of (3.37) in Def. 3.27:

$$J^1(\gamma^{1*}, T^2\gamma^{1*}) = \min_{\gamma^1 \in \Gamma^1} J^1(\gamma^1, T^2\gamma^1) \triangleq J^{1*}. \tag{3.38}$$

Here J^{1*} is no longer only a secured equilibrium cost level for the leader (**P1**), but it is the cost level that is actually attained. □

From the follower's point of view, the equilibrium strategy in a Stackelberg game is any optimal response to the announced Stackelberg strategy of the leader. More precisely, we have the following.

Definition 3.28 *Let $\gamma^{1*} \in \Gamma^1$ be a Stackelberg strategy for the leader* (**P1**). *Then, any element $\gamma^{2*} \in R^2(\gamma^{1*})$ is an* optimal strategy *for the follower* (**P2**) *that is* in equilibrium *with γ^{1*}. The pair $\{\gamma^{1*}, \gamma^{2*}\}$ is a* Stackelberg solution *for the game with* **P1** *as the leader, and the cost pair $(J^1(\gamma^{1*}, \gamma^{2*}), J^2(\gamma^{1*}, \gamma^{2*}))$ is the corresponding* Stackelberg equilibrium outcome.

Remark 3.19 In the preceding definition, the cost level $J^1(\gamma^{1*}, \gamma^{2*})$ could in fact be lower than the Stackelberg cost J^{1*}—a feature which has already been observed within the context of the bimatrix game (3.35). However, if $R^2(\gamma^{1*})$ is a singleton, then these two cost levels have to coincide. □

For a given two-person finite game, let J^{1*} again denote the Stackelberg cost of the leader (**P1**), and J^1_N denote any Nash equilibrium cost for the same player. We have already seen within the context of the bimatrix game (3.35) that J^{1*} is not necessarily lower than J^1_N, in particular, when the optimal response of the follower is not unique. The following proposition now provides one sufficient condition under which the leader never does worse in a "Stackelberg game" than in a "Nash game".

Proposition 3.16 *For a given two-person finite game, let* J^{1*} *and* J^1_N *be as defined before. If* $R^2(\gamma^1)$ *is a singleton for each* $\gamma^1 \in \Gamma^1$, *then*

$$J^{1*} \leq J^1_N.$$

Proof. Under the hypothesis of the proposition, assume to the contrary that there exists a Nash equilibrium solution $\{\gamma^{1\circ} \in \Gamma_1, \gamma^{2\circ} \in \Gamma^2\}$ whose corresponding cost to **P1** is lower than J^{1*}, i.e.,

$$J^{1*} > J^1(\gamma^{1\circ}, \gamma^{2\circ}). \qquad (i)$$

Since $R^2(\gamma^1)$ is a singleton, let $T^2 : \Gamma^2 \to \Gamma^1$ be the unique mapping introduced in Remark 3.18. Then, clearly, $\gamma^{2\circ} = T^2\gamma^{1\circ}$, and if this is used in (i), together with the RHS of (3.38), we obtain

$$\min_{\gamma^1 \in \Gamma^1} J^1(\gamma^1, T^2\gamma^1) = J^{1*} > J^1(\gamma^{1\circ}, T^2\gamma^{1\circ}),$$

which is a contradiction. □

Remark 3.20 One might be tempted to think that if a nonzero-sum game admits a unique Nash equilibrium solution and a unique Stackelberg strategy (γ^{1*}) for the leader, and further if $R^2(\gamma^{1*})$ is a singleton, then the inequality of Prop. 3.16 still should hold. This, however, is not true as the following bimatrix game demonstrates:

$A =$ (P2 columns L, R; rows L, R for P1)

	L	R
L	0^{S_1}	1
R	-1^{N}	2

P1, $B =$

	L	R
L	0^{S_1}	2
R	1^{N}	1

P1.

Here, there exists a unique Nash equilibrium solution, as indicated, and a unique Stackelberg strategy $\gamma^{1*} = L$ for the leader (**P1**). Furthermore, the follower's optimal response to $\gamma^{1*} = L$ is unique (which is $\gamma^2 = L$). However, $0 = J^{1*} > J^1_N = -1$. This counterexample indicates that the sufficient condition of Prop. 3.16 cannot be relaxed any further in any satisfactory way. □

Since the Stackelberg equilibrium concept is nonsymmetric as to the roles of the players, there is no recursive procedure that can be developed (as in Section 3.5) to determine the Stackelberg solution of dynamic games. The only possibility is the brute-force method, that is, to convert the original two-person finite dynamic game in extensive form into an equivalent normal form (which is basically a bimatrix game), to exhaustively work out the possible outcomes of the game for each strategy choice of the leader, and to see which of these is the most favorable one for the leader. Since this is a standard application of Def. 3.27 and is no different from the approach adopted in solving the bimatrix games (3.34) and (3.35), we do not provide any further examples here. However, the reader is referred to Section 3.8 for problems that involve derivation of Stackelberg equilibria in single-act and multi-act dynamic games.

The feedback (stagewise) Stackelberg solution

Within the context of multi-act dynamic games, the Stackelberg equilibrium concept is suitable for the class of decision problems in which the leader has the ability to announce his decisions at all of his possible information sets ahead of time—an attitude which naturally forces the leader to commit himself irrevocably to the actions dictated by these strategies. Hence, in a sense, the Stackelberg solution is of the prior commitment type from the leader's point of view. There are yet other types of hierarchical multi-act (multi-stage) decision problems, however, in which the leader does not have the ability to announce and enforce his strategy at all levels of play prior to the start of the game, but can instead enforce his strategy on the follower(s) at every level (stage) of the game. Such a hierarchical equilibrium solution, which has the "Stackelberg" property at every level of play (but not globally), is called a *feedback Stackelberg solution.*[22] We now provide below a precise definition of this equilibrium solution within the context of two-person nonzero-sum feedback games in extensive form (cf. Def. 3.21). To this end, let us adopt the terminology and notation of Section 3.5, and consider **I** to be a two-person nonzero-sum feedback game in extensive form with K levels of play and with **P1** being the first-acting player at each level of play. Let a typical strategy $\gamma^i \in \Gamma^i$ of $\mathbf{P}i$ be decomposed as $(\gamma_1^i \in \Gamma_1^i, \ldots, \gamma_K^i \in \Gamma_K^i)$, where γ_j^i stands for the corresponding strategy of $\mathbf{P}i$ at the jth level of play. Furthermore, for a given strategy $\gamma^i \in \Gamma^i$, let $\gamma^{i,k}$ and $\gamma_{i,k}$ denote two truncated versions of γ^i, defined as

$$\gamma^{i,k} = (\gamma_k^i, \ldots, \gamma_K^i) \tag{3.39a}$$

and

$$\gamma_{i,k} = (\gamma_1^i, \ldots, \gamma_{k-1}^i). \tag{3.39b}$$

Then, we have the following.

Definition 3.29 *Using the preceding notation and terminology, a pair* $\{\beta^1 \in \Gamma^1, \beta^2 \in \Gamma^2\}$ *constitutes a* feedback Stackelberg solution *for* **I** *with* **P1** *as the leader if the following two conditions are fulfilled:*

(i)

$$\begin{aligned} &J^1(\gamma_{1,k}, \beta^{1,k}; \gamma_{2,k}, \beta^{2,k}) \\ &= \min_{\gamma_k^1 \in \Gamma_k^1} \max_{\gamma_k^2 \in R_k^2(\beta;\gamma_{1,k+1})} J^1(\gamma_{1,k}, \gamma_k^1, \beta^{1,k+1}; \gamma_{2,k}, \gamma_k^2, \beta^{2,k+1}) \end{aligned} \tag{3.40a}$$

for all $\gamma_j^i \in \Gamma_j^i$, $j = 1, \ldots, k-1$; $i = 1, 2$; *and with* $k = K, K-1, \ldots, 1$, *where* $R_k^2(\beta; \gamma_{1,k+1})$ *is the optimal response set of the follower at level* k,

[22] We shall henceforth refer to the Stackelberg equilibrium solution introduced through Defs. 3.26-3.28 as the *global Stackelberg* equilibrium solution whenever it is not clear from the context which type of equilibrium solution we are working with.

defined by

$$R_k^2(\beta;\gamma_{1,k+1}) \triangleq \{\mathring{\gamma}_k^2 \in \Gamma_k^2 : J^2(\gamma_{1,k},\gamma_k^1,\beta^{1,k+1};\gamma_{2,k},\mathring{\gamma}_k^2,\beta^{2,k+1}) = \min_{\gamma_k^2\in\Gamma_k^2} J^2(\gamma_{1,k},\gamma_k^1,\beta^{1,k+1};\gamma_{2,k},\gamma_k^2,\beta^{2,k+1})\}. \quad (3.40b)$$

(ii) $R_k^2(\beta;\beta_k^1,\gamma_{1,k})$ *is a singleton, with its only element being* β_k^2, $k = 1,\ldots,K$.

The quantity $J^1(\beta^1,\beta^2)$ *is the corresponding* feedback Stackelberg cost *of the leader.*

Relations (3.40a) and (3.40b) now indicate that the feedback Stackelberg solution features the "Stackelberg" property at every level of play, and this can in fact be verified recursively—thus leading to a recursive construction of the equilibrium solution. Before we go into a discussion of this recursive procedure, let us first prove a result concerning the structures of $R_k^2(\cdot)$ and $J^1(\cdot)$.

Proposition 3.17 *For a two-person feedback game that admits a feedback Stackelberg solution with* **P1** *as the leader,*

(i) $R_k^2(\beta;\gamma_{1,k+1})$ *is not a variant of* $\gamma_{1,k}$, *and hence it can be written as* $R_k^2(\beta;\gamma_k^1)$,

(ii) $J^1(\gamma_{1,k},\beta^{1,k};\gamma_{2,k},\beta^{2,k})$ *depends on* $\gamma_{1,k}$ *and* $\gamma_{2,k}$ *only through the singleton information sets* η_k^1 *of* **P1** *at level* k.

Proof. Both of these results follow from the "feedback" nature of the two-person dynamic game under consideration, which enables one to decompose the K-level game recursively into a number of single-act games each of which corresponds to a singleton information set of **P1** (the leader). In accordance with this observation, let us start with $k = K$, in which case the proposition is trivially true. Relation (3.40a) then defines the (global) Stackelberg solution of the single-act game obtained for each information set η_k^1 of **P1**, and the quantity on the RHS of (3.40a) defines, for $k = K$, the Stackelberg cost for the leader in each of these single-act games, which is actually realized because of condition (ii) of Def. 3.29 (see Remark 3.19). Then, the cost pair transferred to level $k = K - 1$ is well defined (not ambiguous) for each information set η_K^1 of **P1**. Now, with the Kth level crossed out, we repeat the same analysis and first observe that $J^2(\cdot)$ in (3.40) depends only on the information set η_{K-1}^1 of **P1** and on the strategies γ_{K-1}^1 and γ_{K-1}^2 of **P1** and **P2**, respectively, at level $k = K-1$, since we again have essentially single-act games. Consequently, $R_{K-1}^2(\cdot)$ depends only on γ_{K-1}^1. If the cost level of **P1** at $k = K-1$ is minimized subject to that constraint, β_{K-1}^1 is by hypothesis a solution to that problem, to which a unique response of the follower corresponds. Hence, again a well-defined cost pair is determined for each game corresponding to the singleton information sets η_{K-1}^1 of **P1**, and these are attached to these nodes to provide cost pairs for the remaining $(K-2)$-level game. An iteration on this analysis then proves the result. □

The foregoing proof already describes the recursive procedure involved to obtain a feedback Stackelberg solution of a two-person feedback game. The requirement (ii) in Def. 3.29 ensures that the outcomes of the Stackelberg games considered at each level are unambiguously determined, and it imposes an indirect restriction on the problem under consideration, which cannot be tested *a priori*. This then leads to the conclusion that a two-person nonzero-sum feedback game does not necessarily admit a feedback Stackelberg solution.

There are also cases in which a feedback game admits more than one feedback Stackelberg solution. Such a situation might arise if, in the recursive derivation, one or more of the single-act Stackelberg games admit multiple Stackelberg strategies for the leader whose corresponding costs to the follower are not the same. The leader is, of course, indifferent to these multiple equilibria at that particular level where they arise, but this nonuniqueness might affect the (overall) feedback Stackelberg cost of the leader. Hence, if $\{\beta^1, \beta^2\}$ and $\{\xi^1, \xi^2\}$ are two feedback Stackelberg solutions of a given game, it could very well happen that $J^1(\beta^1, \beta^2) < J^1(\xi^1, \xi^2)$, in which case the leader definitely prefers β^1 over ξ^1 and attempts to enforce the components of β^1 at each level of the game. This argument then leads to a total ordering among different feedback Stackelberg solutions of a given game, which we formalize below.

Definition 3.30 *A feedback Stackelberg solution $\{\beta^1, \beta^2\}$ of a two-person nonzero-sum feedback game is* admissible *if there exists no other feedback Stackelberg solution $\{\xi^1, \xi^2\}$ with the property $J^1(\xi^1, \xi^2) < J^1(\beta^1, \beta^2)$.*

It should be noted that if a feedback game admits more than one feedback Stackelberg solution, then the leader can definitely enforce the admissible one on the follower since he leads the decision process at each level. A unique feedback Stackelberg solution is clearly admissible. We now provide below an example of a 2-level two-person nonzero-sum feedback game which admits a unique feedback Stackelberg solution. This example also illustrates recursive derivation of this equilibrium solution.

Example 3.13 Consider the multi-act feedback nonzero-sum game of Fig. 3.9. To determine its feedback Stackelberg solution with **P1** acting as the leader, we first attempt to solve the Stackelberg games corresponding to each of the four information sets of **P1** at level 2. The first one (counting from the left) is a trivial game, admitting the unique equilibrium solution

$$\gamma_2^{1*}(\eta_2^1) = L, \quad \gamma_2^{2*}(\eta_2^2) = \begin{cases} R & \text{if } u_2^1 = L, \\ L & \text{otherwise}, \end{cases} \tag{i}$$

with the corresponding outcome being $(1, 0)$. The next two admit, respectively, the following bimatrix representations whose unique Stackelberg equilibria are indicated on the matrices:

$$A = \begin{array}{c|c|c|} \multicolumn{3}{c}{\mathbf{P2}} \\ \cline{2-3} L & 3 & 4 \\ \cline{2-3} R & \textcircled{1} & 5 \\ \cline{2-3} \multicolumn{1}{c}{} & \multicolumn{1}{c}{L} & \multicolumn{1}{c}{R} \end{array} \ \mathbf{P1}, \qquad B = \begin{array}{c|c|c|} \multicolumn{3}{c}{\mathbf{P2}} \\ \cline{2-3} L & 1 & 2 \\ \cline{2-3} R & \textcircled{1} & 2 \\ \cline{2-3} \multicolumn{1}{c}{} & \multicolumn{1}{c}{L} & \multicolumn{1}{c}{R} \end{array} \ \mathbf{P1}, \qquad (ii)$$

$$A = \begin{array}{c|c|c|} \multicolumn{3}{c}{\mathbf{P2}} \\ \cline{2-3} L & \textcircled{0} & 2 \\ \cline{2-3} R & 2 & 2 \\ \cline{2-3} \multicolumn{1}{c}{} & \multicolumn{1}{c}{L} & \multicolumn{1}{c}{R} \end{array} \ \mathbf{P1}, \qquad B = \begin{array}{c|c|c|} \multicolumn{3}{c}{\mathbf{P2}} \\ \cline{2-3} L & \textcircled{$-\frac{1}{2}$} & 1 \\ \cline{2-3} R & 1 & 0 \\ \cline{2-3} \multicolumn{1}{c}{} & \multicolumn{1}{c}{L} & \multicolumn{1}{c}{R} \end{array} \ \mathbf{P1}. \qquad (iii)$$

Finally, the fourth one admits the trivial bimatrix representation

$$A = \begin{array}{c|c|c|} \multicolumn{3}{c}{\mathbf{P2}} \\ \cline{2-3} L & \textcircled{0} & \text{-2} \\ \cline{2-3} \multicolumn{1}{c}{} & \multicolumn{1}{c}{L} & \multicolumn{1}{c}{R} \end{array} \ \mathbf{P1}, \qquad B = \begin{array}{c|c|c|} \multicolumn{3}{c}{\mathbf{P2}} \\ \cline{2-3} L & \textcircled{-1} & 1 \\ \cline{2-3} \multicolumn{1}{c}{} & \multicolumn{1}{c}{L} & \multicolumn{1}{c}{R} \end{array} \ \mathbf{P1} \qquad (iv)$$

together with the 2×2 bimatrix representation

$$A = \begin{array}{c|c|c|} \multicolumn{3}{c}{\mathbf{P2}} \\ \cline{2-3} M & 3 & 0 \\ \cline{2-3} R & 2 & \textcircled{-1} \\ \cline{2-3} \multicolumn{1}{c}{} & \multicolumn{1}{c}{M} & \multicolumn{1}{c}{R} \end{array} \ \mathbf{P1}, \qquad B = \begin{array}{c|c|c|} \multicolumn{3}{c}{\mathbf{P2}} \\ \cline{2-3} L & 2 & 3 \\ \cline{2-3} R & 1 & \textcircled{0} \\ \cline{2-3} \multicolumn{1}{c}{} & \multicolumn{1}{c}{L} & \multicolumn{1}{c}{R} \end{array} \ \mathbf{P1} \qquad (v)$$

with the unique Stackelberg solutions being as indicated, in each case. By a comparison of the leader's cost levels at (iv) and (v), we conclude that he plays R at the fourth of his information sets at level 2. Hence, in going from the second to the first level, we obtain from (i), (ii), (iii) and (v) the unique cost pairs at each of **P1**'s information sets to be (counting from the left): $(1,0)$, $(1,1)$, $(0,-1/2)$ and $(-1,0)$. Then, the Stackelberg game at the first level has the bimatrix representation

$$A = \begin{array}{c|c|c|} \multicolumn{3}{c}{\mathbf{P2}} \\ \cline{2-3} L & 1 & 1 \\ \cline{2-3} R & \textcircled{0} & -1 \\ \cline{2-3} \multicolumn{1}{c}{} & \multicolumn{1}{c}{L} & \multicolumn{1}{c}{R} \end{array} \ \mathbf{P1}, \qquad B = \begin{array}{c|c|c|} \multicolumn{3}{c}{\mathbf{P2}} \\ \cline{2-3} L & 0 & 1 \\ \cline{2-3} R & \textcircled{$-\frac{1}{2}$} & 0 \\ \cline{2-3} \multicolumn{1}{c}{} & \multicolumn{1}{c}{L} & \multicolumn{1}{c}{R} \end{array} \ \mathbf{P1} \qquad (vi)$$

whose unique Stackelberg solution with **P1** as leader is as indicated.

To recapitulate, the leader's unique feedback Stackelberg strategy in this feedback game is

$$\gamma_1^{1*}(\eta_1^1) = R, \qquad \gamma_2^{1*}(\eta_2^1) = u_1^2,$$

and the follower's unique optimal response strategy is

$$\gamma_1^{2*}(\eta_1^2) = L, \quad \gamma_2^{2*}(\eta_2^2) = \begin{cases} R & \text{if } u_1^1 = L \text{ and } u_2^1 = L, \text{ or } u_1^1 = R \\ & \qquad \text{and } u_1^2 = R \text{ and } u_2^1 \neq L, \\ L & \text{otherwise} . \end{cases}$$

The unique feedback Stackelberg outcome of the game is $(0, -1/2)$, which follows from (vi) by inspection. □

Before concluding our discussion on the feedback Stackelberg equilibrium solution, we should mention that the same concept has potential applicability to multi-act decision problems in which the roles of the players change from one level of play to another, that is, when different players act as leaders at different stages. An extension of Def. 3.29 is possible so that such classes of decision problems are also covered; but since this involves only an appropriate change of indices and an interpretation in the right framework, it will not be covered here.

Mixed and behavioral Stackelberg equilibria

The motivation behind introducing mixed strategies in the investigation of saddle-point equilibria (in Chapter 2) and Nash equilibria (in Section 3.2) was that such equilibria do not always exist in pure strategies, whereas within the enlarged class of mixed strategies one can ensure existence of noncooperative equilibria. In the case of the Stackelberg solution of two-person finite games, however, an equilibrium always exists (cf. Thm. 3.3), and thus, at the outset, there seems to be no need to introduce mixed strategies. Besides, since the leader dictates his strategy on the follower, in a Stackelberg game, it might at first seem to be unreasonable to imagine that the leader would ever employ a mixed strategy. Such an argument, however, is not always valid, and there are cases in which the leader can actually do better (in the average sense) with a proper mixed strategy, than the best he can do within the class of pure strategies. As an illustration of such a possibility, consider the bimatrix game (A, B) displayed below:

$$A = \begin{array}{c} \mathbf{P2} \\ \begin{array}{c|c|c|} & L & R \\ \hline L & 1 & 0 \\ \hline R & 0 & 1 \\ \hline \end{array} \end{array} \mathbf{P1}, \qquad B = \begin{array}{c} \mathbf{P2} \\ \begin{array}{c|c|c|} & L & R \\ \hline L & 1/2 & 1 \\ \hline R & 1 & 1/2 \\ \hline \end{array} \end{array} \mathbf{P1}. \tag{3.41}$$

If **P1** acts as the leader, then the game admits two pure-strategy Stackelberg equilibrium solutions, which are $\{L, L\}$ and $\{R, R\}$, the Stackelberg outcome in each case being $(1, 1/2)$. However, if the leader (**P1**) adopts the mixed strategy which is to pick L and R with equal probability $1/2$, then the average cost incurred to **P1** will be equal to $1/2$, quite independent of the follower's (pure or mixed) strategy. This value $\bar{J}^1 = 1/2$ is clearly lower than the leader's Stackelberg cost in pure strategies, which can further be shown to be the unique Stackelberg cost of the leader in mixed strategies, since any deviation from $(1/2, 1/2)$ for the leader results in higher values for $\bar{J}^1$, by taking into account the optimal responses of the follower.

The preceding result then establishes the significance of mixed strategies in the investigation of Stackelberg equilibria of two-person nonzero-sum games, and demonstrates the possibility that a proper mixed-strategy Stackelberg solution could lead to a lower cost level for the leader than the Stackelberg cost level in pure strategies. To introduce the concept of mixed-strategy Stackelberg equilibrium in mathematical terms, we take the two-person nonzero-sum finite game to be in normal form (without any loss of generality) and associate with it a bimatrix game (A, B). Abiding by the notation and terminology of Section 3.2, we let Y and Z denote the mixed-strategy spaces of **P1** and **P2**, respectively, with their typical elements designated as y and z. Then, we have the following.

Definition 3.31 *For a bimatrix game (A, B), the set*

$$\bar{R}^2(y) = \{z^\circ \in Z : y'Bz^\circ \leq y'Bz, \forall z \in Z\} \tag{3.42}$$

is the optimal response *(*rational reaction*)* set *of* **P2** in mixed strategies *to the mixed strategy $y \in Y$ of* **P1**.

Definition 3.32 *In a bimatrix game (A, B) with* **P1** *acting as the leader, a mixed strategy $y^* \in Y$ is called a* mixed Stackelberg equilibrium strategy *for the leader if*

$$\max_{z \in \bar{R}^2(y^*)} y^{*\prime}Az = \inf_{y \in Y} \max_{z \in \bar{R}^2(y)} y'Az \triangleq \bar{J}^{1*}. \tag{3.43}$$

The quantity $\bar{J}^{1*}$ is the Stackelberg cost *of the leader in mixed strategies.*

It should be noted that the "maximum" in (3.43) always exists since, for each $y \in Y$, $y'Az$ is continuous in z, and $\bar{R}^2(y)$ is a closed and bounded subset of Z (which is a finite dimensional simplex). Hence, $\bar{J}^{1*}$ is a well-defined quantity. The "infimum" in (3.43), however, cannot always be replaced by a "minimum", unless the problem admits a mixed Stackelberg equilibrium strategy for the leader. The following example now demonstrates the possibility that a two-person finite game might not admit a mixed-strategy Stackelberg strategy even though $\bar{J}^{1*} < J^{1*}$.

Example 3.14 Consider the following modified version of the bimatrix game of (3.41):

$A =$ (**P2** columns L, R; **P1** rows L, R)

	L	R
L	1	0
R	0	1

P1, $B =$ (**P2** columns L, R; **P1** rows L, R)

	L	R
L	1/2	1
R	1	1/3

P1.

With **P1** as the leader, this bimatrix game also admits two pure-strategy Stackelberg equilibria, which are $\{L, L\}$ and $\{R, R\}$, the Stackelberg cost for the leader being $J^{1*} = 1$. Now, let the leader adopt the mixed strategy $y = (y_1, (1 - y_1))'$, under which $\bar{J}^2$ is

$$\bar{J}^2(y, z) = y'Bz = \left(-\frac{7}{6}y_1 + \frac{2}{3}\right) z_1 + \frac{2}{3}y_1 + \frac{1}{3},$$

where $z = (z_1, (1 - z_1))'$ denotes any mixed strategy of **P2**. Then, the mixed-strategy optimal response set of **P2** can readily be determined as

$$\bar{R}^2(y) = \begin{cases} \{z = (1, 0)\} & \text{if } y_1 > 4/7, \\ \{z = (0, 1)\} & \text{if } y_1 < 4/7, \\ Z & \text{if } y_1 = 4/7. \end{cases}$$

Hence, for $y_1 > 4/7$, the follower chooses "column 1" with probability 1, and this leads to an average cost of $\bar{J}^1 = y_1$ for **P1**. For $y_1 < 4/7$, on the other hand, **P2** chooses "column 2" with probability 1, which leads to an average cost level of $\bar{J}^1 = (1-y_1)$ for **P1**. Then, clearly, the leader will prefer to stay in this latter region; in fact, if he employs the mixed strategy $y = (4/7-\epsilon, 3/7+\epsilon)'$ where $\epsilon > 0$ is sufficiently small, his realized average cost will be $\bar{J}^1 = 3/7+\epsilon$, since then **P2** will respond with the unique pure-strategy $\gamma^2 = R$. Since $\epsilon > 0$ can be taken as small as possible, we arrive at the conclusion that $\bar{J}^{1*} = \frac{3}{7} < 1 = J^{1*}$. In spite of this fact, the leader does not have a mixed Stackelberg strategy since, for the only candidate $y^\circ = (4/7, 3/7)$, $\bar{R}^2(y^\circ) = Z$, and therefore $\max_{z\in\bar{R}^2(y^\circ)} y^{\circ\prime} Az = 4/7$ which is higher than $\bar{J}^{1*}$. □

The preceding example thus substantiates the possibility that a mixed Stackelberg strategy might not exist for the leader, but he can still do better than his pure Stackelberg cost J^{1*} by employing some sub-optimal mixed strategy (such as the one $y = (4/7-\epsilon, 3/7+\epsilon)'$ in Example 3.14, for sufficiently small $\epsilon > 0$). In fact, whenever $\bar{J}^{1*} < J^{1*}$, there will always exist such an approximating mixed strategy for the leader. If $\bar{J}^{1*} = J^{1*}$, however, it is, of course, reasonable to employ the pure Stackelberg strategy which always exists by Thm. 3.3. The following proposition now verifies that $\bar{J}^{1*} < J^{1*}$ and $\bar{J}^{1*} = J^{1*}$ are the only two possible relations we can have between $\bar{J}^{1*}$ and J^{1*}; in other words, the inequality $\bar{J}^{1*} > J^{1*}$ never holds.

Proposition 3.18 *For every two-person finite game, we have*

$$\bar{J}^{1*} \leq J^{1*}. \tag{3.44}$$

Proof. Let Y_0 denote the subset of Y consisting of all one-point distributions. Analogously, define Z_0 as comprised of one-point distributions in Z. Note that Y_0 is equivalent to Γ^1, and Z_0 is equivalent to Γ^2. Then, for each $y \in Y_0$

$$\min_{z\in Z} y'Bz = \min_{z\in Z_0} y'Bz$$

since any minimizing solution in Z can be replaced by an element of Z_0. This further implies that, for each $y \in Y_0$, elements of $\bar{R}^2(y)$ are probability distributions on $R^2(y)$, where the latter set is defined by (3.42) with Z replaced by Z_0. Now, since $Y_0 \subset Y$,

$$\bar{J}^{1*} = \min_{y\in Y} \max_{z\in\bar{R}^2(y)} y'Az \leq \min_{y\in Y_0} \max_{z\in\bar{R}^2(y)} y'Az,$$

and further, because of the cited relation between $\bar{R}^2(\cdot)$ and $R^2(\cdot)$, the latter quantity is equal to

$$\min_{y\in Y_0} \max_{z\in R^2(y)} y'Az$$

which, by definition, is J^{1*}, since $R^2(y)$ is equivalent to the pure-strategy optimal response set of the follower, as defined by (3.36). Hence, $\bar{J}^{1*} \leq J^{1*}$. □

Computation of a mixed-strategy Stackelberg equilibrium (whenever it exists) is not as straightforward as in the case of pure-strategy equilibria, since the spaces Y and Z are not finite. The standard technique is first to determine the minimizing solution(s) of

$$\min_{z \in Z} y' B z$$

as functions of $y \in Y$. This will lead to a decomposition of Y into subsets (regions), on each of which a reaction set for the follower is defined. (Note that in the analysis of Example 3.14, Y has been decomposed into three regions.) Then, one has to minimize $y'Az$ over $y \in Y$, subject to the constraints imposed by these reaction sets, and under the stipulation that the same quantity is maximized on these reaction sets whenever they are not singletons. This brute-force approach also provides approximating strategies for the leader, whenever a mixed Stackelberg solution does not exist, together with the value of $\bar{J}^{1*}$.

If the two-person finite game under consideration is a dynamic game in extensive form, then it is more reasonable to restrict attention to behavioral strategies (cf. Def. 2.9). Stackelberg equilibria within the class of behavioral strategies can be introduced as in Defs. 3.31 and 3.32, by replacing the mixed-strategy sets with the behavioral-strategy sets. Hence, using the terminology and notation of Def. 2.9, we have the following counterparts of Defs. 3.31 and 3.32, in behavioral strategies.

Definition 3.33 *Given a two-person finite dynamic game with behavioral-strategy sets $(\hat{\Gamma}^1, \hat{\Gamma}^2)$ and average cost functions $(\hat{J}^1, \hat{J}^2)$ the set*

$$\hat{R}^2(\hat{\gamma}^1) = \{\hat{\gamma}^{2o} \in \hat{\Gamma}^2 : \hat{J}^2(\hat{\gamma}^1, \hat{\gamma}^{2o}) \leq \hat{J}^2(\hat{\gamma}^1, \hat{\gamma}^2), \forall \hat{\gamma}^2 \in \hat{\Gamma}^2\}, \tag{3.45}$$

is the optimal response (rational reaction) set *of* **P2** *in* behavioral strategies *to the behavioral strategy $\hat{\gamma}^1 \in \hat{\Gamma}^1$ of* **P1**.

Definition 3.34 *In a two-person finite dynamic game with* **P1** *acting as the leader, a behavioral strategy $\hat{\gamma}^{1*} \in \hat{\Gamma}^1$ is called a* behavioral Stackelberg equilibrium *strategy for the leader if*

$$\sup_{\hat{\gamma}^2 \in \hat{R}^2(\hat{\gamma}^{1*})} \hat{J}^1(\hat{\gamma}^{1*}, \hat{\gamma}^2) = \inf_{\hat{\gamma}^1 \in \hat{\Gamma}^1} \sup_{\hat{\gamma}^2 \in \hat{R}^2(\hat{\gamma}^1)} \hat{J}^1(\hat{\gamma}^1, \hat{\gamma}^2) \triangleq \hat{J}^{1*}. \tag{3.46}$$

The quantity $\hat{J}^{1*}$ is the Stackelberg cost *of the leader in* behavioral strategies.

Remark 3.21 The reason why we use "supremum" in (3.46), instead of "maximum", is because the set $\hat{R}^2(\hat{\gamma}^1)$ is not necessarily compact. The behavioral Stackelberg cost for the leader is again a well-defined quantity, but a behavioral Stackelberg strategy for the leader does not necessarily exist. □

We now conclude our discussion on the notion of behavioral Stackelberg equilibrium by presenting an example of a single-act dynamic game that admits a Stackelberg solution in behavioral strategies.

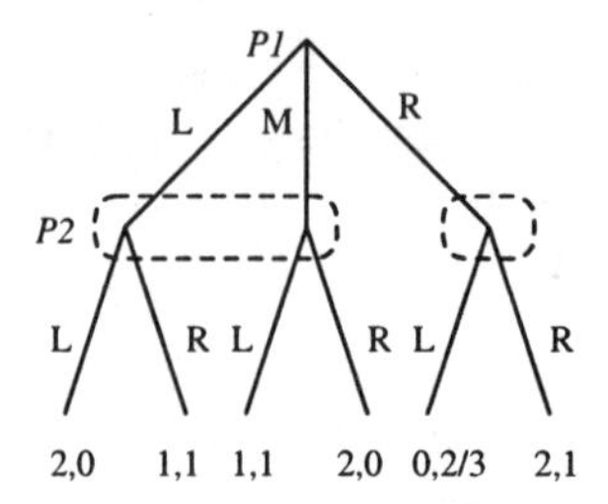

Figure 3.10: The two-person single-act game of Example 3.15.

Example 3.15 Consider the two-person single-act game whose extensive form is depicted in Fig. 3.10. With **P2** acting as the leader, it admits two pure-strategy Stackelberg solutions:

$$\gamma^{2*}(\eta^2) = \begin{cases} L & \text{if } u^1 = R, \\ R & \text{otherwise} \end{cases}, \quad \gamma^{1*}(\eta^1) = R,$$

and

$$\gamma^{2*}(\eta^2) = L, \quad \gamma^{1*}(\eta^1) = R, v$$

with the unique Stackelberg outcome in pure strategies being $(0, \frac{2}{3})$. Now, let us allow the leader to use behavioral strategies. The bimatrix game at the first (counting from the left) of the leader's information sets is

$$A = \begin{array}{c} \\ L \\ M \\ \end{array} \overset{\mathbf{P2}}{\begin{array}{|c|c|} \hline 2 & 1 \\ \hline 1 & 2 \\ \hline \end{array}} \ \mathbf{P1}, \qquad B = \begin{array}{c} \\ L \\ M \\ \end{array} \overset{\mathbf{P2}}{\begin{array}{|c|c|} \hline 0 & 1 \\ \hline 1 & 0 \\ \hline \end{array}} \ \mathbf{P1},$$
$$\qquad\quad L \quad R \qquad\qquad\qquad\qquad L \quad R$$

for which $\bar{J}^{2*} = 1/2$. The mixed Stackelberg strategy exists for the leader, given as

$$\bar{\gamma}^2 = \begin{cases} L & \text{w.p.} \quad 1/2, \\ R & \text{w.p.} \quad 1/2, \end{cases}$$

and any mixed strategy is an optimal response strategy for **P1** (the follower) yielding him an average cost level of $\bar{J}^1 = 3/2$.

We thus observe that if the leader can force the follower to play L or M, his average cost can be pushed down to $1/2$ which is lower than $J^{2*} = 2/3$, and the follower's cost is then pushed up to $3/2$ which is higher than the cost level of $J^1 = 0$ which would have been attained if **P2** were allowed to use only pure strategies. There indeed exists such a (behavioral) strategy for **P2**, which is given as

$$\hat{\gamma}^{2*}(\eta^2) = \begin{cases} \left.\begin{matrix} L & \text{w.p. } 1/2 \\ R & \text{w.p. } 1/2 \end{matrix}\right\} & \text{if } u^1 = L \text{ or } M, \\ R \quad \text{w.p. } 1 & \text{if } u^1 = R, \end{cases} \qquad (i)$$

and the behavioral Stackelberg cost for the leader is

$$\hat{J}^{2*} = 1/2. \qquad (ii)$$

It should be noted that the follower (**P1**) does not dare to play $u^1 = R$ since, by announcing (i) ahead of time, the leader threatens him with an undesirable cost level of $J^1 = 2$.

We now leave it to the reader to verify that the behavioral strategy (i) is indeed a Stackelberg strategy for the leader and the quantity (ii) is his Stackelberg cost in behavioral strategies. This verification can be accomplished by converting the original game in extensive form into equivalent normal form and then applying Def. 3.34. Another alternative method is to show that (ii) is the mixed-strategy Stackelberg cost for the leader, and since (i) attains that cost level, it is also a behavioral Stackelberg strategy, by the result of Problem 21 (Section 3.8). □

Many-player games

Heretofore, we have confined our investigation on the Stackelberg concept in nonzero-sum games to two-person games wherein one of the players is the leader and the other one is the follower. Extension of this investigation to N-player games could be accomplished in a number of ways; here we discuss the possible extensions to the case when $N = 3$, with the further generalizations to $N > 3$ being conceptually straightforward.

For three-person nonzero-sum games, we basically have three different possible modes of play among the players within the framework of noncooperative decision making.

(1) There are two levels of hierarchy in decision making—one leader and two followers. The followers react to the leader's announced strategy by playing according to a specific equilibrium concept among themselves (for instance, Nash).

(2) There are still two levels of hierarchy in decision making—now two leaders and one follower. The leaders play according to a specific equilibrium concept among themselves, by taking into account possible optimal responses of the follower.

(3) There are three levels of hierarchy. First **P1** announces his strategy, then **P2** determines his strategy by also taking into account possible responses of **P3** and enforces this strategy on him, and finally **P3** optimizes his cost function in view of the announced strategies of **P1** and **P2**.

Let us now provide precise definitions for the three types of hierarchical equilibria outlined above. To this end, let $J^i(\gamma^1, \gamma^2, \gamma^3)$ denote the cost function of **P**i corresponding to the strategy triple $\{\gamma^1 \in \Gamma^1, \gamma^2 \in \Gamma^2, \gamma^3 \in \Gamma_3\}$. Then, we have the following.

Definition 3.35 *For the three-person finite game with one leader* (**P1**) *and two followers (***P2** *and* **P3***),* $\gamma^{1*} \in \Gamma^1$ *is a* hierarchical equilibrium strategy *for the leader if*

$$\max_{(\gamma^2,\gamma^3)\in R^F(\gamma^{1*})} J^1(\gamma^{1*}, \gamma^2, \gamma^3) = \min_{\gamma^1\in\Gamma^1} \max_{(\gamma^2,\gamma^3)\in R^F(\gamma^1)} J^1(\gamma^1, \gamma^2, \gamma^3), \quad (3.47a)$$

where $R^F(\gamma^1)$ *is the* optimal response set *of the* followers' group *and is defined for each* $\gamma^1 \in \Gamma^1$ *by*

$$R^F(\gamma^1) = \{(\xi^2,\xi^3) \in \Gamma^2 \times \Gamma^3 : J^2(\gamma^1,\xi^2,\xi^3) \le J^2(\gamma^1,\gamma^2,\xi^3) \quad and \\ J^3(\gamma^1,\xi^2,\xi^3) \le J^3(\gamma^1,\xi^2,\gamma^3),\ \forall \gamma^2 \in \Gamma^2, \gamma^3 \in \Gamma^3\}. \tag{3.47b}$$

Any $(\gamma^{2*},\gamma^{3*}) \in R^F(\gamma^{1*})$ *is a corresponding optimal strategy pair for the followers' group.*

Definition 3.36 *For the three-person finite game with two leaders* (**P**1 *and* **P**2) *and one follower* (**P**3), $\gamma^{i*} \in \Gamma^i$ *is a* hierarchical equilibrium strategy *for* **P***i* $(i = 1, 2)$ *if*

$$\max_{\gamma^3 \in R^3(\gamma^{1*};\gamma^{2*})} J^1(\gamma^{1*},\gamma^{2*},\gamma^3) = \min_{\gamma^1\in\Gamma^1} \max_{\gamma^3 \in R^3(\gamma^1;\gamma^{2*})} J^1(\gamma^1,\gamma^{2*},\gamma^3), \tag{3.48a}$$

$$\max_{\gamma^3 \in R^3(\gamma^{*};\gamma^{2*})} J^2(\gamma^{1*},\gamma^{2*},\gamma^3) = \min_{\gamma^2\in\Gamma^2} \max_{\gamma^3 \in R^3(\gamma^{1*};\gamma^2)} J^2(\gamma^{1*},\gamma^2,\gamma^3), \tag{3.48b}$$

where $R^3(\gamma^1;\gamma^2)$ *is the* optimal response set *of the* follower *and is defined for each* $(\gamma^1,\gamma^2) \in \Gamma^1 \times \Gamma^2$ *by*

$$R^3(\gamma^1;\gamma^2) = \{\xi \in \Gamma^3; J^3(\gamma^1,\gamma^2,\xi) \le J^3(\gamma^1,\gamma^2,\gamma^3), \forall \gamma^3 \in \Gamma^3\}. \tag{3.48c}$$

Any strategy $\gamma^{3*} \in R^3(\gamma^{1*};\gamma^{2*})$ *is a corresponding* optimal strategy *for the follower* (**P**3).

Definition 3.37 *For the three-person finite game with three levels of hierarchy in decision making (with* **P**1 *enforcing his strategy on* **P**2 *and* **P**3, *and* **P**2 *enforcing his strategy on* **P**3*),* $\gamma^{1*} \in \Gamma^1$ *is a* hierarchical equilibrium strategy *for* **P**1 *if*

$$\max_{\gamma^2\in S^2(\gamma^{1*})} \max_{\gamma^3\in S^3(\gamma^{1*};\gamma^2)} J^1(\gamma^{1*},\gamma^2,\gamma^3) \\ = \min_{\gamma^1\in\Gamma^1} \max_{\gamma^2\in S^2(\gamma^1)} \max_{\gamma^3\in S^3(\gamma^1;\gamma^2)} J^1(\gamma^1,\gamma^2,\gamma^3), \tag{3.49a}$$

where

$$S^2(\gamma^1) \triangleq \{\xi \in \Gamma^2 : \max_{\gamma^3\in S^3(\gamma^1;\xi)} J^2(\gamma^1,\xi,\gamma^3) \\ \le \max_{\gamma^3\in S^3(\gamma^1;\gamma^2)} J^2(\gamma^1,\gamma^2,\gamma^3), \forall\gamma^2 \in \Gamma^2\}, \tag{3.49b}$$

$$S^3(\gamma^1;\gamma^2) \triangleq \{\xi \in \Gamma^3 : J^3(\gamma^1,\gamma^2,\xi) \le J^3(\gamma^1,\gamma^2,\gamma^3), \forall\gamma^3 \in \Gamma^3\}. \tag{3.49c}$$

Any $\gamma^{2*} \in S^2(\gamma^{1*})$ *is a corresponding* optimal strategy *for* **P**2, *and any* $\gamma^3 \in S^3(\gamma^{1*},\gamma^{2*})$ *is an* optimal strategy *for* **P**3 *corresponding to the pair* $(\gamma^{1*},\gamma^{2*})$.

Several simplifications in these definitions are possible if the optimal response sets are singletons. In the case of Def. 3.35, for example, we would have the following.

Proposition 3.19 *In the three-person finite game of Def. 3.35, if $R^F(\gamma^1)$ is a singleton for each $\gamma^1 \in \Gamma^1$, then there exist unique mappings $T^2 : \Gamma^1 \to \Gamma^2$ and $T^3 : \Gamma^1 \to \Gamma^3$ such that*

$$J^1(\gamma^{1*}, T^2\gamma^{1*}, T^3\gamma^{1*}) = \min_{\gamma^1 \in \Gamma^1} J^1(\gamma^1, T^2\gamma^1, T^3\gamma^1) \tag{3.50a}$$

and

$$J^2(\gamma^1, T^2\gamma^1, T^3\gamma^1) \leq J^2(\gamma^1, \gamma^2, T^3\gamma^1), \forall \gamma^2 \in \Gamma^2, \tag{3.50b}$$

$$J^3(\gamma^1, T^2\gamma^1, T^3\gamma^1) \leq J^3(\gamma^1, T^2\gamma^1, \gamma^3), \forall \gamma^3 \in \Gamma^3 \tag{3.50c}$$

and for each $\gamma^1 \in \Gamma^1$.

Analogous results can be obtained in the cases of Defs. 3.36 and 3.37, if the optimal response sets are singletons. The following example now illustrates all these cases with singleton optimal response sets.

Example 3.16 Using the notation of Section 3.3, consider a three-person game with possible outcomes given as

$$\begin{aligned}
(a^1_{1,1,1}, a^2_{1,1,1}, a^3_{1,1,1}) &= (1,0,0),\\
(a^1_{1,2,1}, a^2_{1,2,1}, a^3_{1,2,1}) &= (2,1,-1),\\
(a^1_{2,1,1}, a^2_{2,1,1}, a^3_{2,1,1}) &= (0,0,1),\\
(a^1_{2,2,1}, a^2_{2,2,1}, a^3_{2,2,1}) &= (2,-1,1),\\
(a^1_{1,1,2}, a^2_{1,1,2}, a^3_{1,1,2}) &= (1,-1,1),\\
(a^1_{1,2,2}, a^2_{1,2,2}, a^3_{1,2,2}) &= (0,1,0),\\
(a^1_{2,1,2}, a^2_{2,1,2}, a^3_{2,1,2}) &= (-1,2,0),\\
(a^1_{2,2,2}, a^2_{2,2,2}, a^3_{2,2,2}) &= (0,1,2),
\end{aligned}$$

and determine its hierarchical equilibria (of the three different types discussed above).

1. *One leader* (**P**1) *and two followers* (**P**2 *and* **P**3).

For the first alternative of **P**1 (i.e., $n_1 = 1$), the bimatrix game faced by **P**2 and **P**3 is

P2

$(1,0,0)^*$	(2, 1, −1)	**P3**
(1, −1, 1)	(0,1,0)	

which clearly admits the unique Nash equilibrium solution $\{n_2^* = 1, n_3^* = 1\}$ whose cost to the leader is $J^1 = 1$. For $n_1 = 2$, the bimatrix game is

P2

(0,0,1)	$(2,-1,1)^*$	**P3**
(−1, 2, 0)	(0,1,2)	

with a unique Nash equilibrium solution $\{n_2^* = 2, n_3^* = 1\}$ whose cost to **P**1 is $J^1 = 2$. This determines the mappings T^2 and T^3 of Prop. 3.19 uniquely. Now, since $1 < 2$, the leader has a unique hierarchical equilibrium strategy that satisfies (3.50a), which is γ^{1*} = "alternative 1", and the corresponding unique optimal strategies for **P2** and **P3** are γ^{2*} = "alternative 1", γ^{3*} = "alternative 1", respectively. The unique equilibrium outcome is $(1,0,0)$.

2. *Two leaders* (**P1** *and* **P2**) *and one follower* (**P3**).
There are four pairs of possible strategies for the leaders' group, which are listed below together with the corresponding unique optimal responses of the follower and the cost pairs incurred to **P1** and **P2**:

$$\begin{aligned}
\{n_1 = 1, n_2 = 1\} \quad &: \quad n_3^* = 1 \to (1,0),\\
\{n_1 = 1, n_2 = 2\} \quad &: \quad n_3^* = 1 \to (2,1),\\
\{n_1 = 2, n_2 = 1\} \quad &: \quad n_3^* = 2 \to (-1,2),\\
\{n_1 = 2, n_2 = 2\} \quad &: \quad n_3^* = 1 \to (2,-1).
\end{aligned}$$

The two leaders are faced with the bimatrix game

	P2		
	(1,0)	(2,1)	**P1**
	(−1,2)	(2, −1)*	

whose Nash solution is uniquely determined as $\{n_1^* = 2, n_2^* = 2\}$. Hence, the hierarchical equilibrium strategies of the leaders are unique: γ^{1*} = "alternative 2", γ^{2*} = "alternative 2". The corresponding unique optimal strategy of **P3** is γ^{3*} = "alternative 1", and the unique equilibrium outcome is $(2,-1,1)$.

3. *Three levels of hierarchy* (*first* **P1**, *then* **P2** *and finally* **P3**).
We first observe that $S^2(\gamma^1)$ is a singleton and is uniquely determined by the mapping $P^2 : \Gamma^1 \to \Gamma^2$ defined by

$$P^2 n_1 = \begin{cases} 1 & \text{if } n_1 = 1, \\ 2 & \text{if } n_1 = 2. \end{cases}$$

Hence, if **P1** chooses $n_1 = 1$, his cost will be $J^1 = 1$, and if he chooses $n_1 = 2$, his cost will be $J^1 = 2$. Since $1 < 2$, this leads to the conclusion that **P1**'s unique hierarchical equilibrium strategy is γ^{1*} = "alternative 1" and the corresponding unique optimal strategies of **P2** and **P3** are γ^{2*} = "alternative 1" and γ^{3*} = "alternative 1", respectively. The unique equilibrium outcome is $(1,0,0)$. □

3.7 Nonzero-Sum Games with Chance Moves

In this section, we briefly discuss an extension of the general analysis of the previous sections to the class of nonzero-sum games which incorporate chance moves, that is, games wherein the final outcome is determined not only by the actions of the players, but also by the outcome of a chance mechanism. A special class of such decision problems was already covered in Section 2.6 within the context of zero-sum games, and the present analysis in a way builds on the material of Section 2.6 and extends that analysis to nonzero-sum games under the Nash and Stackelberg solution concepts. To this end, let us first provide a precise definition of an extensive form description of a nonzero-sum finite game with chance moves.

Definition 3.38 *An* extensive form *of an N-person nonzero-sum finite game with chance moves is a tree structure with*

(i) a specific vertex *indicating the starting point of the game,*

(ii) N cost functions, *each one assigning a real number to each terminal vertex of the tree, where the ith cost function determines the loss incurred to* $\mathbf{P}i$ *for each possible set of actions of the player together with the possible choices of nature,*

(iii) a partition of the nodes of the tree into $N+1$ player sets *(to be denoted by* $\bar{N}^i$ *for* $\mathbf{P}i$*,* $i \in \mathbf{N}$*, and by* $\bar{N}^0$ *for nature),*

(iv) a probability distribution, *defined at each node of* $\bar{N}^0$*, among the immediate branches (alternatives) emanating from that node,*

(v) a subpartition of each player set $\bar{N}^i$ *into* information sets $\{\eta_j^i\}$ *such that the same number of immediate branches emanates from every node belonging to the same information set and no node follows another node in the same information set.*

For such a game, let us again denote the (pure) strategy space of $\mathbf{P}i$ by Γ^i, with a typical element designated as γ^i. For a given N-tuple of strategies $\{\gamma^j \in \Gamma^j; j \in \mathbf{N}\}$ the cost function $J^i(\gamma^1, \ldots, \gamma^N)$ of $\mathbf{P}i$ is in general a random quantity, and hence the real quantity of interest is the one obtained by taking the average value of J^i over all relevant actions of nature and under the *a priori* known probability distribution. Let us denote that quantity by L^i $(\gamma^1, \ldots, \gamma^N)$ which in fact provides us with a normal form description of the original nonzero-sum game with chance moves in extensive form. Apart from a difference in notation, there is no real difference between this description and the normal form description studied heretofore in this chapter, and this observation readily leads to the conclusion that all concepts introduced and general results obtained in the previous sections on the normal form description of nonzero-sum games are also equally valid in the present context. In particular, the concept of Nash equilibrium for N-person nonzero-sum finite games with chance moves can be introduced through Def. 3.12 by merely replacing J^i with L^i $(i = 1, \ldots, N)$. Analogously, the Stackelberg equilibrium solution in two-person nonzero-sum finite games with chance moves can be introduced by Defs. 3.26-3.29. Rather than list all these apparent analogies, we shall consider, in the remaining part of this section, two examples which illustrate derivation of Nash and Stackelberg equilibria in static and dynamic games that incorporate chance moves.

Example 3.17 Consider the two-person static nonzero-sum game depicted in Fig. 3.11. If nature picks the left branch, then the bimatrix game faced by the

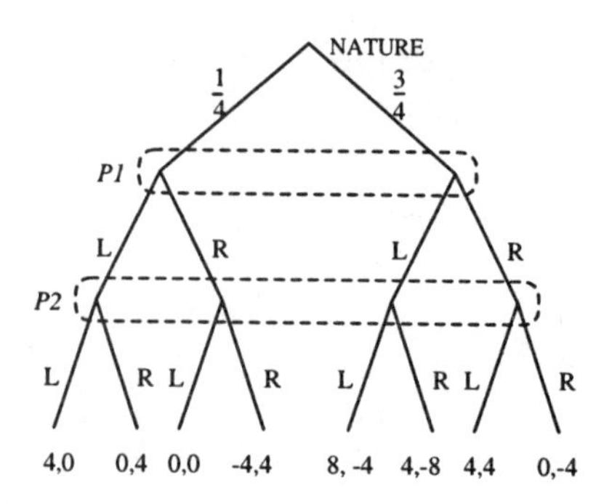

Figure 3.11: A single-act static nonzero-sum game that incorporates a chance move.

players is

$$\begin{array}{c} \quad \mathbf{P2} \\ \begin{array}{c} L \\ R \end{array} \begin{array}{|c|c|} \hline (4,0) & (0,4) \\ \hline (0,0) & (-4,4) \\ \hline \end{array} \; \mathbf{P1}, \\ \quad L \qquad R \end{array} \qquad (i)$$

whereas if nature picks the right branch, the relevant bimatrix game is

$$\begin{array}{c} \quad \mathbf{P2} \\ \begin{array}{c} L \\ R \end{array} \begin{array}{|c|c|} \hline (8,-4) & (4,-8) \\ \hline (4,4) & (0,-4) \\ \hline \end{array} \; \mathbf{P1}. \\ \quad L \qquad R \end{array} \qquad (ii)$$

Since the probabilities with which nature chooses the left and right branches are 1/4 and 3/4, respectively, the final bimatrix game of average costs is

$$\begin{array}{c} \quad \mathbf{P2} \\ \begin{array}{c} L \\ R \end{array} \begin{array}{|c|c|} \hline (7,-3) & (3,-5) \\ \hline (3,3) & (-1,-2) \\ \hline \end{array} \; \mathbf{P1}. \\ \quad L \qquad R \end{array} \qquad (iii)$$

This game admits a unique Nash equilibrium solution in pure strategies, which is

$$\gamma^{1*} = R, \quad \gamma^{2*} = R$$

with an average outcome of $(-1, -2)$.

To determine the Stackelberg solution with **P1** as the leader, we again focus our attention on the bimatrix game (iii) and observe that the Stackelberg solution is in fact the same as the unique Nash solution determined above. □

Example 3.18 Let x_0 be a discrete random variable taking the values

$$x_0 = \begin{cases} -1 & \text{w.p. } 1/4, \\ 0 & \text{w.p. } 1/2, \\ 1 & \text{w.p. } 1/4. \end{cases}$$

We view x_0 as the random initial state of some system under consideration. **P1**, whose decision variable (denoted by u^1) assumes the values -1 or $+1$, acts first, and as a result the system is transferred to another state (x_1) according to the equation

$$x_1 = x_0 + u^1.$$

Then **P2**, whose decision variable u^2 can take the same two values (-1 and $+1$), acts and transfers the system to some other state (x_2) according to

$$x_2 = x_1 + u^2.$$

It is further assumed that, in this process, **P1** does not make any observation (i.e., he does not know what the real outcome of the chance mechanism (value of x_0) is), whereas **P2** observes sgn (x_1), where

$$\text{sgn } (x) \triangleq \begin{cases} -1, & x < 0, \\ 0, & x = 0, \\ +1, & x > 0. \end{cases}$$

This observation provides **P2** with partial information concerning the value of x_0 and the action of **P1**. The cost structure of the problem is such that **P**i wishes to minimize J^i, $i = 1, 2$, where

$$J^1 = g(x_2) + u^1, \qquad J^2 = g(x_2) - u^2$$

with

$$g(x) \triangleq \begin{cases} -1, & \text{if } x = -3, -1, 2, \\ 0, & \text{if } x = 0, 3, \\ 1, & \text{if } x = -2, 1. \end{cases}$$

We now seek to obtain Nash equilibria and also Stackelberg equilibria of this single-act game with either player as the leader.

To this end, let us first note that this is a finite nonzero-sum game which incorporates a chance move, but it has not been cast in the standard framework. Though, by spelling out all possible values of J^1 and J^2, and the actions of the players leading to those values, an extensive tree formulation can be obtained for the game, which is the one depicted in Fig. 3.12. This extensive form also clearly displays the information sets and thereby the number of possible strategies for each player. Since **P1** gains no information, he has two possible strategies:

$$\gamma_1^1 = -1, \quad \gamma_2^1 = +1.$$

P2, on the other hand, has eight possible strategies which can be listed as

$$\gamma_1^2(\eta^2) = -1, \qquad \gamma_2^2(\eta^2) = +1, \qquad \gamma_3^2(\eta^2) = \begin{cases} -1, \text{if } \eta^2 = \eta_1^2, \\ +1, \text{ otherwise }, \end{cases}$$

$$\gamma_4^2(\eta^2) = \begin{cases} +1, \text{if } \eta^2 = \eta_1^2, \\ -1, \text{ otherwise }, \end{cases} \qquad \gamma_5^2(\eta^2) = \begin{cases} -1, \text{if } \eta^2 = \eta_2^2, \\ +1, \text{ otherwise }, \end{cases}$$

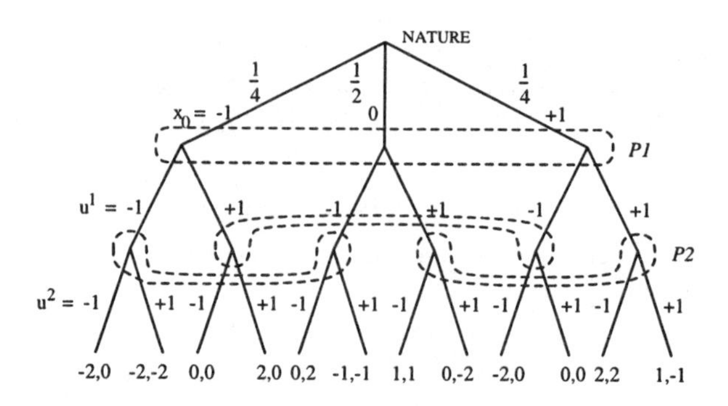

Figure 3.12: Extensive tree formulation for the single-act two-person game of Example 3.18.

$$\gamma_6^2(\eta^2) = \begin{cases} +1, \text{if } \eta^2 = \eta_2^2, \\ -1, \text{ otherwise} , \end{cases} \qquad \gamma_7^2(\eta^2) = \begin{cases} -1, \text{if } \eta^2 = \eta_3^2, \\ +1, \text{ otherwise} , \end{cases}$$

$$\gamma_8^2(\eta^2) = \begin{cases} +1, \text{if } \eta^2 = \eta_3^2, \\ -1, \text{ otherwise} . \end{cases}$$

Then, the equivalent bimatrix representation in terms of average costs is

$A =$ (P2 columns, P1 rows)

	1	2	3	4	5	6	7	8
1	-1	-1	$-1/2$	$-3/2$	$-3/2$	$-1/2$	-1	-1
2	1	$3/4$	$3/4$	1	$1/4$	$3/2$	$3/2$	$1/4$

P1,

$B =$ (P2 columns, P1 rows)

	1	2	3	4	5	6	7	8
1	1	-1	1	-1	-1	1	-1	1
2	1	$-5/4$	$-5/4$	1	$-3/4$	1	1	$-3/4$

P1.

Since the first row of matrix A strictly dominates the second row, **P1** clearly has a unique *(permanent) optimal* strategy $\gamma^{1*} = -1$ regardless of whether it is a Nash or a Stackelberg game. **P2** has four strategies that are in Nash or Stackelberg equilibrium with γ^{1*}, which are γ_2^2, γ_4^2, γ_5^2 and γ_7^2. Hence, we have four sets of Nash equilibrium strategies, with the equilibrium outcome being either $(-1, -1)$ or $(-3/2, -1)$. If **P1** is the leader, his Stackelberg strategy is $\gamma^{1*} = -1$ and the Stackelberg average cost (as defined in Def. 3.27) is $L^{1*} = -1$. He may, however, also end up with the more favorable average cost level of $L^1 = -3/2$, depending on the actual play of **P2**. If **P2** is the leader, he has four Stackelberg strategies (γ_2^2, γ_4^2, γ_5^2 and γ_7^2), with his unique Stackelberg average cost being $L^{2*} = -1$.

To summarize the results in terms of the terminology of the original problem, regardless of the specific mode of play (i.e., whether it is Nash or Stackelberg), **P1** always chooses $u^1 = -1$, and **P2** chooses $u^2 = +1$ if $x_1 < 0$, and is indifferent between his two alternatives (-1 and $+1$) if $x_1 \geq 0$. □

3.8 Problems

1. Determine the Nash equilibrium solutions of the following bimatrix game, and investigate the admissibility properties of these equilibria:

$$A = \overset{\mathbf{P2}}{\begin{array}{|c|c|c|}\hline 8 & -4 & 8 \\ \hline 6 & 1 & 6 \\ \hline 12 & 0 & 0 \\ \hline 0 & 0 & 12 \\ \hline 0 & 4 & 0 \\ \hline \end{array}}\ \mathbf{P1}, \qquad B = \overset{\mathbf{P2}}{\begin{array}{|c|c|c|}\hline 4 & 0 & 0 \\ \hline 0 & 0 & 4 \\ \hline 0 & 4 & 0 \\ \hline 0 & 4 & 0 \\ \hline 0 & 1 & 0 \\ \hline \end{array}}\ \mathbf{P1}.$$

2. Verify that minimax values of a bimatrix game are not lower (in an ordered way) than the pair of values of any Nash equilibrium outcome. Now, construct a bimatrix game wherein minimax strategies of the players are also their Nash equilibrium strategies, but in which the minimax values are not the same as the ordered values of the Nash equilibrium outcome.

3. Obtain the mixed Nash equilibrium solution of the following bimatrix game:

$$A = \overset{\mathbf{P2}}{\begin{array}{|c|c|}\hline 10 & -2 \\ \hline -1 & 1 \\ \hline \end{array}}\ \mathbf{P1}, \qquad B = \overset{\mathbf{P2}}{\begin{array}{|c|c|}\hline -5 & 2 \\ \hline 1 & -1 \\ \hline \end{array}}\ \mathbf{P1}.$$

What is the mixed Nash equilibrium solution of the game $(-A, -B)$?

4. If a bimatrix game (A, B) admits multiple minimax strategies for a player, then these can be ordered according to the notion of *admissibility* (which is different from the admissibility of Nash equilibrium à la Def. 3.3). A minimax strategy for **P1**, say "row i", is said to be admissible if there exists no other minimax strategy, say "row j", with the property

$$a_{ik} \geq a_{jk} \qquad \forall k = 1, \ldots, n$$

with the inequality being strict for at least one k. An admissible minimax strategy for **P2** can be defined analogously.
(i) Does every bimatrix game admit admissible minimax strategies for **P1** and **P2**? If they do, are they necessarily unique?
(ii) Obtain the set of all admissible minimax strategies for **P1** and **P2** in the following bimatrix game:

$$A = \overset{\mathbf{P2}}{\begin{array}{|c|c|c|}\hline 1 & 2 & 3 \\ \hline 3 & 4 & 0 \\ \hline 1 & 1 & 3 \\ \hline \end{array}}\ \mathbf{P1}, \qquad B = \overset{\mathbf{P2}}{\begin{array}{|c|c|c|}\hline 4 & 0 & 0 \\ \hline 0 & 4 & 3 \\ \hline 0 & 0 & 4 \\ \hline \end{array}}\ \mathbf{P1}.$$

5. Let (A, B) be an $(m \times n)$ bimatrix game admitting a unique Nash equilibrium solution in mixed strategies, say (y^*, z^*). Prove that the number

of nonzero entries of y^* and z^* are equal. How would this statement have to be modified if the number of mixed-strategy equilibrium solutions is greater than one?

6. Prove (or disprove) the following conjecture: Every bimatrix game that admits more than one pure Nash equilibrium solution with different equilibrium outcomes necessarily also admits a Nash equilibrium solution in proper mixed strategies (that is, mixed strategies that are not pure).

7. Determine all Nash equilibria of the three-person nonzero-sum game, with $m_1 = m_2 = m_3 = 2$, whose possible outcomes are given below:

$$\begin{aligned}
(a^1_{1,1,1}, a^2_{1,1,1}, a^3_{1,1,1}) &= (0,0,1),\\
(a^1_{1,2,1}, a^2_{1,2,1}, a^3_{1,2,1}) &= (-2,1,2),\\
(a^1_{2,1,1}, a^2_{2,1,1}, a^3_{2,1,1}) &= (0,-3,1),\\
(a^1_{2,2,1}, a^2_{2,2,1}, a^3_{2,2,1}) &= (-1,1,-2),\\
(a^1_{1,1,2}, a^2_{1,1,2}, a^3_{1,1,2}) &= (-2,1,1),\\
(a^1_{1,2,2}, a^2_{1,2,2}, a^3_{1,2,2}) &= (-2,0,1),\\
(a^1_{2,1,2}, a^2_{2,1,2}, a^3_{2,1,2}) &= (-1,-2,2),\\
(a^1_{2,2,2}, a^2_{2,2,2}, a^3_{2,2,2}) &= (1,0,0).
\end{aligned}$$

8. Obtain the minimax strategies of the players in the three-person game of Problem 7. What are the security levels of the players?

9. Consider the following bimatrix game, which admits three pure-strategy Nash equilibria:

$$A = \overset{\textbf{P2}}{\begin{array}{|c|c|}\hline 1 & 3 \\ \hline 1 & 2 \\ \hline 2 & 2 \\ \hline \end{array}}\ \textbf{P1}, \qquad B = \overset{\textbf{P2}}{\begin{array}{|c|c|}\hline 1 & 1 \\ \hline 1 & 2 \\ \hline 2 & 1 \\ \hline \end{array}}\ \textbf{P1}.$$

Which of these three Nash equilibria are (i) perfect, (ii) proper? (**Note:** See the discussion in subsection 3.5.5.)

10. This problem is concerned with a specific N-person zero-sum game[23] ($N \geq 2$) which is called Mental Poker after Epstein (1967). Each of the N players chooses independently an integer out of the set $\{1, 2, \ldots, n\}$. The player who has chosen the lowest unique integer is declared the winner and he receives $N-1$ units from the other players each of whom loses one unit. If there is no player who has chosen a unique integer, then no one wins the game and everybody receives zero units. If for instance $N = 3$, $n = 5$

[23] In terms of the notation of Section 3.3, the game is zero-sum if $\sum_{i \in \mathbf{N}} a^i_{n_1,\ldots,n_N} = 0$, $\forall n_j \in \mathbf{M}_j$, $j \in \mathbf{N}$.

and the integers chosen are 2, 2 and 5, then the player who has chosen 5 is the winner. Show that for $N = 2$, n arbitrary, there exists a unique saddle-point strategy for both players, which is to choose integer 1 with probability 1.

Next consider the case $N = 3$. A mixed strategy for $\mathbf{P}i$ will be denoted by $y^i = (y_1^i, \ldots, y_n^i)$, $i = 1, 2, 3$, with $y_j^i \geq 0$ and $\sum_{j=1}^n y_j^i = 1$. We seek a Nash equilibrium solution $\{y^{1*}, y^{2*}, y^{3*}\}$ with the property $y^{1*} = y^{2*} = y^{3*}$. Show that there exists a unique such Nash equilibrium solution, which is given by $y_j^{i*} = (1/2)^j$, $j = 1, \ldots, n-1$, $y_n^{i*} = (1/2)^{n-1}$. Generalize these results to $N > 3$.

11. Obtain all pure-strategy Nash equilibria of the following single-act games in extensive form. Which of these equilibria are (i) admissible, (ii) of the delayed commitment type?

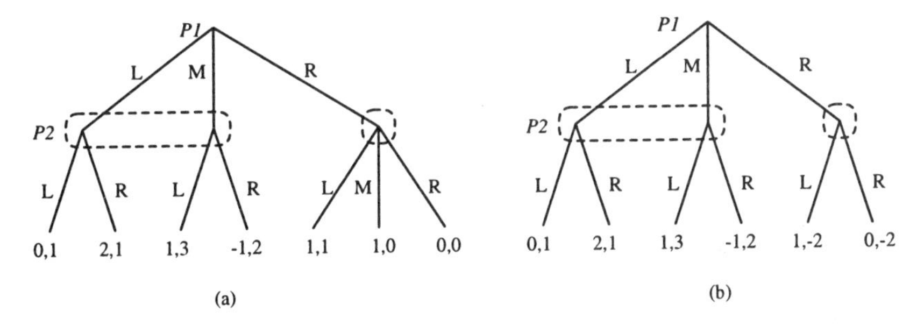

12. Obtain all pure-strategy Nash equilibria of the following three-player single-act game wherein the order in which $\mathbf{P}2$ and $\mathbf{P}3$ act is a variant of $\mathbf{P}1$'s strategy. (Hint: A recursive procedure still applies here for each fixed strategy of $\mathbf{P}1$.)

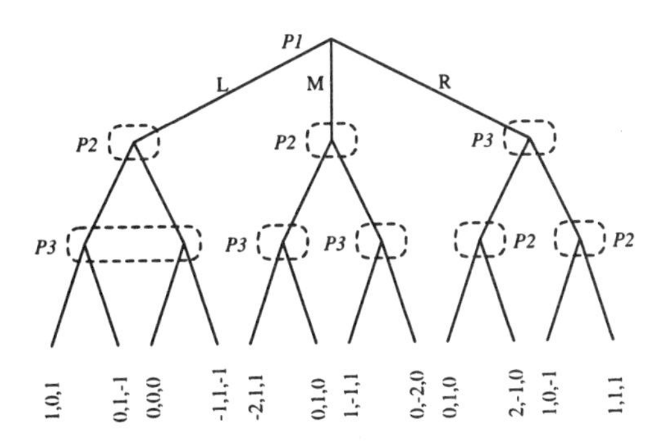

13. Investigate whether the three-person single-act game in nested extensive form depicted in Fig. 3.13 admits a Nash equilibrium of the delayed commitment type.

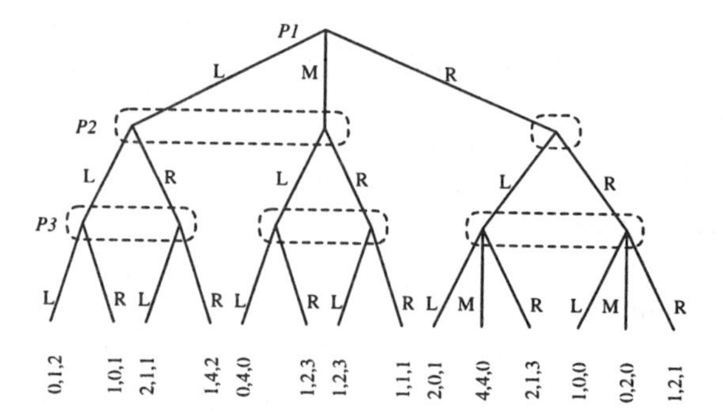

Figure 3.13: Single-act game of Problem 13.

14. Determine all multi-act nonzero-sum feedback games which are informationally inferior to the feedback game whose extensive form is depicted in Fig. 3.9. Obtain their delayed commitment type feedback Nash equilibria.

15. Investigate whether the multi-act game of Fig. 3.14 admits pure-strategy Nash equilibria, by transforming it into equivalent normal form.

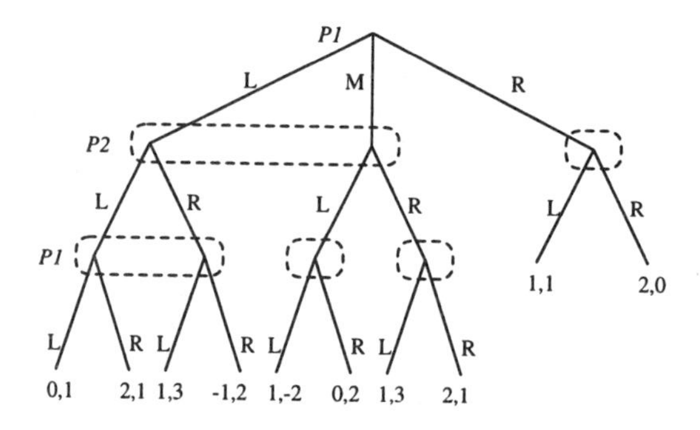

Figure 3.14: Multi-act game of Problem 15.

16. Determine the perfect Nash equilibria (in pure and mixed strategies) of the multi-act game of Problem 15.

17. Determine the Stackelberg strategies for the leader (**P1**) and also his Stackelberg costs in the single-act games of Problem 11. Solve the same game problems under the Stackelberg mode of play with **P2** acting as the leader.

18. Let I be a two-person single-act finite game in extensive form in which **P2** is the first-acting player, and denote the Stackelberg cost for the leader (**P1**) as J_{I}^{1*}. Let II be another two-person single-act game that is informationally inferior to I, and whose Stackelberg cost for the leader (**P1**) is J_{II}^{1*}. Prove the inequality $J_{\mathrm{I}}^{1*} \leq J_{\mathrm{II}}^{1*}$.

19. Determine the Stackelberg solution of the two-person dynamic game of Problem 15 with (i) **P1** as the leader, (ii) **P2** as the leader.

20. Obtain the pure, behavioral and mixed Stackelberg costs with **P2** as the leader for the single-act game of Fig. 3.15. Does the game admit a behavioral Stackelberg strategy for the leader (**P2**)?

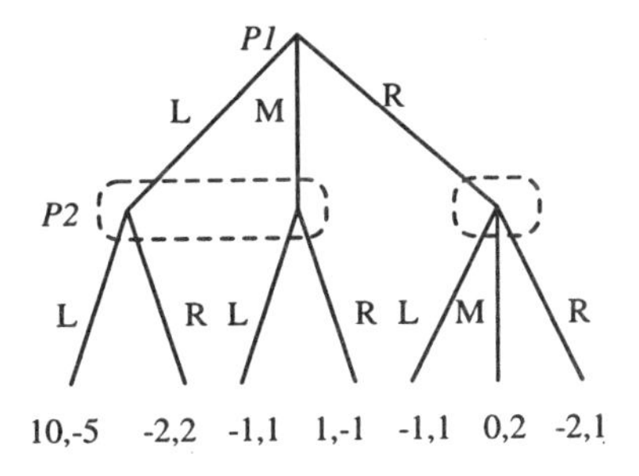

Figure 3.15: Single-act game of Problem 20.

21. Let $\bar{\gamma}^{1*} \in \bar{\Gamma}^1$ be a mixed Stackelberg strategy for the leader (**P1**) in a two-person finite game, with the further property that $\bar{\gamma}^{1*} \in \hat{\Gamma}^1$, i.e., it is a behavioral strategy. Prove that $\bar{\gamma}^{1*}$ also constitutes a behavioral Stackelberg strategy for the leader.

22. Prove that the hierarchical equilibrium strategy γ^{1*} introduced in Def. 3.37 always exists. Discuss the reasons why γ^{1*} introduced in Def. 3.35, and γ^{1*} and γ^{2*} introduced in Def. 3.36, do not necessarily exist.

23. Determine hierarchical equilibria in pure or behavioral strategies of the single-act game of Fig. 3.4(b) when (i) **P2** is the leader, **P1** and **P3** are followers, (ii) **P1** and **P2** are leaders, **P3** is the follower, (iii) there are three levels of hierarchy with the ordering being **P2**, **P1**, **P3**.

24. In the four-person single-act game of Fig. 3.5, let **P3** and **P4** form the leaders' group, and **P1** and **P2** form the followers' group. Stipulating that the noncooperative Nash solution concept is adopted within each group and the Stackelberg solution concept between the two groups, determine the corresponding hierarchical equilibrium solution of the game in pure or behavioral strategies.

25. Determine the pure-strategy Nash equilibria of the three-person single-act game in extensive form depicted in Fig. 3.16, which incorporates a chance move. Also determine its (pure-strategy) Stackelberg equilibria with (i) **P1** as leader, **P2** and **P3** as followers (i.e., two levels of hierarchy), (ii) **P2** as leader, **P1** as the first follower, **P3** as the second follower (i.e., three levels of hierarchy).

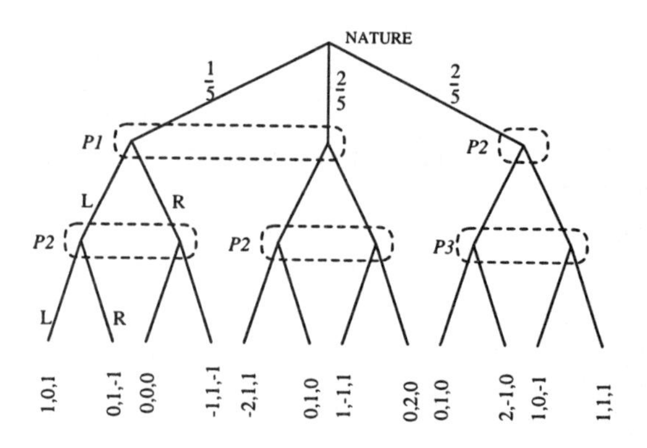

Figure 3.16: Three-person single-act game of Problem 25.

26. Let x_0 be a discrete random variable taking the values

$$x_0 = \begin{cases} -1 & \text{w.p. } 1/3, \\ 0 & \text{w.p. } 1/3, \\ 1 & \text{w.p. } 1/3. \end{cases}$$

Consider the two-act game whose evolution is described by

$$\begin{aligned} x_1 &= x_0 + u_0^1 + u_0^2, \\ x_2 &= x_1 + u_1^1 + u_1^2, \end{aligned}$$

where u_0^i (respectively, u_1^i) denotes the decision (action) variable of **P**i at level 1 (respectively, level 2), $i = 1, 2$, and they take values out of the set $\{-1, 0, 1\}$. At level 1 none of the players make any observation (i.e., they do not acquire any information concerning the actual outcome of the chance mechanism (x_0)). At level 2, however, **P2** still makes no observation, while **P1** observes the sign of x_1. At this level, both players also recall their own past actions. Finally, let

$$J^i = (-1)^{i+1} \text{ mod } (x_2) + \text{ sgn } (u_0^i) + \text{ sgn } (u_1^i), i = 1, 2,$$

denote the cost function of **P**i which he desires to minimize, where

$$\text{mod } (x) = \begin{cases} -1, & \text{if } x = -4, -1, 2, 5 \\ 0, & \text{if } x = -3, 0, 3 \\ -1, & \text{if } x = -5, -2, 1, 4 \end{cases} \quad ; \text{ sgn } (0) = 0.$$

(i) Formulate this dynamic multi-act game in the standard extensive form and obtain its pure-strategy Nash equilibrium.

(ii) Determine its pure-strategy Stackelberg equilibria with **P1** as leader.

3.9 Notes

Sections 3.2 and 3.3. The first formulation of nonzero-sum finite games dates back to the pioneering work of Von Neumann and Morgenstern (1947). But the noncooperative equilibrium solution discussed in these two sections within the context of N-person finite games in normal form was first introduced by Nash (1951) who also gave a proof of Thm. 3.2. For interesting discussions on the reasons and implications of occurrence of nonunique Nash equilibria, see Howard (1971). A nontechnical treatment of the "prisoner's dilemma" phenomenon can be found in the recent book by Poundstone (1992). Some other selected texts on nonzero-sum (as well as zero-sum) games are Binmore (1992), Fudenberg and Tirole (1991), Kreps (1990), Myerson (1991) and Szép and Forgó (1985).

Section 3.4. For further discussion on the relationship between bimatrix games and nonlinear programming, see Section 7.4 of Parthasarathy and Raghavan (1971). Section 7.3 of the same reference also presents the Lemke-Howson algorithm which is further exemplified in Section 7.6 of that reference.

Sections 3.5 and 3.7. A precise formulation for finite nonzero-sum dynamic games in extensive form was first given by Kuhn (1953), who also proved that every N-person nonzero-sum game with perfect information (i.e., with singleton information sets) admits a pure-strategy Nash equilibrium solution (note that this is a special case of Prop. 3.11 given in subsection 3.5.4). He also showed that in every game in extensive form with perfect recall, every behavioral-strategy mixture (and hence a mixed strategy) for a player can be replaced by a realization-equivalent behavioral strategy, and hence equilibrium can be sought in the class of behavioral strategies. The notion of ladder-nested extensive forms and the recursive derivation of the Nash equilibrium solution of the delayed commitment type are introduced here for the first time. The "informational nonuniqueness" property of the Nash equilibrium solution in finite dynamic games is akin to a similar feature in infinite nonzero-sum games observed by Başar (1974, 1976a, 1977b); see also Section 6.3. A special version of this feature was observed earlier by Starr and Ho (1969a,b) who showed that, in nonzero-sum finite feedback games, the feedback Nash equilibrium outcome could be different (for all players) from the open-loop Nash equilibrium outcome. The result of Prop. 3.10 was also proved in Dubey and Shubik (1981), and later included in the text by Shubik (1983). This latter reference calls the *static (open-loop)* version of an extensive form the *most coarsened form*, and its (open-loop) Nash equilibria *information-insensitive* equilibria, as they also constitute equilibria for all refined versions. The material of subsection 3.5.5 is based on Selten (1975) and Myerson (1978). Further refinements on Nash equilibrium can be found in Kreps and Wilson (1982), Kohlberg and Mertens (1986) and van Damme (1987, 1989).

A detailed study of N-player nonzero-sum games on finite graphs with cycles, viewed as "nonzero-sum extensions" of the zero-sum theory treated in subsection 2.7.1, can be found in Alpern (1991). The cost N-tuples corresponding to an infinite path on the graph in such dynamic games are the long term averages of the local cost N-tuples. Results have been obtained for the existence of Nash equilibria confined to certain classes of pure strategies.

If a static game (such as the game of the prisoners' dilemma) is played repeatedly, one speaks of *repeated games*. In such games one gets to know about the past behavior of the other player(s) which may therefore influence the current decision to be made.

With this role of information about the past actions of the players, the repeated game must be viewed as a dynamic one. It is possible to derive conditions for the existence of a Nash equilibrium for such a dynamic game, which, if viewed at the individual time steps, converges to a Pareto solution of the underlying static ("one shot") game as time evolves. The reason why a "sequence of Pareto-like solutions" can form a (stable) Nash equilibrium is the ability to perform threats, which, in its turn, is possible by applying strategies which include history. Many results exist in this direction. One can for instance consult Radner (1981, 1985), Mertens (1985, 1991) and the references therein.

Sections 3.6 and 3.7. The Stackelberg solution in nonzero-sum static games was first introduced by H. Von Stackelberg (1934) within the context of economic competition and it was later extended to dynamic games primarily in the works of Chen and Cruz, Jr (1972) and Simaan and Cruz, Jr (1973a,b) who also introduced the concept of a feedback Stackelberg solution, but their formulation is restricted by the assumption that the follower's response is unique for every strategy of the leader. The present analysis is free from such a restriction. The feature of the mixed-strategy Stackelberg solution exhibited in (Counter-) Example 3.14 is reported here for the first time. Its counterpart in the context of infinite games was first observed in Başar and Olsder (1980a); see also Section 4.4. For a discussion on a unified solution concept, from which Nash, Stackelberg, minimax solutions (and also cooperative solutions) follow as special cases, see Blaquière (1976).

Chapter 4

Static Noncooperative Infinite Games

4.1 Introduction

This chapter deals with static zero-sum and nonzero-sum infinite games, i.e., static noncooperative games wherein at least one of the players has at his disposal an infinite number of alternatives to choose from. In Section 4.2, the concept of ϵ equilibrium solutions is introduced, and several of its features are discussed. In Section 4.3, continuous-kernel games defined on closed and bounded subsets of finite dimensional spaces are treated, and general questions of the existence and uniqueness of Nash and saddle-point equilibrium solutions in such games are addressed; furthermore, some iterative algorithms are developed for their computation.

Section 4.4 is devoted to the Stackelberg solution of continuous-kernel games, in both pure and mixed strategies; it presents some existence and uniqueness results, together with illustrative examples. Section 4.5 focuses on the consistent conjectural variations (*CCV*) equilibrium in continuous-kernel games, and introduces different levels of approximation to the CCV solution. Finally, Section 4.6 deals with quadratic games and economic applications. Here, the theory developed in the previous sections is specialized to games whose kernels are quadratic functions of the decision variables, and in that context the equilibrium strategies are obtained in closed form. These results are then applied to different price and quantity models of oligopoly. Section 4.7 deals with the Braess paradox, which shows that the Nash equilibrium solution concept could lead to rather counter-intuitive results in certain situations.

4.2 ϵ Equilibrium Solutions

Abiding by the notation and terminology of the previous chapters, we have N players, **P**1,...,**P**N, where **P**i has the cost function J^i whose value depends not only on his action but also on the actions of some or all of the other players. **P**i's action is denoted by u^i, which belongs to his action space U^i.[24] For infinite games, the action space of at least one of the players has infinitely many elements.

In Chapters 2 and 3, we have only dealt with finite games in normal and extensive form, and in that context we have proven that noncooperative equilibrium solutions always exist in the extended space of mixed strategies (cf. Thm. 3.2). The following example now shows that this property does not necessarily hold for infinite games in normal form; it will then motivate the introduction of a weaker equilibrium solution concept — the so-called ϵ equilibrium solution.

Example 4.1 Consider the zero-sum matrix game A represented by

	P2					
P1	2	1/2	1/3	1/4	1/5	......
	0	1/2	2/3	3/4	4/5	......

Here, **P1** has two pure strategies, whereas **P2** has countably many alternatives. As in finite zero-sum games, let us denote the mixed strategies of **P1** and **P2** by y_i, $i = 1, 2$; and z_j, $j = 1, 2, \ldots$, respectively; that is, y_i denotes the probability with which **P1** chooses row i, and z_j denotes the probability with which **P2** chooses column j. Obviously we should have $y_i \geq 0$, $z_j \geq 0$, $y_1 + y_2 = 1$, $\sum_{j=1}^{\infty} z_j = 1$.

If we employ a natural extension of the graphical solution method described in Section 2.3, to determine the mixed security strategy of **P1**, we arrive at Fig. 4.1. The striking feature here is that the upper envelope is not well defined.

Suppose that the matrix A is truncated to A_k, where A_k comprises the first k columns of A. The mixed security strategy for **P1** is then $y_1^* = (k-1)/(3k-2)$; $y_2^* = (2k-1)/(3k-2)$, and the average security level of **P1** is $\bar{V}_{\rm m}(A_k) = 2(k-1)/(3k-2)$. If we let $k \to \infty$, we obtain $y_1^* = \frac{1}{3}$, $y_2^* = \frac{2}{3}$, $\bar{V}_{\rm m}(A) = \frac{2}{3}$. Thus it is easily seen that $(y_1^* = \frac{1}{3}, y_2^* = \frac{2}{3})$ is a mixed security strategy for **P1** in the matrix game A. The mixed security strategy for **P2** in the truncated game A_k is $(z_1^* = (k-2)/(3k-2), z_k^* = 2k/(3k-2))$ and the average security level of **P2** is $\underline{V}_{\rm m}(A_k) = 2(k-1)/(3k-2)$. In the limit as $k \to \infty$, we obtain $(z_1^* = \frac{1}{3}, z_\infty^* = \frac{2}{3})$ and $\underline{V}_{\rm m}(A) = \frac{2}{3}$. But, **P2** cannot guarantee for himself the average lower value $\underline{V}_{\rm m}(A)$, since he cannot choose column "∞". We observe, however, that if he chooses column k with probability $\frac{2}{3}$, k being sufficiently large, then he can secure for himself an average lower value arbitrarily close to $\underline{V}_{\rm m}(A) = \frac{2}{3}$. In accordance with this observation, if **P2** chooses k such that

[24] In this chapter, we deal with static problems only, and in that context we do not distinguish between strategy space and action space, since they are identical.

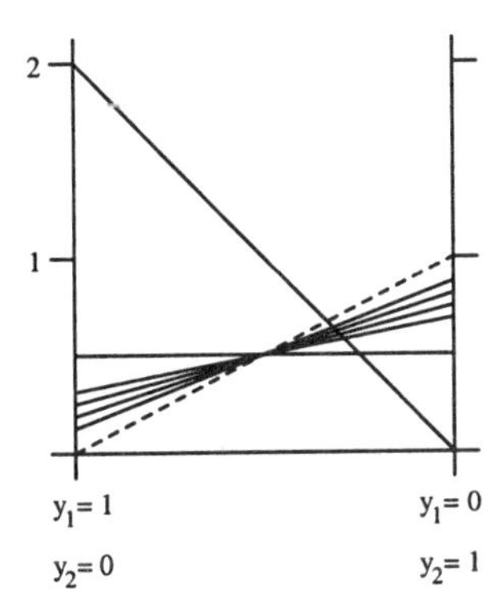

Figure 4.1: Graphical solution to the $(2 \times \infty)$ matrix game of Example 4.1.

$\underline{V}_{\mathrm{m}}(A_k) > \frac{2}{3} - \epsilon$, then the corresponding mixed security strategy is called an *ϵ-mixed security strategy*, which is also an *ϵ-mixed saddle-point strategy* for **P**2.

The reason for the observed disparity between finite and infinite games lies in the fact that the set of mixed strategies of **P**2 is not compact in the latter case. Consequently, an $(m \times \infty)$ matrix game might not admit a saddle point in mixed strategies, in spite of the fact that the outcome is continuous with respect to mixed strategies of **P**2. □

The preceding example has displayed the non-existence of a saddle point within the class of mixed strategies. It is even possible to construct examples of zero-sum (semi-infinite) matrix games which admit ϵ saddle-point strategies within the class of pure strategies. This happens, for instance, in the degenerate matrix game A, where $A = [0\ \frac{1}{2}\ \frac{2}{3}\ \frac{3}{4} \ldots]$. Similar features are also exhibited by nonzero-sum matrix games under the Nash solution concept. We now provide below a precise definition of an ϵ equilibrium solution in N-person games within the context of pure strategies. Extension to the class of mixed strategies is then immediate, and the last part of this section is devoted to a discussion on that extension, as well as to some results on existence of mixed ϵ equilibrium solutions.

Definition 4.1 *For a given $\epsilon \geq 0$, an N-tuple $\{u_\epsilon^{1^*}, \ldots, u_\epsilon^{N^*}\}$, with $u_\epsilon^{i^*} \in U^i$, $i \in \mathbf{N}$, is called a (pure) ϵ* Nash equilibrium solution *for an N-person nonzero-sum infinite game if*

$$J^i(u_\epsilon^{1^*}, \ldots, u_\epsilon^{N^*}) \leq \inf_{u^i \in U^i} J^i(u_\epsilon^{1^*}, \ldots, u_\epsilon^{i-1^*}, u^i, u_\epsilon^{i+1^*}, \ldots, u_\epsilon^{N^*}) + \epsilon, i \in \mathbf{N}.$$

For $\epsilon = 0$, one simply speaks of "equilibrium" instead of "0 equilibrium" solution, in which case we denote the equilibrium strategy of **P***i by u^{i^*}.*

For zero-sum two-person games the terminology is somewhat different. For $N = 2$ and $J^1 \equiv -J^2 \triangleq J$, we have the following definition.

Definition 4.2 *For a given $\epsilon \geq 0$, the pair $\{u_\epsilon^{1^*}, u_\epsilon^{2^*}\} \in U^1 \times U^2$ is called an ϵ* saddle point *if*

$$J(u_\epsilon^{1^*}, u^2) - \epsilon \leq J(u_\epsilon^{1^*}, u_\epsilon^{2^*}) \leq J(u^1, u_\epsilon^{2^*}) + \epsilon$$

for all $\{u^1, u^2\} \in U^1 \times U^2$. For $\epsilon = 0$ one simply speaks of a "saddle point".

By direct analogy with the treatment given in Section 2.2 for finite games, the *lower value* of a zero-sum two-person infinite game is defined by

$$\underline{V} = \sup_{u^2 \in U^2} \inf_{u^1 \in U^1} J(u^1, u^2),$$

which is also the security level of **P2** (the maximizer). Furthermore, the *upper value* is defined by

$$\bar{V} = \inf_{u^1 \in U^1} \sup_{u^2 \in U^2} J(u^1, u^2),$$

which is also the security level of **P1**. Since the strategy spaces are fixed, and in particular the structure of U^i does not depend on u^j, $i \neq j$, $i, j = 1, 2$, it can easily be shown, as in the case of Thm. 2.1(iii), that $\underline{V} \leq \bar{V}$. If $\underline{V} = \bar{V}$, then $V = \underline{V} = \bar{V}$ is called the *value* of the game. We should note that existence of the value of a game does not necessarily imply existence of (pure) equilibrium strategies; it, however, implies existence of an ϵ saddle point—a property that is verified in Thm. 4.1 below, which proves, in addition, that the converse statement is also true.

Theorem 4.1 *A two-person zero-sum (infinite) game has a finite value if, and only if, for every $\epsilon > 0$, an ϵ saddle point exists.*

Proof. First, suppose that the game has a finite value ($V = \underline{V} = \bar{V}$). Then, given an $\epsilon > 0$, one can find $u_\epsilon^{1^*} \in U^1$ and $u_\epsilon^{2^*} \in U^2$ such that

$$J(u^1, u_\epsilon^{2^*}) > \underline{V} - \frac{1}{2}\epsilon, \quad \forall u^1 \in U^1, \tag{i}$$

$$J(u_\epsilon^{1^*}, u^2) < \bar{V} + \frac{1}{2}\epsilon, \quad \forall u^2 \in U^2, \tag{ii}$$

which follow directly from the definitions of $\underline{V}$ and $\bar{V}$, respectively. Now, since $\underline{V} = \bar{V} = V$, by adding ϵ to both sides of the first inequality we obtain

$$J(u^1, u_\epsilon^{2^*}) + \epsilon > V + \frac{1}{2}\epsilon > J(u_\epsilon^{1^*}, u_\epsilon^{2^*}), \quad \forall u^1 \in U^1, \tag{iii}$$

where the latter inequality follows from (ii) by letting $u^2 = u_\epsilon^{2^*}$. Analogously, if we now add $-\epsilon$ to both sides of (ii), and also make use of (i) with $u^1 = u_\epsilon^{1^*}$, we obtain

$$J(u_\epsilon^{1^*}, u^2) - \epsilon < V - \frac{1}{2}\epsilon < J(u_\epsilon^{1^*}, u_\epsilon^{2^*}), \quad \forall u^2 \in U^2. \tag{iv}$$

If (iii) and (iv) are collected together, the result is the set of inequalities

$$J(u_\epsilon^{1^*}, u^2) - \epsilon < J(u_\epsilon^{1^*}, u_\epsilon^{2^*}) < J(u^1, u_\epsilon^{2^*}) + \epsilon, \qquad (v)$$

for all $u^1 \in U^1$, $u^2 \in U^2$, which verifies the sufficiency part of the theorem, in view of Def. 4.2.

Second, suppose that for every $\epsilon > 0$, an ϵ saddle point exists, that is, a pair $\{u_\epsilon^{1^*} \in U^1, u_\epsilon^{2^*} \in U^2\}$ can be found satisfying (v) for all $u^1 \in U^1$, $u^2 \in U^2$. Let the middle term be denoted as J_ϵ. We now show that the sequence $\{J_{\epsilon_1}, J_{\epsilon_2}, \ldots\}$, with $\epsilon_1 > \epsilon_2 > \cdots > 0$ and $\lim_{i\to\infty} \epsilon_i = 0$, is Cauchy.[25] Toward this end, let us first take $\epsilon = \epsilon_k$ and $\epsilon = \epsilon_j$, $j > k$, in subsequent order in (v) and add the resulting two inequalities to obtain

$$J(u_{\epsilon_k}^{1^*}, u^2) - J(u^1, u_{\epsilon_j}^{2^*}) + J(u_{\epsilon_j}^{1^*}, u^2) - J(u^1, u_{\epsilon_k}^{2^*}) \le 2(\epsilon_k + \epsilon_j) < 4\epsilon_k.$$

Now, substituting first $\{u^1 = u_{\epsilon_j}^{1^*}, u^2 = u_{\epsilon_k}^{2^*}\}$ and then $\{u^1 = u_{\epsilon_k}^{1^*}, u^2 = u_{\epsilon_j}^{2^*}\}$ in the preceding inequality, we get

$$-4\epsilon_k < J_{\epsilon_k} - J_{\epsilon_j} < 4\epsilon_k$$

for any finite k and j, with $j > k$, which proves that $\{J_{\epsilon_k}\}$ is indeed a Cauchy sequence. Hence, it has a limit in $\mathbf{R}$, which is the value of the game.

It should be noted that, although the sequence $\{J_{\epsilon_k}\}$ converges, the sequences $\{u_{\epsilon_k}^{1^*}\}$ and $\{u_{\epsilon_k}^{2^*}\}$ need not have limits. □

Extension to mixed strategies

All definitions, as well as the result of Thm. 4.1, of this section have been presented for pure strategies. We now outline possible extensions to mixed strategies.

We define a mixed strategy for $\mathbf{P}i$ as a probability distribution μ^i on U^i, and denote the class of all such probability distributions by M^i. For the case in which U^i is a finite set, we have already seen in Chapters 2 and 3 that M^i is a simplex of an appropriate dimension. For an infinite U^i, however, its description will be different. Let us consider the simplest case in which U^i is the unit interval $[0, 1]$. Then, elements of M^i are defined as mappings $\mu : [0, 1] \to [0, 1]$ with the properties

$$\left.\begin{array}{ll} \mu(0) = 1 - \mu(1) = 0, & \\ \mu(u) \ge \mu(\tilde{u}) & \text{whenever } u > \tilde{u}, \\ \mu(u) = \mu(u+) & \text{if } u \ne 0, \end{array}\right\} \qquad (4.1)$$

that is, M^i is the class of all probability distribution functions defined on $[0, 1]$. More generally, if U^i is a closed and bounded subset of a finite dimensional space, a mixed strategy for $\mathbf{P}i$ can likewise be defined since every such U^i has the cardinality of the continuum and can therefore be mapped into the unit

[25] For a definition of a Cauchy sequence, the reader is referred to Appendix A.

interval in a one-to-one manner. However, such a definition of a mixed strategy leads to difficulties in the construction of the average cost function, and therefore it is convenient to define μ^i as a probability measure (see Ash (1972)) on U^i, which is valid even if U^i is not closed and bounded. The average cost function of $\mathbf{P}i$ is then defined as a mapping $\bar{J}^i : M^1 \times \cdots \times M^N \to \mathrm{R}$ by

$$\bar{J}^i(\mu^1, \ldots, \mu^N) = \int_{U^1} \cdots \int_{U^N} J^i(u^1, \ldots, u^N)\, \mathrm{d}\mu^1(u^1) \cdots \mathrm{d}\mu^N(u^N) \qquad (4.2)$$

whenever this integral exists, which has to be interpreted in the Lebesgue-Stieltjes sense (see Appendix B.3).

With the average cost functions defined as in (4.2) the *mixed ϵ Nash equilibrium solution* concept can be introduced as in Def. 4.1, by merely replacing J^i by $\bar{J}^i$, U^i by M^i, u^i by μ^i and $u_\epsilon^{i^*}$ by $\mu_\epsilon^{i^*}$. Analogously, Def. 4.2 can be extended to mixed strategies, and Thm. 4.1 finds a direct counterpart in mixed strategies, with $\underline{V}$ and $\bar{V}$ replaced by

$$\underline{V}_{\mathrm{m}} = \sup_{\mu^2 \in M^2} \inf_{\mu^1 \in M^1} \bar{J}(\mu^1, \mu^2)$$

and

$$\bar{V}_{\mathrm{m}} = \inf_{\mu^1 \in M^1} \sup_{\mu^2 \in M^2} \bar{J}(\mu^1, \mu^2),$$

respectively, and with V replaced by V_{m}.

We devote the remainder of this section to the mixed ϵ Nash equilibrium solution concept within the context of semi-infinite bimatrix games. The two matrices, whose entries characterize the possible costs to be incurred to **P1** and **P2**, will be denoted by $A = \{a_{ij}\}$ and $B = \{b_{ij}\}$, respectively, and both will have size $(m \times \infty)$. If $m < \infty$, then we speak of a semi-infinite bimatrix game, and if $m = \infty$, we simply have an infinite bimatrix game. The case when the matrices both have the size $(\infty \times n)$ follows the same lines and will not be treated separately.

A mixed strategy for **P1**, in this case, is an m-tuple $\{y_1, y_2, \ldots, y_m\}$ satisfying $\sum_{i=1}^m y_i = 1$, $y_i \geq 0$. The class of such m-tuples is denoted by Y. The class Z is defined, similarly, as the class of all sequences $\{z_1, z_2, \ldots\}$ satisfying $\sum_{i=1}^\infty z_i = 1$, $z_i \geq 0$. Each element of Z defines a mixed strategy for **P2**. Now, for each pair $\{y \in Y, z \in Z\}$, the average costs incurred to **P1** and **P2** are defined, respectively, by

$$\bar{J}^1(y, z) = \sum_i \sum_j y_i a_{ij} z_j; \quad \bar{J}^2(y, z) = \sum_i \sum_j y_i b_{ij} z_j,$$

assuming that the infinite series corresponding to these sums are absolutely convergent. If the entries of A and B are bounded, for example, then such a requirement is fulfilled for every $y \in Y$, $z \in Z$. The following theorem now verifies the existence of mixed ϵ equilibrium strategies for the case $m < \infty$.

Theorem 4.2 *For each $\epsilon > 0$, the semi-infinite bimatrix game (A, B), with $A = \{a_{ij}\}_{i=1}^{m}{}_{j=1}^{\infty}, B = \{b_{ij}\}_{i=1}^{m}{}_{j=1}^{\infty}, m < \infty$ and the entries a_{ij} and b_{ij} being bounded, admits an ϵ equilibrium solution in mixed strategies.*

Proof. Let $\underline{\mathbf{N}}$ denote the class of all positive integers. For each $j \in \underline{\mathbf{N}}$, define c_j as the jth column of B, i.e., $c_j = Be_j$. The set $C = \{c_j, j \in \underline{\mathbf{N}}\}$ is a bounded subset of $\mathbf{R}^m$. Let $\epsilon > 0$ be given. Then, a finite subset $\tilde{C} \subset C$ exists such that to each $c \in C$ a $\tilde{c} \in \tilde{C}$ corresponds with the property $\|c - \tilde{c}\|_\infty < \epsilon$, which means that the magnitude of each component of $(c - \tilde{c})$ is less than ϵ. Without loss of generality, we assume that $\tilde{C} = \{c_1, c_2, \ldots, c_n\}$ with $n < \infty$. If necessary, the pure strategies of **P2** can be rearranged so that $\tilde{C}$ consists of the first n columns of B.

Define $A_n = \{a_{ij}\}_{i=1}^{m}{}_{j=1}^{n}, B_n = \{b_{ij}\}_{i=1}^{m}{}_{j=1}^{n}$. By Thm. 3.1, the "truncated" bimatrix game (A_n, B_n) has at least one equilibrium point in mixed strategies, to be denoted by $(\hat{y}, \hat{z})$. We now prove that $(\hat{y}, \bar{z})$ is a mixed ϵ equilibrium solution for (A, B), where $\bar{z}$ is defined as $(\hat{z}', 0, 0, \ldots)'$. For all $y \in Y$ we have

$$y' A\bar{z} = y' A_n \hat{z} \geq \hat{y}' A_n \hat{z} = \hat{y}' A\bar{z}. \qquad (i)$$

For each $r \in \underline{\mathbf{N}}$, we choose a $b(r) \in \{1, \ldots, n\}$ such that $\|c_r - c_{b(r)}\|_\infty < \epsilon$. For each $j \in \{1, \ldots, n\}$, let $R(j) \triangleq \{r \in \underline{\mathbf{N}} : b(r) = j\}$. Next, for each $z \in Z$, we define $\tilde{z} = (\tilde{z}_1, \ldots, \tilde{z}_n)$ by $\tilde{z}_j = \sum_{r \in R(j)} z_r$. Then $\tilde{z}_j \geq 0$ and $\sum \tilde{z}_j = 1$. Subsequently,

$$Bz = \sum_{r=1}^{\infty} z_r c_r \geq \sum_{j=1}^{\infty} \left(\sum_{r \in R(j)} z_r \right) c_j - \epsilon 1_m = B_n \tilde{z} - \epsilon 1_m,$$

where $1_m = (1, \ldots, 1)' \in \mathbf{R}^m$. Hence

$$\hat{y}' Bz \geq \hat{y}' B_n \tilde{z} - \epsilon \geq \hat{y}' B_n \hat{z} - \epsilon = \hat{y}' B\bar{z} - \epsilon \qquad (ii)$$

for all $z \in Z$. Inequalities (i) and (ii) verify that $(\hat{y}, \bar{z})$ constitutes a mixed ϵ equilibrium solution for (A, B). □

Corollary 4.1 *Every bounded semi-infinite zero-sum matrix game has a value in the class of mixed strategies.*

Proof. This result is a direct consequence of Thm. 4.2, and constitutes a natural extension of Thm. 4.1 to mixed strategies. □

For $(\infty \times \infty)$ bimatrix games such powerful results do not exist as shown in the following (counter-) example.

Example 4.2 Each one of two players chooses a natural number, independently of the other. The player who has chosen the highest number wins and receives one unit from the other player. If both players choose the same number, then the outcome is a draw. The matrix corresponding to this zero-sum game is

$$\begin{bmatrix} 0 & 1 & 1 & 1 & \dots \\ -1 & 0 & 1 & 1 & \dots \\ -1 & -1 & 0 & 1 & \dots \\ -1 & -1 & -1 & 0 & \dots \\ \vdots & \vdots & \vdots & \vdots & \\ \vdots & \vdots & \vdots & \vdots & \end{bmatrix}$$

whose size is $(\infty \times \infty)$. For this game, $\underline{V} = -1$ and $\bar{V} = +1$ and consequently the average value V_{m} does not exist, in spite of the fact that the matrix entries are bounded. □

4.3 Continuous-Kernel Games: Reaction Curves, and Existence and Uniqueness of Nash and Saddle-Point Equilibria

This section deals with static games in which the number of alternatives available to each player is a continuum and the cost functionals are continuous. In particular, we shall consider the class of games for which a pure strategy of each player can be represented as an element of a finite-dimensional space, and hence take $U^i \subset \mathrm{R}^{m_i}$, where m_i is an integer denoting the dimension of the decision vector of $\mathbf{P}i$. We will sometimes speak of "games on the square", which will refer to two-player games where $U^i = [0, 1]$, $i = 1, 2$.

We first discuss the role of reaction curves in the construction of pure-strategy Nash equilibria in continuous-kernel games, and present some classification of Nash equilibria based on the topological classifications of the reaction curves. Then, we state and prove some results on the existence of Nash equilibria, and as a special case the existence of saddle-point solutions in continuous-kernel games, in both pure- and mixed-strategy spaces. Finally, we discuss some computational algorithms motivated by the notion of *stability* of Nash equilibria.

Reaction curves and classification of Nash equilibria

Pure-strategy Nash equilibrium solution in infinite static games can be obtained as the common intersection point of the reaction curves of the players. The concept of a reaction set was already introduced in Chapter 3 (Def. 3.26) within the context of finite games, and its counterpart in infinite games describes a curve, or a family of curves, with some properties like continuity and differentiability, depending on the structure of the action sets and the cost functionals. We now make this notion precise for N-person games.

Definition 4.3 *In an N-person nonzero-sum game, let the minimum of the cost function of* **P1**, $J^1(u^1, \cdots, u^N)$, *with respect to* $u^1 \in U^1$ *be attained for each* $u_{-1} \in U_{-1}$, *where* $u_{-1} \triangleq \{u^2, \ldots, u^N\}$ *and* $U_{-1} \triangleq U^2 \times \cdots \times U^N$. *Then, the set* $R^1(u_{-1}) \subset U^1$ *defined by*

$$R^1(u_{-1}) = \{\xi \in U^1 : J^1(\xi, u_{-1}) \leq J^1(u^1, u_{-1}), \quad \forall u^1 \in U^1\}$$

is called the optimal response *or* rational reaction set *of* **P1**. *If* $R^1(u_{-1})$ *is a singleton for every* $u_{-1} \in U_{-1}$, *then it is called the* reaction curve *or* reaction function *of* **P1**, *and is denoted by* $l_1(u_{-1})$. *The reaction sets and curves of* **P***i*, $i = 2, \ldots, N$ *are similarly defined (simply by replacing the index* 1 *by* i*).*

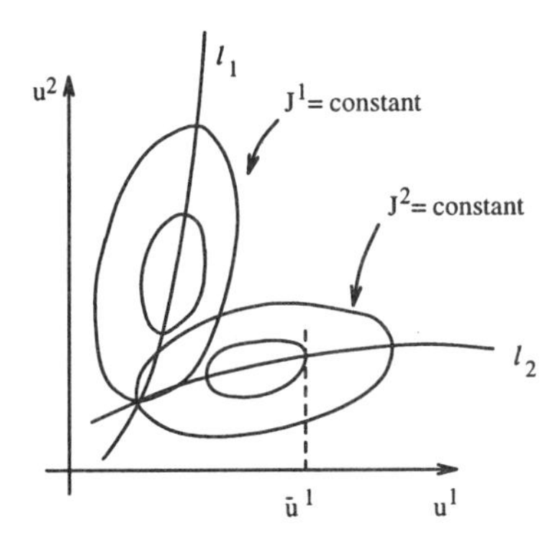

Figure 4.2: Constant level curves for J^1 and J^2, and the corresponding reaction curves (l_1 and l_2) of **P1** and **P2**, respectively.

To illustrate the role of reaction curves in the derivation of Nash equilibria, we have drawn in Fig. 4.2, the "constant level" or *iso-cost* curves corresponding to two cost functions J^1 and J^2 for a specific two-person game with $U^1 = U^2 =$ R. For fixed u^1, say $u^1 = \bar{u}^1$, the best **P2** can do is to minimize J^2 along the line $u^1 = \bar{u}^1$. Assuming that this minimization problem admits a unique solution, the said optimal response of **P2** is determined, in the figure, as the point where the line $u^1 = \bar{u}^1$ is tangent to iso-cost curve $J^2 =$ constant. For each different $\bar{u}^1$, a possibly different unique optimal response can thus be found for **P2**, and the collection of all these points forms the reaction curve of **P2**, indicated by l_2 in the figure. The reaction curve of **P1** is similarly constructed: it is the collection of all points (u^1, u^2) where horizontal lines are tangent to the iso-cost curves of J^1, and it is indicated by l_1 in the figure. By definition, the Nash solution must lie on both reaction curves, and therefore, if these curves have only one point of intersection, as in the figure, the Nash solution exists and is unique.

The configuration of the reaction curves is not always as depicted in Fig. 4.2. In Fig. 4.3, some other possibilities have been displayed. In all these cases it has been assumed that $U^1 = U^2 =$ **R**. In Fig. 4.3a, l_1 and l_2 are parallel straight lines and neither a Nash nor an ϵ Nash solution exists. In Fig. 4.3b, l_1 and l_2

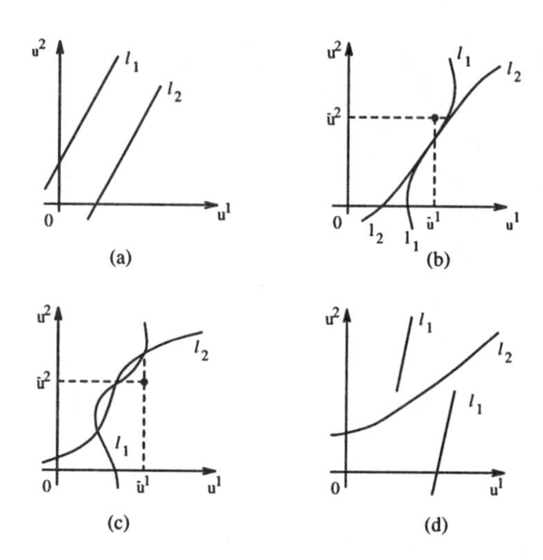

Figure 4.3: Possible configurations of the reaction curves.

partly coincide and there exists a continuum of Nash solutions. In Fig. 4.3c, there is a finite number of Nash solutions. In Fig. 4.3d, one of the reaction curves is not connected and neither a Nash solution nor an ϵ Nash solution exists.

The nonzero-sum games associated with Figs. 4.3b and 4.3c both exhibit multiple Nash equilibria, in which case, as discussed earlier in Section 3.2, the Nash concept is weak as an equilibrium solution concept. Unless other criteria are imposed, there is no reason why players should prefer one particular equilibrium solution over the other(s). In fact, since the players make their decisions independently, it could so happen that the outcome of their joint (but noncooperative) choices is not an equilibrium point at all. Such nonequilibrium outcomes of unilateral equilibrium strategies are shown in Figs. 4.3b and 4.3c, where **P1** picks a Nash strategy $\bar{u}^1$ and **P2** chooses a Nash strategy $\bar{u}^2$.

In the case of nonunique Nash equilibrium solutions, we can classify them in a number of ways. One such classification is provided by the notion of "robustness", which is introduced below for (two-person) games on the square, which, however, easily extends to general N-person games covered by Def. 4.3.

Definition 4.4 *Given two connected curves $u^2 = l_2(u^1)$ and $u^1 = l_1(u^2)$ on the square, denote their weak δ-neighborhoods by N_δ^2 and N_δ^1, respectively.*[26] *Then, a point P of intersection of these two curves is said to be* robust *if, given $\epsilon > 0$, there exists a $\delta_0 > 0$ so that every ordered pair selected from $N_{\delta_0}^2$ and $N_{\delta_0}^1$ has an intersection in an ϵ-neighborhood of P.*

As specific examples, consider the two-person nonzero-sum games for which the reaction functions are depicted in Figs. 4.4a and 4.4b. In Fig. 4.4a, the

[26] A weak δ-neighborhood N_δ, of $l : [0,1] \to [0,1]$ is defined as the set of all maps $\tilde{l} : [0,1] \to [0,1]$ such that $|l(\xi) - \tilde{l}(\xi)| < \delta$, $\forall \xi \in [0,1]$.

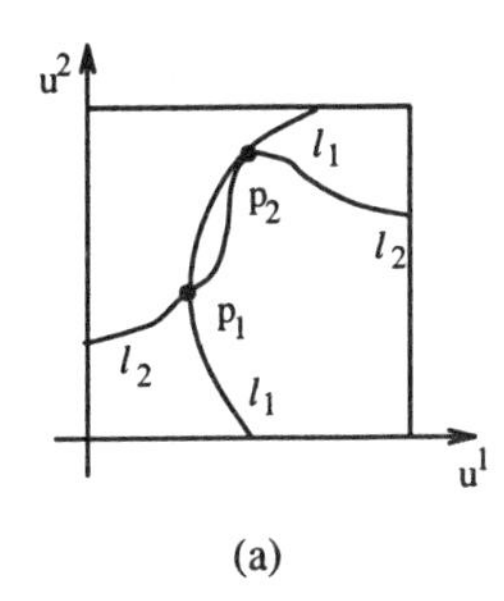

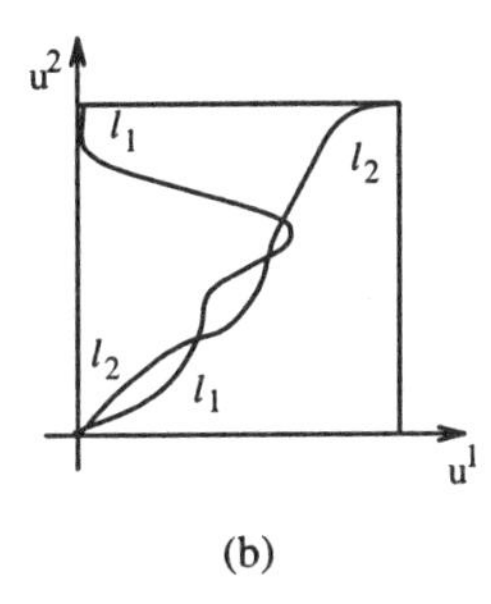

Figure 4.4: Games on the square, illustrating robust Nash solutions.

point P_1, constitutes a robust Nash solution, while point P_2 is not robust. In Fig. 4.4b, however, all the Nash equilibrium solutions are robust. It should be noted that the number of robust intersections of two connected curves is actually odd, provided that possible points of intersections at the corners of the square are excluded. Hence, we can say that nonzero-sum two-person static games on the square, with well-defined reaction functions for both players, essentially admit an odd number of Nash equilibrium solutions. An exception to this rule is the case when the two reaction curves intersect only at the corners of the square.

Yet another classification within the class of multiple Nash equilibrium solutions of a two-person nonzero-sum static game is provided by the notion of "stability" of the solution(s) of the fixed point equation. Given a Nash equilibrium solution, consider the following sequence of moves: (i) One of the players (say **P1**) deviates from his corresponding equilibrium strategy, (ii) **P2** observes this and minimizes his cost function in view of the new strategy of **P1**, (iii) **P1** now optimally reacts to that (by minimizing his cost function), (iv) **P2** optimally reacts to that optimum reaction, etc. Now, if this infinite sequence of moves converges to the original Nash equilibrium solution, and this being so regardless of the nature of the initial deviation of **P1**, we say that the Nash equilibrium solution is *stable*. If convergence is valid only under small initial deviations, then we say that the Nash equilibrium solution is *locally stable*. Otherwise, the Nash solution is said to be *unstable*. A nonzero-sum game can of course admit more than one locally stable equilibrium solution, but a stable Nash equilibrium solution has to be unique.

The reaction functions of two different nonzero-sum games on the square are depicted in Fig. 4.5. In the first one (Fig. 4.5a) the equilibrium solution is stable, whereas in the second one it is unstable.

The notion of *stability*, as introduced above for two-person games, brings in a refinement to the concept of Nash equilibrium, which finds natural extensions to the N-player case. Essentially, we have to require that the equilibrium be "restorable" under any rational readjustment scheme when there is a deviation from it by any player. For $N > 2$ this will depend on the specific scheme adopted, which brings us to the following formal definition of a stable Nash equilibrium.

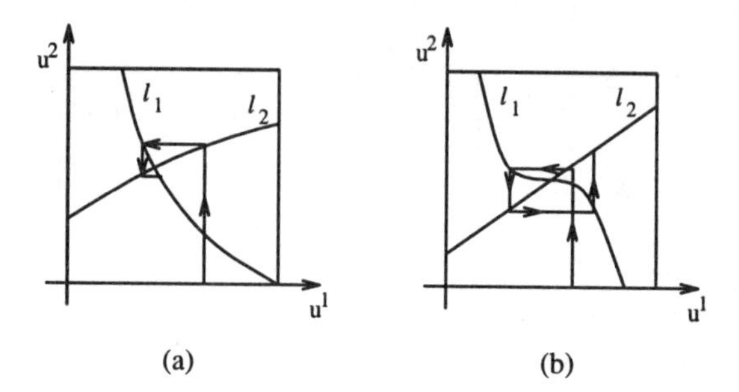

Figure 4.5: Illustration of (a) a stable and (b) an unstable Nash solution.

Definition 4.5 *A Nash equilibrium $u^{i^*}, i \in \mathbf{N}$, is (*globally*) *stable *with respect to an adjustment scheme $\mathcal{S}$ if it can be obtained as the limit of the iteration:*

$$u^{i^*} = \lim_{k\to\infty} u^{i(k)}, \tag{4.3}$$

$$u^{i(k+1)} = \arg \min_{u^i \in U^i} J^i(u_{-i}^{(\mathcal{S}_k)}, u^i), \quad u^{i(0)} \in U^i, \quad i \in \mathbf{N}, \tag{4.4}$$

where the superscript $\mathcal{S}_k$ indicates that the precise choice of $u_{-i}^{(\mathcal{S}_k)}$ depends on the readjustment scheme selected.

One possibility for the scheme above is

$$u_{-i}^{(\mathcal{S}_k)} = u_{-i}^{(k)} \tag{4.5}$$

which corresponds to the situation where the players update (readjust) their actions simultaneously, in response to the most recently determined actions of the other players. Yet another possibility is

$$u_{-i}^{(\mathcal{S}_k)} = \left(u^{1(k+1)}, \ldots, u^{i-1(k+1)}, u^{i+1(k)}, \ldots, u^{N(k)}\right), \tag{4.6}$$

where the players do the update in a predetermined (in this case numerical) order. A third possibility is

$$u_{-i}^{(\mathcal{S}_k)} = \left(u^{1 m_{1,k}^i}, \ldots, u^{i-1(m_{i-1,k}^i)}, u^{i+1(m_{i+1,k}^i)}, \ldots, u^{N(m_{N,k}^i)}\right), \tag{4.7}$$

where $m_{j,k}^i$ is an integer-valued random variable, satisfying the bounds

$$\max(0, k-d) \leq m_{j,k}^i \leq k+1, \quad j \neq i, j \in \mathbf{N}, i \in \mathbf{N},$$

which corresponds to a situation where $\mathbf{P}i$ receives action update information from $\mathbf{P}j$ at random times, with the delay not exceeding d time units.

Clearly, if the iteration of Def. 4.5 converges under any one of the readjustment schemes above (or any other readjustment scheme where a player receives update information from every other player infinitely often), then the Nash equilibrium solution is *unique*. Every unique Nash equilibrium, however, is not necessarily *stable*, nor is a Nash equilibrium that is stable with respect to a particular readjustment scheme necessarily stable with respect to some other scheme. Hence *stability* is generally given with some qualification (such as "stable with respect to scheme $\mathcal{S}$" or "with respect to a given class of schemes"), except when the number of players is two, in which case all schemes (with at most a finite delay in the transmission of update information) lead to the same condition of stability, as one then has the simplified recursions

$$u^{i(r_{k+1,i})} = \tilde{l}_i(u^{i(r_{k,i})}), \quad k = 0, 1, \ldots; \quad i = 1, 2,$$

where $r_{1,i}$, $r_{2,i}$, $r_{3,i}$, ... denote the time instants when $\mathbf{P}i$ receives new action update information from $\mathbf{P}j$, $j \neq i$, i, $j = 1, 2$.

Existence of Nash equilibria

In view of the discussion in the previous subsection, the existence of pure-strategy Nash equilibria in N-person continuous-kernel games can be established by proving existence of well-defined reaction functions with a common point of intersection. The following theorem does precisely that, and provides a set of sufficient conditions under which N-person nonzero-sum games admit pure-strategy Nash equilibria.

Theorem 4.3 *For each $i \in \mathbf{N}$, let U^i be a closed, bounded and convex subset of a finite-dimensional Euclidean space, and the cost functional $J^i : U^1 \times \cdots \times U^N \to \mathbf{R}$ be jointly continuous in all its arguments and strictly convex in u^i for every $u^j \in U^j$, $j \in \mathbf{N}$, $j \neq i$. Then, the associated N-person nonzero-sum game admits a Nash equilibrium in pure strategies.*

Proof. Let us take $i = 1$. By strict convexity, there exists a unique mapping $l_1 : U_{-1} \to U^1$ such that $u^1 = l_1(u^2, \ldots, u^N)$ uniquely minimizes $J^1(u^1, \ldots, u^N)$ for any given $(N-1)$-tuple $\{u^2, \ldots, u^N\}$.

The mapping l_1 is actually the reaction function of $\mathbf{P}1$ in this N-person game. Similarly, reaction functions l_i, $i = 2, \ldots, N$, can be defined as unique mappings from U_{-i} into U^i. Using vector notation, these relations can be written in compact form as $u = L(u)$, where $u = (u^1, \ldots, u^N) \in U \triangleq U^1 \times \cdots \times U^N$, and $L = (l_1, \ldots, l_N)$. It will be shown in the sequel that the individual reaction functions l_i are continuous in their arguments, and hence L is a continuous mapping. Since L maps a closed and bounded subset U of a finite-dimensional space into the same subset, this then implies, by utilization of Brouwer's fixed point theorem (see Appendix C), that there exists a $u^* \in U$ such that $u^* = L(u^*)$ (that is, u^* is a fixed point of L). Obviously, the individual components of u^* constitute a Nash equilibrium solution.

To complete the proof of the theorem, what remains to be shown is the continuity of l_i. Let us take $i = 1$ and assume that, to the contrary, l_1 is discontinuous at $(u_0^2, \ldots, u_0^N)$. Further, let $l_1(u_0^2, \ldots, u_0^N) = u_0^1$. Then there exists a sequence of vectors $\{\bar{u}_j \triangleq (u_j^2, \ldots, u_j^N)'; \; j = 1, 2, \ldots\}$ such that $(u_0^2, \ldots, u_0^N)'$ is the limit of this sequence but u_0^1 is not the limit of $l_1(u_j^2, \ldots, u_j^N)$ as $j \to \infty$. Because of compactness of the action spaces, there exists a subsequence of $\{\bar{u}_j\}$, say $\{\bar{u}_{j_k}\}$, such that $l_1(\bar{u}_{j_k})$ converges to a limit $\tilde{u}_0^1 \neq u_0^1$, and simultaneously the following inequality holds:

$$J^1(l_1(\bar{u}_{j_k}), \bar{u}_{j_k}) < J^1(u_0^1, \bar{u}_{j_k}).$$

Now, by taking limits with respect to the subsequence of indices $\{j_k\}$, one obtains the inequality

$$J^1(\tilde{u}_0^1, u_0^2, \ldots, u_0^N) \leq J^1(u_0^1, u_0^2, \ldots, u_0^N),$$

which, together with $\tilde{u}_0^1 \neq u_0^1$, forms a contradiction to the initial hypothesis that u_0^1 is the unique u^1 which minimizes $J^1(u^1, u_0^2, \ldots, u_0^N)$. Hence, l_1 is continuous. The continuity of l_i, $i > 1$, can analogously be proven. □

One of the conditions of Thm. 4.3 was the compactness of the action spaces U^i, which was needed in the proof of the theorem to ensure that for each $u_{-i} \in U_{-i}$ there exists a $u^i \in U^i$ that minimizes J^i, and hence that the reaction curve of each player is well defined. If the U^i are unbounded (such as $U^i = \mathbf{R}^{m_i}$), then the same end result can be obtained provided that the cost function of the corresponding player becomes arbitrarily large (positive) as its action variable becomes arbitrarily large in norm. Under such a condition, the player's action variable can be restricted to a closed and bounded set, without any loss of generality, and hence the theorem again applies. This reasoning then leads to the following corollary to Thm. 4.3.

Corollary 4.2 *For each $i \in \mathbf{N}$, let $U^i = \mathbf{R}^{m_i}$, the cost functional $J^i : U^1 \times \cdots \times U^N \to \mathbf{R}$ be jointly continuous in all its arguments and strictly convex in u^i for every $u^j \in U^j$, $j \in \mathbf{N}$, $j \neq i$. Furthermore, let*

$$J^i(u^i, u_{-i}) \to \infty \quad \text{as } |u^i| \to \infty, \quad \forall u_{-i} \in U_{-i}, \quad i \in \mathbf{N}\,.$$

Then, the associated N-person nonzero-sum game admits a Nash equilibrium in pure strategies.

The notion of *stability* introduced in the previous subsection (cf. Def. 4.5) can be used to sharpen the result above for two-person games, so as to establish unicity of equilibria under some additional conditions. Toward that end, let J^1 and J^2 be twice continuously differentiable on $\mathbf{R}^{m_1} \times \mathbf{R}^{m_2}$, with J^1 strictly convex on $\mathbf{R}^{m_1}$ for each $u^2 \in \mathbf{R}^{m_2}$, and J^2 strictly convex on $\mathbf{R}^{m_2}$ for each $u^1 \in \mathbf{R}^{m_1}$. Consider the parallel readjustment scheme whereby each player responds (optimally) to the previously selected action of the other player, which can be written as

$$u^{1^{(k+1)}} = l_1(u^{2^{(k)}}); \qquad u^{2^{(k+1)}} = l_2(u^{1^{(k)}}),$$

where l_1 and l_2 are continuous (reaction) functions, uniquely solvable from

$$\nabla_{u^1} J^1(l_1(u^2), u^2) = 0, \qquad \nabla_{u^2} J^2(u^1, l_2(u^1)) = 0.$$

Hence, if (u^{1^*}, u^{2^*}) is a stable Nash equilibrium, then

$$u^{1^*} = \lim_{k\to\infty} u^{1^{(k)}},$$

where $u^{1^{(k)}}$ is generated from

$$u^{1^{(k+1)}} = l_1 \circ l_2(u^{1^{(k)}}).$$

Now introduce the matrix

$$\begin{aligned} T(u,w,v) \triangleq\ & \left[\nabla^2_{u^1} J^1(u,v)\right]^{-1} \nabla_{u^1u^2} J^1(u,v) \\ & \cdot \left[\nabla^2_{u^2} J^2(w,v)\right]^{-1} \nabla_{u^2u^1} J^2(w,v), \end{aligned} \tag{4.8}$$

where w is an arbitrary m_1-dimensional vector, $v = l_2(w)$, and $u = l_1(v)$. If the operator norm of T (equivalently, its spectral radius) is bounded above by α, $0 \le \alpha < 1$, for all $w \in \mathbf{R}^{m_1}$, it can be shown (see Li and Başar , 1987) that for some $\bar{\alpha}$, $\alpha < \bar{\alpha} < 1$,

$$|u^{1^{(k+1)}} - u^{1^*}| \le \bar{\alpha}|u^{1^{(k)}} - u^{1^*}|, \quad k = 0, 1, 2, \ldots$$

which implies (by a contraction mapping argument) that $u^{1^{(k)}} \to u^{1^*}$, and hence $u^{2^{(k)}} = l_2(u^{1^{(k)}}) \to u^{2^*}$—which proves existence of a stable Nash equilibrium, which by definition is also unique. Hence, we have the following.

Proposition 4.1 *For a two-person nonzero-sum game, in addition to the hypotheses of Corollary 4.2, assume that J^i is twice continuously differentiable in u^i for each $u^j \in \mathbf{R}^{m_j}$, $j \ne i$, $i,j = 1,2$. Further assume that the matrix function (4.8) has operator norm strictly less than 1. Then the game admits a unique Nash equilibrium, which is also stable.*

Remark 4.1 One class of games for which the last condition of Prop. 4.1 (that is, the one involving the operator norm of (4.8)) is satisfied is that with *weakly coupled* players, that is one with cost functions:

$$J^i(u^1, u^2; \epsilon) = J^i_i(u^i) + \epsilon J^i_j(u^1, u^2), \quad j \ne i, \quad i,j = 1,2,$$

where ϵ is a sufficiently small scalar. More precisely, there exists an $\epsilon_1 > 0$ such that for all $\epsilon \in (-\epsilon_1, \epsilon_1)$, this class of convex-kernel games admits a unique stable Nash equilibrium, which can be obtained as the limit point of the parallel update scheme introduced earlier (see Srikant and Başar , 1992). □

One important class of games not covered by Thm. 4.3 are those where the action spaces (or the constraint sets) of the players are not rectangular. As an example of such a scenario consider the two-person scalar game on the unit circle (instead of the unit square), where the scalar decision variables of the two players are coupled through the inequality $(u^1)^2 + (u^2)^2 \le 1$. Note that in such a case, there does not exist a U^1 and a U^2 from which the actions of the players can be chosen independently, but rather a single set $U = \{(u^1, u^2) : (u^1)^2 + (u^2)^2 \le 1\}$, to which the pair (u^1, u^2) belongs. Extending this to the N-person case, we have the *coupled constraint set* $U \subset \mathbf{R}^m$, where the N-tuple $u \triangleq (u^1, u^2, \ldots, u^N)$ belongs, with $m \triangleq \sum_{i=1}^N m_i$. The concept of *Nash equilibrium* is still well defined in this general framework, with the obvious modification that $u \in U$ is a Nash equilibrium if

$$J^i(u) \le J^i(u_{-i}, v^i), \quad \forall v^i \in U^i(u_{-i}), \quad i \in \mathbf{N}\,, \tag{4.9}$$

where $U^i(u_{-i})$ is a subset of $\mathbf{R}^{m_i}$, obtained by the projection

$$U^i(u_{-i}) \triangleq \{v^i \in \mathbf{R}^{m_i} : (u_{-i}, v^i) \in U\}\,.$$

Note that (4.9) can equivalently be written as

$$J(u; u) \le J(u; v), \quad \forall v \in U\,, \tag{4.10}$$

where

$$J(u; v) \triangleq \sum_{i=1}^N J^i(u_{-i}, v^i),$$

a form that we will have occasion to use in the sequel. The following theorem now generalizes the result of Thm. 4.3 to such games.[27]

Theorem 4.4 *Let U be a closed, bounded and convex subset of $\mathbf{R}^m$, and for each $i \in \mathbf{N}$ the cost functional $J^i : U \to \mathbf{R}$ be continuous on U and convex in u^i for every $u^j \in U^j$, $j \in \mathbf{N}$, $j \neq i$. Then, the associated N-person nonzero-sum game admits a Nash equilibrium in pure strategies.*

Proof. First note that under the hypotheses of the theorem the function $J(u; v)$ defined prior to the statement of the theorem is continuous in u and v and is convex in v for every fixed v, with $(u, v) \in U \times U$. Introduce the reaction set for the game:

$$Tu = \{v \in U : J(u; v) \le J(u; w) \quad \forall w \in U\},$$

which, by the continuity and convexity property of J cited above, is an upper semicontinuous mapping that maps each point u in the convex and compact set U into a closed convex subset of U. Then, by the Kakutani fixed point theorem (see Appendix C), there exists a point $u^* \in U$ such that $u^* \in Tu^*$, or equivalently such that it minimizes $J(u^*; v)$ over $v \in U$. Such a point indeed

[27] This theorem is of course also valid for the special case in which the action constraint sets are rectangular, and for this class it relaxes the assumption of strict convexity of the cost functions to simple convexity.

constitutes a Nash equilibrium, because if it does not, then this would imply that for some $i \in \mathbf{N}$ there would be a $\bar{u}^i \in U^i(u_{-i}^*)$ such that

$$J^i(u_{-i}^*, \bar{u}^i) < J^i(u^*) ,$$

which would in turn imply (by adding $J^j(u^*)$ to both sides and summing over $j \in \mathbf{N}, j \neq i$) the strict inequality

$$J(u^*; \bar{u}) < J(u^*; u^*), \quad \bar{u} \triangleq (u_{-i}^*, \bar{u}^i) ,$$

contradicting the initial hypothesis that u^* minimizes $J(u^*; v)$ over $v \in U$. □

Since two-person zero-sum games are special types of (two-person) nonzero-sum games, the statements of Thms. 4.3 and 4.4 are equally valid for saddle-point solutions. The first one can even be sharpened to some extent. We first recall the *ordered interchangeability* property of multiple saddle points (viewed as a direct extension of Corollary 2.1), which we will need in the sequel.

Property 4.1 *For a given two-person zero-sum game with rectangular action sets U^1 and U^2 for the players, multiple saddle points satisfy the* ordered interchangeability *property; that is, if (u^{1*}, u^{2*}) and $(\bar{u}^1, \bar{u}^2)$ are two saddle-point equilibrium solutions, so are the pairs $(u^{1*}, \bar{u}^2)$ and $(\bar{u}^1, u^{2*})$.*

Theorem 4.5 *Consider a two-person zero-sum game on convex finite-dimensional action sets $U^1 \times U^2$, defined by the continuous kernel $J(u^1, u^2)$. Suppose that $J(u^1, u^2)$ is strictly convex in u^1 for each $u^2 \in U^2$ and strictly concave in u^2 for each $u^1 \in U^1$. Suppose that either*

(i) U^1 and U^2 are closed and bounded, or

(ii) $U^i = \mathrm{R}^{m_i}, i = 1, 2,$ and $J(u^1, u^2) \to \infty\ \forall u^2$ as $|u^1| \to \infty$, and $J(u^1, u^2) \to -\infty\ \forall u^1$ as $|u^2| \to \infty$.

Then, the game admits a unique pure-strategy saddle-point equilibrium.

Proof. Existence of a saddle-point equilibrium is a direct consequence of Thm. 4.3 and Corollary 4.2. Furthermore, by strict convexity and concavity, there can be no saddle-point solutions outside the class of pure strategies. Hence only uniqueness within the class of pure strategies remains to be proven, which, however, follows readily from the ordered interchangeability property of multiple saddle points (cf Property 4.1). □

The next theorem, which follows as a special case of Thm. 4.4, provides a generalization of the preceding theorem to coupled constraint sets, but in the absence of uniqueness of equilibria.

Theorem 4.6 *Let $J(u^1, u^2)$ be a functional defined on the (not necessarily rectangular) convex and compact action set U. If J is continuous on U, concave in u^2 for each $u^1 \in U^1(u^2)$ and convex in u^1 for each $u^2 \in U^2(u^1)$, then a saddle point exists in pure strategies, but it is not necessarily unique.*

We now focus attention on nonzero-sum games whose cost functionals are continuous but not necessarily convex. For such games one cannot, in general, hope to obtain pure-strategy Nash equilibria; however, in the enlarged class of mixed strategies, the Nash equilibrium solution exists as it is stated in the following theorem.

Theorem 4.7 *An N-person nonzero-sum game in which the finite-dimensional action spaces U^i ($i \in \mathbf{N}$) are compact and the cost functionals J^i ($i \in \mathbf{N}$) are continuous on $U^1 \times \cdots \times U^N$ admits a Nash equilibrium in mixed strategies.*

Proof. A proof of this theorem can be found in Owen (1974). The underlying idea is to make the kernels J^i discrete so as to obtain an N-person finite matrix game that suitably approximates the original game in the sense that a mixed-strategy solution of the latter (which always exists by Thm. 3.1) is arbitrarily close to a mixed equilibrium solution of the former. Compactness of the action spaces ensures that a limit to the sequence of solutions obtained for approximating finite matrix games exists. For yet another proof of this result, see Glicksberg (1950). □

As a special case of Thm. 4.7 we now have the following.

Corollary 4.3 *Every continuous-kernel two-person zero-sum game with compact action (strategy) spaces admits a saddle point in mixed strategies.*

The question of the weakest conditions under which a mixed-strategy saddle point exists naturally arises. Corollary 4.3 says that continuity is a sufficient condition. But it is by no means necessary, and, in fact, it can be replaced by semi-continuity conditions (see Glicksberg, 1950). However, it is not possible to relax the semi-continuity conditions any further and still retain existence of a mixed-strategy saddle point within a general enough framework (see Sion and Wolfe, 1957).

We conclude this section with an example of a zero-sum game whose cost functional is continuous but not convex-concave, and which admits a mixed saddle-point equilibrium.

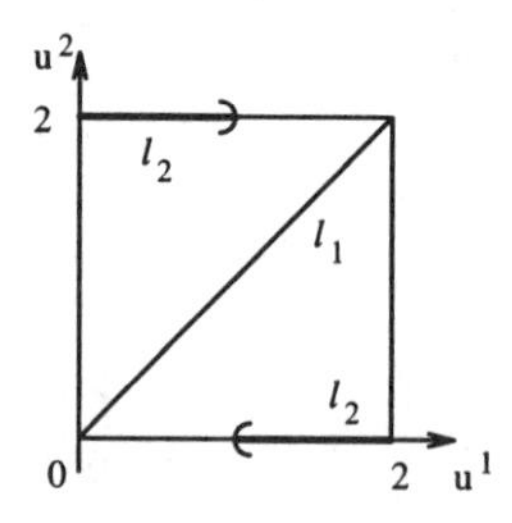

Figure 4.6: Reaction curves for the zero-sum game with kernel $J = (u^1 - u^2)^2$.

Example 4.3 Consider the two-person zero-sum game on the square $[0,2]^2$ characterized by the kernel $J = (u^1 - u^2)^2$. In Fig. 4.6, the two reaction curves

l_1 and l_2 have been drawn, which clearly do not intersect. It can readily be verified that the *upper value* $\bar{V} = 1$ and the *lower value* $\underline{V} = 0$, i.e., a pure-strategy saddle point does not exist. In the extended class of mixed strategies, however, a candidate saddle-point solution directly follows from Fig. 4.6, which is

$$u^{1^*} = 1 \quad \text{w.p. } 1; \quad u^{2^*} = \begin{cases} 0 & \text{w.p. } 1/2, \\ 2 & \text{w.p. } 1/2. \end{cases}$$

It can readily be verified that this pair of strategies indeed provides a mixed saddle-point solution. □

4.4 Stackelberg Solution of Continuous-Kernel Games

This section is devoted to the Stackelberg solution of static nonzero-sum game problems in which the number of alternatives available to each player is not a finite set and the cost functions are described by continuous kernels. For the sake of simplicity and clarity in exposition, we shall deal primarily with two-person static games. A variety of possible extensions of the Stackelberg solution concept to N-person static games with different levels of hierarchy will be briefly mentioned towards the end of the section, with the details left to the interested reader as exercises.

Let $u^i \in U^i$ denote the action variable of $\mathbf{P}i$, where his action set U^i is assumed to be a subset of an appropriate metric space and, the cost function J^i of $\mathbf{P}i$ be defined as a continuous function on the product space $U^1 \times U^2$. Then we can give the following general definition of a Stackelberg equilibrium solution, which is the counterpart of Def. 3.27 for infinite games.

Definition 4.6 *In a two-person game, with* **P1** *as the leader, a strategy* $u^{1^*} \in U^1$ *is called a* Stackelberg equilibrium strategy *for the leader if*

$$J^{1^*} \triangleq \sup_{u^2 \in R^2(u^{1^*})} J^1(u^{1^*}, u^2) \leq \sup_{u^2 \in R^2(u^1)} J^1(u^1, u^2) \tag{4.11}$$

for all $u^1 \in U^1$. *Here,* $R^2(u^1)$ *is the rational reaction set of the follower as introduced in Def. 4.3.*

Remark 4.2 If $R^2(u^1)$ is a singleton for each $u^1 \in U^1$, in other words, if it is described completely by a reaction curve $l_2 : U^1 \to U^2$, then inequality (4.11) in the above definition can be replaced by

$$J^{1^*} \triangleq J^1(u^{1^*}, l_2(u^{1^*})) \leq J^1(u^1, l_2(u^1)) \tag{4.12}$$

for all $u^1 \in U^1$. □

If a Stackelberg equilibrium strategy exists for the leader, then the LHS of inequality (4.11) is known as the *leader's Stackelberg cost*, and is denoted by J^{1^*}. A more general definition for J^{1^*} is, in fact,

$$J^{1^*} = \inf_{u^1 \in U^1} \sup_{u^2 \in R^2(u^1)} J^1(u^1, u^2), \tag{4.13}$$

which also covers the case when a Stackelberg equilibrium strategy does not exist. It follows from this definition that the Stackelberg cost of the leader is a well-defined quantity, and that there will always exist a sequence of strategies for the leader which will ensure him a cost value arbitrarily close to J^{1^*}. This observation brings us to the following definition of ϵ Stackelberg strategies.

Definition 4.7 *Let $\epsilon > 0$ be a given number. Then, a strategy $u_\epsilon^{1^*} \in U^1$ is called an ϵ* Stackelberg strategy *for the leader* (**P1**) *if*

$$\sup_{u^2 \in R^2(u_\epsilon^{1^*})} J^1(u_\epsilon^{1^*}, u^2) \le J^{1^*} + \epsilon.$$

The following two properties of ϵ Stackelberg strategies now readily follow.

Property 4.2 *In a two-person game, let J^{1^*} be a finite number. Then, given an arbitrary $\epsilon > 0$, an ϵ Stackelberg strategy for the leader necessarily exists.*

Property 4.3 *Let $\{u_{\epsilon_i}^{1^*}\}$ be a given sequence of ϵ Stackelberg strategies in U^1, and with $\epsilon_i > \epsilon_j$ for $i < j$ and $\lim_{j\to\infty} \epsilon_j = 0$. Then, if there exists a convergent subsequence $\{u_{\epsilon_{ik}}^{1^*}\}$ in U^1 with its limit denoted as u^{1^*}, and further if $\sup_{u^2 \in R^2(u^1)} J^1(u^1, u^2)$ is a continuous function of u^1 in an open neighborhood of $u^{1^*} \in U^1$, u^{1^*} is a Stackelberg strategy for* **P1**.

The equilibrium strategy for the follower, in a Stackelberg game, would be any strategy that constitutes an optimal response to the one adopted (and announced) by the leader. Mathematically speaking, if u^{1^*} (respectively, $u_\epsilon^{1^*}$) is adopted by the leader, then any $u^2 \in R^2(u^1)$ (respectively, $u^2 \in R^2(u_\epsilon^{1^*})$) will be referred to as an *optimal strategy* for the follower that is in *equilibrium* with the Stackelberg (respectively, ϵ Stackelberg) strategy of the leader. This pair is referred to as a *Stackelberg* (respectively, ϵ Stackelberg) solution of the two-person game with **P1** as the leader (see Def. 3.28). The following theorem now provides a set of sufficient conditions for two-person nonzero-sum games to admit a Stackelberg equilibrium solution.

Theorem 4.8 *Let U^1 and U^2 be compact metric spaces, and J^i be continuous on $U^1 \times U^2$, $i = 1, 2$. Further let there exist a finite family of continuous mappings $l^{(i)} : U^1 \to U^2$, indexed by a parameter $i \in I \triangleq \{1, \ldots, M\}$, so that $R^2(u^1) = \{u^2 \in U^2 : u^2 = l^{(i)}(u^1), i \in I\}$. Then, the two-person nonzero-sum static game admits a Stackelberg equilibrium solution.*

Proof. It follows from the hypothesis of the theorem that J^{1^*}, as defined by (4.13), is finite. Hence, by Property 4.2, a sequence of Stackelberg strategies exists for the leader, and it admits a convergent subsequence whose limit lies in U^1, due to compactness of U^1. Now, since $R^2(\cdot)$ can be constructed from a finite family of continuous mappings (by hypothesis),

$$\sup_{u^2 \in R^2(u^1)} J^1(u^1, u^2) = \max_{i \in I} J^1(u^1, l^{(i)}(u^1)),$$

and the latter function is continuous on U^1. Then, the result follows from Property 4.3. □

Remark 4.3 The assumption of Thm. 4.8, concerning the structure of $R^2(\cdot)$, imposes some severe restrictions on J^2; but such an assumption is inevitable as the following example demonstrates. Take $U^1 = U^2 = [0, 1]$, $J^1 = -u^1u^2$ and $J^2 = (u^1 - \frac{1}{2})u^2$. Here, $R^2(\cdot)$ is determined by a mapping $l(\cdot)$ which is continuous on the half-open intervals $[0, \frac{1}{2}), (\frac{1}{2}, 1]$, but is multivalued at $u^1 = \frac{1}{2}$. The Stackelberg cost of the leader is clearly $J^{1^*} = -\frac{1}{2}$, but a Stackelberg strategy does not exist because of the "infinitely multivalued" nature of l. □

If $R^2(u^1)$ is a singleton for every $u^1 \in U^1$, the hypothesis of Thm. 4.8 can definitely be made less restrictive. One such set of conditions is provided in the following corollary to Thm. 4.8 under which there exists a unique l which is continuous (by an argument similar to the one used in the proof of Thm. 4.3).

Corollary 4.4 *Every two-person nonzero-sum continuous-kernel game on the square, for which $J^2(u^1, \cdot)$ is strictly convex for all $u^1 \in U^1$ and* **P1** *acts as the leader, admits a Stackelberg equilibrium solution.*

It should be noted that the Stackelberg equilibrium solution for a two-person game exists under a set of sufficiency conditions which are much weaker than those required for existence of Nash equilibria (cf. Thm. 4.3). It should further be noted, however, that the statement of Thm. 4.8 does not also rule out the existence of a mixed-strategy Stackelberg solution which might provide the leader with a lower average cost. We have already observed occurrence of such a phenomenon within the context of matrix games, in Section 3.6, and we now investigate to what extent such a result could remain valid in continuous-kernel games.

If mixed strategies are also allowed, then permissible strategies for **P***i* will be probability measures μ^i on the space U^i. Let us denote the collection of all such probability measures for **P***i* by M^i. Then, the quantity replacing J^i will be the average cost function

$$\bar{J}^i(\mu^1, \mu^2) = \int_{U^2} \int_{U^1} J^i(u^1, u^2)\, \mathrm{d}\mu^1(u^1)\, \mathrm{d}\mu^2(u^2), \tag{4.14}$$

and the reaction set R^2 will be replaced by

$$\bar{R}^2(\mu^1) \triangleq \{\hat{\mu}^2 \in M^2 : \bar{J}^2(\mu^1, \hat{\mu}^2) \leq \bar{J}^2(\mu^1, \mu^2), \forall \mu^2 \in M^2\}. \tag{4.15}$$

Hence, we have the following.

Definition 4.8 *In a two-person game with* **P1** *as the leader, a mixed strategy* $\mu^{1^*} \in M^1$ *is called a* mixed Stackelberg equilibrium strategy *for the leader if*

$$\bar{J}^{1^*} \triangleq \sup_{\mu^2 \in \bar{R}^2(\mu^{1^*})} \bar{J}^1(\mu^{1^*}, \mu^2) \leq \sup_{\mu^2 \in \bar{R}^2(\mu^1)} \bar{J}^1(\mu^1, \mu^2)$$

for all $\mu^1 \in M^1$*, where* $\bar{J}^{1^*}$ *is known as the* average Stackelberg cost *of the leader in* mixed strategies.

Proposition 4.2

$$\bar{J}^{1^*} \leq J^{1^*}. \tag{4.16}$$

Proof. Since M^i also includes all one-point measures, we have (by an abuse of notation) $U^i \subset M^i$. Then, for each $u^1 \in U^1$, considered as an element of M^1,

$$\begin{aligned} \inf_{\mu^2 \in M^2} \bar{J}^2(u^1, \mu^2) &\equiv \inf_{\mu^2 \in M^2} \int_{U^2} J^2(u^1, u^2)\, \mathrm{d}\mu^2(u^2) \\ &= \inf_{u^2 \in U^2} J^2(u^1, u^2), \end{aligned}$$

where the last equality follows since any infimizing sequence in M^2 can be replaced by a subsequence of one-point measures. This implies that, for one point measures in $M^1, \bar{R}^2(\mu^1)$ coincides with the set of all probability measures defined on $R^2(u^1)$. Now, since $M^1 \supset U^1$,

$$\bar{J}^{1^*} = \inf_{M^1} \sup_{\bar{R}^2(\mu^1)} \bar{J}^1(\mu^1, \mu^2) \leq \inf_{U^1} \sup_{\bar{R}^2(\mu^1)} \bar{J}^1(\mu^1, \mu^2),$$

and because of the cited relation between $\bar{R}^2(\mu^1)$ and $R^2(u^1)$, the last expression can be written as

$$\inf_{U^1} \sup_{R^2(u^1)} J^1(u^1, u^2) = J^{1^*}.$$

□

We now show, by a counter-example, that, even under the hypothesis of Thm. 4.8, it is possible to have strict inequality in (4.16).

Example 4.4 Consider a two-person continuous-kernel game with $U^1 = U^2 = [0, 1]$, and with cost functions

$$J^1 = \epsilon(u^1)^2 + u^1\sqrt{u^2} - u^2; \quad J^2 = (u^2 - (u^1)^2)^2,$$

where $\epsilon > 0$ is a sufficiently small parameter. The unique Stackelberg solution of this game, in pure strategies, is $u^{1^*} = 0$, $u^{2*} = (u^1)^2$, and the Stackelberg cost for the leader is $J^{1^*} = 0$. We now show that the leader can actually do better by employing a mixed strategy.

First note that the follower's unique reaction to a mixed strategy of the leader is $u^2 = E[(u^1)^2]$ which, when substituted into $\bar{J}^1$, yields the expression

$$\bar{J}^1 = \epsilon E[(u^1)^2] + E[u^1]\sqrt{\{E[(u^1)^2]\}} - E[(u^1)^2].$$

Now, if the leader uses the uniform probability distribution on $[0, 1]$, his average cost becomes

$$\bar{J}^1 = \frac{\epsilon - 1}{3} + \frac{1}{2}\sqrt{\frac{1}{3}}$$

which clearly indicates that, for ϵ sufficiently small,

$$\bar{J}^{1^*} < 0 = J^{1^*}.$$

□

The preceding example has displayed the fact that even Stackelberg games with *strictly convex* cost functionals may fail to admit only pure-strategy solutions, and the mixed Stackelberg solution may in fact be preferable.[28] However, if we further restrict the cost structure to be quadratic, it can then be shown that only pure-strategy Stackelberg equilibria exist.

Proposition 4.3 *Consider the two-person nonzero-sum game with* $U^1 = \mathbf{R}^{m_1}$, $U^2 = \mathbf{R}^{m_2}$, *and*

$$J^i = \frac{1}{2}u^{i'}R^i_{ii}u^i + u^{1'}R^i_{ij}u^j + \frac{1}{2}u^{j'}R^i_{jj}u^j + u^{i'}r^i_i + u^{j'}r^i_j; i, j = 1, 2, i \neq j,$$

where $R^i_{ii} > 0$, R^i_{ii}, R^i_{ij}, R^i_{jj} *are matrices of appropriate dimensions, and* r^i_i, r^i_j *are vectors of appropriate dimensions. This "quadratic" game can only admit a pure-strategy Stackelberg solution, with either* **P1** *or* **P2** *as the leader.*

Proof. Without any loss of generality take **P1** as the leader, and assume, to the contrary, that the game admits a mixed-strategy Stackelberg solution, and denote the leader's optimal mixed strategy by μ^{1^*}. Furthermore, denote the expectation operation under μ^{1^*} by $E[\cdot]$. If the leader announces this mixed strategy, then the follower's reaction is unique and is given by

$$u^2 = -R^2_{22}{}^{-1}R^2_{21}E[u^1] - R^2_{22}{}^{-1}r^2_2.$$

By substituting this in $\bar{J}^{1^*} = E[J^1]$, we obtain

$$\bar{J}^{1^*} = \frac{1}{2}E[u^{1'}R^1_{11}u^1] + E[u^1]'KE[u^1] + E[u^1]'k + c,$$

[28]In retrospect, this should not be surprising since for the special case of zero-sum games (without pure-strategy saddle points) we have already seen that the minimizer could further decrease his guaranteed expected cost by playing a mixed strategy; here, however, it holds even if $J^1 \not\equiv -J^2$.

where

$$K \triangleq \frac{1}{2}R_{21}^{2}{}'R_{22}^{2}{}^{-1}R_{22}^{1}R_{22}^{2}{}^{-1}R_{21}^{2} - R_{12}^{1}R_{22}^{2}{}^{-1}R_{21}^{2},$$

$$k \triangleq r_1^1 - R_{21}^{2}{}'R_{22}^{2}{}^{-1}r_2^1 - R_{12}^{1}R_{22}^{2}{}^{-1}r_2^2 + R_{21}^{2}{}'R_{22}^{2}{}^{-1}R_{22}^{1}R_{22}^{2}{}^{-1}r_2^2,$$

$$c \triangleq \frac{1}{2}r_2^{2}{}'R_{22}^{2}{}^{-1}R_{22}^{1}R_{22}^{2}{}^{-1}r_2^2 - r_2^{2}{}'R_{22}^{2}{}^{-1}r_2^1.$$

Now, applying the Cauchy–Schwarz inequality (see Appendix B.4) on the first term of $\bar{J}^{1^*}$, we further obtain the bound

$$\bar{J}^{1^*} \geq \frac{1}{2}E[u^1]'R_{11}^1E[u^1] + E[u^1]'KE[u^1] + E[u^1]'k + c, \tag{i}$$

which depends only on the mean value of u^1. Hence,

$$J^{1^*} = \inf_{u^1 \in U^1} \left\{ \frac{1}{2}u^{1'}R_{11}^1u^1 + u^{1'}Ku^1 + u^{1'}k + c, \right\} \leq \bar{J}^{1^*}.$$

This implies that enlargement of the strategy space of the leader, so as to include mixed strategies as well, does not yield him any better performance. In fact, since $E[u^{1'}u^1] > E[u^1]'E[u^1]$, whenever the probability distribution is not one-point, it follows that the inequality in (i) is actually strict for the case of a proper mixed strategy. This then implies that, outside the class of pure strategies, there can be no Stackelberg solution. □

Graphical display of Stackelberg solutions and the notion of relative leadership

If U^1 and U^2 are one-dimensional spaces, and if, in addition, $R^2(u^1)$ is singleton for each $u^1 \in U^1$, then the Stackelberg equilibrium solution can easily be visualized on a graph. In Fig. 4.7, iso-cost curves for J^1 and J^2 have been drawn, together with the reaction curves l_1 and l_2. With **P1** as the leader, the Stackelberg solution will be situated on l_2, at the point where it is tangent to the

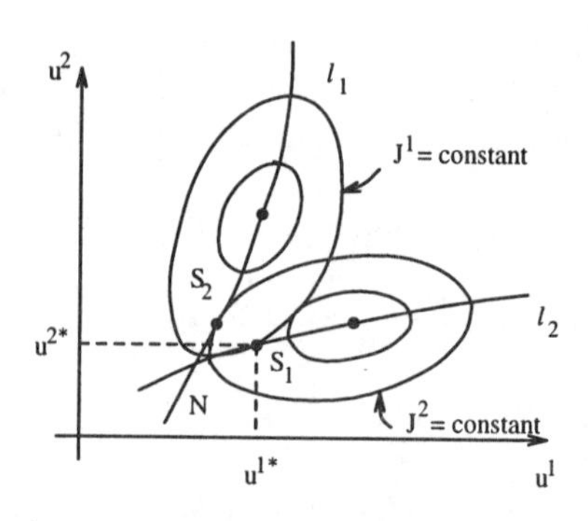

Figure 4.7: Graphical construction of the Stackelberg solution in a two-person game—a "stalemate" solution.

appropriate iso-cost curve of J^1. This point is designated as S_1 in Fig. 4.7. The coordinates of S_1 now correspond to the Stackelberg solution of this game with **P1** as the leader. The point of intersection of l_1 and l_2 (which is denoted by N in the figure) denotes the unique Nash equilibrium solution of this game. It should be noted that the Nash costs of both players are higher than their corresponding equilibrium costs in the Stackelberg game with **P1** as the leader. This is, though, only a specific example, since it would be possible to come up with situations in which the follower is worse off in the Stackelberg game than under the associated "Nash game". However, when the reaction set $R^2(u^1)$ is a singleton for every $u^1 \in U^1$, the leader cannot do worse in the "Stackelberg game" than his best performance in the associated "Nash game" since he can, at the worst, play the strategy corresponding to the most favorable (from the leader's point of view) Nash equilibrium solution pair, as discussed earlier in Section 3.6 (see, in particular, Prop. 3.14). To summarize, we have the following.

Proposition 4.4 *For a two-person nonzero-sum game, let V_N^1 denote the infimum of all the Nash equilibrium costs of* **P1**. *Then, if $R^2(u^1)$ is a singleton for every $u^1 \in U^1$, $V_N^1 \geq J^{1^*}$.*

Hence, in two-person nonzero-sum games with unique follower responses, the leader never prefers to play the "Nash game" instead of the "Stackelberg game", whereas the follower could prefer to play the "Nash game", if such an option is open to him. In some games, even the option of who should be the leader and who should follow might be open to the players, and in such situations there is the question of whether it is profitable for either player to act as the leader rather than be the follower. For the two-person game depicted in Fig. 4.7, for example, both players would prefer to act as the follower (in the figure, point S_2 characterizes the Stackelberg solution with **P2** as the leader). There are other cases, however, in which both players prefer the leadership of only one of the players (see Fig. 4.8a, where leadership of **P2** is more advantageous to both players) or in which each player wants to be the leader himself (Fig. 4.8b).

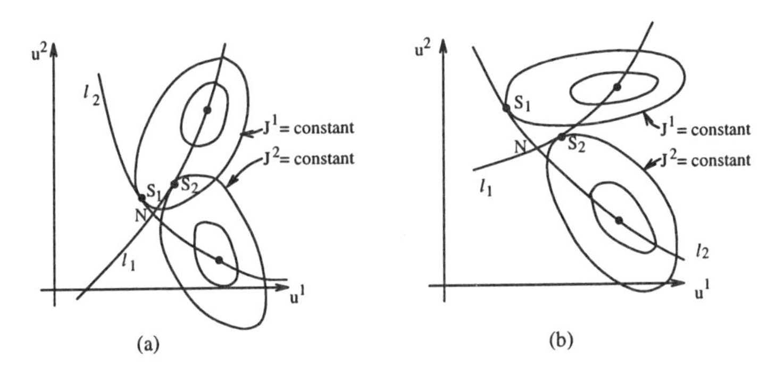

Figure 4.8: Different types of Stackelberg equilibrium solutions. (a) Concurrent solution. (b) Nonconcurrent solution.

When the players mutually benefit from the leadership of one of them, then this constitutes a stable situation since there is no reason for either player to deviate from the corresponding Stackelberg solution which was computed under mutual agreement—such a Stackelberg equilibrium solution is called a "*concurrent*" solution.

If each player prefers to be the leader himself, then the Stackelberg solution is called "*nonconcurrent*". In such a situation each player will try to dictate to the other player his own Stackelberg strategy and thus force him to play the "Stackelberg game" under his leadership. In this case, the one who can process his data faster will certainly be the leader and announce his strategy first. However, if the slower player does not actually know that the other player can process his data faster than he does, and/or if there is a delay in the information exchange between the players, then he might tend to announce a Stackelberg strategy under his own leadership quite unaware of the announcement of the other player—which certainly results in a nonequilibrium situation.

Finally, for the case in which neither player wants to be the leader (as in Fig. 4.7), the "Stackelberg game" is said to be in a "*stalemate*", since each player will wait for the other one to announce his strategy first.

4.5 Consistent Conjectural Variations Equilibrium

A third noncooperative solution concept for nonzero-sum games is the *consistent conjectural variations* (CCV) equilibrium, which we introduce here for the two-player case. This solution concept is in a sense a "double-sided Stackelberg equilibrium", where an equilibrium is sought in the class of reaction functions (or, more generally, reaction sets). For a precise mathematical formulation, let $\mathcal{T}_1 \times \mathcal{T}_2$ be the class of all mappings (T_1, T_2), $T_i : U^j \to U^i$, $j \neq i$, $i,j = 1,2$, with the property that the composite maps $T_1 \circ T_2$ and $T_2 \circ T_1$ have unique fixed points. For a given pair $(T_1^c, T_2^c) \in \mathcal{T}_1 \times \mathcal{T}_2$, let $(u^{1^c}, u^{2^c}) \in U^1 \times U^2$ denote the unique fixed points, satisfying

$$u^{1^c} = T_1^c \circ T_2^c(u^{1^c}), \qquad u^{2^c} = T_2^c \circ T_1^c(u^{2^c}). \tag{4.17}$$

Introduce the unique pair $(\hat{u}^1, \hat{u}^2) \in U^1 \times U^2$, defined by

$$\hat{u}^1 = \arg \min_{u^1 \in U^1} J^1(u^1, T_2^c(u^1)); \;\; \hat{u}^2 = \arg \min_{u^2 \in U^2} J^2(T_1^c(u^2), u^2). \tag{4.18}$$

Let

$$\Delta T_i^c(u^j) \triangleq T_i^c(u^j) - T_i^c(u^{j^c}), \quad j \neq i, i, j = 1,2; \tag{4.19a}$$

$$g_i^{\Delta T_j^c}(u^j) \triangleq \arg \min_{u^i \in U^i} J^i(u^i, u^j + \Delta T_j^c(u^i)), j \neq i, i, j = 1,2. \tag{4.19b}$$

Note that $\Delta T_i^c(u^j)$ is the "differential" reaction of $\mathbf{P}i$ under T_i^c to a deviation from u^{jc} to u^j. Furthermore, $g_i^{\Delta T_j^c}$ can be interpreted as a reaction function of

$\mathbf{P}i$ when $\mathbf{P}j$ uses a policy that is additively composed of a predetermined policy (u^j) and a policy determined by the policy choice of $\mathbf{P}i$ under the reaction function ΔT_j^c. Clearly, $\hat{u}^i$ introduced by (4.18) can also be obtained from

$$\hat{u}^i = g_i^{\Delta T_j^c}(u^{jc}). \tag{4.20}$$

This now brings us to the following definition.

Definition 4.9 *A pair of reaction functions $(T_1^c, T_2^c) \in \mathcal{T}_1 \times \mathcal{T}_2$, along with their unique fixed point $(u^{1^c}, u^{2^c}) \in U^1 \times U^2$, constitute a* consistent conjectural variations equilibrium *(CCVE) if we have the consistency of reaction functions:*

$$g_1^{\Delta T_2^c} = T_1^c, \qquad g_2^{\Delta T_1^c} = T_2^c \tag{4.21}$$

and consistency of policies

$$\hat{u}^1 = u^{1^c}, \qquad \hat{u}^2 = u^{2^c}. \tag{4.22}$$

Remark 4.4 In the most general case, it may be impossible to validate the consistency of the reaction functions, though this may be possible to any "order" under appropriate smoothness conditions on J^1 and J^2 (this will be further discussed in the sequel). Consistency of policies, on the other hand, is more readily testable. □

To gain more insight into CCVE, let us take $U^1 = U^2 = \mathbf{R}$, and J^1 and J^2 to be continuously differentiable and (jointly) strictly convex in their arguments. Furthermore, let $\mathcal{T}_1 \times \mathcal{T}_2$ be chosen such that $J^1(\cdot, T_2(\cdot))$ and $J^2(T_1(\cdot), \cdot)$ are strictly convex in their arguments, for every pair $(T_1, T_2) \in \mathcal{T}_1 \times \mathcal{T}_2$. To simplify the notation, let $u^1 = u$, $u^2 = v$. Then, two reaction functions $(T_1^c, T_2^c) \in \mathcal{T}_1 \times \mathcal{T}_2$ are in CCV equilibrium if (and only if)

$$\frac{\partial J_1(u,v)}{\partial u} + \frac{\partial J_1(u,v)}{\partial v} \cdot \frac{\partial T_2^c(u)}{\partial u} \equiv 0, \quad \text{for } u = T_1^c(v), \tag{i}$$

$$\frac{\partial J_2(u,v)}{\partial v} + \frac{\partial J_2(u,v)}{\partial u} \cdot \frac{\partial T_1^c(v)}{\partial v} \equiv 0, \quad \text{for } v = T_2^c(u). \tag{ii}$$

Note that these are two coupled partial differential equations which are, in general, difficult to solve. To gain some further insight, we expand these around the CCV solution (u^c, v^c) and perform a local analysis. First, to third order in v:

$$T_1^c(v) \cong u^c + \frac{dT_1^c(v^c)}{dv}(v - v^c) + \frac{d^2T_1^c(v^c)}{2\,dv^2}(v - v^c)^2 + \frac{d^3T_1^c(v^c)}{6\,dv^3}(v - v^c)^3$$

and likewise, to third order in u:

$$T_2^c(u) \cong v^c + \frac{dT_2^c(u^c)}{du}(u - u^c) + \frac{d^2T_2^c(u^c)}{2\,du^2}(u - u^c)^2 + \frac{d^3T_2^c(u^c)}{6\,du^3}(u - u^c)^3.$$

Now, rewriting (i):

$$\frac{\partial J^1(T_1^c(v),v)}{\partial u} + \frac{\partial J^1(T_1^c(v),v)}{\partial v} \cdot \frac{\partial T_2^c(T_1^c(v))}{\partial u} \equiv 0, v \in \mathbf{R},$$

and using the above expansion for $T_1^c(v)$ around $v = v^c$, we arrive at the following:

zeroth order

$$\frac{\partial J^1(u^c,v^c)}{\partial u} + \frac{\partial J^1(u^c,v^c)}{\partial v} \cdot \frac{\partial T_2^c(u^c)}{\partial u} = 0, \tag{0}$$

first order (arguments at $u = u^c$, $v = v^c$)

$$\frac{\partial^2 J^1}{\partial u^2} \cdot \frac{dT_1^c}{dv} + \frac{\partial^2 J^1}{\partial u \partial v} \left[1 + \frac{dT_1^c}{dv} \cdot \frac{dT_2^c}{du}\right] + \frac{\partial^2 J^1}{\partial v^2} \cdot \frac{dT_2^c}{du} + \frac{\partial J^1}{\partial v} \cdot \frac{d^2 T_2^c}{du^2} \cdot \frac{dT_1^c}{dv} = 0, \tag{1}$$

second order (arguments at $u = u^c$, $v = v^c$)

$$\begin{aligned} &\frac{\partial^3 J^1}{\partial u^3}\left(\frac{dT_1^c}{dv}\right)^2 + \frac{\partial^3 J^1}{\partial u^2 \partial v}\frac{dT_1^c}{dv} + \left[\frac{\partial^3 J^1}{\partial u^2 \partial v}\frac{dT_1^c}{dv} + \frac{\partial^3 J^1}{\partial u \partial v^2}\right]\left[1 + \frac{dT_1^c}{dv}\frac{dT_2^c}{du}\right] \\ &+ \frac{\partial^2 J^1}{\partial u^2}\frac{d^2 T_1^c}{dv^2} + \frac{\partial^2 J^1}{\partial u \partial v}\left[\frac{d^2 T_1^c}{dv^2}\frac{dT_2^c}{\partial u} + \left(\frac{dT_1^c}{dv}\right)^2 \frac{d^2 T_2^c}{du^2}\right] + \frac{dT_2^c}{du}\left[\frac{\partial^3 J^1}{\partial v^3} + \frac{\partial^3 J^1}{\partial v^2 \partial u}\frac{dT_1^c}{dv}\right] \\ &+ \left[2\frac{\partial^2 J^1}{\partial v^2} + \frac{\partial^2 J^1}{\partial u \partial v}\frac{dT_1^c}{dv}\right]\frac{d^2 T_2^c}{du^2}\frac{dT_1^c}{dv} + \frac{\partial J^1}{\partial v}\frac{d^3 T_2^c}{du^3}\left(\frac{dT_1^c}{dv}\right)^2 + \frac{\partial J^1}{\partial v}\frac{d^2 T_2^c}{du^2}\frac{d^2 T_1^c}{dv^2} = 0, \end{aligned} \tag{2}$$

where we assume that derivatives of all required orders exist. Likewise, rewriting (ii)

$$\frac{\partial J^2(u,T_2^c(u))}{\partial v} + \frac{\partial J^2(u,T_2^c(u))}{\partial u} \cdot \frac{dT_1^c(T_2^c(u))}{dv} \equiv 0, u \in \mathbf{R},$$

and using the above expansion for $T_2^c(u)$ around $u = u^c$, we have the following:

zeroth order

$$\frac{\partial J^2(u^c,v^c)}{\partial v} + \frac{\partial J^2(u^c,v^c)}{\partial u} \cdot \frac{dT_1^c(v^c)}{dv} = 0, \tag{0'}$$

first order (arguments at $u = u^c$, $v = v^c$)

$$\frac{\partial^2 J^2}{\partial v^2} \cdot \frac{dT_2^c}{du} + \frac{\partial^2 J^2}{\partial u \partial v} \left[1 + \frac{dT_2^c}{du} \cdot \frac{dT_1^c}{dv}\right] + \frac{\partial^2 J^2}{\partial u^2} \cdot \frac{dT_1^c}{dv} + \frac{\partial J^2}{\partial u} \cdot \frac{d^2 T_1^c}{dv^2} \cdot \frac{dT_2^c}{du} = 0, \tag{1'}$$

second order (arguments at $u = u^c$, $v = v^c$)

$$\begin{aligned} &\frac{\partial^3 J^2}{\partial v^3}\left(\frac{dT_2^c}{du}\right)^2 + \frac{\partial^2 J^2}{\partial v^2}\frac{d^2 T_2^c}{du^2} + \left[\frac{\partial^3 J^2}{\partial v \partial u^2} + \frac{\partial^3 J^2}{\partial u \partial v^2}\frac{dT_2^c}{du}\right]\left[1 + \frac{dT_1^c}{dv}\frac{dT_2^c}{du}\right] \\ &+ \frac{\partial^3 J^2}{\partial v^2 \partial u}\frac{dT_2^c}{du} + \frac{\partial^2 J^2}{\partial u \partial v}\left[\frac{d^2 T_2^c}{du^2}\frac{dT_1^c}{dv} + \left(\frac{dT_2^c}{du}\right)^2 \frac{d^2 T_1^c}{dv^2}\right] + \left[\frac{\partial^3 J^2}{\partial u^3} + \frac{\partial^3 J^2}{\partial u^2 \partial v}\frac{dT_2^c}{du}\right]\frac{dT_1^c}{dv} \\ &+ \left[2\frac{\partial^2 J^2}{\partial u^2} + \frac{\partial^2 J^2}{\partial u \partial v}\frac{dT_2^c}{du}\right]\frac{d^2 T_1^c}{dv^2}\frac{dT_2^c}{du} + \frac{\partial J^2}{\partial u}\frac{d^3 T_1^c}{dv^3}\left(\frac{dT_2^c}{du}\right)^2 + \frac{\partial J^2}{\partial u}\frac{d^2 T_1^c}{dv^2}\frac{d^2 T_2^c}{du^2} = 0. \end{aligned} \tag{2'}$$

In view of these relationships, we now refine the definition of CCVE for thrice continuously differentiable cost functions J^1 and J^2.

Definition 4.10 *A pair of conjectured response functions* (T_1^c, T_2^c) *is in CCV equilibrium to* zeroth order *if* (0) *and* (0′) *are satisfied. It is CCV equilibrium to* first order *if* (0)-(1) *and* (0′)-(1′) *are satisfied, and is in CCV equilibrium to* second order *if* (0)-(2) *and* (0′)-(2′) *are satisfied.*

It should now be clear from the above that this refinement of CCVE can be extended to any order, provided that J^1 and J^2 are continuously differentiable up to that order. An nth-order expansion (of CCVE) involves $2n+2$ separate, recursively solvable equations, which yield derivatives of T_1^c and T_2^c up to order n, evaluated at (u^c, v^c). If a pair of response functions (T_1^c, T_2^c) satisfy these $2n+2$ equations, then we say that it is "in CCVE to nth-order", which provides an nth-order approximation to the true (if it exists) CCVE. In some cases one need not go to higher orders, as they may vanish beyond a certain "n", in which case we clearly have the true CCVE. Games with quadratic cost functions constitute one such class of problems, as we shall see in the next section. Another point worth mentioning here is that the definition of an nth-order CCVE can be given also if U^1 and U^2 are higher (than one) dimensional Euclidean spaces, and even if they are (infinite-dimensional) Banach spaces. In the latter case one deals with Fréchet derivatives (Luenberger, 1969) instead of regular (Euclidean) differentials.

An observation we can make from the analysis given above in the context of scalar two-player games is that the Nash equilibrium (u^*, v^*), where each player takes the policy of the other player as given (and fixed), is a zeroth-order CCVE, with

$$T_1^c(v) \equiv u^*, \qquad T_2^c(u) \equiv v^*.$$

It is also a first-order CCVE under the set of restrictive conditions

$$\frac{\partial^2 J^1(u^*, v^*)}{\partial u \partial v} = 0, \qquad \frac{\partial^2 J^2(u^*, v^*)}{\partial u \partial v} = 0.$$

For (u^*, v^*) to be a second-order CCVE, also the set of conditions

$$\frac{\partial^3 J^1(u^*, v^*)}{\partial u \partial v^2} = 0, \qquad \frac{\partial^3 J^2(u^*, v^*)}{\partial v \partial u^2} = 0$$

has to be satisfied. Since the latter two conditions are overly restrictive, we can say that generically the Nash equilibrium solution is a zeroth-order CCVE. Clearly this holds not only for the scalar game, and hence we have the following.

Proposition 4.5 *If a two-player nonzero-sum game admits a unique Nash equilibrium, say* (u^{1^*}, u^{2^*}), *then this pair is a zeroth-order CCVE, where* u^{1^*} *and* u^{2^*} *are the constant reaction functions. The Nash equilibrium is not necessarily a higher-order CCVE.*

Remark 4.5 The notion of a CCVE can be extended from two- to many-player games, at least at the conceptual level. Computationally, however, the difficulties would be compounded here, as a player's response now depends on the policies or actions of more than one player (cf. Def. 4.3). □

4.6 Quadratic Games with Applications in Microeconomics

In this section, we obtain explicit expressions for the Nash, Stackelberg and consistent conjectural variations equilibrium solutions of static nonzero-sum games in which the cost functions of the players are quadratic in the decision variables—the so-called *quadratic games.* The action (strategy) spaces will be taken as Euclidean spaces of appropriate dimensions, but the results are also equally valid (under the right interpretation) if the strategy spaces are taken as infinite-dimensional Hilbert spaces. In that case, the Euclidean inner products will have to be replaced by the inner product of the underlying Hilbert space, and the positive-definiteness requirements on some of the matrices will have to be replaced by *strong positive definiteness* of the corresponding self-adjoint operators. This section will also include some discussion on iterative algorithms for the computation of Nash equilibria, and an illustration of the quadratic model and its Nash and Stackelberg solutions within the context of noncooperative economic equilibrium behavior of firms in oligopolistic markets.

A general quadratic cost function for $\mathbf{P}i$, which is strictly convex in his action variable, can be written

$$J^i = \frac{1}{2}\sum_{j=1}^{N}\sum_{k=1}^{N} u^{j\prime} R^i_{jk} u^k + \sum_{j=1}^{N} u^{j\prime} r^i_j + c^i, \tag{4.23}$$

where $u^j \in U^j = \mathbf{R}^{m_j}$ is the m_j-dimensional decision variable of $\mathbf{P}j$, R^i_{jk} is an $(m_j \times m_k)$-dimensional matrix with $R^i_{ii} > 0$, r^i_j is an m_j-dimensional vector and c^i is a constant. Without loss of generality, we may assume that, for $j \neq k$, $R^i_{jk} = R^{i\,\prime}_{kj}$, since if this were not the case, the corresponding two quadratic terms could be written as

$$u^{j\prime} R^i_{jk} u^k + u^{k\prime} R^i_{kj} u^j = u^{j\prime}\left(\frac{R^i_{jk}+R^{i\,\prime}_{kj}}{2}\right) u^k + u^{k\prime}\left(\frac{R^i_{jk}+R^{i\,\prime}_{kj}}{2}\right) u^j \tag{4.24}$$

and redefining R^i_{jk} as $(R^i_{jk} + R^{i\,\prime}_{jk})/2$, a symmetric matrix could be obtained. By an analogous argument, we may take R^i_{jj} to be symmetric, without any loss of generality.

Quadratic cost functions are of particular interest in game theory, firstly because they constitute second-order approximation to other types of nonlinear cost functions, and secondly because they are analytically tractable, admitting, in general, closed-form equilibrium solutions which provide insight into the properties and features of the equilibrium solution concept under consideration.

To determine the Nash equilibrium solution in strictly convex quadratic games, we differentiate J^i with respect to u^i $(i \in \mathbf{N})$, set the resulting expressions equal to zero, and solve the set of equations thus obtained. This set of equations, which also provides a sufficient condition because of strict convexity, is

$$R_{ii}^i u^i + \sum_{j \neq i} R_{ij}^i u^j + r_i^i = 0 \quad (i \in \mathbf{N}), \tag{4.25}$$

which can be written in compact form as

$$Ru = -r, \tag{4.26a}$$

where

$$R \triangleq \begin{bmatrix} R_{11}^1 & R_{12}^1 & \cdots & R_{1N}^1 \\ R_{21}^2 & R_{22}^2 & \cdots & R_{N2}^2 \\ \cdot & \cdot & & \cdot \\ \cdot & \cdot & & \cdot \\ \cdot & \cdot & & \cdot \\ R_{N1}^N & R_{N2}^N & \cdots & R_{NN}^N \end{bmatrix}, \tag{4.26b}$$

$$u' \triangleq (u^1, u^2, \ldots, u^N), \tag{4.26c}$$

$$r' \triangleq (r_1^1, r_2^2, \ldots, r_N^N). \tag{4.26d}$$

This then leads to the following proposition.

Proposition 4.6 *The quadratic N-player nonzero-sum static game defined by the cost functions (4.23) and with $R_{ii}^i > 0$, admits a Nash equilibrium solution if, and only if, (4.26a) admits a solution, say u^*; this Nash solution is unique if the matrix R defined by (4.26b) is invertible, in which case it is given by*

$$u^* = -R^{-1} r. \tag{4.27}$$

Remark 4.6 Since each player's cost function is strictly convex and continuous in his action variable, quadratic nonzero-sum games of the type discussed above cannot admit a Nash equilibrium solution in mixed strategies. Hence, in strictly convex quadratic games, the equilibrium analysis can be confined to the class of pure strategies. □

We now investigate the stability properties of the unique Nash solution of quadratic games, where the notion of stability was introduced in Section 4.3. Assuming $N = 2$, and directly specializing Prop. 4.1 to the quadratic case, we arrive at the following iteration:

$$u^{1(k+1)} = C_1 u^{2(k)} + d_1, \quad u^{2(k+1)} = C_2 u^{1(k+1)} + d_2, \quad k = 0, 1, \ldots \tag{4.28}$$

with an arbitrary starting choice $u^{2^{(0)}}$, where

$$C_i = -(R_{ii}^i)^{-1}R_{ij}^i, \quad d_i = -(R_{ii}^i)^{-1}r_i^i, \quad j \neq i, \ i,j = 1,2.$$

This iteration corresponds to the sequential (Gauss–Seidel) update scheme where **P**1 responds to the most recent past action of **P**2, whereas **P**2 responds to the current action of **P**1. The alternative to this is the parallel (Jacobi) update scheme where (4.28) is replaced by

$$u^{1^{(k+1)}} = C_1 u^{2^{(k)}} + d_1, \quad u^{2^{(k+1)}} = C_2 u^{1^{(k)}} + d_2, \quad k = 0,1,\ldots \tag{4.29}$$

starting with arbitrary initial choices $(u^{1^{(0)}}, u^{2^{(0)}})$. Then, the question of stability of the Nash solution (4.27), with $N = 2$, reduces to the question of stability of the fixed point of either (4.28) or (4.29). Note that, apart from a relabeling of indices, stability of these two iterations is equivalent to the stability of the single iteration:

$$u^{1^{(k+1)}} = C_1 C_2 u^{1^{(k)}} + C_1 d_2 + d_1.$$

Since this is a linear difference equation, a necessary and sufficient condition for it to converge (to the actual Nash strategy of **P**1) is that the eigenvalues of the matrix C_1C_2, or equivalently those of C_2C_1, should be in the unit circle, i.e.,

$$\rho(C_1C_2) \equiv \rho(C_2C_1) < 1, \tag{4.30}$$

where $\rho(A)$ is the spectral radius of the matrix A. This spectral radius condition is precisely the one given in Prop. 4.1, which shows that the condition of Prop. 4.1 is tight for the quadratic case.

Note that the condition of stability is considerably more stringent than the condition of existence of a unique Nash equilibrium, which is

$$\det(I - C_1C_2) \neq 0. \tag{4.31}$$

The question we address now is whether, in the framework of Gauss–Seidel or Jacobi iterations, this gap between (4.30) and (4.31) could be shrunk or even totally eliminated, by allowing players to incorporate memory into the iterations. While doing this, it would be desirable for the players to need to know as little as possible regarding the reaction functions of each other (note that no such information is necessary in the Gauss–Seidel or Jacobi iterations given above).

To study this issue, consider the Gauss–Seidel iteration (4.28), but with a one-step memory for (only) **P**1. Then, the "relaxed" algorithm will be (using the simpler notation $u^{1^{(k)}} = u_k$, $u^{2^{(k)}} = v_k$):

$$\left.\begin{array}{rcl} u_{k+1} & = & C_1 v_k + d_1 + A(u_k - C_1 v_k - d_1), \\ v_{k+1} & = & C_2 u_{k+1} + d_2, \end{array}\right\} \tag{4.32}$$

where A is a gain matrix, yet to be chosen. Substituting the second (for v_k) into the first, we obtain the single iteration

$$u_{k+1} \quad = \quad [C + A(I - C)]u_k + (I - A)[d_1 + C_1 d_2],$$

where

$$C \triangleq C_1C_2.$$

By choosing

$$A = -C(I-C)^{-1}, \tag{4.33}$$

where the required inverse exists because of (4.31), we obtain an immediate convergence, assuming that the true value of C_2 is known to **P1**. If the true value of C_2 is not known, but a nominal value is given in a neighborhood of which the true value lies, the scheme (4.32) along with the choice (4.33) and using the nominal value, still leads to convergence (but not in a finite number of steps) provided that the neighborhood is sufficiently small (see Başar , 1987).

Now, if the original scheme is instead the parallel (Jacobi) scheme, then a one-step memory for **P1** will not be sufficient to obtain a finite-step convergence result as above. In this case we replace (4.32) by

$$\left.\begin{aligned} u_{k+1} &= C_1v_k + d_1 + B(u_{k-1} - C_1v_k - d_1), \\ v_{k+1} &= C_2u_k + d_2, \end{aligned}\right\} \tag{4.34}$$

where B is another gain matrix. Note that here **P1** uses, in the computation of u_{k+1}, not u_k but rather u_{k-1}. Now, substituting for v_k from the second into the first equation of (4.34), we arrive at the iteration

$$u_{k+1} = [C + B(I-C)]u_{k-1} + (I-B)[d_1 + C_1d_2],$$

which again shows immediate convergence, with B chosen as

$$B = -C(I-C)^{-1}. \tag{4.35}$$

Again, there is a certain neighborhood of nominal C_2 or equivalently of the nominal C, where the iteration (4.34) is convergent.

In general, however, the precise scheme according to which **P2** responds to **P1**'s policy choices may not be common information, and hence one would like to develop relaxation-type algorithms for **P1** which would converge to the true equilibrium solution regardless of what particular scheme **P2** adopts (for example, Gauss–Seidel or Jacobi). Consider, for example, the scheme where **P2**'s responses for different k are modeled by (see also (4.7))

$$v_{k+1} = C_2u_{k+1-i_k} + d_2, \tag{4.36}$$

where $i_k \geq 0$ is an integer denoting the delay in the receipt of current policy information by **P2** from **P1**. The choice $i_k = 0$ for all k, would correspond to the Gauss–Seidel iteration, and the choice $i_k = 1$ for all k, to the Jacobi iteration — assuming that u_{k+1} is still determined according to (4.28). An extreme case would be the totally asynchronous communication where $\{i_k\}_{k\geq 0}$ could be any sequence of positive integers. Under the assumptions that **P1** communicates new policy choices to **P2** *infinitely often*, and he uses the simple ("nonrelaxed") iteration

$$u_{k+1} = C_1 v_k + d_1, \tag{4.37}$$

it is known from the work of Chazan and Miranker (1969) that such a scheme converges if, and only if,

$$\rho(|C|) < 1 \tag{4.38}$$

where $|C|$ is the matrix derived from C by multiplying all its negative entries by -1.

This condition can be improved upon, however, by incorporating relaxation terms in (4.37), as well as in (4.36), such as

$$\left.\begin{aligned} u_{k+1} &= \alpha C_1 v_k + (1-\alpha) C_1 v_{k-1} + d_1, \\ v_{k+1} &= \beta C_2 u_k + (1-\beta) C_2 u_{k-1} + d_2, \end{aligned}\right\} \tag{4.39}$$

where α and β are some scalars. A sufficient condition for convergence of any asynchronously implemented version of (4.39) is (see Başar (1987) for details)

$$\rho(\bar{A}(\alpha)) < 1, \tag{4.40}$$

where

$$\bar{A}(\alpha) \triangleq \begin{pmatrix} 0 & \max(\alpha, 1-\alpha)|C_1| \\ \max(\beta, 1-\beta)|C_2| & 0 \end{pmatrix}. \tag{4.41}$$

Clearly, there are values of $\alpha \neq 0$, $\beta \neq 0$, for which (4.40) requires a less stringent condition (on C_1 and C_2) than (4.38). For example, if C_1 and C_2 are positive scalars, and $\alpha = \beta = \frac{1}{2}$, inequality (4.40) dictates

$$C_1 C_2 < 4$$

while (4.38) requires that $C_1 C_2 < 1$.

From a game-theoretic point of view, each of the iteration schemes discussed above corresponds to a game with a sufficiently large number of stages and with a particular mode of play among the players. Moreover, the objective of each player is to minimize a kind of an average long horizon cost, with costs at each stage contributing to this average cost. Viewing this problem overall as a multi-act nonzero-sum game, we observe that the behavior of each player at each stage of the game is rather "myopic", since at each stage the players minimize their cost functions only under past information, and quite in ignorance of the possibility of any future moves. If the possibility of future moves is also taken into account, then the rational behavior of each player at a particular stage could be quite different. For an illustration of this possibility the reader is referred to Example 4.5 which is given later in this section. Such myopic decision making could make sense, however, if the players have absolutely no idea as to how many stages the game comprises, in which case there is the possibility that at any stage a particular player could be the last one to act in the game. In such a situation, risk-aversing players would definitely adopt "myopic" behavior, minimizing their current cost functions under only the past information, whenever given the opportunity to act.

Two-person zero-sum games

Since zero-sum games are special types of two-person nonzero-sum games with $J_1 = -J_2$ (**P1** minimizing and **P2** maximizing), in which case the Nash equilibrium solution concept coincides with the concept of saddle-point equilibrium, a special version of Prop. 4.6 will be valid for quadratic zero-sum games. To this end, we first note that the relation $J_1 = -J_2$ imposes in (4.23) the restrictions

$$R_{12}^{1\,\prime} = -R_{21}^2, R_{11}^2 = -R_{11}^1, R_{22}^1 = -R_{22}^2, r_1^2 = -r_1^1, r_2^1 = -r_2^2, c^1 = -c^2,$$

under which matrix R defined by (4.26b) can be written as

$$R = \begin{pmatrix} R_{11}^1 & R_{12}^1 \\ -R_{12}^{1\,\prime} & R_{22}^2 \end{pmatrix}$$

which has to be nonsingular for existence of a saddle point. Since R can also be written as the sum of two matrices

$$R = \begin{pmatrix} R_{11}^1 & 0 \\ 0 & R_{22}^2 \end{pmatrix} + \begin{pmatrix} 0 & R_{12}^1 \\ -R_{12}^{1\,\prime} & 0 \end{pmatrix}$$

the first one being positive definite and the second one skew-symmetric, and since eigenvalues of the latter are always imaginary, it readily follows that R is a nonsingular matrix. Hence we arrive at the conclusion that every quadratic strictly convex-concave zero-sum game admits a unique saddle-point equilibrium in pure strategies—a result that also follows as a special case of Thm. 4.5.

Corollary 4.5 *The strictly convex-concave quadratic zero-sum game with cost function*

$$\begin{aligned} J &= \frac{1}{2}u^{1\prime}R_{11}^1u^1 + u^{1\prime}R_{12}^1u^2 - \frac{1}{2}u^{2\prime}R_{22}^2u^2 + u^{1\prime}r_1^1 + u^{2\prime}r_2^1 + c^1; \\ & \quad R_{11}^1 > 0, R_{22}^2 > 0, \end{aligned}$$

admits a unique saddle-point equilibrium in pure strategies, which is given by

$$\begin{aligned} u^{1^*} &= -[R_{11}^1 + R_{12}^1(R_{22}^2)^{-1}R_{12}^{1\,\prime}]^{-1}[r_1^1 + R_{12}^1(R_{22}^2)^{-1}r_2^1], \\ u^{2^*} &= [R_{22}^2 + R_{12}^{1\,\prime}(R_{11}^1)^{-1}R_{12}^1]^{-1}[r_2^1 - R_{12}^{1\,\prime}(R_{11}^1)^{-1}r_1^1]. \end{aligned}$$

Remark 4.7 The positive-definiteness requirements on R_{11}^1 and R_{22}^2 in Corollary 4.5 are necessary and sufficient for the game kernel to be strictly convex-strictly concave, but this structure is clearly not necessary for the game to admit a saddle point. If the game is simply convex-concave (that is, if the matrices above are nonnegative definite, with a possibility of zero eigenvalues), then in view of Thm. 4.6 a saddle point will still exist provided that the upper and lower values are bounded.[29] If the quadratic game is not convex-concave, however, then either the upper or the lower value (or both) will be unbounded, implying that a saddle point will not exist. □

[29] For a convex-concave quadratic game, the upper value will not be bounded if, and only if, there exists a $v \in \mathrm{R}^{m_2}$ such that $v'R_{22}^2v = 0$ while $v'r_2^1 \neq 0$. A similar result applies to the lower value.

Team problems

Yet another special class of nonzero-sum games is the team problems in which the players (or equivalently, members of the team) share a common objective. Within the general framework, this corresponds to the case $J^1 \equiv J^2 \equiv \cdots \equiv J^N \triangleq J$, and the objective is to minimize this cost function over all $u^i \in U^i$, $i = 1, \ldots, N$. The resulting solution N-tuple $(u^{1^*}, u^{2^*}, \ldots, u^{N^*})$ is known as the *team-optimal solution.* The Nash solution, however, corresponds to a weaker solution concept in team problems, the so-called *person-by-person (pbp) optimality.* In a two-member team problem, for example, a pbp optimal solution (u^{1^*}, u^{2^*}) dictates satisfaction of the pair of inequalities

$$\begin{aligned} J(u^{1^*}, u^{2^*}) &\leq J(u^1, u^{2^*}), \quad \forall u^1 \in U^1, \\ J(u^{1^*}, u^{2^*}) &\leq J(u^{1^*}, u^2), \quad \forall u^2 \in U^2, \end{aligned}$$

whereas a team-optimal solution (u^{1^*}, u^{2^*}) requires satisfaction of a single inequality

$$J(u^{1^*}, u^{2^*}) \leq J(u^1, u^2), \quad \forall u^1 \in U^1, u^2 \in U^2.$$

A team-optimal solution always implies pbp optimality, but not vice versa. Of course, if J is quadratic and strictly convex on the product space $U^1 \times \cdots \times U^N$, then a unique pbp optimal solution exists, and it is also team-optimal.[30] However, for a cost function that is strictly convex only on individual spaces U^i, but not on the product space, this latter property may not be true. Consider, for example, the quadratic cost function

$$J = (u^1)^2 + (u^2)^2 + 10u^1u^2 + 2u^1 + 3u^2$$

which is strictly convex in u^1 and u^2, separately. The matrix corresponding to R defined by (4.26b) is

$$\begin{pmatrix} 2 & 10 \\ 10 & 2 \end{pmatrix}$$

which is clearly nonsingular. Hence a unique pbp optimal solution will exist. However, a team-optimal solution does not exist since the said matrix (which is also the Hessian of J) has one positive and one negative eigenvalue. By cooperating along the direction of the eigenvector corresponding to the negative eigenvalue, the members of the team can make the value of J as small as possible. In particular, taking $u^2 = -\frac{2}{3}u^1$ and letting $u^1 \to +\infty$, drives J to $-\infty$.

The Stackelberg solution

We now elaborate on the Stackelberg solutions of quadratic games of type (4.23) but with $N = 2$, and **P1** acting as the leader. We first note that since the

[30]This result may fail to hold true for team problems with strictly convex but nondifferentiable kernels (see Problem 10, Section 4.8).

quadratic cost function J^i is strictly convex in u^i, by Prop. 4.3 we can confine our investigation of an equilibrium solution to the class of pure strategies. Then, to every announced strategy u^1 of **P1**, the follower's unique response will be as given by (4.25) with $N = 2$, $i = 2$:

$$u^2 = -(R_{22}^2)^{-1}[R_{21}^2 u^1 + r_2^2]. \tag{4.42}$$

Now, to determine the Stackelberg strategy of the leader, we have to minimize J^1 over U^1 and subject to the constraint imposed by the reaction of the follower. Since the reaction curve gives u^2 uniquely in terms of u^1, this constraint can best be handled by substitution of (4.42) in J^1 and by minimization of the resulting functional (to be denoted by $\tilde{J}^1$) over U^1. To this end, we first determine $\tilde{J}^1$:

$$\begin{aligned}\tilde{J}^1(u^1) &= \frac{1}{2}u^{1\prime}R_{11}^1 u^1 + \frac{1}{2}[R_{21}^2 u^1 + r_2^2]'(R_{22}^2)^{-1}R_{22}^1(R_{22}^2)^{-1}[R_{21}^2 u^1 + r_2^2] \\ &\quad -u^{1\prime}R_{12}^1(R_{22}^2)^{-1}[R_{21}^2 u^1 + r_2^2] + u^{1\prime}r_1^1 \\ &\quad -[R_{21}^2 u^1 + r_2^2]'(R_{22}^2)^{-1}r_2^1 + c^1.\end{aligned}$$

For the minimum of $\tilde{J}^1$ over U^1 to be unique, we have to impose a strict convexity condition on $\tilde{J}^1$. Because of the quadratic structure of $\tilde{J}^1$, this condition amounts to having the coefficient matrix of the quadratic term in u^1 positive definite, which is

$$\begin{aligned}R_{11}^1 + R_{21}^{2\prime}(R_{22}^2)^{-1}R_{22}^1(R_{22}^2)^{-1}R_{21}^2 - R_{12}^1(R_{22}^2)^{-1}R_{21}^2 \\ -R_{21}^{2\prime}(R_{22}^2)^{-1}R_{12}^{1\prime} > 0.\end{aligned} \tag{4.43}$$

Under this condition, the unique minimizing solution can be obtained by setting the gradient of $\tilde{J}^1$ equal to zero, which yields

$$\begin{aligned}u^{1^*} &= -[R_{11}^1 + R_{21}^2(R_{22}^2)^{-1}R_{22}^1(R_{22}^2)^{-1}R_{21}^2 - R_{12}^1(R_{22}^2)^{-1}R_{21}^2 \\ &\quad -R_{21}^{2\prime}(R_{22}^2)^{-1}R_{12}^{1\prime}]^{-1}[R_{21}^{2\prime}(R_{22}^2)^{-1}R_{22}^1(R_{22}^2)^{-1}r_2^2 \\ &\quad -R_{12}^1(R_{22}^2)^{-1}r_2^2 + r_1^1 - R_{21}^{2\prime}(R_{22}^2)^{-1}r_2^1].\end{aligned} \tag{4.44}$$

Proposition 4.7 *Under condition (4.43), the two-person version of the quadratic game (4.23) admits a unique Stackelberg strategy for the leader, which is given by (4.44). The follower's unique response is then given by (4.42).*

Remark 4.8 The reader can easily verify that a sufficient condition for condition (4.43) is strict convexity of J^1 on the product space $U^1 \times U^2$. □

The consistent conjectural variations equilibrium

Again working with the two-player version of the quadratic game defined by the cost functions (4.23), but with $R_{11}^1 = I$ and $R_{22}^2 = I$ for simplicity in notation (and without any loss of generality), let us "conjecture" linear reaction functions:

$$T_i^c(u^j) = K_i u^j + k_i, \quad j \neq i, i, j = 1, 2, \tag{4.45}$$

where K_i is an $m_i \times m_j$ dimensional matrix, and $k_i \in IR^{m_i}$. Substituting this into (4.21) we arrive at

$$\begin{aligned}
\tilde{u}^1 + R^1_{12}u^2 + r^1_1 + K'_2(R^1_{12}{}'\tilde{u}^1 + R^1_{22}u^2 + r^1_2) &= 0, \quad \text{for } \tilde{u}^1 = K_1u^2 + k_1, \\
\tilde{u}^2 + R^2_{21}u^1 + r^2_2 + K'_1(R^2_{21}{}'\tilde{u}^2 + R^2_{11}u^1 + r^2_1) &= 0, \quad \text{for } \tilde{u}^2 = K_2u^1 + k_2.
\end{aligned}$$

These subsequently lead to the pair of equations

$$\begin{aligned}
(K_1 + K'_2R^1_{12}{}'K_1 + R^1_{12} + K'_2R^1_{22})u^2 + k_1 + r^1_1 + K'_2r^1_2 + K'_2R^1_{12}{}'k_1 &= 0, \\
(K_2 + K'_1R^2_{21}{}'K_2 + R^2_{21} + K'_1R^2_{11})u^1 + k_2 + r^2_2 + K'_1r^2_1 + K_1R^2_{21}{}'k_2 &= 0,
\end{aligned}$$

which are required to hold for all $(u^1, u^2) \in \mathrm{R}^{m_1} \times \mathrm{R}^{m_2}$. A sufficient condition would be the existence of K_1, K_2, k_1, k_2 that satisfy

$$\begin{aligned}
(I + K'_2R^1_{12}{}')K_1 + K'_2R^1_{22} &= -R^1_{12}, &(4.46)\\
(I + K'_1R^2_{21}{}')K_2 + K'_1R^2_{11} &= -R^2_{21}, &(4.47)\\
(I + K'_2R^1_{12}{}')k_1 &= -r^1_1 - K'_2r^1_2, &(4.48)\\
(I + K'_1R^2_{21}{}')k_2 &= -r^2_2 - K'_1r^2_1. &(4.49)
\end{aligned}$$

Note that what we have here is a pair of quadratic equations for K_1 and K_2, and after K_1 and K_2 are determined the next two linear equations yield the corresponding choices for k_1 and k_2. This shows that a CCVE equilibrium in the form of linear reaction functions is plausible in the quadratic case, but existence is by no means guaranteed. For some numerical results on the scalar (R^1) version of this problem we refer the reader to Başar , Turnovsky and D'Orey (1986), and Turnovsky, Başar and D'Orey (1988).

Applications in microeconomics

We now consider specific applications of the quadratic nonzero-sum game theory in microeconomics, in establishing economic equilibria in certain types of markets. The markets we have in mind are defined with respect to a specific good or a collection of goods, and an individual participant (player) in these markets is either a buyer or a seller. The traded goods are produced by the seller(s) and are consumed by the buyer(s).

An important characteristic of such markets is the number of participants involved. Markets in which there are a large number of both buyers and sellers, and in which each participant is involved with only a negligible fraction of the transactions, are called (bilaterally) *competitive*. A key characteristic of such competitive markets is that no single participant can, by his own actions, have a noticeable effect on the prices at which the transactions take place. There is a market price, and no seller needs to sell for less, and no buyer needs to pay more.

Our interest in the sequel will be on the so-called *oligopoly* markets, which comprise a few sellers and many competitive buyers. In the special case of two sellers this market is known as a *duopoly*. In an oligopoly, the buyers cannot influence the price or quantities offered, and it is assumed that the collective behavior of the buyers is fixed and known. The sellers, however, have an influence via the price they ask and the output (production) they realize.

Example 4.5 This example refers to what is known as a Cournot duopoly situation. There are two firms with identical products, which operate in a market in which the market demand function is known. The demand function relates the unit price of the product to the total quantity offered by the firms. It is then assumed that the whole production is sold at the prevailing price dictated by the demand function. Let q^i denote the production level of firm i (**P**i), $i = 1, 2$, and p denote the price of the commodity. Then, assuming a linear structure for the market demand curve, we have the relation

$$p = \alpha - \beta(q^1 + q^2),$$

where α and β are positive constants known to both firms. Furthermore, if we assign quadratic production costs to both firms, the profit functions are given by

$$\begin{aligned} P^1 &= q^1 p - k_1 (q^1)^2, \\ P^2 &= q^2 p - k_2 (q^2)^2, \end{aligned}$$

for **P1** and **P2**, respectively, where k_1, k_2 are nonnegative constants. **P1** wants to maximize P^1 and **P2** wants to maximize P^2. Under the stipulation that the firms act rationally and that there is no explicit collusion, the Nash equilibrium concept seems to fit rather well within this framework. Formulated as a nonzero-sum game, the Nash equilibrium solution of this decision problem is

$$q^1 = \frac{\alpha(\beta + 2k_2)}{3\beta^2 + 4k_1k_2 + 4\beta(k_1 + k_2)}; \qquad q^2 = \frac{\alpha(\beta + 2k_1)}{3\beta^2 + 4k_1k_2 + 4\beta(k_1 + k_2)},$$

which has been obtained by solving equation (4.25). The corresponding (Nash equilibrium) level of price is

$$p = \frac{\alpha(\beta^2 + 2\beta(k_1 + k_2) + 4k_1k_2)}{3\beta^2 + 4k_1k_2 + 4\beta(k_1 + k_2)}.$$

To obtain the Stackelberg solution with, for instance, **P1** as the leader, we first compute the reaction curve of **P2**,

$$q^2 = (\alpha - \beta q^1)/[2(\beta + k_2)],$$

and then substitute this into P^1 and maximize the resulting expression over q^1. The result is the following quantity level for the leader:

$$q^{1^*} = \frac{\alpha(\beta + 2k_2)}{2[\beta(\beta + 2k_2) + 2(\beta + k_2)k_1]}.$$

The corresponding quantity level for the follower, and the price level are, respectively,

$$q^{2^*} = \frac{\alpha}{4(\beta+k_2)} \cdot \frac{\beta^2 + 2\beta k_2 + 4k_1(\beta+k_2)}{\beta(\beta+2k_2)+2(\beta+k_2)k_1},$$
$$p^* = \frac{\alpha(\beta+2k_2)}{4(\beta+k_2)} \cdot \frac{4(\beta+k_2)k_1 + \beta(\beta+2k_2)}{\beta(\beta+2k_2)+2(\beta+k_2)k_1}.$$

The CCV solution can be obtained by making direct use of (4.46)-(4.49). For simplicity in computation, let us consider the symmetric case where $k_1 = k_2 = k$, and let $\bar{\beta} \triangleq \beta/k$. Then, (4.46)-(4.47) admit two pairs of solutions:

$$K_1 = K_2 = \frac{\pm\sqrt{1+2\bar{\beta}} - (1+\bar{\beta})}{\bar{\beta}}$$

with the corresponding values of k_1 and k_2 from (4.48)-(4.49) being

$$k_1 = k_2 = \frac{\alpha}{\bar{\beta}^2 k}\left[1 + \bar{\beta} \mp \sqrt{1+2\bar{\beta}}\right].$$

These two determine completely the optimum reaction functions (4.45) (in the sense of CCVE). Now letting

$$q^1 = K_1 q^2 + k_1, \qquad q^2 = K_2 q^1 + k_2$$

and solving for q^1 and q^2, we obtain the following unique quantity levels for this symmetric duopoly under the CCVE:

$$q^1 = q^2 = q^{\mathrm{CCV}} = \frac{\alpha(1+\bar{\beta})(1+2\bar{\beta} \mp \sqrt{1+2\bar{\beta}})}{2\bar{\beta}(2\bar{\beta}+1)}.$$

□

Example 4.6 This is a somewhat more sophisticated, and extended version of Example 4.5. There will be N firms or players in the oligopoly problem to be defined, and it will be possible for the price of the product to vary from one firm to another. Let us first introduce the following notation:

P^i = the profit for the ith firm (**P**i),
p^i = the price per unit product charged by **P**i,
$\bar{p}$ = average price = $\left(\sum_i p^i\right)/N$,
β = price sensitivity of overall demand,
V = the price at which demand is zero,
c = fixed costs per unit product of **P**i,
d^i = demand for the product of the ith firm.

Suppose that demand and price of the product of **P**i are related through the relation

$$d^i = \frac{1}{N}\beta[V - p^i - \gamma(p^i - \bar{p})], \tag{4.50}$$

where $\gamma > 0$ is a constant, or equivalently through

$$p^i = \left(1 + \gamma\left(1 - \frac{1}{N}\right)\right)^{-1} \left(V + \frac{\gamma}{N}\sum_{j \neq i} p^j - \frac{Nq^i}{\beta}\right), \tag{4.51}$$

where we have replaced d^i with q^i since the quantity produced by each firm is assumed to be equal to the demand for the product of that firm under the price dictated by this relation. The profit of $\mathbf{P}i$ can then be written as

$$P^i = (p^i - c)d^i,$$

where d^i is to be interpreted as being equal to q^i if (4.51) is used. Note that in this example all firms enter symmetrically into the model. One could incorporate an asymmetry by including weighting factors in the determination of the average price $\bar{p}$ or by making c firm-dependent. This modified model would definitely be more realistic; however, the structural properties of equilibria are not that sensitive to such changes, and therefore we have avoided such a more complicated model for the sake of simplicity in exposition.

For this example, we again seek the Nash equilibrium solution; but, from the model, it is not clear at the outset which set of variables should be taken as the decision (action) variables: the prices or the quantities? We shall elaborate here on both possibilities. In the so-called "*price game*" the Nash solution must satisfy the set of equations

$$\frac{\partial P^i}{\partial p^i} = 0, \quad i \in \mathbf{N}, \tag{4.52}$$

and in the so-called "*quantity game*" the first-order conditions for Nash equilibrium are

$$\frac{\partial P^i}{\partial q^i} = 0, \quad i \in \mathbf{N}. \tag{4.53}$$

In the price game it is assumed that the demands are determined via the prices according to (4.50), whereas in the quantity game it is assumed that the prices satisfy (4.51). In both cases the profit function of each firm is quadratic in the decision variables, and strictly concave in his own decision variable, and therefore it suffices to consider only the first order conditions.

For the price game, the set of equations (4.52) becomes

$$d^i(p^1, \ldots, p^N) + (p^i - c)\frac{\partial d^i(p^1, \ldots, p^N)}{\partial p^i} = 0, \quad i \in \mathbf{N},$$

and since the p^i's appear symmetrically in this set of equations, we may substitute $p \triangleq p^1 = p^2 = \cdots = p^N$ and solve for p, to obtain

$$p^i = \tilde{p} \triangleq \frac{V + c\left(1 + \frac{N-1}{N}\gamma\right)}{2 + \frac{N-1}{N}\gamma}, \quad i \in \mathbf{N}. \tag{4.54}$$

If this solution is substituted into (4.50), the following equilibrium values for the demands are obtained:

$$d^1 = \cdots = d^N \triangleq \tilde{d} = \frac{\beta}{N} \cdot \frac{1 + \frac{N-1}{N}\gamma}{2 + \frac{N-1}{N}\gamma} \cdot (V - c). \tag{4.55}$$

Now, for the quantity game, the set of equations (4.53) can be rewritten as

$$p^i(q^1, \ldots, q^N) - c + q^i \frac{\partial p^i(q^1, \ldots, q^N)}{\partial q^i} = 0, \quad i \in \mathbf{N}. \tag{4.56}$$

In order to express p^i as a function of $q^1, \ldots, q^N$ only, we must solve (4.51), which as it stands is an implicit equation in the variables p^i. Because of symmetry, the solution should have the form $p^i = \alpha_1 + \alpha_2 d^i + \alpha_3 \sum_{j \neq i} d^j$, where the coefficients α_k, $k = 1, 2, 3$, do not depend on i. Some analysis then leads to

$$p^i = V - \frac{1}{\beta(1+\gamma)}[(N+\gamma)d^i + \gamma \sum_{j \neq i} d^j]. \tag{4.57}$$

Because of symmetry reasons, in the equilibrium situation we have $\hat{p} \triangleq p^1 = \cdots = p^N$ and $\hat{q} \triangleq q^1 = \cdots = q^N$ and hence

$$\hat{p} = p^i = V - \frac{N\hat{q}}{\beta}, \quad i \in \mathbf{N}. \tag{4.58}$$

Now evaluating the quantity $\partial p^i / \partial q^i$ from (4.57), we obtain

$$\frac{\partial p^i}{\partial q^i} = -\frac{N}{\beta} \cdot \frac{N+\gamma}{N(1+\gamma)}, \quad i \in \mathbf{N}. \tag{4.59}$$

Substitution of (4.58) and (4.59) into (4.56) yields the following unique equilibrium value for q^i:

$$q^i = \hat{q} \triangleq \beta(V - c) \cdot \frac{1+\gamma}{2N + \gamma(1+N)}, \quad i \in \mathbf{N}. \tag{4.60}$$

If this value is further substituted into (4.58), we obtain

$$p^i = \hat{p} \triangleq \frac{V(N+\gamma) + cN(1+\gamma)}{2N + \gamma(1+N)}, \quad i \in \mathbf{N}. \tag{4.61}$$

A comparison of (4.54) with (4.61) readily yields the conclusion that the Nash equilibria of price and quantity games are not the same. For the trivial game with only one firm, i.e., $N = 1$, however, we obtain in both cases the same result, which is

$$p^* = \frac{1}{2}(V + c), \quad d^* = \frac{1}{2}\beta(V - c).$$

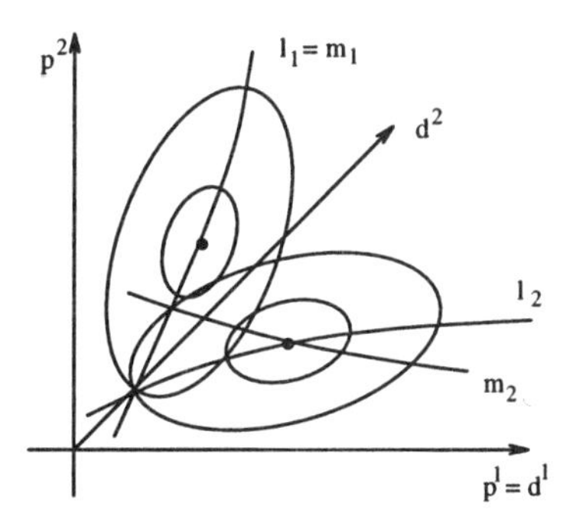

Figure 4.9: Graphical illustration of the effect of a coordinate transformation on Nash equilibria.

Also, in the limit as $N \to \infty$, the corresponding limiting equilibrium values of p and $q(d)$ in both games are again the same

$$p \to \frac{V + c(1+\gamma)}{2+\gamma}, \quad q \downarrow 0.$$

For $1 < N < \infty$, however, the Nash equilibria differ, in spite of the fact that the economic background in both models is the same. Therefore, it is important to know *a priori* in this market as to which variable is going to be adopted as the decision variable. The price game with its equilibrium is sometimes named after Edgeworth (1925), whereas the quantity game is named after Cournot who first suggested it in the nineteenth century (Cournot, 1838).

We now present a graphical explanation for occurrence of different Nash equilibria when different variables are taken as decision variables in the same model. Suppose that the iso-cost curves for a nonzero-sum game are as depicted in Fig. 4.2, which have been redrawn in Fig. 4.9, where the decision variables are denoted by p^1 and p^2, and the reaction curves corresponding to these decision variables are shown as l_1 and l_2, respectively. If a transformation to a new set of decision variables $d^1(p^1, p^2)$, $d^2(p^1, p^2)$ is made, then it should be apparent from the figure that we obtain different reaction curves m_1 and m_2 and hence a different Nash solution. As a specific example, in Fig. 4.9 we have chosen $d^1 = p^1$, $d^2 = p^1 + p^2$, in which case $m_1 = l_1$; but since $m_2 \neq l_2$ Nash equilibrium becomes a variant of the coordinate transformation. It should be noted, however, that if the transformation is of the form $d^1(p^1)$, $d^2(p^2)$, then the Nash equilibrium remains invariant. □

4.7 Braess Paradox

In this section we present a nonzero-sum game for which the Nash equilibrium solution leads to a surprising phenomenon, called the "Braess paradox", after Dietrich Braess who was the first to publish about it (Braess, 1968). Braess' field of application concerned traffic behavior, but the Braess paradox has since been observed in other fields of application as well (Cohen and Horowitz, 1991). We will more or less follow here Braess' original paper.

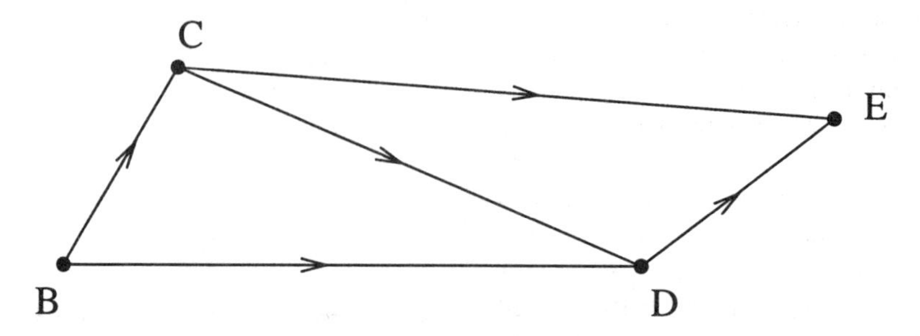

Figure 4.10: The routes of the Braess paradox.

Consider a network of roads as given in Fig. 4.10. All these roads allow only one-way traffic, as indicated by the arrows. The drivers on this network all want to go from point B to point E. They have to choose one of the three possible routes:

- route 1 consists of two segments: from B via C to E;
- route 2 consists of two segments: from B via D to E;
- route 3 consists of three segments: from B via C and D to E.

There are many cars and this leads to congestion on the roads. The time needed to drive along a segment depends on the intensity i, i.e., the number of cars per time unit that choose this segment. Measurements have shown that the time duration t needed to traverse each segment is as follows:

- along segment BC: $t = 10 \times i$;
- along segment BD: $t = 50 + i$;
- along segment CE: $t = 50 + i$;
- along segment CD: $t = 10 + i$;
- along segment DE: $t = 10 \times i$.

Suppose that x_j drivers choose route j, $j = 1, 2, 3$. The total time needed to go from B to E is then

- along route 1: $10 \times (x_1 + x_3) + 50 + x_1$;
- along route 2: $50 + x_2 + 10 \times (x_2 + x_3)$;
- along route 3: $10 \times (x_1 + x_3) + 10 + x_3 + 10 \times (x_2 + x_3)$.

Each driver will individually (and independently) choose that route for which the total driving time is the smallest. At equilibrium, this will lead to all three total driving times being equal. If we pick an arbitrary normalization, say $x_1 + x_2 + x_3 = 6$ (this normalization leads automatically to integer-valued solutions), then these driving times being equal leads to the unique Nash solution $x_1 = x_2 = x_3 = 2$, and the total driving time along each of the three routes equals 92 time units.

Now reconsider the problem described above with one change: segment CD is not available (due to roadworks) and the drivers must now choose between route 1 and route 2 only. The unique Nash solution in this new game turns out to be $x_1 = x_2 = 3$ ($x_3 = 0$) and the total driving time along each of the two routes equals 83 time units. Apparently, reducing the number of possibilities has led to a uniformly better result (the driving time for each driver being now 83 rather than 92). Or, if we reason in the other direction, increasing capacity leads to a worse equilibrium—a rather counter-intuitive result (Cohen, 1988).

If in the original problem, with three routes available, all drivers together would agree upon minimizing the maximum of the three total driving times, i.e., minimizing the maximum of the three expressions above for the total driving time, then the result turns out to be (the elementary calculation is left to the reader) $x_1 = x_2 = 3$, $x_3 = 0$. Hence in this *group optimum*, equivalently *Pareto optimum* (see Chapter 1), route 3 is not used at all even though it is available. However, this solution is cheating-prone as each individual driver would be tempted to use route 3, since if he were the only one doing so, his total driving time would be reduced to approximately 70. This, of course, is not an equilibrium situation.

4.8 Problems

1. Solve the following zero-sum semi-infinite matrix games:

P2

P1					
	0	1	1	1	...
	0	−3/2	−4/3	−5/4	...

P2

P1					
	3/4	8/9	15/16	24/25	...
	1	0	0	0	...

2. The kernel of a zero-sum game on the unit square is given by $J(u^1, u^2) = (u^1)^3 - 3u^1u^2 + (u^2)^3$. Determine the pure or mixed saddle-point strategies. (Hint: Note that J is strictly convex in u^1, and that for any u^1 the maximum value of J with respect to u^2 is attained either at $u^2 = 0$ or at $u^2 = 1$.)

3. Consider the zero-sum continuous-kernel game on the unit square, defined by $J(u^1, u^2) = (u^1 - u^2)^2 - \alpha(u^2)^2$, where α is a scalar parameter. Determine

its pure or mixed saddle-point solution when (i) $1 < \alpha \leq 2$ and (ii) $0 \leq \alpha \leq 1$.

4. Consider the following zero-sum game, known as a *duel*. Two men, starting at $t = 0$, walk toward each other at such a speed that they meet each other at $t = 1$ if nothing intervenes. Each one has a pistol with exactly one bullet, and may fire it at any time $t \in [0, 1]$. If one of them hits the other, the duel is immediately over, and the one who has fired successfully is declared the winner. If neither one fires successfully or if both fire simultaneously and successfully, then the duel becomes a stand-off. The probability of a hit after firing is inversely proportional with the distance between the two duelists, and assuming a uniform distribution we can let this probability (of hitting the other player) to be t at time $t \in [0, 1]$. This is a continuous-time version of the duel described in Section 2.8. Consider now the following two versions of the duel:

   (i) *Silent duel.* Assuming that the players do not know whether their opponents have fired (unless, of course, a particular player is hit), first show that an appropriate zero-sum game modeling this (silent) duel is one with the kernel

$$J(u^1, u^2) = \begin{cases} -u^1u^2 - u^1 + u^2, & u^1 < u^2, \\ 0, & u^1 = u^2, \\ +u^1u^2 - u^1 + u^2, & u^1 > u^2, \end{cases}$$

   where u^1 and u^2 are the instants at which the pistols are fired by **P1** and **P2**, respectively. Then, noting that this is a skew-symmetric game (à la Problem 9 of Section 2.9), show that its value in mixed strategies (V_{m}) is *zero*, and the corresponding mixed saddle-point policies are:

$$\mu^{1^*}(u) = \mu^{2^*}(u) = \begin{cases} 0, & 0 \leq u \leq \frac{1}{3} \\ \frac{1}{8}(9 - (\frac{1}{u})^2), & \frac{1}{3} \leq u \leq 1 \end{cases}.$$

   (ii) *Noisy duel.* Now consider the case where the pistols are noisy, i.e., a player knows whether his opponent has already fired or not. Show that this version of the duel game can be described by the kernel

$$J(u^1, u^2) = \begin{cases} 1 - 2u^1, & u^1 < u^2, \\ 0, & u^1 = u^2, \\ 2u^2 - 1, & u^1 > u^2, \end{cases}$$

   and that it admits a saddle point in pure strategies, given by $(u^{1^*} = \frac{1}{2}, u^{2^*} = \frac{1}{2})$.

5. Consider the two-person nonzero-sum game with cost functions

$$J^1(u, v) = [u(x - u - v^{\alpha_1})^{\tau\beta_1}]^{-1}, \quad J^2(u, v) = [v(x - v - u^{\alpha_2})^{\tau\beta_2}]^{-1},$$

where u and v are the action variables of **P1** and **P2**, respectively, belonging to the coupled constraint set

$$U = \{u \geq 0.01, v \geq 0.01, u + v^{\alpha_1} \leq x, v + u^{\alpha_2} \leq x\},$$

and x, α_1, α_2 are positive constants, and τ, β_1, β_2 are each positive and smaller than 1.

(i) Using Thm. 4.4, show that the game admits a pure-strategy Nash equilibrium.

(ii) Using Prop. 4.1, obtain a set of sufficient conditions on the six parameters characterizing the game, under which the pure-strategy Nash equilibrium is unique. Show that this condition is satisfied for the set of values:

$$x = 1.259,\ \alpha_1 = 1.1,\ \alpha_2 = 1.2,\ \beta_1 = 0.8,\ \beta_2 = 0.48,\ \tau = 0.2852.$$

(iii) For the same set of parameter values given above, obtain the unique Nash equilibrium solution. Is it (globally) stable? If yes, can you obtain the solution using an iterative procedure?

Partial answer for part (iii): The Nash equilibrium solution is $(u^* = 0.3, v^* = 0.9)$.

6. Prove the statement of Remark 4.1, and find the range of values of the scalar parameter ϵ for which the Nash equilibrium solution of the following game (with weakly coupled players) can be obtained using the Gauss–Seidel readjustment scheme

$$\begin{aligned} J^1(u^1, u^2) &= u^{1\prime}Qu^1 + u^{1\prime}r + \epsilon(u^{1\prime}u^1 + u^{1\prime}u^2), \\ J^2(u^1, u^2) &= u^{2\prime}Qu^2 + u^{2\prime}r - \epsilon(u^{2\prime}u^2 + u^{1\prime}u^2), \end{aligned}$$

where

$$Q = \begin{pmatrix} 2 & 1 & 0 \\ 1 & 1 & -1 \\ 0 & -1 & 3 \end{pmatrix}, \quad r = \begin{pmatrix} 1 \\ 2 \\ -1 \end{pmatrix}.$$

7. Consider the two-person Stackelberg game defined by the cost functions

$$J^i = \alpha_i(1 + u^1 + u^2)^2 + (u^i)^2, \quad i = 1, 2,$$

where α_i, $i = 1, 2$, are scalar parameters; the decision variables u^1 and u^2 are also scalars. For what values of (α_1, α_2) is the game (i) concurrent, (ii) nonconcurrent, or (iii) stalemate?

8. Consider the two-person Stackelberg game, with **P1** as the leader, defined on the unit square, and with the cost functions

$$\begin{aligned} J^1 &= \beta(1 - u^1)u^2(u^2 - 1) + u^1, \\ J^2 &= ((u^1)^k - u^2)^2, \end{aligned}$$

where β and k are scalars. For what values of β and k does a mixed Stackelberg solution for the leader yield a lower average cost for him than the pure Stackelberg solution?

9. In a three-person game with three levels of hierarchy, first **P1** announces his strategy and dictates it on both **P2** and **P3**; subsequently **P2** announces his strategy and dictates it on **P3**. The cost functions are given by

$$\begin{aligned} J^1 &= (x-2)^2 + (u^1)^2, \quad J^2 = (x-1)^2 + (u^2)^2, \\ J^3 &= (x-3)^2 + (u^3)^2, \end{aligned}$$

where $x = u^1 + u^2 + u^3$ and $u^i \in \mathrm{R}$, $i = 1, 2, 3$.

Determine the hierarchical equilibrium solution of this game. Compare this with the Nash equilibrium solution of a nonzero-sum game having the same cost structure. In which one (the three-level hierarchy or the Nash game) does **P1** attain a better performance?

10. Consider a two-person game with $J^1 \equiv J^2 \triangleq J$. Show that the cost function $J : \mathrm{R}^2 \to \mathrm{R}$ defined by

$$J = \begin{cases} (1-u^2)^2 + (u^1)^2, & u^1 \geq u^2, \\ (1-u^1)^2 + (u^2)^2, & u^1 < u^2, \end{cases}$$

is strictly convex on R^2. What are the reaction sets $R^2(u^1)$ and $R^1(u^2)$? Prove that infinitely many person-by-person (pbp) optimal solutions exist, but that the team solution is unique.

Now prove that if J is quadratic and strictly convex, the reaction curves can have at most one point in common, and therefore the pbp optimal solution and the team solution are unique and identical.

11. The purpose of this problem is to show that the notion of stability of Nash equilibrium, as introduced in Def. 4.5, indeed depends on the particular scheme used when the number of players is three or more. Consider the scalar three-person game with the cost functions

$$\begin{aligned} J^1 &= (u^1)^2 + 2\epsilon u^1 u^2 + 3\epsilon u^1 u^3 - 2u^1, \\ J^2 &= \epsilon u^2 u^1 + (u^2)^2 + \epsilon u^2 u^3 - 2u^2, \\ J^3 &= \epsilon u^3 u^1 + \epsilon u^3 u^2 + (u^3)^2 - 2u^3, \end{aligned}$$

where ϵ is a scalar parameter. The following two numerical schemes for the determination of Nash equilibrium correspond to (4.5) and (4.6), respectively:

$$\text{Scheme 1}: \quad \begin{aligned} u^1_{k+1} &= \arg \min_u J^1(u, u^2_k, u^3_k), \\ u^2_{k+1} &= \arg \min_u J^2(u^1_k, u, u^3_k), \\ u^3_{k+1} &= \arg \min_u J^3(u^1_k, u^2_k, u), \end{aligned}$$

$$\text{Scheme 2}: \quad \begin{aligned} u_{k+1}^1 &= \arg \min_u J^1(u, u_k^2, u_k^3), \\ u_{k+1}^2 &= \arg \min_u J^2(u_{k+1}^1, u, u_k^3), \\ u_{k+1}^3 &= \arg \min_u J^3(u_{k+1}^1, u_{k+1}^2, u), \end{aligned}$$

where, by a slight abuse of notation, u_k^i denotes the value of u^i at the kth step of iteration.

(i) Prove that, for $\epsilon = 0.36$, if u_k^i is calculated according to scheme 1 (which is the Cournot–Jacobi iteration), the sequences $\{u_k^i\}$, $i = 1, 2, 3$, do not converge, whereas they do converge if calculated according to scheme 2 (which is a particular Gauss–Seidel iteration).

(ii) Obtain the complete range of values of ϵ for which (a) scheme 1 is convergent, (b) scheme 2 is convergent.

(iii) Repeat (ii) for the following variation of scheme 2:

$$\text{Scheme 3}: \quad \begin{aligned} u_{k+1}^1 &= \arg \min_u J^1(u, u_{k+1}^2, u_k^3), \\ u_{k+1}^2 &= \arg \min_u J^2(u_k^1, u, u_k^3), \\ u_{k+1}^3 &= \arg \min_u J^3(u_{k+1}^1, u_{k+1}^2, u), \end{aligned}$$

12. Consider the following "relaxed" version of scheme 1 in Problem 11, where α, β, γ are the *relaxation* parameters (à la Section 4.6) yet to be chosen:

$$\textit{Relaxed}\ \text{Scheme 1}: \quad \begin{aligned} u_{k+1}^1 &= \alpha u_k^1 + (1-\alpha) \arg \min_u J^1(u, u_k^2, u_k^3), \\ u_{k+1}^2 &= \beta u_k^2 + (1-\beta) \arg \min_u J^2(u_k^1, u, u_k^3), \\ u_{k+1}^3 &= \gamma u_k^3 + (1-\gamma) \arg \min_u J^3(u_k^1, u_k^2, u). \end{aligned}$$

Let $\epsilon = 0.36$, and show that there exist nonzero values of α, β, γ such that the sequences generated by the scheme above converge.

13. Relaxation algorithms of the type used in Problem 12 above can also be used in the computation of saddle-point solutions in zero-sum games, not only to assure convergence of a particular scheme, but also to improve the speed of convergence to equilibrium. To explore this possibility, consider the scalar zero-sum two-person game with reaction functions

$$l^1(v) = \frac{v}{v^2 + 0.1}, \quad l^2(u) = \frac{1 - 0.608u}{1 - 0.608u^2}.$$

(i) First consider the Gauss–Seidel update scheme, written as:

$$u_{k+1} = l^1(v_k), \ v_{k+1} = l^2(u_{k+1});$$
$$v_0 \text{ arbitrarily chosen}, \quad k = 0, 1, \ldots.$$

Verify that with the starting choice picked as $v_0 = 0$, the sequence generated by this algorithm converges (to the nearest six figures) to the unique saddle-point solution ($u^* = 0.9547778$, $v^* = 0.94110625$) in approximately 120 iterations.

(ii) Now consider the "relaxed" version:

$$u_{k+1} = \alpha u_k + (1-\alpha)l^1(v_k), \; v_{k+1} = \beta v_k + (1-\beta)l^2(u_{k+1});$$
$$v_0 \text{ arbitrarily chosen}, \quad k = 0, 1, \ldots$$

and show by numerical experimentation that by choosing the values of α and β appropriately the speed of convergence can be improved significantly. Consider, for example, the values of $\alpha = 1, \beta = 0.67$, for which the improvement is by a magnitude of 10.

14. Consider the "price" and "quantity" models of oligopoly as introduced in Example 4.6, but with the roles of the players such that **P1** is the leader and **P2**, ..., **P**N are the followers. Further assume that the followers play according to the Nash solution concept among themselves. Obtain the solution of this "leader-followers" game, and study its asymptotic properties as the number of followers goes to infinity.

15. Consider a two-person nonzero-sum game with the cost functions defined by $J^1 = (u^1)^2 + (u^2)^2$; $J^2 = (u^1-1)^2 + (u^2-1)^2$.

(i) Determine the Nash solution, the consistent conjectural variations solution and the Stackelberg solution with **P1** as the leader.

(ii) Now assume that **P1** has access to the action variable u^2 of **P2**, and can announce a strategy in the form

$$u^1 = \alpha u^2 + \beta,$$

where α and β are constants yet to be determined. Assuming that **P2** is a rational player seeking to minimize his own cost function J^2, how should **P1** choose the values for α and β? Show that the resulting value of J^1 is its global minimum, i.e., *zero*. Provide a graphical interpretation for this result.

4.9 Notes

Section 4.2. The concept of ϵ equilibrium solution is as old as the concept of equilibrium solution itself. The results presented in this section are standard, and our proof of Thm. 4.2 follows closely the one given in Burger (1966) where additional results on equilibrium solutions can be found. Corollary 4.1, which follows from Thm. 4.2 in our case, was originally proven by Wald (1945) in a different way. More recently, the concept of ϵ equilibrium solution has received attention in the works of Tijs (1975) and Rupp (1977).

Section 4.3. The notion of a reaction curve (also called reaction function) was first introduced in an economics context and goes as far back as 1838, to the original work of Cournot. There are several results in the literature, known as minimax theorems, which establish existence of the (saddle-point) value in pure strategies under different conditions on U^1, U^2 and J. Thoerem 4.5 provides only one set of conditions which are sufficient for existence of a value. Some other minimax theorems have been given (in more general spaces) by Sion (1958), Fan (1953), Blackwell and Girshick (1954), Balakrishnan (1976) and Bensoussan (1971). The result on the existence of Nash equilibria in pure strategies (i.e., Thm. 4.3) is due to Nikaido and Isoda (1955), and extensions to mixed strategies were first considered by Ville (1938). The extension of Thm. 4.3 to coupled constraint sets (i.e., Thm. 4.4) is due to Rosen (1965), who also introduces a dynamic (differential equation-based) model for the computation of unique stable equilibrium. For extensions of these existence results to more general topological spaces, with applications to economics, see Aubin (1980). The discretization procedure alluded to in the proof of Thm. 4.7, in the text, suggests a possible method for obtaining mixed Nash equilibria of continuous-kernel games in an approximate way, if an explicit derivation is not possible. In this case, one has to solve for mixed Nash equilibria of only finite matrix games, with the size of these matrices determined by the degree of approximation desired.

Section 4.4. The Stackelberg solution concept was first introduced in economics through the work of H. Von Stackelberg in the early 1900s (cf. Von Stackelberg, 1934). But the initial formulation involved a definition in terms of reaction curves, with the generalization to nonunique reaction curves (or the so-called reaction sets) being relatively recent (cf. Başar and Selbuz, 1979a; Leitmann, 1978). A different version of Thm. 4.8, which accounts only for unique reaction curves, was given by Simaan and Cruz, Jr (1973b), but the more general version given here is due to Başar and Olsder (1980a). The latter reference is also the first source for extension to mixed strategies. A classification of different static Stackelberg solutions as given here is from Başar (1973), but the concept of "relative leadership" goes as far back as the original work of Von Stackelberg (1934). See also Ho and Olsder (1981).

Section 4.5. The material in this section is based on Başar (1986a). Other, more restrictive definitions of CCVE can be found in Bresnahan (1981), and Kamien and Schwartz (1983). See also Olsder (1987) for another perspective on CCVE.

Section 4.6. Quadratic games have attracted considerable attention in the literature partly because of analytic tractability (reaction functions are always affine). For further discussion on asynchronous implementation of Gauss–Seidel and Jacobi algorithms, and their generalized versions, for two as well as three-player games we refer the reader to Başar (1987). Some other selective references which deal with asynchronous algorithms in other contexts are Bertsekas and Tsitsiklis (1989a,b, 1991) and Tsitsiklis (1987, 1989). Two good sources for economic applications of (quadratic) games are Friedman (1977) and Case (1979). Example 4.5 is the so-called Cournot's quantity model and can be found in the original work of Cournot (1838). Several later publications have used this model in one way or another (see e.g., Intriligator (1971) and Başar and Ho (1974)). Example 4.6 is a simplified version of a model presented in Levitan and Shubik (1971), where the authors also provide a simulation study on some data obtained from the automobile industry in the USA.

Section 4.7. The Braess paradox is named after Braess who apparently was the first to write about a paradoxal phenomenon in traffic control (Braess, 1968). This phenomenon can also occur in other fields of application such as electric networks and mechanical constructions (Cohen and Horowitz, 1991).

Section 4.8. Problem 4 can be found in Karlin (1959). Extensions of the game of timing involve the cases of **P1** having m bullets and **P2** having n bullets, where m and n are positive integers. The general silent (m, n) case was solved by Restrepo (1957) and the general noisy duel with m and n bullets was solved by Fox and Kimeldorf (1969). The case of **P1** having m noisy bullets and **P2** having n silent bullets remains unsolved for $m > 1$. Problem 5 is taken from Li and Başar (1987), which the reader can consult with motivation and more theoretical as well as numerical results. A theoretical development for weakly coupled games, in which context Problems 11 and 12 arise, can be found in Srikant and Başar (1992). Problem 15 (ii) exhibits a dynamic information structure and can therefore also be viewed as a dynamic game; for more details on such Stackelberg games the reader is referred to Chapter 7 of this book, and in particular to Section 7.4.

Part II

Chapter 5

General Formulation of Infinite Dynamic Games

5.1 Introduction

This chapter provides a general formulation and some background material for the class of infinite dynamic games to be studied in the remaining chapters of the book. In these games, the action sets of the players comprise an infinite number (in fact a continuum) of elements (alternatives), and the players gain some dynamic information throughout the decision process. Moreover, such games are defined either in discrete time, in which case there exists a finite (or at most a countable) number of levels of play, or in continuous time, which corresponds to a continuum of levels of play.

Quite analogous to finite dynamic games, infinite dynamic games can also be formulated in extensive form which, however, does not lead to a (finite) tree structure, firstly because the action sets of the players are not finite, and secondly because of existence of a continuum of levels of play if the game is defined in continuous time. Instead, the extensive form of an infinite dynamic game involves a difference (in discrete time) or a differential (in continuous time) equation which describes the evolution of the underlying decision process. In other words, possible paths of action are not determined by following the branches of a tree structure (as in the finite case) but by the solutions of these functional equations which we also call "state equations". Furthermore, the information sets, in such an extensive formulation, are defined as subsets of some infinite sets on which some further structure is imposed (such as Borel subsets). A precise formulation which takes into account all these extensions as well as the possibility of chance moves is provided in Section 5.2 for discrete-time problems and in Section 5.3 for game problems defined in continuous time. In a discrete-time dynamic game, every player acts only at discrete instants of time, whereas in the continuous-time formulation the players act throughout a time interval which is either fixed *a priori* or determined by the rules of the game and the actions of the players. For finite games, the case in which the duration of the game is finite, but not fixed *a priori*, has already been included in the extensive tree formulation of Chapters 2 and 3. These chapters, in fact, included examples of multi-act games wherein the number of times a player acts is explicitly dependent on the actions of some other player(s)—and these examples have indicated that such decision problems do not require a separate treatment, and they can be handled by making use of the theory developed for multi-act games with an *a priori* fixed number of levels. In infinite dynamic games, however, an *a priori* fixed upper bound on the duration may sometimes be lacking, in which case "termination" of the play becomes a more delicate issue, as already shown in Section 2.7. Therefore, we devote a portion of Section 5.3 to a discussion on this topic, which will be useful later in Chapter 8.

Even though we shall not be dealing, in this book, with mixed and behavioral strategies in infinite dynamic games, we devote Section 5.4 to a brief discussion on this topic, mainly to familiarize the reader with the difficulties encountered in extending these notions from the finite to the infinite case, and to point out the direction along which such an extension would be possible.

Section 5.5 deals with certain standard techniques of one-person single-goal optimization and, more specifically, with optimal control problems. Our objective in including such a section is twofold; first, to introduce the reader to our notation and terminology for the remaining portions of the book, and second, to summarize some of the tools for one-person optimization problems which will be heavily employed in subsequent chapters in the derivation of noncooperative equilibria of infinite dynamic games. Section 5.6 deals with the notion of "representations of strategies on trajectories", and the issue of "time consistency", both of which are of prime importance in infinite dynamic games, as it will become apparent in Chapters 6 and 7.

Finally, Section 5.7 deals with so-called viscosity solutions. An important standard technique for continuous-time problems, as described in Section 5.5, is the solution of a partial differential equation. Among all possible solutions of this equation, the viscosity solution is a particular one; it is quite often unique and has a clear interpretation in terms of a generalized optimal control problem to which some stochastic perturbations have been added.

5.2 Discrete-Time Infinite Dynamic Games

In order to motivate the formulation of a discrete-time infinite dynamic game in extensive form, we now first introduce an alternative to the tree model of finite dynamic games—the so-called *loop model*. To this end, consider an N-person finite dynamic game with strategy sets $\{\Gamma^i; i \in \mathbf{N}\}$ and with decision (action) vectors $\{u^i \in U^i; i \in \mathbf{N}\}$, where the kth block component (u^i_k) of u^i designates $\mathbf{P}i$'s action at the kth level (stage) of the game and no chance moves are allowed. Now, if such a finite dynamic game is defined in extensive form (cf. Def. 3.10), then for any given N-tuple of strategies $\{\gamma^i \in \Gamma^i; i \in \mathbf{N}\}$ the actions of the players are completely determined by the relations

$$u^i = \gamma^i(\eta^i), \quad i \in \mathbf{N}, \tag{5.1}$$

where η^i denotes the information set of $\mathbf{P}i$. Moreover, since each η^i is completely determined by the actions of the players, there exists a point-to-set mapping[31] $g^i : U^1 \times U^2 \times \cdots \times U^N \to N^i$ $(i \in \mathbf{N})$ such that $\eta^i = g^i(u^1, \ldots, u^N)$ $(i \in \mathbf{N})$ and when substituted into (5.1), yields the "loop" relation

$$u^i = \gamma^i(g^i(u^1, \ldots, u^N)) \triangleq p^i(u^1, \ldots, u^N), \qquad i \in \mathbf{N}. \tag{5.2}$$

This clearly indicates that, for a given strategy N-tuple, the actions of the players are completely determined through the solution of a set of simultaneous equations which admits a unique solution under the requirements of Kuhn's extensive form (Kuhn, 1953). If the cost function of $\mathbf{P}i$ is defined on the action spaces, as $L^i : U^1 \times U^2 \times \cdots \times U^N \to \mathbf{R}$, then substitution of the solution of (5.2) into L^i yields $L^i(u^1, \ldots, u^N)$ as the cost incurred to $\mathbf{P}i$ under the strategy N-tuple $\{\gamma^i \in \Gamma^i; i \in \mathbf{N}\}$. If the cost function is instead defined on the strategy spaces, as $J^i : \Gamma^1 \times \Gamma^2 \times \cdots \times \Gamma^N \to \mathbf{R}$, then we clearly have the relation $J^i(\gamma^1, \ldots, \gamma^N) = L^i(u^1, \ldots, u^N)$ $(i \in \mathbf{N})$. Now, if a finite dynamic game is described through a set of equations of the form (5.2), with $\{p^i; i \in \mathbf{N}\}$ belonging to an appropriately specified class, together with an N-tuple of cost functions $\{L^i; i \in \mathbf{N}\}$, then we call this a *loop model* for the dynamic game (Witsenhausen, 1971a, 1975b). The above discussion has already displayed the steps involved in going from a tree model to a loop model in finite deterministic dynamic games. Conversely, it is also possible to start with a loop model and derive a tree model out of that, provided that the functions $p^i : U^1 \times U^2 \times \cdots \times U^N \to U^i$ $(i \in \mathbf{N})$ are restricted to a suitable class satisfying conditions like causality, unique solvability of the loop equation (5.2), etc. In other words, if the sets P^i $(i \in \mathbf{N})$ are chosen properly, then the set of equations

$$u^i = p^i(u^1, \ldots, u^N), \quad p^i \in P^i, \quad i \in \mathbf{N}, \tag{5.3}$$

together with the cost structure $\{L^i; i \in \mathbf{N}\}$ leads to a tree structure that satisfies the requirements of Def. 3.10.

If the finite dynamic game under consideration also incorporates a chance move, with possible alternatives for nature being $\omega \in \Omega$, then the loop equation (5.3) will accordingly have to be replaced by

$$u^i = p^i(u^1, \ldots, u^N, \omega), \quad p^i \in P^i, \quad \omega \in \Omega, \quad i \in \mathbf{N}, \tag{5.4}$$

where, by an abuse of notation, we again let P^i be the class of all permissible mappings $p^i : U^1 \times U^2 \times \cdots \times U^N \times \Omega \to U^i$. The solution of (5.4) will now be a function of ω; and when this is substituted into the cost function $L^i : U^1 \times \cdots \times U^N \times \Omega \to \mathbf{R}$ and expectation is taken over the statistics of ω, the resulting quantity determines the corresponding expected loss to $\mathbf{P}i$.

---

[31] Here N^i denotes the set of all η^i as in Def. 3.11.

Now, for infinite dynamic games defined in discrete time, the tree model of Def. 3.10 is not suitable since it cannot accommodate infinite action sets. However, the loop model defined by relations such as (5.3) does not impose such a restriction on the action sets; in other words, in the loop model we can take each U^i, to which the action variable u^i belongs, to be an infinite set. Hence, let us now start with a set of equations of the form (5.3) with U^i ($i \in \mathbf{N}$) taken as infinite sets. Furthermore let us decompose u^i into K blocks such that, considered as a column vector,

$$u^i = [u_1^{i\prime}, u_2^{i\prime}, \ldots, u_K^{i\prime}]', \qquad i \in \mathbf{N}, \tag{5.5}$$

where K is an integer denoting the maximum number of stages (levels) in the game,[32] and u_k^i denotes the action (decision or control) variable of $\mathbf{P}i$ corresponding to his move during the kth stage of the game.

In accordance with this decomposition, let us decompose each $p^i \in P^i$ in its range space, so that (5.3) can equivalently be written as

$$u_k^i = p_k^i(u^1, \ldots, u^N), \quad p_k^i \in P_k^i, \quad i \in \mathbf{N},\, k \in \mathbf{K}. \tag{5.6}$$

Under the causality assumption which requires u_k^i to be a function of only the past actions of the players, (5.6) can equivalently be written as (by an abuse of notation)

$$u_k^i = p_k^i(u_1^1, \ldots, u_{k-1}^1; \ldots; u_1^N, \ldots, u_{k-1}^N),\ p_k^i \in P_k^i,\ i \in \mathbf{N},\, k \in \mathbf{K}. \tag{5.7}$$

By following an analysis parallel to the one used in system theory in going from input-output relations for systems to state space models (Zadeh, 1969), we now assume that P_k^i is structured in such a way that there exist sets Γ_k^i, Y_k^i, X ($i \in \mathbf{N}, k \in \mathbf{K}$), with the latter two being finite dimensional, and functions $f_k : X \times U_k^1 \times \cdots \times U_k^N \to X$, $h_k^i : X \to Y_k^i$ ($i \in \mathbf{N}, k \in \mathbf{K}$), and for each $p_k^i \in P_k^i$ ($i \in \mathbf{N}, k \in \mathbf{K}$) there exists a $\gamma_k^i \in \Gamma_k^i$ ($i \in \mathbf{N}, k \in \mathbf{K}$), such that (5.7) can equivalently be written as

$$u_k^i = \gamma_k^i(\eta_k^i), \quad i \in \mathbf{N},\, k \in \mathbf{K}, \tag{5.8}$$

where η_k^i is a subcollection of

$$\{y_1^1, \ldots, y_k^1; \ldots; y_1^N, \ldots, y_k^N; u_1^1, \ldots, u_{k-1}^1; \ldots; u_1^N, \ldots, u_{k-1}^N\},$$

and

$$y_k^i = h_k^i(x_k), \quad y_k^i \in Y_k^i, \tag{5.9a}$$

$$x_{k+1} = f_k(x_k, u_k^1, \ldots, u_k^N), \quad x_{k+1} \in X \tag{5.9b}$$

for some $x_1 \in X$.

[32]In other words, no player can make more than K moves in the game, regardless of the strategies picked.

Adopting the system theory terminology, we call x_k the state of the game, y_k^i the (deterministic) *observation* of $\mathbf{P}i$ and η_k^i the *information* available to $\mathbf{P}i$, all at stage k; and we shall take relations (5.8), (5.9a)-(5.9b) as the starting point in our formulation of discrete-time infinite dynamic games. More precisely, we define an N-person discrete-time infinite dynamic game of prespecified fixed duration as follows.

Definition 5.1 *An N-person discrete-time deterministic infinite dynamic game*[33] *of prespecified fixed duration involves*

(i) An index set $\mathbf{N} = \{1, \ldots, N\}$ called the players' set.

(ii) An index set $\mathbf{K} = \{1, \ldots, K\}$ denoting the stages *of the game, where K is the maximum possible number of moves a player is allowed to make in the game.*

(iii) An infinite set X with some topological structure, called the state set (space) *of the game, to which the state of the game (x_k) belongs for all $k \in \mathbf{K} \cup \{K+1\}$.*

(iv) An infinite set U_k^i with some topological structure, defined for each $k \in \mathbf{K}$ and $i \in \mathbf{N}$, which is called the action (control) *set of $\mathbf{P}i$ at stage k. Its elements are the permissible* actions *u_k^i of $\mathbf{P}i$ at stage k.*

(v) A function $f_k : X \times U_k^1 \times \cdots \times U_k^N \to X$, defined for each $k \in \mathbf{K}$, so that

$$x_{k+1} = f_k(x_k, u_k^1, \ldots, u_k^N), k \in \mathbf{K} \qquad (i)$$

for some $x_1 \in X$ which is called the initial state *of the game. This difference equation is called the* state equation *of the dynamic game, describing the evolution of the underlying decision process.*

(vi) A set Y_k^i with some topological structure, defined for each $k \in \mathbf{K}$ and $i \in \mathbf{N}$, and called the observation set *of $\mathbf{P}i$ at stage k, to which the observation y_k^i of $\mathbf{P}i$ belongs at stage k.*

(vii) A function $h_k^i : X \to Y_k^i$, defined for each $k \in \mathbf{K}$ and $i \in \mathbf{N}$, so that

$$y_k^i = h_k^i(x_k), \quad k \in \mathbf{K}, \quad i \in \mathbf{N},$$

which is the state-measurement (-observation) *equation of $\mathbf{P}i$ concerning the value of x_k.*

(viii) A finite set η_k^i, defined for each $k \in \mathbf{K}$ and $i \in \mathbf{N}$ as a subset of $\{y_1^1, \ldots, y_k^1; \ldots; y_1^N, \ldots, y_k^N; u_1^1, \ldots, u_{k-1}^1; \ldots; u_1^N, \ldots, u_{k-1}^N\}$, which determines the information gained and recalled by $\mathbf{P}i$ at stage k of the game. Specification of η_k^i for all $k \in \mathbf{K}$ characterizes the information structure (pattern) *of $\mathbf{P}i$, and the collection (over $i \in \mathbf{N}$) of these information structures is the* information structure *of the game.*

[33] Also known as an "N-person deterministic multi-stage game".

(ix) A set N_k^i*, defined for each* $k \in \mathbf{K}$ *and* $i \in \mathbf{N}$ *as an appropriate subset of* $\{(Y_1^1 \times \cdots \times Y_k^1) \times \cdots \times (Y_1^N \times \cdot \times Y_k^N) \times (U_1^1 \times \cdots \times U_{k-1}^1) \times \cdots \times (U_1^N \times \cdots \times U_{k-1}^N)\}$*, compatible with* η_k^i*.* N_k^i *is called the* information space *of* **P***i at stage k, induced by his information* η_k^i*.*

(x) A prespecified class Γ_k^i *of mappings* $\gamma_k^i : N_k^i \to U_k^i$ *which are the* permissible strategies *of* **P***i at stage k. The aggregate mapping* $\gamma^i = \{\gamma_1^i, \gamma_2^i, \ldots, \gamma_K^i\}$ *is a* strategy *for* **P***i in the game, and the class* Γ^i *of all such mappings* γ^i *so that* $\gamma_k^i \in \Gamma_k^i$*,* $k \in \mathbf{K}$*, is the* strategy set *(*space*) of* **P***i.*

(xi) A functional $L^i : (X \times U_1^1 \times \cdots \times U_1^N) \times (X \times U_2^1 \times \cdots \times U_2^N) \times \cdots \times (X \times U_K^1 \times, \ldots, \times U_K^N) \to \mathbf{R}$ *defined for each* $i \in \mathbf{N}$*, and called the* cost functional *of* **P***i in the game of fixed duration.*

The preceding definition of a deterministic discrete-time infinite dynamic game is clearly not the most general one that could be given, first because the duration of the game need not be fixed, but be a variant of the players' strategies, and second because a "quantitative" measure might not exist to reflect the preferences of the players among different alternatives. In other words, it is possible to relax and/or modify the restrictions imposed by items (ii) and (xi) in Def. 5.1, and still retain the essential features of a dynamic game. A relaxation of the requirement of (ii) would involve introduction of a *termination set* $\Lambda \subset X \times \{1, 2, \ldots\}$, in which case we say that *the game terminates, for a given N-tuple of strategies, at stage k, if k is the smallest integer for which* $(x_k, k) \in \Lambda$.[34] Such a more general formulation clearly also covers fixed duration game problems in which case $\Lambda = X \times \{K\}$, where K denotes the number of stages involved. A modification of (xi), on the other hand, might for instance involve a "qualitative" measure (instead of the "quantitative" measure induced by the cost functional), thus giving rise to the so-called *qualitative games* (as opposed to "*quantitative games*" covered by Def. 5.1). Any qualitative game (also called *game of kind*) can, however, be formulated as a quantitative game (also known as *game of degree*) by assigning a fixed cost of zero to paths and strategies leading to preferred states, and positive cost to the remaining paths and strategies. For example, in a two-player game, if **P1** wishes to reach a certain subset Λ of the state set X after K stages and **P2** wishes to avoid it, we can choose

$$L^1 = \begin{cases} 0 & \text{if } x_{K+1} \in \Lambda \\ 1 & \text{otherwise} \end{cases}, \qquad L^2 = \begin{cases} 0 & \text{if } x_{K+1} \in \Lambda, \\ -1 & \text{otherwise} \end{cases}$$

and thus consider it as a zero-sum quantitative game.

Now, returning to Def. 5.1, we note that it corresponds to an *extensive form description* of a dynamic game, since the evolution of the game, the information gains and exchanges of the players throughout the decision process, and the interactions of the players among themselves are explicitly displayed in such a

[34]It is, of course, implicit here that x_k is determined by the given N-tuple of strategies, and the strategies are defined as in Def. 5.1, but by taking K sufficiently large.

formulation. It is, of course, also possible to give a normal form description of such a dynamic game, which in fact readily follows from Def. 5.1. More specifically, for each fixed initial state x_1 and for each fixed N-tuple permissible strategies $\{\gamma^i \in \Gamma^i; i \in \mathbf{N}\}$ the extensive form description leads to a *unique* set of vectors $\{u_k^i \triangleq \gamma_k^i(\eta_k^i), x_{k+1}; i \in \mathbf{N}, k \in \mathbf{K}\}$ because of the causal nature of the information structure and because the state evolves according to a difference equation. Then, substitution of these quantities into L^i $(i \in \mathbf{N})$ clearly leads to a unique N-tuple of numbers reflecting the corresponding costs to the players. This further implies existence of a composite mapping $J^i : \Gamma^1 \times \cdots \times \Gamma^N \to \mathbf{R}$, for each $i \in \mathbf{N}$, which is also known as the *cost functional* of $\mathbf{P}i$ $(i \in \mathbf{N})$. Hence, the permissible strategy spaces of the players (i.e., $\Gamma^1, \ldots, \Gamma^N$) together with these cost functions $(J^1, \ldots, J^N)$ constitute the *normal form* description of the dynamic game for each fixed initial state vector x_1.

It should be noted that, under the normal form description, there is no essential difference between infinite discrete-time dynamic games and finite games (the complex structure of the former being disguised in the strategy spaces and the cost functionals), and this permits us to adopt all the noncooperative equilibrium solution concepts introduced in Chapters 2 and 3 directly in the present framework. In particular, the reader is referred to Defs 2.8, 3.12 and 3.27 which introduce the *saddle-point*, *Nash* and *Stackelberg* equilibrium solution concepts, respectively, which are equally valid for infinite dynamic games (in normal form). Furthermore the *feedback Nash* (cf. Def. 3.22) and *feedback Stackelberg* (cf. Def. 3.29) solution concepts are also applicable to discrete-time infinite dynamic games under the right type of interpretation, and these are discussed in Chapters 6 and 7, respectively.

Before concluding our discussion on the ingredients of a discrete-time dynamic game as presented in Def. 5.1 we now finally classify possible information structures that will be encountered in the following chapters, and also introduce a specific class of cost functions—the so-called *stage-additive* cost functions.

Definition 5.2 *In an N-person discrete-time deterministic dynamic game of prespecified fixed duration, we say that $\mathbf{P}i$'s information structure is a(n)*

(i) open-loop *(OL) pattern if* $\eta_k^i = \{x_1\}$, $k \in \mathbf{K}$,

(ii) closed-loop perfect state information *(CLPS) pattern if* $\eta_k^i = \{x_1, \ldots, x_k\}$, $k \in \mathbf{K}$,

(iii) closed-loop imperfect state information *(CLIS) pattern if* $\eta_k^i = \{y_1^i, \ldots, y_k^i\}$, $k \in \mathbf{K}$,

(iv) memoryless perfect state information *(MPS) pattern if* $\eta_k^i = \{x_1, x_k\}$, $k \in \mathbf{K}$,

(v) feedback (perfect state) information *(FB) pattern if* $\eta_k^i = \{x_k\}$, $k \in \mathbf{K}$,

(vi) feedback imperfect state information *(FIS) pattern if* $\eta_k^i = \{y_k^i\}$, $k \in \mathbf{K}$,

(*vii*) one-step delayed CLPS *(1DCLPS) pattern if* $\eta_k^i = \{x_1, \ldots, x_{k-1}\}$, $k \in \mathbf{K}, k \neq 1$,

(*viii*) one-step delayed observation sharing *(1DOS) pattern if* $\eta_k^i = \{y_1, \ldots, y_{k-1}, y_k^i\}$, $k \in \mathbf{K}$, *where* $y_j \triangleq \{y_j^1, y_j^2, \ldots, y_j^N\}$.

Definition 5.3 *In an N-person discrete-time deterministic dynamic game of prespecified fixed duration (i.e., K stages),* **P***i's cost functional is said to be* stage-additive *if there exist* $g_k^i : X \times X \times U_k^1 \times \cdots \times U_k^N \to \mathbf{R}$, $(k \in \mathbf{K})$, *so that*

$$L^i(u^1, \ldots, u^N) = \sum_{k=1}^{K} g_k^i(x_{k+1}, u_k^1, \ldots, u_k^N, x_k), \tag{5.10}$$

where

$$u^j = (u_1^{j\prime}, \ldots, u_K^{j\prime})'.$$

Furthermore, if $L^i(u^1, \ldots, u^N)$ *depends only on* x_{K+1} *(the terminal state), then we call it a* terminal cost functional.

Remark 5.1 It should be noted that every stage-additive cost functional can be converted into a terminal cost functional, by introducing an additional variable z_k $(k \in \mathbf{K})$ through the recursive relation

$$z_{k+1} = z_k + g_k(f_k(x_k, u_k^1, \ldots, u_k^N), u_k^1, \ldots, u_k^N, x_k), z_1 = 0,$$

and by adjoining z_k to the state vector x_k as the last component. Denoting the new state vector as $\tilde{x}_k \triangleq (x_k', z_k)'$, the stage-additive cost functional (5.10) can then be written as

$$L^i(u^1, \ldots, u^N) = (0, \ldots, 0, 1)\tilde{x}_{K+1}$$

which is a terminal cost functional. □

Games with chance moves: Stochastic games

Infinite discrete-time dynamic games which also incorporate chance moves (the so-called *stochastic games*) can be introduced by modifying Def. 5.1 so that we now have an additional player, called "nature", whose actions influence the evolution of the state of the game. These are, in fact, random actions which obey an *a priori* known probability law, and hence what replaces the state equation (5.9b) in stochastic games is a conditional probability distribution function of the state given the past actions of the players and the past values of the state. An equivalent way of saying this is that there exists a function $F_k : X \times U_k^1 \times \cdots \times U_k^N \times \Theta \to X$, defined for each $k \in \mathbf{K}$, so that

$$x_{k+1} = F_k(x_k, u_k^1, \ldots, u_k^N, \theta_k),\ k \in \mathbf{K},$$

where θ_k is the action variable of nature at stage k, taking values in Θ; the initial state x_1 is also a random variable and the joint probability distribution function of $\{x_1, \theta_1, \ldots, \theta_K\}$ is known. A precise formulation of a stochastic dynamic game in discrete time would then be as follows.

Definition 5.4 *An N-person* discrete-time stochastic infinite dynamic game *of prespecified fixed duration involves all but items (v) and (xi) of Def. 5.1, and in addition*

(0) A finite or infinite set Θ, with some topological structure, which denotes the action set *of the auxiliary (N+1st) player,* nature. *Any permissible action θ_k of nature at stage k is an element of Θ.*

(v) A function $F_k : X \times U_k^1 \times \cdots \times U_k^N \times \Theta \to X$, defined for each $k \in \mathbf{K}$, so that

$$x_{k+1} = F_k(x_k, u_k^1, \ldots, u_k^N, \theta_k), \quad k \in \mathbf{K}, \tag{5.11}$$

where x_1 is a random variable taking values in X, and the joint probability distribution function of $\{x_1, \theta_1, \ldots, \theta_k\}$ is specified.

(xi) A functional $L^i : (X \times U_1^1 \times \cdots \times U_1^N \times \Theta) \times (X \times U_2^1 \times \cdots \times U_2^N \times \Theta) \times \cdots \times (X \times U_K^1 \times \cdots \times U_K^N \times \Theta) \to \mathbf{R}$ defined for each $i \in \mathbf{N}$, and called the cost functional *of* $\mathbf{P}i$ *in the stochastic game of fixed duration.*

To introduce the noncooperative equilibrium solution concepts for stochastic games as formulated in Def. 5.4, we again have to transfer the original game in extensive form into equivalent normal form, by first computing $L^i(\cdot)$ $(i \in \mathbf{N})$ for each N-tuple of permissible strategies $\{\gamma^i \in \Gamma^i, i \in \mathbf{N}\}$ and as a function of the random variables $\{x_1, \theta_1, \ldots, \theta_N\}$, and then by taking expectation of the resulting N-tuple of cost functions over the statistics of these random variables. The resulting deterministic functions $J^i(\gamma^1, \ldots, \gamma^N)$ $(i \in \mathbf{N})$ are known as the *expected (average) cost functionals* of the players, and they characterize the normal form of the original stochastic game together with the strategy spaces $\{\Gamma^i; i \in \mathbf{N}\}$. We now note that, since such a description is free of any stochastic nature of the problem, all the solution concepts applicable to the deterministic dynamic game are also applicable to the stochastic dynamic game formulated in Def. 5.4, and hence the stochastic case need not be treated separately while introducing the equilibrium solution concepts.

Remark 5.2 Definition 5.4 does not cover the most general class of stochastic dynamic games since (i) the state measurements of the players could be of stochastic nature by taking h_k^i to be a mapping: $X \times \Theta \to Y_k^i$, (ii) the duration of the game might not be fixed *a priori*, but be a variant of the players' actions as well as the outcome(s) of the chance move(s), and (iii) the order in which the players act might not be fixed *a priori*, but again depend on the outcome of the chance move as well as on the actions of the players (Witsenhausen, 1971a). Such extensions will, however, not be considered in this book. Yet another possible extension which is also valid in deterministic dynamic games (and which is included in the formulation of finite games in extensive form in Chapters 2 and 3) is the case when the action sets U_k^i $(i \in \mathbf{N}, k \in \mathbf{K})$ of the players are structurally dependent on the history of the evolution of the game. Such games will also not be treated in the following chapters. □

5.3 Continuous-Time Infinite Dynamic Games

Continuous-time infinite dynamic games, also known as *differential games* in the literature, constitute a class of decision problems wherein the evolution of the state is described by a differential equation and the players act throughout a time interval. Hence, as a counterpart of Def. 5.1, we can formulate such games of prespecified fixed duration as follows.

Definition 5.5 *A* quantitative N-person differential game *of prespecified fixed duration involves the following:*

(i) An index set $\mathbf{N} = \{1, \ldots, N\}$ *called the* players' set.

(ii) A time interval $[0, T]$ *which is specified* a priori *and which denotes the duration of the evolution of the game.*

(iii) An infinite set S_0 *with some topological structure, called the* trajectory space *of the game. Its elements are denoted as* $\{x(t), 0 \le t \le T\}$ *and constitute the permissible* state trajectories *of the game. Furthermore, for each fixed* $t \in [0, T]$, $x(t) \in S^0$, *where* S^0 *is a subset of a finite dimensional vector space, say* $\mathbf{R}^n$.

(iv) An infinite set U^i *with some topological structure, defined for each* $i \in \mathbf{N}$ *and which is called the* control (action) space *of* $\mathbf{P}i$, *whose elements* $\{u^i(t), 0 \le t \le T\}$ *are the* control functions *or simply the* controls *of* $\mathbf{P}i$. *Furthermore, there exists a set* $S^i \subseteq \mathbf{R}^{m_i} (i \in \mathbf{N})$ *so that, for each fixed* $t \in [0, T]$, $u^i(t) \in S^i$.

(v) A differential equation

$$\frac{dx(t)}{dt} = f(t, x(t), u^1(t), \ldots, u^N(t)), x(0) = x_0, \tag{5.12}$$

whose solution describes the state trajectory *of the game corresponding to the N-tuple of control functions* $\{u^i(t),\ 0 \le t \le T\}$ $(i \in \mathbf{N})$ *and the given initial state* x_0.

(vi) A set-valued function $\eta^i(\cdot)$ *defined for each* $i \in \mathbf{N}$ *as*

$$\eta^i(t) = \{x(s), \quad 0 \le s \le \epsilon_t^i\}, \quad 0 \le \epsilon_t^i \le t, \tag{5.13}$$

where ϵ_t^i *is nondecreasing in* t, *and* $\eta^i(t)$ *determines the state information gained and recalled by* $\mathbf{P}i$ *at time* $t \in [0, T]$. *Specification of* $\eta^i(\cdot)$ *(in fact,*

ϵ_t^i *in this formulation) characterizes the* information structure (pattern) *of* **Pi**, *and the collection (over* $i \in \mathbf{N}$*) of these information structures is the* information structure *of the game.*

(vii) A sigma-field N_t^i, *in* S_0, *generated for each* $i \in \mathbf{N}$ *by the cylinder sets* $\{x \in S_0, x(s) \in B\}$ *where* B *is a Borel set in* S^0 *and* $0 \leq s \leq \epsilon_t^i$. *The sigma-field* $/N_t^i$, $t \geq t_0$, *is called the* information field *of* **Pi**.

(viii) A prespecified class Γ^i *of mappings* $\gamma^i : [0,T] \times S_0 \to S^i$, *with the property that* $u^i(t) = \gamma^i(t,x)$ *is* N_t^i*-measurable (i.e., it is adapted to the information field* N_t^i*).* Γ^i *is the strategy space of* **Pi** *and each of its elements* γ^i *is a permissible strategy for* **Pi**.

(ix) Two functionals $q^i : S^0 \to \mathbf{R}, g^i : [0,T] \times S^0 \times S^1 \times \cdots \times S^N \to \mathbf{R}$ *defined for each* $i \in \mathbf{N}$, *so that the composite functional*

$$L^i(u^1, \ldots, u^N) \triangleq \int_0^T g^i(t, x(t), u^1(t), \ldots, u^N(t))\, d\, t + q^i(x(T)) \quad (5.14)$$

is well defined [35] *for every* $u^j(t) = \gamma^j(t,x), \gamma^j \in \Gamma^j \quad (j \in \mathbf{N})$, *and for each* $i \in \mathbf{N}$. L^i *is the* cost function *of* **Pi** *in the differential game of fixed duration.*

A differential game, as formulated above, is yet not well defined unless we impose some additional restrictions on some of the terms introduced. In particular, we have to impose conditions on f and $\Gamma^i(i \in \mathbf{N})$, so that the differential equation (5.12) admits a unique solution for every N-tuple $\{u^i(t) = \gamma^i(t,x), i \in \mathbf{N}\}$, with $\gamma^i \in \Gamma^i$. A nonunique solution to (5.12) is clearly not allowed under the extensive form description of a dynamic game, since it corresponds to nonunique state trajectories (or game paths) and thereby to a possible nonuniqueness in the cost functions for a single N-tuple of strategies. We now provide below in Thm. 5.1 a set of conditions under which this uniqueness requirement is fulfilled. But first we list down some information structures within the context of deterministic differential games, as a counterpart of Def. 5.2.

Definition 5.6 *In an* N*-person continuous-time deterministic dynamic game (differential game) of prespecified fixed duration* $[0,T]$, *we say that* **Pi***'s information structure is a(n)*

(i) open-loop *(OL) pattern if* $\eta^i(t) = \{x_0\}, t \in [0,T]$,

(ii) closed-loop perfect state *(CLPS) pattern if*

$$\eta^i(t) = \{x(s), 0 \leq s \leq t\}, \quad t \in [0,T],$$

[35] This term will be made precise in the sequel.

(iii) ϵ-delayed closed-loop perfect state *(ϵDCLPS) pattern if*

$$\eta^i(t) = \begin{cases} \{x_0\}, & 0 \le t \le \epsilon \\ \{x(s), 0 \le s \le t-\epsilon\}, & \epsilon < t \end{cases}$$

where $\epsilon > 0$ is fixed,

(iv) memoryless perfect state *(MPS) pattern if $\eta^i(t) = \{x_0, x(t)\}$, $t \in [0, T]$,*[36]

(v) feedback (perfect state) *(FB) pattern if $\eta^i(t) = \{x(t)\}$, $t \in [0, T]$.*

Theorem 5.1 *Within the framework of Def. 5.5, let the information structure for each player be any one of the information patterns of Def. 5.6. Furthermore, let $S_0 = C^n[0, T]$. Then, if*

(i) $f(t, x, u^1, \ldots, u^N)$ is continuous in $t \in [0, T]$ for each $x \in S^0$, $i \in \mathbf{N}$,

(ii) $f(t, x, u^1, \ldots, u^N)$ is uniformly Lipschitz in $x, u^1, \ldots, u^N$; i.e., for some $k > 0$,[37]

$$\begin{aligned} &|f(t, x, u^1, \ldots, u^N) - f(t, \bar{x}, \bar{u}^1, \ldots, \bar{u}^N)| \\ &\qquad \le k \max_{0 \le t \le T} \{|x(t) - \bar{x}(t)| + \sum_{i \in \mathbf{N}} |u^i(t) - \bar{u}^i(t)|\}, \\ &x(\cdot), \bar{x}(\cdot) \in C^n[0, T]; u^i(\cdot), \bar{u}^i(\cdot) \in U^i \quad (i \in \mathbf{N}), \end{aligned}$$

(iii) for $\gamma^i \in \Gamma^i$ $(i \in \mathbf{N})$, $\gamma^i(t, x)$ is continuous in t for each $x(\cdot) \in C^n[0, T]$ and uniformly Lipschitz in $x(\cdot) \in C^n[0, T]$,

the differential equation (5.12) admits a unique solution (i.e., a unique state trajectory) for every $\gamma^i \in \Gamma^i$ $(i \in \mathbf{N})$, so that $u^i(t) = \gamma^i(t, x)$, and furthermore this unique trajectory is continuous.

Proof. It follows from a standard result on the existence of unique continuous solutions to differential equations. See for instance Coddington and Levinson (1955). □

Remark 5.3 Theorem 5.1 provides only one set of conditions which are sufficient to ensure existence of a unique state trajectory for every N-tuple of strategies $\{\gamma^i \in \Gamma^i; i \in \mathbf{N}\}$, which necessarily implies a well-defined differential game problem within the framework of Def. 5.5. Since these conditions are all related to existence of a unique solution to

$$\frac{\mathrm{d}x(t)}{\mathrm{d}t} = f(t, x(t), \gamma^1(t, x), \ldots, \gamma^N(t, x)), \gamma^i \in \Gamma^i, i \in \mathbf{N}, \tag{5.15}$$

they can definitely be relaxed (but slightly) by making use of the available theory on functional differential equations (see, for example, Hale (1977)). But,

[36] Note that (iv) and (v) are not covered by (5.13) in Def. 5.5.

[37] $|v|$ denotes here the Euclidean norm for the vector v.

inevitably, these conditions all involve some sort of Lipschitz-continuity on the permissible strategies of the players. However, although such a restriction could be reasonable in the extreme case of one-person differential games (i.e., optimal control problems), it might be quite demanding in an N-player ($N \geq 2$) differential game. To illustrate this point, consider, for instance, the one-player game described by the scalar differential equation

$$\frac{dx}{dt} = u^1, \quad x(0) = 0, \tag{i}$$

and adopt the strategy $\gamma^1(t,x) = \text{ sgn } (x(t))$. The solution to (i) with $u^1(t) = \gamma^1(t,x)$ is clearly not unique; a multitude of solutions exists. If we adopt yet another strategy, viz. $\gamma^1(t,x) = - \text{ sgn } (x(t))$, then the solution to (i) does not even exist in the classical sense. Hence, a relaxation of the Lipschitz-continuity condition on the permissible strategies could make an optimal control problem quite ill-defined. In such a problem, the single player may be satisfied with smooth (but sub-optimal) strategies. In differential games, however, it is unlikely that players are willing to restrict themselves to smooth strategies voluntarily. If one player would restrict his strategy to be Lipschitz, the other player(s) may be able to exploit this. Specifically in pursuit evasion games (see Chapter 8) the players play "on the razor's edge", and such a restriction could result in a drastic change in the outcome of the game. Extensive discussions on these issues, and on various different definitions of "solution", can be found in the book by Krasovskii and Subbotin (1988).

In conclusion, non-Lipschitz strategies cannot easily be put into a rigorous mathematical framework. On the other hand, in many games, we do not want the strategy spaces to comprise only smooth mappings. These difficulties may show up especially under the Nash equilibrium solution concept. In two-person Stackelberg games, however, non-Lipschitz strategies can more easily be handled in the general formulation, since one of the player's (follower's) choice of strategy is allowed to depend on the other player's (leader's) announced strategy. □

The saddle-point, Nash, Stackelberg and consistent conjectural variations equilibrium concepts introduced earlier for finite games are equally valid for (continuous-time) differential games if we bring them into equivalent normal form. To this end, we start with the extensive form description of a differential game, as provided in Def. 5.5 and under the hypotheses of Thm. 5.1, and for each fixed N-tuple of strategies $\{\gamma^i \in \Gamma^i; i \in \mathbf{N}\}$ we obtain the unique solution of the functional differential equation (5.15) and determine the corresponding action (control) vectors $u^i(\cdot) = \gamma^i(\cdot, x), i \in \mathbf{N}, x \in S^0$. Substitution of these into (5.14), together with the corresponding unique state trajectory, thus yields an N-tuple of numbers $\{L^i; i \in \mathbf{N}\}$, for each choice of strategies by the players — assuming of course that functions g^i ($i \in \mathbf{N}$) are integrable, so that (5.14) are well defined. Therefore, we have mappings $J^i : \Gamma^1 \times \cdots \times \Gamma^N \to \mathbf{R}$ ($i \in \mathbf{N}$) for each fixed initial state vector x_0, which we call the *cost functional* of $\mathbf{P}i$ in a differential game in normal form. These cost functionals, together with the strategy spaces $\{\Gamma^i; i \in \mathbf{N}\}$ of the players, then constitute the equivalent normal form description of the differential game, which is the right framework to introduce noncooperative equilibrium solution concepts, as we have done earlier for other classes of dynamic games.

Termination

Definition 5.5 covers differential games of fixed prespecified duration; however, as discussed in Section 5.2 within the context of discrete-time games, it is possible to extend such a formulation so that the end point in both state and time is a variable. Let S^0 again denote a subset of $\mathbf{R}^n$, and $\mathbf{R}^+$ denote the half-open interval $[0, \infty)$. Let a closed subset $\Lambda \subset S^0 \times \mathbf{R}^+$ be given, which we call a *terminating (target) set*, so that $(x_0, 0) \notin \Lambda$. Then, we say that the *differential game terminates, for a given N-tuple of strategies, at time $T \in \mathbf{R}^+$ if T is the smallest element of* $\mathbf{R}^+$ with the property $(x(T), T) \in \Lambda$, i.e.,

$$T = \min\{t \in \mathbf{R}^+ : (x(t), t) \in \Lambda\}. \tag{5.16}$$

This positive number T is called the *terminal time* of the differential game, corresponding to the given N-tuple of strategies. Terminal time is sometimes defined in a slightly different way, as the smallest time at which $x(\cdot)$ penetrates, Λ, i.e.,

$$T = \inf\{t \in \mathbf{R}^+ : (x(t), t) \in \mathring{\Lambda}\},$$

where $\mathring{\Lambda}$ denotes the interior of Λ. The two definitions can only differ in situations where the trajectory $x(\cdot)$ belongs to the boundary of Λ for a while or only touches Λ at one instant of time. Unless stated differently, we shall adopt (5.16) as the definition of the terminal time.

The question of whether a given differential game necessarily terminates is one that requires some further discussion. If there is a finite time, say t_1, such that $(x, t_1) \in \Lambda$ for all $x \in S^0$, then the game always terminates, in at most t_1 units of time. Such a finite t_1, however, does not always exist, as elucidated by the following optimal control (one-player differential game) problem.

Example 5.1 Consider the optimal control problem described by the two-dimensional state equation

$$\begin{aligned} \dot{x}_1 &= u, & x_1(0) = 1, \\ \dot{x}_2 &= -x_2 + u, & x_2(0) = 1, \end{aligned}$$

where $u(\cdot)$ is the control (action) variable satisfying the constraint $0 \le u(t) \le 3$ for all $t \in [0, \infty)$. Let the terminal time T be defined as $T = \min\{t \in [0, \infty) : x_2(t) = 2\}$, and a permissible OL strategy γ be defined as a mapping from $[0, \infty)$ into $[0, 3]$. Now, if the cost function is $L(u) = \int_0^T 1\, dt = T$, then the minimizing strategy is $\gamma^*(t) = 3$, $t \ge 0$, with the corresponding terminal time (and thereby the minimum value of L) being $\ln 2$. □

If the cost function to be minimized is $L(u) = x_1(T)$, however, the player may not have an incentive to terminate this "one-player game" since, as long as termination has not been achieved, his cost is not defined. So, on the one hand, he does not want to terminate the game; while on the other hand, if he has an incentive to terminate it, he should do it as soon as possible, which dictates him to employ the strategy $\gamma^*(t) = 3$, $t \geq 0$.

The player is faced with a dilemma here, since the latter cost function is not well defined for all strategies available to the player. This ambiguity can, however, be removed by either (i) restricting the class of permissible strategies to the class of so-called *playable strategies* which are those that terminate the game in finite time, or (ii) extending the definition of the cost functional so that

$$L(u) = \begin{cases} x_1(T), & \text{if } T \text{ is finite,} \\ \infty, & \text{otherwise ,} \end{cases}$$

which eliminates any possible incentive for the player not to terminate the game. In both cases, the optimal strategy will be $\gamma^*(t) = 3$, $t \geq 0$.

The difficulties encountered in the preceding optimal control example, as well as the proposed ways out of the dilemma, are also valid for differential games, and this motivates us to introduce the following concept of "playability".

Definition 5.7 *For a given N-person differential game with a target set Λ, a strategy N-tuple is said to be* playable *(at (t_0, x_0)) if it generates a trajectory $x(\cdot)$ such that $(x(t), t) \in \Lambda$ for finite t. Such a trajectory $x(\cdot)$ is called* terminating.

Differential games with chance moves[38]

Chance moves can be introduced in the formulation of differential games by basically following the same lines as in the discrete-time case (cf. Def. 5.4), but this time one has to be mathematically more precise. In particular, if we assume the chance player (nature) to influence the state trajectory throughout a given time interval, then actions of this additional player will be realizations of a stochastic process $\{\theta_t, t \geq 0\}$ whose statistics are known *a priori*. If we adjoin such a stochastic process to (5.12), then the resulting "differential equation"

$$\frac{dx(t)}{dt} = F(t, x(t), u^1(t), \ldots, u^N(t), \theta_t), \quad x(0) = x_0,$$

might not be well defined, in the sense that even though its solution might be unique for each realization (sample path) of $\{\theta_t, t \geq 0\}$, it might not exist as a stochastic process. To obtain a meaningful formulation and tractable results, one has to impose some restrictions of F and $\{\theta_t, t \geq 0\}$. One particular such

[38] Here it is assumed that the reader has some prior knowledge of stochastic processes. If this is not the case, either this part may be skipped (without much loss, since this formulation is used later only in Section 6.7) or the standard references may be consulted for background knowledge (Fleming and Rishel, 1975; Gikhman and Skorohod, 1972; Wong and Hajek, 1985).

assumption is to consider equations of the form

$$x_t = x_0 + \int_0^t F(s, x_s, u^1(s), \ldots, u^N(s))\, \mathrm{d}s + \int_0^t \sigma(s, x_s)\, \mathrm{d}\theta_s, \qquad (5.17\mathrm{a})$$

where F satisfies the conditions imposed on f in Def. 5.5, the function σ satisfies similar conditions in its arguments s and x_s and $\{\theta_t, t \geq 0\}$ is a special type of a stochastic process called the *Wiener process.* Equation (5.17a) can also be written symbolically as

$$\mathrm{d}x_t = F(t, x_t, u^1(t), \ldots, u^N(t))\mathrm{d}t + \sigma(t, x_t)\mathrm{d}w_t, \; x_t\big|_{t=0} = x_0, \qquad (5.17\mathrm{b})$$

where we have used w_t, instead of θ_t, to denote that it is the Wiener process. It should further be noted that in both (5.17a) and (5.17b), the function $x(\cdot)$ is written as $x_{(\cdot)}$, to indicate explicitly that it now stands for a stochastic process instead of a deterministic function. Equation (5.17b) is known as a *stochastic differential equation*, and existence and uniqueness properties of its solution, whenever $u^i(\cdot) = \gamma^i(\cdot, x), \gamma^i \in \Gamma^i \quad (i \in \mathbf{N})$, are elucidated in the following theorem, whose proof can be found in Gikhman and Skorohod (1972).

Theorem 5.2 *Let $\Gamma^1, \ldots, \Gamma^N$ denote the strategy spaces of the players under any one of the information patterns of Def. 5.6. Furthermore let $S_0 = C^n[0, T]$, F satisfy the requirements imposed on f in Thm. 5.1, and $\gamma^i \in \Gamma^i$ $(i \in \mathbf{N})$ satisfy the restrictions imposed on γ^i in Thm. 5.1 (iii). If, further, σ is a nonsingular $(n \times n)$ matrix, whose elements are continuous in t and uniformly Lipschitz in x, the stochastic differential equation (5.17b) with $u^i(\cdot) = \gamma^i(\cdot, x)$, $\gamma^i \in \Gamma^i$ $(i \in \mathbf{N})$, admits as its solution a unique stochastic process with continuous sample paths, for every such N-tuple of strategies.*

To complete the formulation of a differential game with chance moves (i.e., a *stochastic differential game*), it will now be sufficient to replace L^i in Def. 5.5 (ix) with the expected value of the same expression, where the expectation operation is taken with respect to the statistics of the Wiener process $\{w_t, t \geq 0\}$ and the initial state x_0. The game can then be converted into normal form by determining, in the usual sense, functions $J^i : \Gamma^1 \times \cdots \times \Gamma^N \to \mathbf{R}$ $(i \in \mathbf{N})$, which is now the suitable form to introduce the solution concepts already discussed.

5.4 Mixed and Behavioral Strategies in Infinite Dynamic Games

In Chapter 2, we have defined a mixed strategy for a player as a probability distribution on the space of his pure strategies, or equivalently, as a random variable whose values are the player's pure strategies (cf. Def. 2.2), which was also adopted in Chapters 3 and 4, for finite games and static infinite games, respectively. Defined in this way, a mixed strategy is a mathematically well-established object, mainly because the strategy spaces are finite for the former class of problems, and at most a continuum for the latter class of games, thereby allowing one to introduce (Borel-) measurable subsets of these strategy spaces, on which probability measures can be defined. An attempt to extend this directly to infinite dynamic games, however, is hampered by several measure-theoretic difficulties (Aumann, 1964). To illustrate the extent of these difficulties, let us consider a simple two-stage two-person dynamic game defined by the state equation

$$\begin{aligned} x_2 &= 1 + u^1, \\ x_3 &= x_2 + u^2, \end{aligned}$$

where $-1 \leq u^1 \leq 0$, $0 \leq u^2 \leq 1$, and **P2** has access to closed-loop perfect state information (i.e., he knows the value of x_2). A mixed strategy for **P1** can easily be defined in this case, since his pure strategy space is $[-1, 0]$ which is endowed with a measurable structure, viz. its Borel subsets. However, for **P2**, the permissible (pure) strategies are measurable mappings from $[0, 1]$ into $[0, 1]$, and thus the permissible strategy space of **P2** is $\Gamma^2 = I^I$, where I denotes the unit interval. In order to define a probability distribution on I^I, we have to endow it with a measurable structure, but no such structure exists which is suitable for the problem under consideration (see Aumann, 1961). Intuitively, such a difficulty arises because the set of all probability distributions on I is already an extremely rich class, so that if one wants to define a similar object on a domain I^I whose cardinality is higher than that of I, such an increase in cardinality cannot reflect itself on the set of probability distributions.

An alternative, then, is to adopt the "random variable" definition of a mixed strategy. This involves a sample space, say Ω, and a class of measurable mappings from Ω into I^I, with each of these mappings being a candidate mixed strategy. The key issue here, now, is the choice of the sample space Ω. In fact, since Ω stands for a random device, the question can be rephrased as the choice of a random device whose outcomes are rich enough to be compatible with I^I. But the richest one is the continuous roulette wheel which corresponds to a sample space Ω as a copy of the unit interval I.[39] Such a consideration then leads to the conclusion that the random variable (mixed strategy) should be a measurable mapping f from Ω into I^I. Unfortunately, this approach also leads to difficulties, since one still has to define a measurable structure on the range space I^I. But now, if we note that, to every function f: $\Omega \to I^I$, there corresponds a function g: $\Omega \times I \to I$ defined by $g(\omega, x) = f(\omega)(x)$, then the "measurability" difficulty is resolved since one can define measurable mappings from $\Omega \times I$ into I. This conclusion of course remains valid if the action set I is replaced by any measurable space U, and the information space I is replaced by any measurable space N. Moreover, an extension to multi-stage infinite dynamic games is also possible as the following definition elucidates.

[39]This follows from the intuitive idea of a random device (Aumann, 1964). It is, of course, possible to consider purely abstract random devices which are even richer.

Definition 5.8 *Given a K-stage discrete-time infinite dynamic game within the framework of Def. 5.1, let U_k^i, N_k^i be measurable spaces for each $i \in \mathbf{N}$, $k \in \mathbf{K}$. Then, a* mixed strategy *$\hat{\gamma}^i$ for* $\mathbf{P}i$ *is a sequence $\hat{\gamma}^i = (\hat{\gamma}_1^i, \ldots, \hat{\gamma}_K^i)$ of measurable mappings $\hat{\gamma}_j^i : \Omega \times N_j^i \to U_j^i$ $(j \in \mathbf{K})$, so that $\hat{\gamma}^i(\omega, \cdot) \in \Gamma^i$ for every $\omega \in \Omega$, where Ω is a fixed sample space (taken as a copy of the unit interval).*

A direct extension of this definition to continuous-time systems would be as follows.

Definition 5.9 *Given a differential game that fits the framework of Def. 5.5, let S_0 and $S^i (i \in \mathbf{N})$ be measurable spaces. Then, a* mixed strategy *$\hat{\gamma}^i$ for* $\mathbf{P}i$, *in this game, is a measurable transformation $\hat{\gamma}^i : \Omega \times [0, T] \times S_0 \to S^i$, so that $\hat{\gamma}^i(\omega, \cdot, \cdot) \in \Gamma^i$ for every $\omega \in \Omega$, where Ω is a fixed sample space (taken as a copy of the unit interval). Equivalently, $\hat{\gamma}^i(\cdot, \cdot, x)$ is a stochastic process for each $x \in S_0$.*

A behavioral strategy, on the other hand, has been defined (cf. Def. 2.9) as a collection of probability distributions, one for each information set of the player. In other words, using the notation of Section 2.4, it is a mixed strategy $\hat{\gamma}(\cdot)$ with the property that $\hat{\gamma}(\eta_1)$ and $\hat{\gamma}(\eta_2)$ are independent random variables whenever η_1 and η_2 belong to different information sets. If we extend this notion directly to multi-stage infinite dynamic games (and within the framework of Def. 5.8), we have to define a behavioral strategy as a mixed strategy $\hat{\gamma}^i(\cdot, \cdot)$ with the property that the collection of random variables $\{\hat{\gamma}_j^i(\cdot, \eta_j^i), \eta_j^i \in N_j^i; j \in \mathbf{K}\}$ is mutually independent. But, since N_j^i is in general a non-denumerable set, in infinite dynamic games this would imply that we have a non-denumerable number of mutually independent bounded random variables on the same sample space. This is clearly not possible, since any bounded random variable defined on our sample space Ω should have a countable basis (Loève , 1963). Then, a way out of this difficulty is to define a behavioral strategy as a mixed strategy which is independent from stage to stage, but not necessarily stagewise. Aumann actually discusses that stagewise correlation is quite irrelevant (Aumann, 1964), and the expected cost function is invariant under such a correlation. Therefore, we have the following definition.

Definition 5.10 *Given a K-stage discrete-time infinite dynamic game within the framework of Def. 5.1, let U_k^i, N_k^i be measurable spaces for each $i \in \mathbf{N}$, $k \in \mathbf{K}$. Then, a* behavioral strategy *$\hat{\gamma}^i$ for* $\mathbf{P}i$ *is a mixed strategy (cf. Def. 5.8) with the further property that the sequence of random variables $\{\hat{\gamma}_j^i(\cdot, \eta_j^i),\ j \in \mathbf{K}\}$ is mutually independent for every fixed $\eta_j^i \in N_j^i$ $(j \in \mathbf{K})$.*

This definition of a behavioral strategy can easily be extended to multi-stage games with a (countably) infinite number of stages; however, an extension to continuous-time infinite dynamic games is not possible since the time set $[0, T]$ is not denumerable.

5.5 Tools for One-Person Optimization

Since optimal control problems constitute a special class of infinite dynamic games with one player and one criterion, the mathematical tools available for such problems may certainly be useful in dynamic game theory. This holds particularly true if the players adopt the noncooperative Nash equilibrium solution concept, in which case each player is faced with a single criterion optimization problem (i.e., optimal control problem) with the strategies of the remaining players taken to be fixed at their equilibrium values. Hence, in order to verify whether a given set of strategies is in Nash equilibrium, we inevitably have to utilize the tools of optimal control theory. We, therefore, present in this section some important results on dynamic one-person optimization problems so as to provide an introduction to the theory of subsequent chapters.

The section comprises three subsections. The first two deal with the dynamic programming (DP) technique applied to discrete-time and continuous-time optimal control problems, and the third is devoted to the "minimum principle". For more details on, and a rigorous treatment of, the material presented in these subsections the reader is referred to Fleming and Rishel (1975), Fleming and Soner (1993) and Boltyanski (1978).

5.5.1 Dynamic programming for discrete-time systems

The method of dynamic programming is based on *the principle of optimality* which states that an optimal strategy has the property that, whatever the initial state and time are, all remaining decisions (from that particular initial state and particular initial time onwards) must also constitute an optimal strategy. To exploit this principle, we work backwards in time, starting at all possible final states with the corresponding final times. Such a technique has already been used in this book within the context of finite dynamic (multi-act) games, specifically in Sections 2.5 and 3.5, in the derivation of noncooperative equilibria, though we did not refer explicitly to dynamic programming. We now discuss the principle of optimality within the context of discrete-time systems that fit the framework of Def. 5.1, but with only one player (i.e., $N = 1$). Toward that end, we consider equation (5.9b), assume feedback perfect state information and a stage-additive cost functional of the form (5.10), all for $N = 1$, i.e.,

$$x_{k+1} = f_k(x_k, u_k), \quad u_k \in U_k, k \in \mathbf{K}, \tag{5.18a}$$

$$L(u) = \sum_{k=1}^{K} g_k(x_{k+1}, u_k, x_k), \tag{5.18b}$$

where $u = \{u_k, k \in \mathbf{K}\}$, $u_k = u_k^1 = \gamma_k(x_k)$; $\gamma_k(\cdot)$ denotes a permissible (control) strategy at stage $k \in \mathbf{K}$, and K is a fixed positive integer. In order to determine the minimizing control strategy, we shall need the expression for the *minimum* (or *minimal*) cost from any starting point at any initial time. This is also called the *value function*, and is defined as

$$V(k,x) = \min_{\gamma_k,\dots,\gamma_K} \left[\sum_{i=k}^{K} g_i(x_{i+1}, u_i, x_i) \right]$$

with $u_i = \gamma_i(x_i) \in U_i$ and $x_k = x$. A direct application of the principle of optimality now readily leads to the recursive relation

$$V(k,x) = \min_{u_k \in U_k} [g_k(f_k(x,u_k), u_k, x) + V(k+1, f_k(x,u_k))]. \tag{5.19}$$

If the optimal control problem admits a solution $u^* = \{u_k^*, k \in \mathbf{K}\}$, then the solution $V(1, x_1)$ of (5.19) should be equal to $L(u^*)$, and furthermore each u_k^* should be determined as an argument of the RHS of (5.19).

Affine-quadratic problems

As a specific application, let us consider the so-called *affine-quadratic (or linear-quadratic) discrete-time optimal control* problem, which is described by the state equation

$$x_{k+1} = A_k x_k + B_k u_k + c_k, \tag{5.20a}$$

and cost functional

$$L(u) = \frac{1}{2} \sum_{k=1}^{K} (x'_{k+1} Q_{k+1} x_{k+1} + u'_k R_k u_k). \tag{5.20b}$$

Let us further assume that $x_k \in \mathbf{R}^n$, $u_k \in \mathbf{R}^m$, $c_k \in \mathbf{R}^n$, $R_k > 0$, $Q_{k+1} \geq 0$ for all $k \in \mathbf{K}$, and A_k, B_k are matrices of appropriate dimensions. Here, the corresponding expression for f_k is obvious, but the one for g_k is not uniquely defined; though, it is convenient to take it as

$$g_k(u_k, x_k) = \begin{cases} \frac{1}{2} u'_1 R_1 u_1, k = 1, \\ \frac{1}{2} u'_k R_k u_k + \frac{1}{2} x'_k Q_k x_k, k \neq 1, K+1, \\ \frac{1}{2} x'_{K+1} Q_{K+1} x_{K+1}, k = K+1. \end{cases}$$

We now obtain, by inspection, that $V(k,x)$ should be a general quadratic function of x for all $k \in \mathbf{K}$, and that $V(K+1, x_{K+1}) = \frac{1}{2} x'_{K+1} Q_{K+1} x_{K+1}$. This leads to the structural form $V(k,x) = \frac{1}{2} x' S_k x + x' s_k + q_k$. Substituting this in the recursive relation (5.19) we obtain the unique solution of the optimal control problem (5.20a)-(5.20b) as follows.

Proposition 5.1 *The optimal control problem (5.20a)-(5.20b) admits the unique solution*

$$u_k^* = \gamma_k^*(x_k) = -P_k S_{k+1} A_k x_k - P_k(s_{k+1} + S_{k+1} c_k), \tag{5.21a}$$

for all $k \in \mathbf{K}$, where

$$\left.\begin{array}{rcl} P_k & = & [R_k + B_k' S_{k+1} B_k]^{-1} B_k', \\ S_k & = & Q_k + A_k' S_{k+1}[I - B_k P_k S_{k+1}] A_k; \quad S_{K+1} = Q_{K+1}, \\ s_k & = & A_k'[I - B_k P_k S_{k+1}]'[s_{k+1} + S_{k+1} c_k]; \quad s_{K+1} = 0. \end{array}\right\} \quad (5.21b)$$

Furthermore, the minimum value of (5.20b) is

$$L(u^*) = \frac{1}{2} x_1' S_1 x_1 + x_1' s_1 + q_1, \quad (5.22)$$

where

$$q_1 = \frac{1}{2} \sum_{k=1}^{K} (c_k' S_{k+1} c_k - (s_{k+1} + S_{k+1} c_k)' P_k' B_k' (s_{k+1} + S_{k+1} c_k) + 2 c_k' s_{k+1}).$$

Remark 5.4 If the requirement $Q_k \geq 0$ is not satisfied, a necessary and sufficient condition for (5.21a)-(5.21b) to provide a unique solution to the affine-quadratic discrete-time optimal control problem is

$$R_k + B_k' S_{k+1} B_k > 0, \qquad \forall k \in \mathbf{K}.$$

□

Infinite horizon linear-quadratic problems

To formulate a possibly meaningful linear-quadratic optimal control problem when $K \to \infty$, we take $c_k = 0$, and the matrices A_k, B_k, Q_k and R_k to be independent of k (which will henceforth be written without the index k). Thus the optimization problem is

$$\min_u \frac{1}{2} \sum_{k=1}^{\infty} (x_{k+1}' Q x_{k+1} + u_k' R u_k); \quad Q \geq 0,\, R > 0$$

subject to

$$x_{k+1} = A x_k + B u_k.$$

It is natural to assume here that the problem is well defined, in the sense that there exists at least one control sequence that renders it a finite cost. Conditions which ensure this are stabilizability of the matrix pair (A, B), and detectability of the matrix pair (A, D), where D is a matrix such that $D'D = Q$.[40] These notions of stabilizability and detectability belong to the realm of the theory of linear systems; see for instance Kailath (1980) or Anderson and Moore (1989). The pair (A, B) is stabilizable if an $m \times n$ matrix F exists such that $(A + BF)$ is a stable matrix, i.e., all its eigenvalues lie strictly within the unit circle. The pair (A, D) is detectable if its "dual pair", (A', D') is stabilizable. The following result is now a standard one, which can be found in any textbook on linear control systems.

[40] Using standard terminology, we will also refer to the latter condition as detectability of the pair (A, Q).

Proposition 5.2 *Assume that the pair (A, B) is stabilizable and the pair (A, Q) is detectable. Then, there exists an $n \times n$ nonnegative-definite matrix S such that*

(i) for fixed k, $S_k^{(K)} \to S$ as $K \to \infty$, where for each K, and arbitrary $Q_{K+1} \geq 0$, $S_k^{(K)}$ is recursively defined by (5.21a)-(5.21b);

(ii) S is the unique solution of the algebraic Riccati equation (ARE)

$$S = Q + A'S[I - B(R + B'SB)^{-1}B'S]A$$

within the class of nonnegative-definite matrices;

(iii) the (closed-loop) matrix $A - B(R + B'SB)^{-1}B'S$ is stable, i.e., all its eigenvalues lie strictly within the unit circle;

(iv) the minimum value of the cost functional is $\frac{1}{2}x_1'Sx_1$;

(v) the stationary optimal control law is

$$\gamma^*(x) = -[R + B'SB]^{-1}B'SAx.$$

Under the stronger assumption that (A, D) is observable (a sufficient condition for which is $Q > 0$), the solution of the ARE is positive definite.

5.5.2 Dynamic programming for continuous-time systems

The dynamic programming approach, when applied to optimal control problems defined in continuous time, leads to a partial differential equation (PDE) which is known as the Hamilton–Jacobi–Bellman (HJB) equation. Toward this end we consider the optimal control problem defined by

$$\left.\begin{aligned} \dot{x}(t) &= f(t, x(t), u(t)),\ x(0) = x_0,\ t \geq 0, \\ u(t) &= \gamma(t, x(t)) \in S,\ \gamma \in \Gamma, \\ L(u) &= \textstyle\int_0^T g(t, x(t), u(t))\,\mathrm{d}t + q(T, x(T)), \\ T &= \min_{t \geq 0} \{t : l(t, x(t)) = 0\}, \end{aligned}\right\} \tag{5.23}$$

where l is a scalar function, defining an n-dimensional smooth manifold in the product space $\mathrm{R}^n \times \mathrm{R}^+$, and Γ is taken to be the class of all admissible feedback strategies.

The minimum cost-to-go from any initial state (x) and any initial time (t) is described by the so-called *value function* which is defined by

$$V(t, x) = \min_{\{u(s), t \geq s \leq T\}} \left[\int_t^T g(s, x(s), u(s))\,\mathrm{d}s + q(T, x(T)) \right], \tag{5.24a}$$

satisfying the boundary condition

$$V(T, x) = q(T, x) \quad \text{along} \quad l(T, x) = 0. \tag{5.24b}$$

A direct application of the principle of optimality on (5.24a), under the assumption of continuous differentiability of V, leads to the HJB equation

$$-\frac{\partial V(t,x)}{\partial t} = \min_u \left[\frac{\partial V(t,x)}{\partial x} f(t,x,u) + g(t,x,u)\right], \tag{5.25}$$

which takes (5.24b) as the boundary condition.

In general, it is not easy to compute $V(t,x)$. Moreover, the continuous differentiability assumption imposed on $V(t,x)$ is rather restrictive (see, for instance, Example 5.2 in subsection 5.5.3, for which it does not hold). Nevertheless, if such a function exists, then the HJB equation (5.25) provides a means of obtaining the optimal control strategy. This "sufficiency" result is now proven in the following theorem.

Theorem 5.3 *If a continuously differentiable function $V(t,x)$ can be found that satisfies the HJB equation (5.25) subject to the boundary condition (5.24b), then it generates the optimal strategy through the static (pointwise) minimization problem defined by the RHS of (5.25).*

Proof. If we are given two strategies, $\gamma^* \in \Gamma$ (the optimal one) and $\gamma \in \Gamma$ (an arbitrary one), with the corresponding terminating trajectories x^* and x, and terminal times T^* and T, respectively, then (5.25) reads

$$g(t,x,u) + \frac{\partial V(t,x)}{\partial x} f(t,x,u) + \frac{\partial V(t,x)}{\partial t} \geq 0, \tag{5.26a}$$

$$g(t,x^*,u^*) + \frac{\partial V(t,x^*)}{\partial x} f(t,x^*,u^*) + \frac{\partial V(t,x^*)}{\partial t} \equiv 0, \tag{5.26b}$$

where γ^* and γ have been replaced by the corresponding controls u^* and u, respectively. Integrating (5.26a) from 0 to T and (5.26b) from 0 to T^*, we obtain

$$\int_0^T g(t,x,u)\,\mathrm{d}t + V(T,x(T)) - V(0,x_0) \geq 0, \tag{5.27}$$

$$\int_0^{T^*} g(t,x^*,u^*)\,\mathrm{d}t + V(T^*,x^*(T^*)) - V(0,x_0) = 0. \tag{5.28}$$

Elimination of $V(0,x_0)$ yields

$$\int_0^T g(t,x,u)\,\mathrm{d}t + q(T,x(T)) \geq \int_0^{T^*} g(t,x^*,u^*)\,\mathrm{d}t + q(T^*,x^*(T^*)), \tag{5.29}$$

from which it readily follows that u^* is the optimal control, and therefore γ^* is the optimal strategy. □

Remark 5.5 If, in the problem statement (5.23), the time variable t does not appear explicitly, i.e., if f, g, q and l are time-invariant, the corresponding value function will also be time-invariant. This then implies that $\partial V/\partial t = 0$, and the resulting optimal strategy can be expressed as a function of only $x(t)$, i.e., $u^*(t) = \gamma^*(x(t))$. In such cases, we will write $V(x)$ for $V(t,x)$. □

Remark 5.6 There exists an alternative derivation of the HJB equation (5.25), which is of a geometrical nature. This will not be discussed here, since it is a special case of the more general derivation to be given in Section 2 of Chapter 8 for two-player zero-sum differential games. □

Affine-quadratic problems

We now consider an important class of problems—the so-called *affine-quadratic (or linear-quadratic) continuous-time optimal control* problems—for which $V(t,x)$ is continuously differentiable, so that Thm. 5.3 applies. Toward that end, let the system be described (as a continuous-time counterpart of (5.20a)-(5.20b)) by

$$\dot{x}(t) = A(t)x(t) + B(t)u(t) + c(t); \quad x(0) = x_0, \tag{5.30a}$$

and the cost functional to be minimized be given as

$$L(u) = \frac{1}{2}x'(T)Q_f x(T) + \frac{1}{2}\int_0^T (x'Qx + 2x'p + u'Ru)\,\mathrm{d}t, \tag{5.30b}$$

where $x(t) \in \mathbf{R}^n$, $u(t) \in \mathbf{R}^m$, $0 \le t \le T$ and T is fixed. $A(\cdot)$, $B(\cdot)$, $Q(\cdot) \ge 0$, $R(\cdot) > 0$ are matrices of appropriate dimensions and with continuous entries on $[0,T]$. The matrix Q_f is nonnegative-definite, and $c(\cdot)$ and $p(\cdot)$ are continuous vector-valued functions, taking values in $\mathbf{R}^n$. Furthermore, we adopt the feedback information pattern and take a typical control strategy as a continuous mapping $\gamma : [0,T] \times \mathbf{R}^n \to \mathbf{R}^m$. Denote the space of all such strategies by Γ. Then the optimal control problem is to find a $\gamma^* \in \Gamma$ such that

$$J(\gamma^*) \le J(\gamma), \quad \forall \gamma \in \Gamma, \tag{5.31a}$$

where

$$J(\gamma) \triangleq L(u), \text{ with } u(\cdot) = \gamma(\cdot, x). \tag{5.31b}$$

Several methods exist to obtain the optimal strategy γ^* or the optimal control function $u^*(\cdot) = \gamma^*(\cdot, x)$. We shall derive the former by making use of Thm. 5.3. Simple arguments (see Anderson and Moore, 1989) lead to the conclusion that $\min_\Gamma J(\gamma)$ is quadratic in x_0. Moreover, it can be shown that, if the system is positioned at an arbitrary $t \in [0,T]$ at an arbitrary point $x \in \mathbf{R}^n$, the minimum cost-to-go, starting from this position, is quadratic in x. Therefore, we may assume existence of a continuously differentiable value function of the form

$$V(t,x) = \frac{1}{2}x'S(t)x + k'(t)x + m(t), \tag{5.32}$$

that satisfies (5.25). Here $S(\cdot)$ is a symmetric $(n \times n)$ matrix with continuously differentiable entries, $k(\cdot)$ is a continuously differentiable n-vector and $m(\cdot)$ is a continuously differentiable function. If we can determine such $S(\cdot)$, $k(\cdot)$ and $m(\cdot)$ so that (5.32) satisfies the HJB equation (5.25), then Thm. 5.3 justifies the assumption of the existence of a value function quadratic in x. Substitution of (5.32) into (5.25) leads to

$$\begin{aligned} &-\frac{1}{2}x'\dot{S}x - x'\dot{k} - \dot{m} \\ &\qquad = \min_u \left[(Sx+k)'(Ax+Bu+c) + \frac{1}{2}x'Qx + x'p + \frac{1}{2}u'Ru\right]. \end{aligned} \tag{5.33}$$

Carrying out the minimization on the RHS yields

$$u^*(t) = \gamma^*(t, x(t)) = -R^{-1}B'[S(t)x(t) + k(t)], \tag{5.34}$$

substitution of which into (5.33) leads to an identity relation which is readily satisfied if

$$\dot{S} + SA + A'S - SBR^{-1}B'S + Q = 0, \qquad S(T) = Q_f, \tag{5.35a}$$

$$\dot{k} + (A - BR^{-1}B'S)'k + Sc + p = 0, \qquad k(T) = 0, \tag{5.35b}$$

$$\dot{m} + k'c - \frac{1}{2}k'BR^{-1}B'k = 0, \qquad m(T) = 0. \tag{5.35c}$$

Thus we arrive at the following proposition.

Proposition 5.3 *The linear-quadratic optimal control problem (5.30a)-(5.30b) admits a unique optimum feedback controller γ^* which is given by (5.34), where $S(\cdot)$, $k(\cdot)$ and $m(\cdot)$ uniquely satisfy (5.35a)-(5.35c). The minimum value of the cost functional is*

$$J(\gamma^*) = \frac{1}{2}x_0'S(0)x_0 + k'(0)x_0 + m(0).$$

Proof. Except for existence of a unique $S(\cdot) \geq 0$, $k(\cdot)$ and $m(\cdot)$ that satisfy (5.35a)-(5.35c), the proof has been given prior to the statement of the proposition. Furthermore, if there exists a unique $S(\cdot)$ that satisfies (5.35a), which is known as the *matrix Riccati equation*, existence of unique solutions to the remaining two differential equations of (5.35a)-(5.35c) is assured since they are linear in k and m, respectively. Then, what remains to be proven is that a unique solution $S(\cdot) \geq 0$ to (5.35a) exists on $[0, T]$. This can be verified by either making use of the theory of differential equations (cf. Reid, 1972) or utilizing the specific form of the optimal control problem with Q_f and $Q(\cdot)$ taken to be nonnegative-definite (Anderson and Moore, 1989). □

Remark 5.7 The solution (5.34) can be obtained by other methods as well. Two of these are the minimum principle (to be discussed shortly; see also (Bryson and Ho, 1975), and the "completion of squares" method (see Brockett, 1970). Since the latter will be discussed in Chapter 6 in the context of affine-quadratic differential games, it will not be covered in this chapter. □

Remark 5.8 The nonnegative-definiteness requirements on Q_f and $Q(\cdot)$ may be relaxed, but then we have to assume from the outset the existence of a unique bounded solution to (5.35a) in order to ensure the existence of a unique minimizing control as given by (5.34).[41] Otherwise (5.35a) might not admit a solution; more precisely its solution may exhibit finite escape (depending on the length of the time interval), implying in turn that the "optimal" cost might tend to $-\infty$. To exemplify this situation consider the scalar example:

$$\left.\begin{array}{rcl} \dot{x} &=& x+u, \quad x(0)=x_0, \\ L(u) &=& \frac{1}{2}\int_0^T(-x^2+u^2)\,\mathrm{d}t. \end{array}\right\} \tag{5.36}$$

The Riccati equation (5.35a), for this example, becomes

$$\dot{S}+2S-S^2-1=0, \quad S(T)=0 \tag{5.37}$$

which admits the solution

$$S(t)=\frac{T-t}{T-1-t} \tag{5.38}$$

on the interval $(T-1,T]$. Hence, for $T\geq 1$, a continuously differentiable solution on the interval $[0,T]$ to (5.37) does not exist. This non-existence of a solution to the matrix Riccati equation is directly related to the well-posedness of the optimal control problem (5.36), since it can readily be shown that, for $T\geq 1$, $L(u)$ can be made arbitrarily small (negative) by choosing a proper $u(t)$. In such a case we say that the Riccati equation has a *conjugate point* in $[0,T]$ (Sagan (1969) and Brockett (1970)). This topic will be revisited in Chapter 6 in the context of affine-quadratic zero-sum differential games. □

Infinite horizon linear-quadratic problems

Meaningful continuous-time linear-quadratic optimal control problems where $T\to\infty$ can also be formulated. Toward that end we take $c(t)\equiv 0$, and the matrices $A(t), B(t), Q(t)$ and $R(t)$ to be independent of t. Then the problem becomes

$$\min_u \frac{1}{2}\int_0^\infty (x'Qx+u'Ru)\,\mathrm{d}t\,; \quad Q\geq 0,\ R>0,$$

subject to

$$\dot{x}=Ax+Bu.$$

Conditions which ensure this problem to be well defined (with a finite minimum) are precisely those of the discrete-time analogue of this problem, i.e., the matrix pair (A,B) must be stabilizable and the matrix pair (A,D), where D is a matrix such that $D'D=Q$, must be detectable; see for instance Kailath (1980)

[41] See Chapter 6, and particularly Lemma 6.4 and Remark 6.15 for further elaboration of this point.

or Davis (1977) for discussions on these notions in continuous time.[42] The pair (A, B) is stabilizable if an $m \times n$ matrix F exists such that $(A + BF)$ is stable, i.e., all its eigenvalues lie in the open left half of the complex plane (note the difference of this definition of stability from the one for its discrete-time counterpart). Furthermore, the pair (A, D) is detectable if its "dual pair" (A', D') is stabilizable. The following proposition, which can be found in any textbook on linear control systems, summarizes the main result.

Proposition 5.4 *Assume that the pair (A, B) is stabilizable and the pair (A, Q) is detectable. Then, there exists an $n \times n$ nonnegative definite matrix S such that*

(i) for fixed t, and for every $Q_f \geq 0$, $S^T(t) \to S$ (uniformly in $t < T$) as $T \to \infty$, where $S^T(t)$ is the unique nonnegative definite solution to (5.35a), parameterized in T;

(ii) S is the unique solution of the algebraic Riccati equation (ARE)

$$SA + A'S - SBR^{-1}B'S + Q = 0$$

in the class of nonnegative definite matrices;

(iii) the closed-loop matrix $A - BR^{-1}B'S$ is stable, i.e., all its eigenvalues lie in the open left half of the complex plane;

(iv) the minimum value of the cost functional is $\frac{1}{2}x_0'Sx_0$;

(v) the stationary optimum control law is

$$\gamma^*(x) = -R^{-1}B'Sx.$$

Under the stronger condition that (A, D) is observable (a sufficient condition for which is $Q > 0$), the solution to the ARE is positive definite.

5.5.3 The minimum principle

Continuous-time systems

In this subsection our starting point will be the HJB equation (5.25) under the additional assumption that V is twice continuously differentiable, and we shall convert it into a series of pointwise optimization problems, indexed by the parameter t. This new form is in some cases more convenient to work with and it is closely related to the so-called minimum principle, as also discussed here.

Toward this end we first introduce the function

$$\tilde{H}(t, x, u) = \frac{\partial V(t, x)}{\partial x} f(t, x, u) + g(t, x, u), \tag{5.39}$$

[42] As in the discrete-time case, we will refer to the latter condition also as detectability of the pair (A, Q).

in terms of which equation (5.25) can be written as

$$-\frac{\partial V}{\partial t} = \min_{u} \tilde{H}(t, x, u). \tag{5.40}$$

Being consistent with our earlier convention, the minimizing u will be denoted by u^*. Then

$$\tilde{H}(t, x, u^*) + \frac{\partial V(t, x)}{\partial t} \equiv 0. \tag{5.41}$$

Since this is an identity in x, its partial derivative with respect to x is also zero. This leads to, by also interchanging the orders of second partial derivatives (which is allowed if V is twice continuously differentiable):

$$\frac{\partial g}{\partial x} + \frac{\mathrm{d}}{\mathrm{d}t}\left(\frac{\partial V}{\partial x}\right) + \frac{\partial V}{\partial x}\frac{\partial f}{\partial x} + \frac{\partial \tilde{H}}{\partial u}\frac{\partial u^*}{\partial x} = 0. \tag{5.42}$$

If there are no constraints on u, then $\partial \tilde{H}/\partial u = 0$ for $u = u^*$ according to equation (5.40). If there are constraints on u, and u^* is not an interior point, then it can be shown that $(\partial \tilde{H}/\partial u)\cdot(\partial u^*/\partial x) = 0$ (because of optimality, $\partial \tilde{H}/\partial u$ and $\partial u^*/\partial x$ are orthogonal; for specific problems we may have $\partial u^*/\partial x = 0$). In view of this, equation (5.42) becomes

$$\frac{\partial g}{\partial x} + \frac{\mathrm{d}}{\mathrm{d}t}\left(\frac{\partial V}{\partial x}\right) + \frac{\partial V}{\partial x}\frac{\partial f}{\partial x} = 0. \tag{5.43}$$

By introducing the so-called *costate vector*, $p'(t) = \partial V(t, x^*(t))/\partial x$, where x^* denotes the state trajectory corresponding to u^*, (5.43) can be re-written as

$$\frac{\mathrm{d}p'}{\mathrm{d}t} = -\frac{\partial}{\partial x}[g(t, x^*, u^*) + p'(t) f(t, x^*, u^*)] \equiv \frac{-\partial}{\partial x} H(t, p, x^*, u^*), \tag{5.44}$$

where H is defined by

$$H(t, p, x, u) = g(t, x, u) + p' f(t, x, u). \tag{5.45}$$

The final time T is determined by the scalar relation $l(T, x) = 0$ and hence can be regarded as a function of the state: $T(x)$. The boundary condition for $p(t)$ is determined from

$$p'(T) = \frac{\partial V(T(x^*), x^*)}{\partial x} = \frac{\partial q(T(x^*), x^*)}{\partial x}. \tag{5.46}$$

In conclusion, we have arrived at the following necessary condition for the optimal control $u^*(\cdot)$: under the assumption that the value function $V(t, x)$ is twice continuously differentiable, the optimal control $u^*(t)$ and corresponding trajectory $x^*(t)$ must satisfy the following so-called *canonical equations*

$$\left.\begin{aligned}
\dot{x}^*(t) &= \left(\frac{\partial H}{\partial p}\right)' = f(t,x^*,u^*), x(t_0) = x_0;\\
\dot{p}'(t) &= -\frac{\partial H(t,p,x^*,u^*)}{\partial x},\\
p'(T) &= \frac{\partial q(T,x^*)}{\partial x} \text{ along } l(T,x) = 0;\\
H(t,p,x,u) &\triangleq g(t,x,u) + p'f(t,x,u),\\
u^*(t) &= \arg\min_{u\in S} H(t,p,x^*,u).
\end{aligned}\right\} \tag{5.47}$$

In the derivation of (5.47), the controls have been assumed to be functions of time and state; i.e., $u(t) = \gamma(t, x(t))$. If, instead, one starts with the class of control functions which depend only on time, i.e., $u(t) = \gamma(t)$, the set of necessary conditions (5.47) can be derived in quite a different way (by using perturbation functions, which is a standard procedure in the classical calculus of variations), and under milder assumptions. Following such a derivation, one obtains the following result which can, for instance, be found in Pontryagin, et al. (1962) and which is referred to as the *minimum principle.*

Theorem 5.4 *Consider the optimal control problem defined by (5.23) and under the OL information structure. If the functions f, g, q and l are continuously differentiable in x and continuous in t and u, then relations (5.47) provide a set of necessary conditions for the optimal control and the corresponding optimal trajectory to satisfy.*

Remark 5.9 A particular class of optimal control problems which are not covered by Thm. 5.4 are those with a *fixed terminal condition*, i.e., with the terminal time T and the terminal state $x(T)$ prescribed, since the function l is not differentiable in this case. However, the necessary conditions for the optimal control are still given by (5.47), with the exception that the condition on $p(T)$ is absent and instead the terminal state constraint $x(T) = x_f$ replaces it. We also note, in passing, that in the theory of zero-sum differential games, reasonable problems with state space dimension ≥ 2 will (almost) never be formulated with a fixed prescribed terminal state, since it is not possible to enforce this end-point constraint on both players if they have totally conflicting objectives. □

The following example now illustrates how Thm. 5.4 can be employed to obtain the solution of an optimal control problem for which the value function V is not continuously differentiable. It also exhibits some features which are frequently met in the theory of differential games, and introduces some terminology which will be useful later in Chapter 8.

Example 5.2 Consider an open-loop control problem with systems dynamics described by

$$\frac{\mathrm{d}}{\mathrm{d}t}\begin{pmatrix}x_1\\x_2\end{pmatrix} = \begin{pmatrix}0&1\\0&0\end{pmatrix}\begin{pmatrix}x_1\\x_2\end{pmatrix} + \begin{pmatrix}0\\1\end{pmatrix}u, \tag{5.48}$$

where the scalar control satisfies the constraint $-1 \leq u(t) \leq 1$ for all $t \geq 0$. The objective is to steer the system from an arbitrary but known initial point in the (x_1, x_2) plane to the line $x_1 = x_2$ in minimum time. Hence, the cost function can be written as

$$L(u) = \int_0^T g(t,x,u)\,\mathrm{d}t = \int_0^T 1\,\mathrm{d}t = T,$$

where T is defined as the first instant for which the function

$$l(t, x_1(t), x_2(t)) \equiv x_1(t) - x_2(t)$$

becomes zero. In the derivation to follow, we shall restrict ourselves to initial points satisfying $x_1 - x_2 > 0$. Initial points with the property $x_1 - x_2 < 0$ can be dealt with analogously. Now, application of relations (5.47) yields (with $p \triangleq (p_1, p_2)'$):

$$\left.\begin{array}{rcl} H &=& 1 + p_1 x_2 + p_2 u, \\ \dot{p}_1 &=& 0, \quad \dot{p}_2 = -p_1, \\ u^*(t) &=& -\operatorname{sgn}\{p_2(t)\}, \end{array}\right\} \tag{5.49}$$

where the costate variable $p_2(t)$ here is also known as the *switching function*, since it determines the sign of $u^*(t)$. For points on the line $x_1 = x_2$, which can be reached by optimal trajectories, obviously $l \equiv 0$ and therefore the costate vector $p(T) = (p_1(T), p_2(T))'$ will be orthogonal to the line $x_1 = x_2$ and point in the direction "south-east". Since the magnitude of the costate vector is irrelevant in this problem, we take

$$p_1(T) = 1, \quad p_2(T) = -1, \tag{5.50}$$

which leads to

$$u^*(t) = \begin{cases} -1 & \text{for} \quad t < T-1, \\ +1 & \text{for} \quad T-1 \leq t \leq T, \end{cases} \tag{5.51}$$

assuming, of course, that $T > 1$. Now, by integrating (5.48) backwards in time from an arbitrary terminal condition on the line $x_1 = x_2$ (which we take as $x_1(T) = x_2(T) = \mathbf{a}$ where $\mathbf{a}$ is the parameter), and by replacing $u(t)$ with $u^*(t)$ as given by (5.51), we obtain

$$\begin{array}{rcl} x_1(t) &=& (t-T)\mathbf{a} + \frac{1}{2}(t-T)^2 + \mathbf{a}, \\ x_2(t) &=& \mathbf{a} + (t-T), \end{array}$$

for $T - 1 \leq t \leq T$. For $t = T - 1$ we get $x_1(T-1) = \frac{1}{2}, x_2(T) = \mathbf{a} - 1$. The line $x_1 = \frac{1}{2}, x_2 = \mathbf{a} - 1$ is called, for varying $\mathbf{a}$, the *switching line*. □

Since we only consider the "south-east" region of the state space, the trajectory $(x_1(t), x_2(t))$ should move into this region from $x_1(T) = x_2(T) = \mathbf{a}$ if t goes backwards in time, which is only true for $\mathbf{a} \leq 1$. For $\mathbf{a} > 1$, the trajectories first move into the region $x_2 > x_1$, in retrograde time. Those trajectories cannot be optimal, since we only consider the region $x_2 < x_1$, and, moreover, boundary conditions (5.50) are not valid for the region $x_2 > x_1$. Hence only points $x_1(t) = x_2(T) = \mathbf{a}$ with $\mathbf{a} \leq 1$ can be terminal points. The half line $x_1 = x_2 = \mathbf{a}$, $\mathbf{a} \leq 1$, is called the *usable part* (UP) of the terminal manifold or *target set*. Integrating the state equations backwards in time from the UP, and substituting $t = T-1$, we observe that only the line segment $x_1 = \frac{1}{2}$, $x_2 = \mathbf{a}-1$, $\mathbf{a} \leq 1$, is a switching line. Using the points on this switching line as new terminal state conditions, we can integrate the state equations further backwards in time with $u^*(t) = -1$. The complete picture with the optimal trajectories is shown in Fig. 5.1. These solutions are the only ones which satisfy the necessary conditions of Thm. 5.4 and therefore they must be optimal, provided that an optimal solution exists.

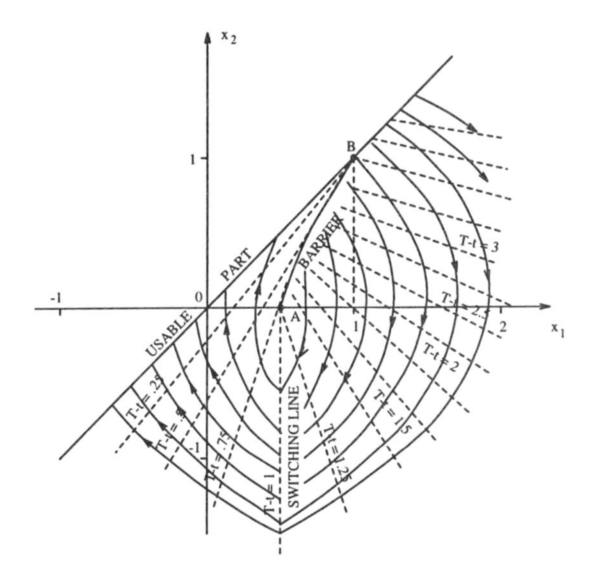

Figure 5.1: The solution of Example 5.2. The dotted lines denote lines of "equal time-to-go".

Once the optimal trajectories are known, it is not difficult to construct $V(x)$. Curves of "equal time-to-go", also called *isochrones*, have been indicated in Fig. 5.1. Such a curve is determined from the equation $V(x) =$ constant. It is seen that along the curve AB the value function $V(x)$ is discontinuous. Curves along which $V(x)$ is discontinuous are called *barriers*. The reason behind the usage of such a terminology is that no optimal trajectory can ever cross a curve along which $V(x)$ is discontinuous in the direction from large V to small V. Note that, because of the discontinuity in $V(x)$, Thm. 5.3 cannot be used in this case.

The corresponding feedback strategy $\gamma^*(t,x)$, which we write as $\gamma^*(x)$ since the system is autonomous (i.e., time-invariant), can readily be obtained from Fig. 5.1: to the right of the switching line and barrier, we have $\gamma^*(x) = -1$, and to the left of the switching line and barrier, and also on the barrier, we have $\gamma^*(x) = +1$.

Discrete-time systems

We now conclude this subsection by stating a counterpart of Thm. 5.4 for discrete-time optimal control problems. Such problems have earlier been formulated by (5.18a) and (5.18b), but now we also assume that f_k is continuously differentiable in x_k, and g_k is continuously differentiable in x_k and x_{k+1} ($k \in \mathbf{K}$). Then, the following theorem, whose derivation can be found in either Canon et al. (1970) or Boltyanski (1978), provides a *minimum principle* for such systems.

Theorem 5.5 *For the discrete-time optimal control problem described by (5.18a) and (5.18b), let*

(i) $f_k(\cdot, u_k)$ *be continuously differentiable on* $\mathbf{R}^n$, $(k \in \mathbf{K})$,

(ii) $g_k(\cdot, u_k, \cdot)$ *be continuously differentiable on* $\mathbf{R}^n \times \mathbf{R}^n$, $(k \in \mathbf{K})$,

(iii) $f_k(\cdot,\cdot)$ *be convex on* $\mathbf{R}^n \times \mathbf{R}^m$, $(k \in \mathbf{K})$.

Then, if $\{u_k^*, k \in \mathbf{K}\}$ *denotes an optimal control sequence, and* $\{x_{k+1}^*, k \in \mathbf{K}\}$ *denotes the corresponding state trajectory, there exists a finite sequence* $\{p_2, \ldots, p_{K+1}\}$ *of n-dimensional costate vectors so that the following relations are satisfied:*

$$\left.\begin{aligned} x_{k+1}^* &= f_k(x_k^*, u_k^*), \quad x_1^* = x_1, \\ u_k^* &= \arg\min_{u_k \in U_k} H_k(p_{k+1}, u_k, x_k^*), \\ p_k &= \frac{\partial}{\partial x_k} f_k(x_k^*, u_k^*) \left[p_{k+1} + \left(\frac{\partial}{\partial x_{k+1}} g_k(x_{k+1}^*, u_k^*, x_k^*) \right)' \right] \\ &\quad + \left(\frac{\partial}{\partial x_k} g_k(x_{k+1}^*, u_k^*, x_k^*) \right)', \quad p_{K+1} = 0, \end{aligned}\right\} \quad (5.52)$$

$$H_k(p_{k+1}, x_k, u_k) \triangleq g_k(f_k(x_k, u_k), u_k, x_k) + p_{k+1}' f_k(x_k, u_k), \; k \in \mathbf{K}. \quad \square$$

The reader should now note that Prop. 5.1 also follows as a special case of this theorem.

5.6 Representations of Strategies Along Trajectories, and Time Consistency of Optimal Policies

In our discussion of some important results of optimal control theory in the previous section, we have confined our treatment to two specific information patterns and thereby to two types of strategies; namely, (i) the open-loop information pattern dictates strategies that depend only on the initial state and time, (ii) the feedback information pattern forces the permissible strategies to depend only on the current value of the state and time. The former class of strategies is known as *open-loop* strategies, while the latter class is referred to as *feedback* strategies.

Referring back to the discrete-time affine-quadratic optimal control problem described by (5.20a)-(5.20b) (and assuming that $c_k = 0$ for all $k \in \mathbf{K}$, for the sake of brevity in the discussion to follow), the solution presented in Prop. 5.1, i.e.,

$$\gamma_k^*(x_k) = -P_k S_{k+1} A_k x_k, \quad k \in \mathbf{K}, \tag{5.53}$$

is a feedback strategy, and in fact it constitutes the *unique* solution of that optimal control problem within the class of feedback strategies. If the optimal control problem had been formulated under the open-loop information pattern, then (5.53) would clearly not be a solution candidate. However, the unique optimal open-loop strategy, for this problem, can readily be obtained from (5.53) as

$$u_k^* = -P_k S_{k+1} A_k x_k^*, \quad k \in \mathbf{K}, \tag{5.54a}$$

where x_k^* is the value of the state at stage k, obtained by substitution of (5.53) in the state equation; in other words, x^* can be solved recursively from the difference equation

$$x_{k+1}^* = (A_k - B_k P_k S_{k+1} A_k) x_k^*, \quad x_1^* = x_1. \tag{5.54b}$$

It should be noted that (5.54a) is an open-loop strategy since it only depends on the discrete-time parameter k and the initial state x_1 (through (5.54b)). We can also interpret (5.54a) as the realization of the feedback strategy (5.53) on the control set U_k, that is to say, (5.54a) is the action dictated by (5.53). For this reason, we also sometimes refer to (5.54a) as the *open-loop realization* or *open-loop value* of the feedback strategy (5.53).

The question now is whether we can obtain other solutions (optimal strategies) for the optimal control problem under consideration by extending the strategy space to conform with some other information structure, say the closed-loop perfect state pattern. If we denote by Γ_k the class of all permissible *closed-loop* strategies (under the CLPS pattern) at stage k (i.e., strategies of the form $\gamma_k(x_k, x_{k-1}, \ldots, x_1)$), then (5.53) is clearly an element of that space, and it constitutes an optimal solution to the optimal control problem (5.20a)-(5.20b) also on that enlarged strategy space. As a matter of fact, any element γ_k of Γ_k ($k \in \mathbf{K}$), which satisfies the boundary condition

$$\gamma_k(x_k^*, x_{k-1}^*, \ldots, x_2^*, x_1^*) = -P_k S_{k+1} A_k x_k^*, \quad k \in \mathbf{K}, \tag{5.55}$$

also constitutes an optimal solution. One such optimal policy is the *one-step memory* strategy

$$\gamma_k(x_k, x_{k-1}) = -P_k S_{k+1} A_k x_k + C_k[x_k - F_{k-1} x_{k-1}],$$

where $F_k \triangleq A_k - B_k P_k S_{k+1} A_k$, and C_k is a matrix of appropriate dimensions. It should be noted that all such strategies have the same open-loop value (which is the RHS of (5.55)) and generate the same state trajectory (5.54b), both of which follow from (5.55). Because of these properties, we call them different *representations* of a single strategy, a precise definition of which is given below. Since the foregoing analysis also applies to continuous-time systems, this definition covers both discrete-time and continuous-time systems.

Definition 5.11 *For a control problem with a strategy space Γ, a strategy $\gamma \in \Gamma$ is said to be a* representation *of another strategy $\tilde{\gamma} \in \Gamma$ if*

(i) they both generate the same unique state trajectory, and

(ii) they both have the same open-loop value on this trajectory.

Remark 5.10 The notion of representation of a strategy, as introduced above, involves no optimality. As a matter of fact, it is a property of the strategy space itself, together with the control system for which it is defined, and no cost functional has to be introduced. □

A significance of the notion of "representations" in control problems is that it enables one to construct equivalence classes (of equal open-loop value control laws) in the general class of closed-loop strategies, so that there is essentially no difference between the elements of each equivalence class. In fact, for an optimal control problem, we can only talk about a *unique optimal* "equivalence class", instead of a *unique optimal* "strategy", since as we have discussed earlier within the context of the optimal control problem (5.20a)-(5.20b), *every representation of the feedback strategy* (5.53) *constitutes an optimal solution* to the optimal control problem, and there are uncountably many such (optimal) strategies. As a converse to the italicized statement, we can say that *every solution of the optimal control problem* (5.20a)-(5.20b) *is a representation of the feedback strategy* (5.53). Both of these statements readily follow from Def. 5.1, and they are valid not only for the specific optimal control problem (5.20a)-(5.20b) but for the general class of control problems defined in either discrete or continuous time.

We now address the problem of constructing equivalence classes of representations of strategies for a control problem with a given space (Γ) of closed-loop strategies, and first for discrete-time systems. The procedure, which follows directly from Def. 5.11, is as follows. For such a system, first determine the set (Γ_{OL}) of all elements of Γ, which are strategies that depend only on the initial state x_1, and the discrete time parameter k. The set Γ_{OL}, thus constructed, is the class of all permissible open-loop controls in Γ. Now, let $\tilde{\gamma} = \{\tilde{\gamma}_k(x_1); k \in \mathbf{K}\}$ be a chosen element of Γ_{OL}, which generates (by substitution into (5.18a)) a (necessarily) unique trajectory $\{\tilde{x}_k, k \in \mathbf{K}\}$. Then, consider all elements $\gamma = \{\gamma_k(\cdot), k \in \mathbf{K}\}$ of Γ with the properties

(i) γ generates the same state trajectory as $\tilde{\gamma}$, and

(ii) $\gamma_k(\tilde{x}_k, \tilde{x}_{k-1}, \ldots, \tilde{x}_2, x_1) \equiv \tilde{\gamma}_k(x_1), k \in \mathbf{K}$.

The subset of Γ thus constructed constitutes an equivalence class of representations which, in this case, have the open-loop value $\tilde{\gamma}$. If this procedure is executed for every element of γ_{OL}, then the construction of all equivalence classes of Γ becomes complete.

For continuous-time systems, essentially the same procedure can be followed to construct equivalence classes of representations of strategies. In both discrete-time and continuous-time infinite dynamic systems such a construction leads, in general, to an uncountable number of elements in each equivalence class which necessarily contains one, and only one, open-loop strategy. The main reason for the "non-denumerable" property of equivalence classes is that in deterministic problems the CLPS information pattern (which involves memory) exhibits redundancy in information—thus giving rise to existence of a plethora of different representations of the same open-loop policy. This "informational nonuniqueness" property of the CLPS pattern is of no real concern to us in optimal control problems, since *every representation of an optimal control strategy also constitutes an optimal solution.* However, in dynamic games, the concept of "representations" plays a crucial role in the characterization of noncooperative equilibria, and the preceding italicized statements are no longer valid in a game situation. We defer details of an investigation in that direction to Chapters 6 and 7, and only provide here an extension of Def. 5.11 to dynamic games.

Definition 5.12 *For an N-person dynamic game with strategy spaces $\{\Gamma^i; i \in \mathbf{N}\}$ let the strategies of all players, except $\mathbf{P}i$, be fixed at $\gamma^j \in \Gamma^j$, $j \in \mathbf{N}$, $j \neq i$. Then, a strategy $\gamma^i \in \Gamma^i$ for $\mathbf{P}i$ is a* representation *of another strategy $\tilde{\gamma}^i \in \Gamma^i$, with $\gamma^j \in \Gamma^j$ $(j \in \mathbf{N}, j \neq i)$ fixed, if*

(i) the N-tuples $\{\gamma^i, \gamma^j; j \in \mathbf{N}, j \neq i\}$ and $\{\tilde{\gamma}^i, \gamma^j; j \in \mathbf{N}, j \neq i\}$ generate the same unique state trajectory, and

(ii) γ^i and $\tilde{\gamma}^i$ have the same open-loop value on this trajectory.

Time consistency

In (one-player) optimal control problems, even though all representations of an optimal control strategy constitute optimal solutions, as discussed above, some of these representations may be preferred over others due to additional appealing properties they carry, such as robustness to modeling and decision errors. One such *refinement* scheme that brings in a selection among different representations of a given optimal strategy is provided by the notion of *time consistency*, which is valid not only for one-player dynamic decision problems (such as optimal control), but also for dynamic games in both discrete and continuous time. This notion manifests itself in two forms—as *weak* and *strong time consistency*, which we introduce below in the general framework of dynamic games. Toward that end, let us first introduce the notation

$$D(\Gamma; [0,T];\ sol) \tag{5.56}$$

to denote an N-person dynamic game, where Γ is the product strategy space, $[0,T]$ is the decision interval,[43] and *sol* stands for any particular solution concept according to which (optimal or equilibrium) policies are determined.[44] Furthermore, let

$$\gamma_{[s,t]} \in \Gamma_{[s,t]}, \qquad \gamma^i_{[s,t]} \in \Gamma^i_{[s,t]} \tag{5.57}$$

denote, respectively, the truncations of $\gamma \in \Gamma$ and $\gamma^i \in \Gamma^i$, to the time interval $[s,t] \subset [0,T]$, and let

$$D^{\beta}_{[s,t]} \triangleq D\left(\{\gamma \in \Gamma:\ \gamma_{[0,s)} = \beta_{[0,s)}, \gamma_{(t,T]} = \beta_{(t,T]}, \gamma_{[s,t]} \in \Gamma_{[s,t]}\}; [0,T]; sol\right) \tag{5.58}$$

denote a version of $D(\Gamma; [0,T];\ sol)$, where the policies of all players in the intervals $[0,s)$ and $(t,T]$ are fixed as $\beta^i_{[0,s)}$, $\beta^i_{(t,T]}$, $i \in \mathbf{N}$. An underlying assumption here is that the cost functions of the players are of additive type, such as (5.10) or (5.14). Then, we have the following two definitions on the "time consistency" of a solution obtained for the original game according to the concept "*sol*".

Definition 5.13 *An N-tuple of policies $\gamma^* \in \Gamma$ solving the dynamic game $D(\Gamma; [0,T];\ sol)$ is* weakly time consistent *(*WTC*) if its truncation to the interval $[s,T]$, $\gamma^*_{[s,T]}$, solves the truncated game $D^{\gamma^*}_{[s,T]}$, this being so for all $s \in (0,T]$. If a solution $\gamma^* \in \Gamma$ is not* WTC, *then it is* time inconsistent.

Definition 5.14 *An N-tuple of policies $\gamma^* \in \Gamma$ solving the dynamic game $D(\Gamma; [0,T];\ sol)$ is* strongly time consistent *(*STC*) if its truncation to the interval $[s,T]$, $\gamma^*_{[s,t_f]}$, solves the truncated game $D^{\beta}_{[s,T]}$, for every $\beta_{[0,s)} \in \Gamma_{[0,s)}$, this being so for every $s \in (0,T]$.*

Both these refinements essentially say that for a solution to be time consistent (in the *weak* or *strong* sense), the players should have no rational reason,

[43] By an abuse of notation, we will let this also denote the discrete interval for the discrete-time problem, in which case $[0,T] = \{1,\ldots,K\}$.

[44] In an optimal control problem, for example, *sol* could stand for minimization or maximization, or reachability to a target set; in a dynamic game it could stand for saddle-point, Nash or Stackelberg equilibrium.

at any future stage of the game, to deviate from the adopted policies. The difference between weak and strong time consistency lies in the consistency (or inconsistency) of the past actions with the adopted policies. In the former (that is, weak time consistency) there is no ground for reneging at future stages only if the past actions are consistent with the original solution, γ^*; whereas in the latter this is true even if there have been deviations in the past from the actions which are dictated by the original solution.

Solutions of optimal control problems of the types discussed in Section 5.5 (and of course also their representations) are weakly time consistent—a fact that easily follows from Def. 5.13. Not every solution is strongly time consistent, however; a case in point being the optimal open-loop control $u^*_{[0,T]} = \{\gamma^*(t), t \in [0,T]\}$ obtained using the minimum principle (cf. subsection 5.5.3). If a non-optimal open-loop policy $u_{[0,s)}$ is applied to the system in the time interval $[0, s)$, which brings the system to a state, say $\bar{x}(s)$, that is not on the optimal trajectory, then the control $u^*_{[s,T]}$ will not necessarily be optimal for the new truncated control problem $D^u_{[s,T]}$. The reason for this is that $u^*_{[s,T]}$ will no longer satisfy the necessary conditions given in Thm. 5.4 or Thm. 5.5 on the interval $[s,T]$, as the initial state at $t = s$ is not on the optimal trajectory. Hence, optimal open-loop control is only *WTC*. The optimal feedback control, obtained from the dynamic programming equation (5.19) or (5.25),[45] however, is *STC*—a fact that follows from Def. 5.14. For the discrete-time linear-quadratic problem (with $c_k \equiv 0$), for example, the policy (5.53), which is obtained from a dynamic programming equation, is *STC*, whereas its open-loop version (5.54a) is only *WTC*. For the more general affine-quadratic problem described by (5.20a), the optimal feedback policy (5.21a) is *STC*.

In a way, an *STC* policy is one that is not only independent of the initial state x_0 but also independent of other past values of the state. Not every optimal policy that depends only on the current value of the state is necessarily *STC*, however, especially if the initial state is a fixed known quantity. To make the point here, consider the optimal feedback policy $\gamma^* \triangleq \{\gamma^*_k,\ k \in \mathbf{K}\}$ obtained from (5.19), for a discrete-time system whose initial state x_1 is fixed and given, under which the optimal trajectory, $\{x^*_{k+1},\ k \in \mathbf{K}\}$, is just a sequence of known vectors. Now, if Γ is a policy space compatible with the feedback (FB) information pattern, and we seek an optimal solution in the class Γ, then not only γ^*, but every $\gamma \in \Gamma$ given by $\gamma_k(x_k) = \psi_k(x_k, x^*_k)$, $k \in \mathbf{K}$, will be an optimal policy, where ψ_k is any function satisfying the boundary condition $\psi_k(x^*_k, x^*_k) = \gamma^*_k(x^*_k)$, $k \in \mathbf{K}$. Different choices of ψ lead to different *representations* of the same (optimal) policy on the optimal trajectory, and they all (in this case) depend only on the current value of the state. Because of this feature, one may be tempted to call all these different representations "optimal feedback policies", but this would not be correct because only one of them, namely γ^*, is *STC*. The others cannot be obtained directly from (5.19) and do not meet the "permanent optimality" property associated with an *STC* optimal solution.

[45]Here, the feedback control is obtained by minimization of the right-hand side of (5.19) or (5.25), depending on whether we have a discrete-time or a continuous-time problem.

There are two points worth mentioning here in connection with the notion of an *STC* optimal policy. The **first** one is that it is quite possible for a constant policy to be *STC*, which would arise if the optimization problem in (5.19) (or (5.25)) admits a solution which is independent of x for every k (or t). In such a case, the optimal feedback and the optimal open-loop solutions would coincide, and hence the open-loop solution would also be *STC*. The **second** point is that a given dynamic optimization problem may not admit an *STC* optimal solution, even under the *CL* information pattern. This would arise if the policy space has additional structural restrictions imposed on the permissible policies. For example, one may require the policies to depend only linearly on the current value of the state, with no such restriction imposed on their dependence on past history. If the unique optimal FB solution obtained from (5.19) (or (5.25)), i.e., γ^*, is not linear in its argument, then the problem would not admit an *STC* solution. However, the problem would still admit a *WTC* optimal solution, such as (in the discrete-time case) any policy of the form $\gamma_k(x_k, x_1) = M_k x_k + \gamma_k^*(x_k^*) - M_k x_k^*$, $k \in \mathbf{K}$, where $\{M_k\}$ is an arbitrary matrix sequence (of compatible dimensions) and $\{x_k^*\}$ is the trajectory generated by γ^*, in general as a function of x_1.

The notions of weak and strong time consistency are of course also valid in systems which incorporate lag variables, either in the state or in the control. Consider, for example, the system where the evolution is governed by (as a counterpart of (5.18a)):

$$x_{k+1} = f_k(x_k, x_{k-1}, u_k), \quad x_1 \text{ and } x_0 \text{ given.}$$

This could be converted into the form (5.18a) by introducing a new state vector ξ_k which corresponds to the pair (x_k, x_{k-1}), thus making our earlier discussion apply here, with x_k replaced by ξ_k. Note that, in view of this, an *STC* solution will exist for the discrete-time optimal control problem (under the cost function (5.18a)) only if one allows the permissible policies at time k to depend not only on x_k but also on x_{k-1}. This can best be seen if one writes down the counterpart of the dynamic programming (DP) equation (5.19) in terms of the original variables:

$$V(k, x_k, x_{k-1}) = \min_{u \in U_k} \{V(k+1, f_k(x_k, x_{k-1}, u), x_k) + g_k(f_k(x_k, x_{k-1}, u), u)\}$$

with the boundary condition being $V(K+1, x_{K+1}, x_K) \equiv 0$. Clearly, an *STC* solution will have to be of the form $u_k = \gamma_k^*(x_k, x_{k-1})$. If the information pattern is FB, then the problem may not admit a solution at all.

This concludes our discussion on *WTC* and *STC* optimal solutions in one-person dynamic decision problems. The topic will be revisited later in Chapters 6 and 7, in the context of saddle-point, Nash and Stackelberg equilibria in dynamic games.

Stochastic systems

We devote the remaining portion of this section to a brief discussion on the representation of closed-loop strategies in stochastic control problems. In particular, we consider, as a special case of Def. 5.4, the class of control problems described by the state equation

$$x_{k+1} = f_k(x_k, u_k) + \theta_k \tag{5.59}$$

where γ_k ($k \in \mathbf{K}$) is a random vector taking values in R^n, and the joint probability distribution function of $\{x_1, \theta_1, \ldots, \theta_k\}$ is known. Then, given a feedback strategy $\{\gamma_k(\cdot), k \in \mathbf{K}\}$, it is in general not possible to write down a representation of this strategy that also involves past values of the state, since now there is no redundancy in the CLPS information, due to existence of the disturbance vector θ_k. In other words, while in deterministic systems x_{k+1} can easily be recovered from x_k if one knows the strategy employed, this is no longer possible in stochastic systems because x_{k+1} also contains some information concerning θ_k. This is particularly true if the random vectors $x_1, \theta_1, \ldots, \theta_K$ are Gaussian, mutually independent and $\mathrm{cov}\,(\theta_k) > 0$[46] for all $k \in \mathbf{K}$. Hence, we have the following proposition.

Proposition 5.5 *If, for the stochastic system (5.59), all random vectors are Gaussian, mutually independent and the covariance of θ_k is of full rank for all $k \in \mathbf{K}$, then every equivalence class of representations is a singleton.*

Remark 5.11 The reason why we do not allow some random vectors to be statistically dependent and/or $\mathrm{cov}\,(\theta_k)$ to have lower rank is because then some components of x_{k+1} can be expressed in terms of some components of x_k, x_{k-1}, etc., independent of θ_k, $\theta_{k-1}, \ldots$, which clearly gives rise to nonunique representations of a strategy. Furthermore, we should note that the requirement of having a Gaussian probability distribution with positive definite covariance can be replaced by having any probability distribution that assigns positive probability mass to every nonempty open subset of R^n. The latter requirement is important because, if any open subset of R^n receives zero probability, then nonunique representations are possible on that particular open set. □

Optimal control of stochastic systems described by (5.59) can be performed by following the basic steps of dynamic programming applied to the deterministic system (5.18a)-(5.18b). Toward this end, we first replace the cost functional (5.18b) by

$$J(\gamma) = E\left[\sum_{k=1}^{K} g_k(x_{k+1}, u_k, x_k)\,|u_k = \gamma_k(x_l, l \leq k), \quad k \in \mathbf{K}\right], \tag{5.60}$$

where $\{\gamma_k; k \in \mathbf{K}\}$ is any permissible closed-loop strategy, and $E[\cdot]$ denotes the expectation operation defined with respect to the underlying statistics of $x_1, \theta_1, \ldots, \theta_K$ which we assume to be mutually independent. Then, the value function $V(k, x)$ is defined by

$$V(k, x) = \min_{\gamma_k, \ldots, \gamma_K} \; E\left[\sum_{i=k}^{K} g_i(x_{i+1}, \gamma_i(x_l, l \leq i), x_i)\right], \quad x_k = x. \tag{5.61a}$$

[46] $\mathrm{cov}\,(\theta_k)$ stands for the covariance matrix of θ_k, i.e., $E[(\theta_k - E[\theta_k])(\theta_k - E[\theta_k])']$.

Following the steps of dynamic programming, and by utilizing the assumption of independence of the random vectors involved, it can readily be shown that the value function satisfies the recursive relation[47]

$$V(k,x) = \min_{u_k \in U_k} E_{\theta_k}[g_k(f_k(x_k,u_k)+\theta_k,u_k,x_k) \\ +V(k+1,f_k(x_k,u_k)+\theta_k)], \quad x_k = x, \tag{5.61b}$$

which also leads to the conclusion that the optimal strategy γ_k ($k \in \mathbf{K}$) depends only on the current value of the state, i.e., is a feedback strategy. This is also a strongly time consistent solution under the *CLPS* information pattern, where the notion (introduced by Def. 5.14) applies equally to stochastic dynamic decision problems, since its definition used strategies, and not actions.

We now apply this result to the so-called linear-quadratic discrete-time stochastic optimal control problem described by the linear state equation

$$x_{k+1} = A_k x_k + B_k u_k + \theta_k \tag{5.62a}$$

and the quadratic cost functional

$$J(\gamma) = E\left[\frac{1}{2}\sum_{k=1}^{K} x'_{k+1}Q_{k+1}x_{k+1} + u'_k R_k u_k \mid u_k = \gamma_k(\cdot), \quad k \in \mathbf{K}\right], \tag{5.62b}$$

where $R_{k+1} > 0$, $Q_{k+1} \geq 0$; $(k \in \mathbf{K})$; θ_k $(k \in \mathbf{K})$ are n-dimensional mutually independent Gaussian vectors with mean value c_k, and covariance $\Lambda_k > 0$, and γ_k $(k \in \mathbf{K})$ are permissible closed-loop strategies. For this special case, $V(k,x)$ again turns out to be a general quadratic function in x, and this leads to the following counterpart of Prop. 5.1.

Proposition 5.6 *For the linear-quadratic stochastic control problem described by (5.62a)-(5.62b), and under the closed-loop perfect state information pattern, there exists a unique minimizing solution (which is a feedback strategy) given by*

$$u_k^* = \gamma_k^*(x_k) = -P_k S_{k+1} A_k x_k - P_k(s_{k+1} + S_{k+1}c_k), \quad k \in \mathbf{K}, \tag{5.63}$$

where all terms are defined in Prop. 5.1.

Remark 5.12 The solution of the stochastic linear-quadratic optimal control problem depends only on the mean value (c_k, $k \in \mathbf{K}$) of the additive disturbance term but not on its covariance ($\Lambda_k, k \in \mathbf{K}$).[48] Moreover, the solution coincides with the optimal feedback solution of the deterministic version of the problem (as presented in Prop. 5.1), which takes the mean value of the Gaussian disturbance term as a constant known input. Finally, in view of Remark 5.11, uniqueness of the solution is valid under the assumption that the covariance matrix Λ_k is positive definite for all $k \in \mathbf{K}$. □

[47] E_{θ_k} denotes the expectation operation with respect to statistics of θ_k.

[48] The minimum expected value of the cost function does, however, depend on $\{\Lambda_k, k \in \mathbf{K}\}$.

Remark 5.13 Optimal solutions to stochastic control problems can be obtained under other deterministic information patterns also, but they are not representations of each other under the full-rank condition of the independent additive noise. Hence, each different information pattern requires a different derivation for the optimal solution. These solutions are not all *STC*, but they are necessarily *WTC*. See Problem 5 in Section 5.8 for an illustrative example. □

5.7 Viscosity Solutions

In this section another solution approach to the HJB equation (5.25) will be given. This approach has its origin in the theory of stochastic optimal control theory (in continuous time) as will be explained soon.

If in (5.25), or in its time-invariant analogue

$$\min_u \left[\frac{\partial V(x)}{\partial x} f(x,u) + g(x,u) \right] = 0, \tag{5.64}$$

the minimizing u, as a function of $\partial V/\partial x$, x (and t for (5.25)), is substituted into the same expressions, the resulting equations will be written formally as

$$\frac{\partial V}{\partial t} + \bar{H}(x, V, \frac{\partial V}{\partial x}, t) = 0, \tag{5.65}$$

and

$$\bar{H}(x, V, \frac{\partial V}{\partial x}) = 0, \tag{5.66}$$

respectively. For both equations we assume appropriate boundary conditions. Sometimes finding a solution of (5.65) is referred to as a Cauchy problem and of (5.66) as a Dirichlet problem. Note that, strictly speaking, the function $\bar{H}$ in (5.65) and (5.66) does not directly depend on the argument V in our case and that the format of (5.65) and (5.66) is more general than strictly necessary. Rather than studying (5.65) it pays sometimes to study

$$\varepsilon \sum_{i,j} \frac{\partial^2 V}{\partial x_i \partial x_j} + \frac{\partial V}{\partial t} + \bar{H}(x, V, \frac{\partial V}{\partial x}, t) = 0, \tag{5.67}$$

where ε is a (small) positive parameter. The first term of this expression is also written compactly as $\varepsilon \Delta V$. In the same way one could add the same term with the second derivatives to (5.66):

$$\varepsilon \sum_{i,j} \frac{\partial^2 V}{\partial x_i \partial x_j} + \bar{H}(x, V, \frac{\partial V}{\partial x}) = 0. \tag{5.68}$$

The theory of partial differential equations tells us that existence and unicity results are easier to obtain for (5.67) than for (5.65). One may hope to get results for (5.67) and then study their limits as ε tends to zero. Hopefully the solution of (5.67) (respectively, (5.68)) will resemble a solution of (5.65) (respectively, (5.66)) in some sense if $\varepsilon \downarrow 0$. Quite often solutions to (5.65), (5.66), viewed as abstract mathematical equations, are not unique, among others depending on the definition of solution that one adopts. With the addition of a proper understanding of the underlying optimal control problem, the solution becomes usually unique. Before continuing along these lines, we will first indicate that (5.67) is not only an auxiliary equation, but that it represents a condition for the value function of a related *stochastic* optimal control problem.

Consider the one-person version of the N-person stochastic differential equation as described by (5.17b):

$$dx_t = F(t, x_t, u(t))dt + \tilde{\sigma}(t, x_t)dw_t, \quad x_t \mid_{t=0} = x_0, \tag{5.69}$$

and the cost functional

$$J(\gamma) = E\left[\int_0^T g(t, x_t, \gamma(t, x_t))\, dt + q(x_T)\right], \tag{5.70}$$

where E stands for the expectation operator (see Appendix B) and where $u(t) = \gamma(t, x_t) \in S$, i.e., only feedback strategies are allowed. The following result from Fleming (1969) (see also Fleming and Rishel (1975)) provides a set of sufficient conditions and a characterization for the optimal solution of this stochastic control problem.

Proposition 5.7 *Let there exist a suitably smooth solution $W(t, x)$ of the semi-linear parabolic partial differential equation*

$$\begin{aligned} &\frac{\partial W}{\partial t} + \frac{1}{2}\sum_{i,j}\sigma^{ij}(t,x)\frac{\partial^2 W}{\partial x_i \partial x_j} \\ &\qquad\qquad + \min_{u \in S}[\nabla_x W \cdot F(t,x,u) + g(t,x,u)] = 0; \\ &W(T,x) = q(T,x), \end{aligned} \tag{5.71}$$

where σ^{ij} denotes the ijth element of the matrix $\tilde{\sigma}\tilde{\sigma}'$, and x_i denotes the ith component of the vector x. If there is a function $u^\circ(t, x, p)$ that minimizes $[p'F(t, x, u) + g(t, x, u)]$ on S, then $\gamma^*(t, x) = u^\circ(t, x, \nabla_x W)$ is an optimal control, with the corresponding expected cost given by $J^* = W(0, x_0)$.

This proposition yields a partial differential equation which the value function of the stochastic optimal control problem should satisfy. The resemblance between (5.67) and (5.71) will be obvious. Thus (5.67), for the time-invariant case, can also be considered as a condition which the value function of an appropriately defined stochastic optimal control problem should satisfy. The formulation of the stochastic optimal control problem and the statement of the proposition have been rather informal. The rigorous underlying mathematics is beyond the scope of this book and can for instance be found in the references already mentioned and in Lions (1982) and in Crandall and Lions (1983).

We now return to (5.67) and will study its solution when $\varepsilon \downarrow 0$. To express the fact that the solution of (5.67) depends on ε, it will be indicated by V_ε. If $\lim_{\varepsilon\downarrow 0} V_\varepsilon(x)$ exists, this limit might be what is called a *viscosity solution* of (5.66). The name "viscosity solution" comes from a celebrated method in fluid dynamics where the term $\varepsilon \sum_{i,j} \frac{\partial^2 V}{\partial x_i \partial x_j}$ represents physically a viscosity. In optimal control theory, however, the term "viscosity solution" is reserved for a particular solution of (5.66), which may equal the limit solution (with $\varepsilon \downarrow 0$) of (5.68) if it exists. More properly, it then could have been called the vanishing viscosity solution.

Definition 5.15 *Suppose that $\bar{H}$ is continuous in its arguments. A continuous function $V(x)$ is called a* viscosity subsolution *of (5.66) provided that for all continuously differentiable $\phi(x)$, which map $\mathbf{R}^n$ or an appropriate subset of $\mathbf{R}^n$ into $\mathbf{R}$ (as does $V(x)$), the following holds: if $V - \phi$ attains a local maximum at x_0, then*

$$\bar{H}(x_0, V(x_0), \frac{\partial \phi(x_0)}{\partial x}) \geq 0. \tag{5.72}$$

A continuous function $V(x)$ is called a viscosity supersolution *of (5.66) provided that for all continuously differentiable $\phi(x)$, which map $\mathbf{R}^n$ or an appropriate subset of $\mathbf{R}^n$ into $\mathbf{R}$ (as does $V(x)$), the following holds: if $V - \phi$ attains a local minimum at x_0, then*

$$\bar{H}(x_0, V(x_0), \frac{\partial \phi(x_0)}{\partial x}) \leq 0. \tag{5.73}$$

A continuous function $V(x)$ is called a viscosity solution *of (5.66) if it is both a viscosity subsolution and a viscosity supersolution.*

An equivalent, somewhat more intuitive, definition of a viscosity subsolution is as follows. A continuous function $V(x)$ is a viscosity subsolution of (5.66) if $\bar{H}(x, V(x), p) \geq 0$ for all $p \in \mathbf{R}^n$ such that

$$\lim_{x \to x_0} \inf (V(x) - V(x_0) - p'(x - x_0)) ||x - x_0||^{-1} \geq 0.$$

Those vectors p which satisfy this latter equation are sometimes referred to as subdifferentials. A similar and somewhat more intuitive definition of a viscosity supersolution can also be given. The equivalences of these definitions have been shown in Crandall, Evans and Lions (1984).

Remark 5.14 It should be noted that, in the definitions of viscosity sub- and supersolutions, in the references cited the inequalities (5.72) and (5.73) are both reversed in sign. The definitions given here, however, are consistent with (5.68), in which it is assumed that ε is positive. In the references cited, the second-order term in (5.68) is preceded by a minus sign. We have not adopted the addition of this minus sign because of the fact that the second-order term is directly related to the covariance of the noise in the system equations (and hence has to be nonnegative). Another way to make the definitions given here and in the references cited fully equivalent is to change $\bar{H}$ into $-\bar{H}$. □

The following simple result establishes the consistency of the notions of viscosity and "classical solutions".

Theorem 5.6 *If the continuously differentiable function V is a classical solution of (5.66), then it is a viscosity solution.*

The following theorem relates the solutions of (5.66) and (5.67).

Theorem 5.7 *Suppose that V_ε is a continuously differentiable solution of (5.67) and that it converges uniformly, as $\varepsilon \downarrow 0$, to some continuous function V. Then V is a viscosity solution of (5.66).*

Proof. Let us check (5.72) first for twice continuously differentiable ϕ. Assuming that $V - \phi$ has a local maximum at x_0, a function ζ is chosen, differentiable a sufficient number of times and with $0 \le \zeta < 1$ for $x \neq x_0$ and $\zeta(x_0) = 1$, such that $V - (\phi - \zeta)$ has a strict local maximum at x_0. Thus for ε sufficiently small, $V_\varepsilon - (\phi - \zeta)$ has a local maximum at some x_ε and $x_\varepsilon \to x_0$ as $\varepsilon \downarrow 0$. Then,

$$\nabla V_\varepsilon(x_\varepsilon) = \nabla(\phi - \zeta)(x_\varepsilon), \quad \Delta V_\varepsilon(x_\varepsilon) \le \Delta(\phi - \zeta)(x_\varepsilon),$$

and therefore

$$\bar{H}(x_\varepsilon, V_\varepsilon(x_\varepsilon), \nabla(\phi - \zeta)(x_\varepsilon)) = -\varepsilon \Delta V_\varepsilon(x_\varepsilon) \ge -\varepsilon \Delta(\phi - \zeta)(x_\varepsilon).$$

Now, if we take the limit $\varepsilon \downarrow 0$, this expression becomes (5.72).

Suppose now that ϕ is only once continuously differentiable. The first part of the proof, as just given, is now repeated for ϕ_k, where ϕ_k is twice continuously differentiable and $\phi_k \to \phi$ as $k \to \infty$. This concludes the part of the proof that V is a viscosity subsolution. The fact that V is also a viscosity supersolution can analogously be shown, which then, together with V being a viscosity subsolution, concludes the proof. □

A weaker version of Thm. 5.7 exists in the sense that the function $\bar{H}$ may also depend on ε, to be indicated by $\bar{H}_\varepsilon$. In the formulation of the extended version of Thm. 5.7 it is then necessary that $\bar{H}_\varepsilon$ converges to $\bar{H}$ uniformly on compact subsets of $\mathrm{R}^n \times \mathrm{R} \times \mathrm{R}^n$, as $\varepsilon \downarrow 0$. This weaker version is useful, for instance, when the continuous-time formulation of the problem is approximated by a discretized one (for numerical purposes, for instance). Such a "discretized version" of $\bar{H}$ could then coincide with $\bar{H}_\varepsilon$; see Lions (1982), Bardi and Sartori (1992) and Pourtallier and Tolwinski (1992).

An issue not dealt with heretofore is that of uniqueness of viscosity solutions. Consider (5.67) again. For special forms of the function $\bar{H}$ the uniqueness has been shown (see Lions (1982) or Crandall, Evans and Lions (1984)). In those cases it is essential that $\bar{H}$ depends explicitly on V, which is for instance the case if $\bar{H}$ has the following form:

$$\bar{H}(x, V, \frac{\partial V}{\partial x}) = V + \alpha(\frac{\partial V}{\partial x}) + \beta(x)$$

for some scalar functions α and β. However, as remarked in the beginning of this section, $\bar{H}$ as introduced does not depend on V. It is, however, possible to create such a dependence by means of the so-called *Kruzkov-transform* of the value function V. This transform of the value function is defined as

$$v(x) \triangleq \begin{cases} \psi(V(x)) & \text{if } V(x) < \infty, \\ 1 & \text{elsewhere,} \end{cases}$$

where $\psi(V) \triangleq 1 - \exp(-V)$. It is important to realize that v is bounded and is itself the value function of an optimal control problem, viz.,

$$v(x) \triangleq \inf_u \psi(L(u)),$$

with L defined in (5.23). The transformed cost function $\psi(L(u))$ can be rewritten in the "standard form" of integral and terminal part as

$$\begin{aligned} \psi(L(u)) &= \int_0^T g(t, x, u) \exp\left(-\int_0^t g(s, x, u)\, \mathrm{d}s\right) \mathrm{d}t \\ &\quad + \exp\left(-\int_0^T g(s, x, u)\, \mathrm{d}s\right)(1 - \exp(-q(T, x(T)))). \end{aligned}$$

As a side remark, this latter cost function resembles cost functions with discount factors, which have the form

$$\int_0^T g(t, x, u) \exp(-\lambda t)\, \mathrm{d}t + q(T, x(T)) \exp(-\lambda T),$$

where the nonnegative constant λ is the so-called discount factor. Such cost functions can be treated in the standard way but offer some advantages if $T = \infty$ and the terminal term is not present. Under some mild conditions, see Bardi and Soravia (1991), the following proposition can be shown to hold.

Proposition 5.8 *If V is a viscosity solution of (5.66), where $\bar{H}$ does not depend on the second argument V, and if there exists a constant g_0 such that $g(x, u) \geq g_0 > 0$, then v is a viscosity solution of*

$$v + \breve{H}(x, v, \frac{\partial v}{\partial x}) = 0, \tag{5.74}$$

where

$$\breve{H}\left(x, v, \frac{\partial v}{\partial x}\right) \triangleq \min_u \left[\frac{\partial v}{\partial x} f(x, u) + g(x, u) - (g(x, u) - g_0)v.\right].$$

This proposition shows how the original equation $\bar{H} = 0$, where H does not depend on the value function V directly, can be transformed into (5.74), which has a linear term in the transformed value function v. For the latter kind of partial differential equations, it is easier to prove uniqueness of solution. It also offers a starting point for numerical procedures to solve for v (and hence for V).

Hitherto in this section, we have taken V to be continuous. For the case in which it is discontinuous the definition of viscosity solution can be modified as follows.

Definition 5.16 *Suppose that $\bar{H}$ is continuous in its arguments. An upper semicontinuous function V is called a* viscosity subsolution *of (5.66) provided that for all continuously differentiable ϕ, which map $\mathbf{R}^n$ or an appropriate subset of $\mathbf{R}^n$ into $\mathbf{R}$ (as does V), the following holds: if $V - \phi$ attains a local maximum at x_0, then*

$$\bar{H}(x_0, V(x_0), \frac{\mathrm{d}\phi(x_0)}{\mathrm{d}x}) \geq 0. \tag{5.75}$$

A lower semicontinuous function V is called a viscosity supersolution *of (5.66) provided that for all continuously differentiable ϕ, which map $\mathbf{R}^n$ or an appropriate subset of $\mathbf{R}^n$ into $\mathbf{R}$ (as does V), the following holds: if $V - \phi$ attains a local minimum at x_0, then*

$$\bar{H}(x_0, V(x_0), \frac{\mathrm{d}\phi(x_0)}{\mathrm{d}x}) \leq 0. \tag{5.76}$$

A function V is called a viscosity solution *of (5.66) if the upper and lower semicontinuous envelopes of V, defined, respectively, by*

$$V^*(x) \triangleq \lim_{y \to x} \sup V(y), \quad V_*(x) \triangleq \lim_{y \to x} \inf V(y),$$

are respectively a viscosity subsolution and a viscosity supersolution.

For value functions with discontinuities relatively few results are available as yet; see the notes section 5.9.

5.8 Problems

1. A company has $x(t)$ customers at time t and makes a total profit of

$$f(t) = \int_0^t (cx(s) - u(s))\,\mathrm{d}s,$$

up to time t, where c is a positive constant. The function $u(t)$, restricted by $u(t) \geq 0$, $f(t) \geq 0$, represents the money put into advertising. The restriction $f(t) \geq 0$ indicates that the company is not allowed to borrow money. Advertisement helps to increase the number of customers according to

$$\dot{x} = u, \qquad x(0) = x_0 > 0.$$

The company wants to maximize the total profit during a given time period $[0, T]$, where T is fixed. Obtain both the open-loop and feedback optimal solutions for this problem.

2. Consider the two-dimensional system

$$\begin{aligned} \dot{x}_1 &= -x_2 u, \\ \dot{x}_2 &= -1 + x_1 u, \end{aligned}$$

where the control variable u is scalar and satisfies $|u(t)| \leq 1$. In polar coordinates, the equations of motion become

$$\begin{aligned} \dot{r} &= -\cos\alpha, \\ r\dot{\alpha} &= -ru + \sin\alpha, \end{aligned}$$

where α is the angle measured clockwise from the positive x_2-axis. The target set Λ is $|\alpha| \geq \bar{\alpha}$, where $0 < \bar{\alpha} < \pi$. (For an interpretation of this system see Example 8.3, later in Chapter 8, with $v_2 = 0$, or Lewin and Olsder (1979).) The decision maker wants the state (x_1, x_2) to stay away from Λ for as long as possible. Determine the time-optimal-control. Show that, for $\bar{\alpha} \geq \pi/2$, there is a barrier.

3. Consider the following discrete-time, discrete state space optimal control problem. The state space consists of the integers $\{0, 1, 2, \ldots\}$. At time t_k the decision maker steers the system positioned at integer $i_k \geq 1$, either to $\{0\}$ or to the integer $\{i_k + 1\}$, where it arrives at time t_{k+1}. The control problem terminates when the system is steered to the origin $\{0\}$. The pay-off is completely determined by the last move. If the system moves from $\{i_k\}$ to $\{0\}$, then the pay-off is 2^{2i_k}, to be maximized by the decision maker.

   Show that each playable pure strategy leads to a finite value, and that the behavioral strategy which dictates both alternatives with equal probability $(\frac{1}{2})$ at each stage leads to an infinite expected pay-off (assuming that $i_0 \geq 1$).

4. Consider the discrete-time system whose state evolves according to the dynamics

$$x_{k+1} = f_k(x_k, u_k, u_{k-1}), \quad u_0 = 0, \ k \in \mathbf{K}.$$

   (i) Show that every *STC* optimal policy for this system, under the cost function (5.18b) and the information pattern that allows the control at stage k to depend on the current and past values of the state and past values of control, has to be in the form $u_k = \gamma_k(x_k, u_{k-1})$.

(ii) Obtain the *STC* solution(s) explicitly for the linear-quadratic (LQ) problem where the cost is given by (5.20b), and the system dynamics (exhibiting lagged dependence on the control) by

$$x_{k+1} = A_k x_k + B_k u_k + C_k u_{k-1}, \quad u_0 = 0, \ k \in \mathbf{K}.$$

(iii) Does the LQ problem above admit (a) an *STC* optimal policy under the CLPS information pattern; (b) a *WTC* optimal policy under the CLPS information pattern? In the latter case obtain one such policy, if there exists one.

5. In Problem 4 above, replace the deterministic linear dynamics by the stochastic difference equation

$$x_{k+1} = A_k x_k + B_k u_k + C_k u_{k-1} + \theta_k, \quad u_0 = 0, \ k \in \mathbf{K},$$

where $x_1, \theta_1, \ldots, \theta_K$ are Gaussian, mutually independent, and have zero mean and positive definite covariances. Further let the cost function be given by (5.62b).

(i) Does the problem admit an *STC* optimal control under the CLPS information pattern?

(ii) Obtain a *WTC* optimal policy for this stochastic control problem under the CLPS information pattern.

(iii) Can you generate an optimal OL control policy from the solution obtained above? Would it be *WTC* under the OL information pattern?

5.9 Notes

Section 5.2. As already noted in Remark 5.2, the state-model description of a discrete-time (deterministic or stochastic) infinite dynamic game as presented here is clearly not the most general extension of Kuhn's finite game model (cf. Chapters 2 and 3) to the infinite case. One such extension which has not been covered in Remark 5.2 is the *infinite-horizon* problem wherein the number of stages K tends to infinity. In such games, the so-called *stationary strategies* (i.e., strategies which do not depend on the discrete-time parameter k) become of real interest, and they have hitherto attracted considerable attention in the literature (see Shapley, 1953; Sobel, 1971; Maitra and Parthasarathy, 1970), but mostly for finite or at most denumerable state and strategy spaces. The general class of discrete-time stochastic dynamic games under the feedback information structure are also referred to as *Markov games* in the literature. For two survey articles on this topic, see Parthasarathy and Stern (1977) and Raghavan and Filar (1991).

Section 5.3. Differential games were first introduced by Isaacs (1954-1956), within the framework of two-person zero-sum games, whose work culminated in the publication of his book about ten years later (see Isaacs, 1975). Nonzero-sum differential games were later introduced in the works of Starr and Ho (1969a,b), Case (1967) and

Friedman (1971), but under specific information structures—namely, the open-loop and feedback perfect state information patterns. Some representative references on stochastic differential games, on the other hand, are Friedman (1972), Elliott (1976) and Başar (1977b). The book by Krasovskii and Subbotin (1988) has an extensive discussions on non-Lipschitzian differential equations. A class of differential games not covered by Def. 5.5 are those whose state dynamics are governed by partial differential equations. These will not be treated in this book; for some examples of practical games which fit into this framework, the reader is referred to Roxin (1977). Yet another class of differential games not to be covered in subsequent chapters are those on which the players are allowed to employ impulsive controls; see Blaquière (1977) for some results on this topic.

Section 5.4. Mixed and behavioral strategies in infinite dynamic games were first introduced by Aumann (1964) who also provided a rigorous extension of Kuhn's extensive form description of finite games to infinite games, and proved in this context that if an N-person nonzero-sum infinite dynamic game is of "perfect recall" for one player, then that player can restrict himself to behavioral strategies. For a further discussion of mixed strategies in differential games and for an elaboration on the relation of this concept with that of "relaxed controls" used in control theory (cf. Warga, 1972), the reader is referred to Wilson (1977), who also proves the existence of mixed-strategy saddle points in suitably structured zero-sum differential games of fixed duration and with open-loop information structure. Some other references which deal with existence of mixed-strategy equilibrium solutions in differential games are Elliott (1976), Pachter and Yavin (1979), Levine (1979) and Kumar and Shiau (1981).

Section 5.5. Some classic texts on the theory of optimal control where the reader can refer to for a more extensive coverage and references are Bryson and Ho (1975), Hermes and LaSalle (1969) and Fleming and Rishel (1975).

Section 5.6. The concept of "representations of a strategy" was introduced by Başar in a series of papers, within the context of infinite dynamic games (see, e.g., Başar, 1974, 1976a, 1977b), and it was shown to be closely related to existence of nonunique Nash equilibria under the CLPS information pattern, as it will be elucidated in Chapter 6. The subsection on time consistency is based on the material in Başar (1989b), which the reader should refer to for extensive discussion on the motivation behind these notions, and some specific results. Also, Chapters 6 and 7 contain further discussion of *time consistency* in the context of Nash and Stackelberg equilibria. The issue of time consistency (or inconsistency) has pervaded the economics literature during the past two decades, following the stimulating paper by Kydland and Prescott (1977).

Section 5.7. Viscosity solutions in the theory of optimal control (and also in the theory of zero-sum differential games) became well known with the publication of the book by Lions (1982). The theory presented in this section is based mainly on Crandall, Evans and Lions (1984). Another reference in the same vein is Crandall and Lions (1983). Viscosity solutions for minimum-time problems have been reported in Bardi (1989) and Staicu (1989). In all these references it was assumed that the value function V is continuous in its arguments, a property not often satisfied (see

Example 5.2 and also Chapter 8). Recently progress has been made on problems with discontinuous value functions, notably for time-optimal control problems; see Ishii (1989), Bardi and Soravia (1991) and Bardi and Staicu (1991). The notion of discontinuous viscosity solutions was introduced by Ishii (1987). For a more recent rigorous treatment of the topic of viscosity solutions in the context of optimal control, see Fleming and Soner (1993).

Chapter 6

Nash and Saddle-Point Equilibria of Infinite Dynamic Games

6.1 Introduction

This chapter discusses properties and derivation of Nash and saddle-point equilibria in infinite dynamic games of prescribed fixed duration. The analysis is first confined to dynamic games defined in discrete time, and with a finite number of stages, and then extended to differential games. Some results for infinite-horizon formulations are also presented, primarily for affine-quadratic structures.

Utilization of the two standard techniques of optimal control theory, viz. the minimum principle and dynamic programming, leads to the so-called open-loop and feedback Nash equilibrium solutions, respectively. These two different Nash equilibria and their derivation and properties (such as existence and uniqueness) are discussed in Section 6.2, and the results are also specialized to affine-quadratic games, as well as to two-person zero-sum games.

When the underlying information structure for at least one player is dynamic and involves memory, a plethora of (so-called *informationally nonunique*) Nash equilibria with different sets of cost values exists, whose derivation entails some rather intricate analysis—not totally based on standard techniques of optimal control theory. This derivation, as well as several important features of Nash equilibria under closed-loop perfect state information pattern, are discussed in Section 6.3, first within the context of a scalar three-person two-stage game (cf. subsection 6.3.1) and then for general dynamic games in discrete time (cf. subsection 6.3.2).

Section 6.4 is devoted to derivation of necessary and sufficient conditions for Nash equilibria in stochastic nonzero-sum dynamic games with deterministic information patterns. Such a stochastic formulation eliminates informational nonuniqueness, thereby making the question of existence of unique Nash equilibrium under closed-loop perfect state information pattern a meaningful one.

Section 6.5 presents the counterparts of the results of Section 6.2 in the continuous time, that is, for differential games with fixed duration, and the next one (Section 6.6) discusses some important applications of this theory to worst-case controller design (so-called H^∞-optimal control), first in continuous and then in discrete time. Finally, Section 6.7 presents the counterparts of the results of Section 6.4 in the continuous time.

6.2 Open-Loop and Feedback Nash and Saddle-Point Equilibria for Dynamic Games in Discrete Time

Within the context of Def. 5.1, consider the class of N-person discrete-time deterministic infinite dynamic games of prescribed fixed duration (K stages) which are described by the state equation

$$x_{k+1} = f_k(x_k, u_k^1, \ldots, u_k^N), \quad k \in \mathbf{K}, \tag{6.1}$$

where $x_k \in X = \mathbf{R}^n$ and x_1 is specified *a priori*. Furthermore, the control sets U_k^i are taken as measurable subsets of $\mathbf{R}^{m_j}$ ($U_k^i \subseteq \mathbf{R}^{m_j}; k \in \mathbf{K}, i \in \mathbf{N}$), and a stage-additive cost functional

$$L^i(u^1, \ldots, u^N) = \sum_{k=1}^{K} g_k^i(x_{k+1}, u_k^1, \ldots, u_k^N, x_k) \tag{6.2}$$

is given for each player $i \in \mathbf{N}$.

In this section, we discuss derivation of Nash equilibria for this class of nonzero-sum games when the information structure of the game is either (i) open-loop or (ii) memoryless perfect state for all the players. In the former case, any permissible strategy for $\mathbf{P}i$ at stage $k \in \mathbf{K}$ is a constant function and therefore can be considered to be an element of U_k^i, i.e., $\Gamma_k^i = U_k^i$, $k \in \mathbf{K}$, $i \in \mathbf{N}$. Under the latter information structure, however, any permissible strategy for $\mathbf{P}i$ at stage $k \in \mathbf{K}$ is a measurable mapping $\gamma_k^i : X \to U^i$, $i \in \mathbf{N}$, $k \in \mathbf{K}$. With Γ_k^i ($k \in \mathbf{K}, i \in \mathbf{N}$) taken as the appropriate strategy space in each case, we recall (from Def. 3.12) that an N-tuple of strategies $\{\gamma^{i*} \in \Gamma^i; i \in \mathbf{N}\}$ constitutes a Nash equilibrium solution if, and only if, the following inequalities are satisfied for all $\{\gamma^i \in \Gamma^i; i \in \mathbf{N}\}$:

$$\left.\begin{aligned}
J^{1*} &\triangleq J^1(\gamma^{1*}; \gamma^{2*}; \ldots; \gamma^{N*}) \le J^1(\gamma^1; \gamma^{2*}; \ldots; \gamma^{N*}),\\
J^{2*} &\triangleq J^1(\gamma^{1*}; \gamma^{2*}; \gamma^{3*}; \ldots; \gamma^{N*}) \le J^2(\gamma^{1*}; \gamma^2; \gamma^{3*}; \ldots; \gamma^{N*}),\\
&\cdots\cdots\cdots\cdots\cdots\cdots\cdots\cdots\\
&\cdots\cdots\cdots\cdots\cdots\cdots\cdots\cdots\\
J^{N*} &\triangleq J^N(\gamma^{1*}; \ldots; \gamma^{N-1*}; \gamma^{N*}) \le J^N(\gamma^{1*}; \ldots; \gamma^{N-1*}; \gamma^N).
\end{aligned}\right\} \tag{6.3}$$

Here, $\gamma^i \triangleq \{\gamma_1^i, \ldots, \gamma_K^i\}$—the aggregate strategy of $\mathbf{P}i$—and the notation $\gamma^i \in \Gamma^i$ stands for $\gamma_k^i \in \Gamma_k^i, \forall k \in \mathbf{K}$. Moreover, $J^i(\gamma^1, \ldots, \gamma^N)$ is equivalent to $L^i(u^1, \ldots, u^N)$ with u_k^i replaced by $\gamma_k^i(\cdot)$ $(i \in \mathbf{N}, k \in \mathbf{K})$.

Under the open-loop information structure, we refer to the Nash solution as "*open-loop Nash equilibrium solution*", which we discuss in the first subsection to follow. For the memoryless perfect state information, the Nash solution will be referred to as "*closed-loop no-memory Nash equilibrium solution*", and under a further restriction, which is the counterpart of Def. 3.22 in the present framework. It will be called "*feedback Nash equilibrium solution*". Both types of equilibria will be discussed in subsection 6.2.2.

6.2.1 Open-loop Nash equilibria

One method of obtaining the open-loop Nash equilibrium solution(s) of the class of discrete-time games formulated above is to view them as static infinite games and directly apply the analysis of Chapter 4. Toward this end we first note that, by backward recursive substitution of (6.1) into (6.2), it is possible to express L^i solely as functions of $\{u^j; j \in \mathbf{N}\}$ and the initial state x_1 whose value is known *a priori*, where u^j is defined as the aggregate control vector $(u_1^{j'}, u_2^{j'}, \ldots, u_K^{j'})'$. This then implies that, to every given set of functions $\{f_k, g_k^i; k \in \mathbf{K}\}$, there corresponds a unique function $\tilde{L}^i : X \times U^1 \times \cdots \times U^N \to \mathbf{R}$, which is the cost functional of $\mathbf{P}i$, $i \in \mathbf{N}$. Here, U^j denotes the aggregate control set of $\mathbf{P}j$, compatible with the requirement that if $u^j \in U^j$ then $u_k^j \in U_k^j$, $\forall k \in \mathbf{K}$, and the foregoing construction leads to a normal form description of the original game, which is no different from the class of infinite games treated in Chapter 4. Therefore, to obtain the open-loop Nash equilibria, we simply have to minimize $\tilde{L}^i(x_1, u^1, \ldots, u^{i-1}, \cdot, u^{i+1}, \ldots, u^N)$ over U^i, for each $i \in \mathbf{N}$, and then determine the intersection point(s) of the resulting reaction curves. In particular, if $\tilde{L}^i(x_1, u^1, \ldots, u^N)$ is continuous on $U^1 \times \cdots \times U^N$, strictly convex in u^i, and further if U^i are closed, bounded and convex, an open-loop Nash equilibrium (in pure strategies) exists (cf. Thm. 4.3).

Such an approach can sometimes lead to quite unwieldy expressions, especially if the number of stages in the game is large. An alternative derivation which partly removes this difficulty is the one that utilizes techniques of optimal control theory, by making explicit use of the stage-additive nature of the cost functionals (6.2) and the specific structure of the extensive form description of the game, as provided by the state equation (6.1). There is in fact a close relationship between derivation of open-loop Nash equilibria and the problem of solving (jointly) N optimal control problems, which can readily be observed from inequalities (6.3) since each one of them (together with (6.1) and (6.2)) describes an optimal control problem whose *structure* is not affected by the remaining players' control vectors. Exploring this relationship a little further, we arrive at the following result.

Theorem 6.1 *For an N-person discrete-time infinite dynamic game, let* *(i)* $f_k(\cdot, u_k^1, \ldots, u_k^N)$ *be continuously differentiable on* $\mathbf{R}^n$, $(k \in \mathbf{K})$,

(ii) $g_k^i(\cdot, u_k^1, \ldots, u_k^N, \cdot)$ *be continuously differentiable on* $\mathbf{R}^n \times \mathbf{R}^n$, $(k \in \mathbf{K}, i \in \mathbf{N})$,

(iii) $f_k(\cdot, \cdot, \ldots, \cdot)$ *be convex on* $\mathbf{R}^n \times \mathbf{R}^{m_1} \times \cdots \mathbf{R}^{m_N}$, $(k \in \mathbf{K})$.

Then, if $\{\gamma^{i*}(x_1) = u^{i*}; i \in \mathbf{N}\}$ *provides an* open-loop Nash equilibrium solution *and* $\{x_{k+1}^*; k \in \mathbf{K}\}$ *is the corresponding state trajectory, there exists a finite sequence of n-dimensional (costate) vectors* $\{p_2^i, \ldots, p_{K+1}^i\}$ *for each* $i \in \mathbf{N}$ *such that the following relations are satisfied:*

$$x_{k+1}^* = f_k(x_k^*, u_k^{1*}, \ldots, u_k^{N*}), \quad x_1^* = x_1, \tag{6.4a}$$

$$\gamma_k^{i*}(x_1) \equiv \arg \min_{u_k^i \in U_k^i} H_k^i(p_{k+1}^i, u_k^{1*}, \ldots, u_k^{i-1*}, u_k^i, u_k^{i+1*}, \ldots, u_k^{N*}, x_k^*), \tag{6.4b}$$

$$\begin{aligned} p_k^i = \frac{\partial}{\partial x_k} f_k(x_k^*, u_k^{1*}, \ldots, u_k^{N*})' \Bigg[p_{k+1}^i + \Bigg(\frac{\partial}{\partial x_{k+1}} g_k^i(x_{k+1}^*, u_k^{1*}, \ldots, \\ u_k^{N*}, x_k^* \Bigg)' \Bigg] + \left[\frac{\partial}{\partial x_k} g_k^i(x_{k+1}^*, u_k^{1*}, \ldots, u_k^{N*}, x_k^*) \right]'; \\ p_{K+1}^i = 0, \quad i \in \mathbf{N}, \quad k \in \mathbf{K}, \end{aligned} \tag{6.4c}$$

where

$$\begin{aligned} H_k^i(p_{k+1}, u_k^1, \ldots, u_k^N, x_k) \triangleq g_k^i(f_k(x_k, u_k^1, \ldots, u_k^N), u_k^1, \ldots, u_k^N, x_k) \\ + p_{k+1}^{i'} f_k(x_k, u_k^1, \ldots, u_k^N); \quad k \in \mathbf{K}, \quad i \in \mathbf{N}. \end{aligned} \tag{6.4d}$$

Every such Nash equilibrium solution is weakly time consistent.

Proof. Consider the ith inequality of (6.3), which says that $\gamma^{i*}(x_1) \equiv u^{i*}$ minimizes $L^i(u^{1*}, \ldots, u^{i-1*}, u^i, u^{i+1*}, \ldots, u^{N*})$ over U^i subject to the state equation

$$x_{k+1} = f_k(x_k, u_k^{1*}, \ldots, u_k^{i-1*}, u_k^i, u_k^{i+1*}, \ldots, u_k^{N*}), \quad k \in \mathbf{K}.$$

This is a standard optimal control problem for $\mathbf{P}i$ since u^{j*} $(j \in \mathbf{K}, j \neq i)$ are open-loop controls and hence do not depend on u^i. The result, then, follows directly from the minimum principle for discrete-time control systems (cf. Thm. 5.5). The "weak time consistency" of the solution is a direct consequence of Def. 5.13, where the "sol" operation is the Nash solution. □

Theorem 6.1 thus provides a set of necessary conditions (solvability of a set of coupled two-point boundary value problems) for the open-loop Nash solution to satisfy; in other words, it produces candidate equilibrium solutions. In principle, one has to determine all solutions of this set of equations and further investigate which of these candidate solutions satisfy the original set of inequalities (6.3). If some further restrictions are imposed on f_k and g_k^i $(i \in \mathbf{N}, k \in \mathbf{K})$ so that the resulting cost functional $\tilde{L}^i$ (defined earlier, in this subsection) is convex in u^i for all $u^j \in U^j$, $j \neq i$, $j \in \mathbf{K}$, the latter phase of verification can clearly be eliminated, since then every solution set of (6.4a)-(6.4c) constitutes an open-loop Nash equilibrium solution. A specific class of problems for which this can be done, and the conditions involved expressed explicitly in terms of the parameters of the game, is the class of so-called "affine-quadratic" games which we first formally introduce below.

Definition 6.1 *An N-person discrete-time infinite dynamic game is of the* affine-quadratic *type if $U_k^i = \mathbf{R}_i^m$ $(i \in \mathbf{N}, k \in \mathbf{K})$, and* [49]

$$f_k(x_k, u_k^1, \ldots, u_k^N) = A_k x_k + \sum_{i \in \mathbf{N}} b_k^i u_k^i + c_k \tag{6.5a}$$

$$g_k^i(x_{k+1}, u_k^1, \ldots, u_k^N, x_k) = \frac{1}{2}\left(x'_{k+1} Q_{k+1}^i x_{k+1} + \sum_{j \in \mathbf{N}} u_k^{j'} R_k^{ij} u_k^j \right), \tag{6.5b}$$

where A_k, B_k, Q_{k+1}^i R_k^{ij} are matrices of appropriate dimensions, Q_{k+1}^i is symmetric, $R_k^{ii} > 0$, $c_k \in \mathbf{R}^n$ is a fixed vector sequence, and $k \in \mathbf{K}$, $i \in \mathbf{N}$.
An affine-quadratic game is of the linear-quadratic *type if $c_k \equiv 0$.*

Theorem 6.2 *For an N-person affine-quadratic dynamic game with $Q_{k+1}^i \geq 0$ $(i \in \mathbf{N}, k \in \mathbf{K})$, let Λ_k, M_{k+1}^i $(k \in \mathbf{K}, i \in \mathbf{N})$ be matrices of appropriate dimensions, defined by*

$$\Lambda_k = I + \sum_{i \in \mathbf{N}} B_k^i [R_k^{ii}]^{-1} B_k^{i'} M_{k+1}^i, \tag{6.6a}$$

$$M_k^i = Q_k^i + A'_k M_{k+1}^i \Lambda_k^{-1} A_k; \quad M_{K+1}^i = Q_{K+1}^i. \tag{6.6b}$$

If the matrices Λ_k $(k \in \mathbf{K})$, thus recursively defined, are invertible, the game admits a unique open-loop Nash equilibrium solution *given by*

$$\gamma_k^{i*}(x_1) \equiv u_k^{i*} = -[R_k^{ii}]^{-1} B_k^{i'} [M_{k+1}^i \Lambda_k^{-1} A_k x_k^* + \xi_k^i], \ (k \in \mathbf{K}, i \in \mathbf{N}), \tag{6.7a}$$

where $\{x_{k+1}^*; k \in \mathbf{K}\}$ is the associated state trajectory determined from

$$x_{k+1}^* = \Lambda_k^{-1}[A_k x_k^* + \eta_k]; \quad x_1^* = x_1, \tag{6.7b}$$

and ξ_k^i, η_k are defined by

$$\xi_k^i = M_{k+1}^i \Lambda_k^{-1} \eta_k + m_{k+1}^i, \tag{6.7c}$$

[49]The stagewise cost functions g_k^i can also be taken to depend on x_k instead of x_{k+1}, as in the proof of Prop. 5.1; but we prefer here the present structure (without any loss of generality) for convenience in the analysis to follow.

$$\eta_k = c_k - \sum_{i\in\mathbf{N}} B_k^i [R_k^{ii}]^{-1} B_k^{i'} m_{k+1}^i , \tag{6.7d}$$

with m_k^i recursively generated by

$$m_k^i = A_k' [m_{k+1}^i + M_{k+1}^i \Lambda_k^{-1} \eta_k] , \; m_{K+1}^i = 0, \; i \in \mathbf{N}, k \in \mathbf{K}. \tag{6.7e}$$

Proof. Since $Q_{k+1} \geq 0$, and $R_k^{ii} > 0$, $\tilde{L}^i(x_1, u^1, \ldots, u^N)$ is a strictly convex function of u^i for all $u^j \in \mathbf{R}^{m_j K}$ $(j \neq i, j \in \mathbf{N})$ and for all $x_1 \in \mathbf{R}^n$. Therefore, every solution set of (6.4a)-(6.4c) provides an open-loop Nash solution. Hence, the proof will be completed if we can show that (6.7a) is the only candidate solution. First note that

$$\begin{aligned} H_k^i &= \frac{1}{2}\left(A_k x_k + c_k + \sum_{j\in\mathbf{N}} B_k^j u_k^j\right)' Q_{k+1}^i \left(A_k x_k + c_k + \sum_{j\in\mathbf{N}} B_k^j u_k^j\right) \\ &\quad + \frac{1}{2}\sum_{j\in\mathbf{N}} u_k^{j'} R_k^{ij} u_k^j + p_{k+1}^{i'}\left(A_k x_k + c_k + \sum_{j\in\mathbf{N}} B_k^j u_k^j\right), \end{aligned}$$

and since $Q_{k+1}^i \geq 0$, $R_k^{ii} > 0$, minimization of this "Hamiltonian" over $u_k^i \in \mathbf{R}^{m_i}$ yields the unique relation

$$u_k^{i*} = -[R_k^{ii}]^{-1} B_k^{i'} [p_{k+1}^i + Q_{k+1}^i x_{k+1}^*], \tag{i}$$

where

$$x_{k+1}^* = A_k x_k^* + c_k + \sum_{i\in\mathbf{N}} B_k^i u_k^{i*}; \quad x_1^* = x_1. \tag{ii}$$

Furthermore, the costate (difference) equation in (6.4c) reads

$$p_k^i = A_k' [p_{k+1}^i + Q_{k+1}^i x_{k+1}^*]; \quad p_{K+1}^i = 0. \tag{iii}$$

Let us start with $k = K$ in which case (i) becomes

$$u_K^{i*} = -[R_K^{ii}]^{-1} B_K^{i'} M_{K+1}^i x_{K+1}^*, \tag{iv}$$

and if both sides are first premultiplied by B_K^i and then summed over $i \in \mathbf{N}$ we obtain, by also making use of (ii) and (6.6a),

$$x_{K+1}^* - A_K x_K^* = (I - \Lambda_K) x_{K+1}^* + c_K$$

which further yields the unique relation

$$x_{K+1}^* = \Lambda_K^{-1} [A_K x_K^* + c_K]$$

which is precisely (6.7b) for $k = K$. Substitution of this relation into (iv) then leads to (6.7a) for $k = K$.

We now prove by induction that the unique solution set of (i)-(iii) is given by (6.7a)-(6.7b) and $p_k^i = A_k^{'}[M_{k+1}^i x_{k+1}^* + m_{k+1}^i]$ $(i \in \mathbf{N}, k \in \mathbf{K})$. Let us assume that this is true for $k = l+1$ (already verified for $l = K-1$) and prove its validity for $k = l$.

First, using the solution $p_{l+1}^i = A_{l+1}^{'}[M_{l+2}^i x_{l+2}^* + m_{l+2}^i]$ in (i) with $k = l$, we obtain, after several algebraic manipulations,

$$u_l^{i*} = -[R_l^{ii}]^{-1} B_l^{i'} [M_{l+1}^i x_{l+1}^* + m_{l+1}^i]. \qquad (v)$$

Again, premultiplying this expression by B_l^i, and summing it over $i \in \mathbf{N}$ leads to, also in view of (ii), (6.6a) and (6.7d),

$$x_{l+1}^* = \Lambda_l^{-1}[A_l x_l^* + \eta_l]$$

which is (6.7b). If this relation is used in (v), we obtain the unique control vectors (6.7a), for $k = l$, and if it is further used in (iii) we obtain, in view of (6.7e),

$$p_l^i = A_l'[M_{l+1}^i x_{l+1}^* + m_{l+1}^i].$$

This then closes the induction argument, and thereby completes the proof of the theorem. □

Remark 6.1 An alternative derivation for the open-loop Nash equilibrium solution of the affine-quadratic game is (as discussed earlier in this subsection in a general context) to convert it into a standard static quadratic game and then to make use of the available results on such games (cf. Prop. 4.6). By backward recursive substitution of the state vector from the state equation into the quadratic cost functionals, it is possible to bring the cost functional of $\mathbf{P}i$ into the structural form as given by (4.23), which further is strictly convex in $u^i = (u_1^{i'}, \ldots, u_K^{i'})'$ because of assumptions $Q_k^i \geq 0$, $R_k^{ii} > 0$, $(k \in \mathbf{K})$.[50] Consequently, each player has a unique reaction curve, and the condition for existence of a unique Nash solution becomes equivalent to the condition for unique intersection of these reaction curves (cf. Prop. 4.6).[51] The existence condition of Thm. 6.2, i.e., nonsingularity of Λ_k, $(k \in \mathbf{K})$, is precisely that condition, but expressed in a different (more convenient, recursive) form. As in Prop. 4.6, it is of course possible for an affine-quadratic dynamic game to have multiple Nash equilibria, which would happen if Λ_k is singular for some k, which corresponds to R being singular in (4.26a), and r being in its range space. For yet another derivation of the result of Thm. 6.2, by making use of Prop. 5.1, the reader is referred to Problem 1, Section 6.8. □

[50] The condition $Q_{k+1}^i \geq 0$ is clearly sufficient (along with $R_k^{ii} > 0$) to make $\tilde{L}^i$ strictly convex in u^i, but is by no means necessary. It can be replaced by weaker conditions (which ensure convexity) under which the statements of Thm. 6.2 and this remark are still valid. One such condition, that is in fact tight, is given in Lemma 6.1 later, which should be interpreted for the present context.

[51] For a derivation of open-loop Nash solution in two-person linear-quadratic games, and along these lines, the reader is referred to Başar (1976a) and Olsder (1977a).

Zero-sum dynamic games

We now turn our attention to another special class of nonzero-sum dynamic games—the two-person zero-sum games—in which case (assuming, in accordance with our earlier convention, that **P1** is the minimizer and **P2** is the maximizer),

$$g_k^1 \equiv -g_k^2 \triangleq g_k \tag{6.8a}$$

and the "Nash" inequalities (6.3) reduce to the saddle-point inequality

$$J(\gamma^{1*}, \gamma^2) \leq J(\gamma^{1*}, \gamma^{2*}) \leq J(\gamma^1, \gamma^{2*}), \gamma^1 \in \Gamma^1, \gamma^2 \in \Gamma^2, \tag{6.8b}$$

where $J \triangleq J^1 \equiv -J^2$. Directly applying Thm. 6.1 in the present context, we first have from (6.4a)-(6.4c) (in view of (6.8a)) the relation $p_k^1 = -p_k^2 \triangleq p_k$, $(k \in \mathbf{K})$, and therefore from (6.4d), $H_k^1 \equiv -H_k^2 \triangleq H_k$, $(k \in \mathbf{K})$. Hence, we have arrived at the following conclusion.

Theorem 6.3 *For a two-person zero-sum discrete-time infinite dynamic game, let*

(i) $f_k(\cdot, u_k^1, u_k^2)$ *be continuously differentiable on* $\mathbf{R}^n$, $(k \in \mathbf{K})$,

(ii) $g_k(\cdot, u_k^1, u_k^1, \cdot)$ *be continuously differentiable on* $\mathbf{R}^n \times \mathbf{R}^n$, $(k \in \mathbf{K})$,

(iii) $f_k(\cdot, \cdot, \cdot)$ *be convex on* $\mathbf{R}^n \times U^1 \times U^2$, $(k \in \mathbf{K})$.

Then, if $\{\gamma^{1*}(x_1) = u^{i*}; i = 1, 2\}$ *provides an* open-loop saddle-point solution, *and* $\{x_{k+1}^*; k \in \mathbf{K}\}$ *is the corresponding state trajectory, there exists a finite sequence of n-dimensional (costate) vectors* $\{p_2, \ldots, p_{K+1}\}$ *such that the following relations are satisfied:*

$$x_{k+1}^* = f_k(x_k^*, u_k^{1*}, u_k^{2*}), x_1^* = x_1, \tag{6.9a}$$

$$\begin{aligned} H_k(p_{k+1}, u_k^{1*}, u_k^2, x_k^*) \quad &\leq \quad H_k(p_{k+1}, u_k^{1*}, u_k^{2*}, x_k^*) \leq H_k(p_{k+1}, u_k^1, u_k^{2*}, x_k^*), \\ &\qquad \forall u_k^1 \in U_k^1, u_k^2 \in U_k^2, \end{aligned} \tag{6.9b}$$

$$\begin{aligned} p_k \quad = \quad & \frac{\partial}{\partial x_k} f_k(x_k^*, u_k^{1*}, u_k^{2*})' \left[p_{k+1} + \left(\frac{\partial}{\partial x_{k+1}} g_k(x_{k+1}^*, u_k^{1*}, u_k^{2*}, x_k^*) \right)' \right] \\ & + \left[\frac{\partial}{\partial x_k} g_k \left(x_{k+1}^*, u_k^{1*}, u_k^{2*}, x_k^* \right) \right]', \qquad p_{K+1} = 0, \end{aligned} \tag{6.9c}$$

where

$$\begin{aligned} H_k(p_{k+1}, u_k^1, u_k^2, x_k) \quad \triangleq \quad & g_k(f_k(x_k, u_k^1, u_k^2), u_k^1, u_k^2, x_k) \\ & + p_{k+1}' f_k(x_k, u_k^1, u_k^2), k \in \mathbf{K}. \end{aligned} \tag{6.9d}$$

Proof. The result follows from Thm. 6.1, as discussed earlier. □

As a specific application of this theorem, we now consider the class of affine-quadratic two-person zero-sum dynamic games (cf. Def. 6.1) described by the state equation

$$x_{k+1} = A_k x_k + B_k^1 u_k^1 + B_k^2 u_k^2 + c_k, \quad k \in \mathbf{K}, \tag{6.10a}$$

and the objective functional

$$L(u^1, u^2) = \frac{1}{2} \sum_{k=1}^{K} (x'_{k+1} Q_{k+1} x_{k+1} + u_k^{1'} u_k^1 - u_k^{2'} u_k^2) \tag{6.10b}$$

which **P1** wishes to minimize and **P2** attempts to maximize. It should be noted that we have taken the weighting matrices for the controls in (6.10b) as unity, without any loss of generality, since otherwise they can be absorbed in B_k^1 and B_k^2 provided, of course, that $R_k^{11} > 0$ and $R_k^{22} > 0$ which was our *a priori* assumption in Def. 6.1. Let us also assume that $Q_{k+1} \geq 0$ ($k \in \mathbf{K}$) which essentially makes $L(u^1, u^2)$ strictly convex in u^1. In order to formulate a meaningful problem, we also have to require $L(u^1, u^2)$ to be (strictly) concave in u^2, since otherwise **P2** can make the value of L unboundedly large. The following lemma now provides a necessary and sufficient condition for (6.10b) to be strictly concave in u^2.

Lemma 6.1 *For the affine-quadratic two-person zero-sum dynamic game introduced above, the objective functional $L(u^1, u^2)$ is* strictly concave *in u^2 (for all $u^1 \in \mathrm{R}^{Km_1}$) if, and only if,*

$$I - B_k^{2'} S_{k+1} B_k^2 > 0 \quad (k \in \mathbf{K}), \tag{6.11a}$$

where S_k is given by

$$\begin{aligned} S_k \;&=\; Q_k + A'_k S_{k+1} A_k + A'_k S_{k+1} B_k^2 [I - B_k^{2'} S_{k+1} B_k^2]^{-1} B_k^{2'} S_{k+1} A_k; \\ &\qquad S_{K+1} = Q_{K+1}. \end{aligned} \tag{6.11b}$$

Proof. Since $L(u^1, u^2)$ is a quadratic function of u^2, the requirement of strict concavity is equivalent to existence of a unique solution to the optimal control problem

$$\min_{u^2 \in \mathrm{R}^{Km_2}} [-L(u^1, u^2)]$$

subject to (6.10a) and for each $u^1 \in \mathrm{R}^{Km_1}$. Furthermore, since the Hessian matrix (matrix of second partials) of L with respect of u^2 is independent of u^1 as well as of c, we can instead consider the optimal control problem

$$\min_{u^2 \in \mathrm{R}^{Km_2}} \sum_{k \in \mathbf{K}} (u_k^{2'} u_k^2 - x'_{k+1} Q_{k+1} x_{k+1})$$

subject to

$$x_{k+1} = A_k x_k + B_k^2 u_k^2;$$

that is, we can take $u_k^1 \equiv 0, c_k \equiv 0$ without any loss of generality. Then, the result follows from the dynamic programming technique outlined in subsection 5.5.1, and in particular from (5.19), which admits the unique solution $V(k,x) = -x'S_k x + x'Q_k x$, if, and only if, (6.11a) holds. □

We are now in a position to present the open-loop saddle-point solution of the affine-quadratic two-person zero-sum dynamic game.

Theorem 6.4 *For the two-person affine-quadratic zero-sum dynamic game described by (6.10a)-(6.10b) and with $Q_{k+1} \geq 0\,(k \in \mathbf{K})$, let condition (6.11a) be satisfied and Λ_k, M_k ($k \in \mathbf{K}$) be matrices of appropriate dimensions, defined through*

$$\Lambda_k = [I + (B_k^1 B_k^{1'} - B_k^2 B_k^{2'})M_{k+1}], \tag{6.12a}$$

$$M_k = Q_k + A_k' M_{k+1} \Lambda_k^{-1} A_k; \quad M_{K+1} = Q_{K+1}. \tag{6.12b}$$

Then,

(i) the matrices Λ_k ($k \in \mathbf{K}$), thus recursively defined, are invertible,

(ii) the game admits a unique open-loop saddle-point solution *given by*

$$\gamma_k^{1*}(x_1) \equiv u_k^{1*} = -B_k^{1'}[M_{k+1}\Lambda_k^{-1}A_k x_k^* + \xi_k], \tag{6.13a}$$

$$\gamma_k^{2*}(x_1) \equiv u_k^{2*} = B_k^{2'}[M_{k+1}\Lambda_k^{-1}A_k x_k^* + \xi_k], \quad k \in \mathbf{K}, \tag{6.13b}$$

where $\{x_{k+1}^*; k \in \mathbf{K}\}$ is the corresponding state trajectory determined from

$$x_{k+1}^* = \Lambda_k^{-1}[A_k x_k^* + \eta_k], \; x_1^* = x_1, \tag{6.13c}$$

and ξ_k, η_k are given by

$$\xi_k = M_{k+1}\Lambda_k^{-1}\eta_k + m_{k+1}, \tag{6.13d}$$

$$\eta_k = c_k - (B_k^1 B_k^{1'} - B_k^2 B_k^{2'})m_{k+1}, \tag{6.13e}$$

with m_k generated by the difference equation

$$m_k = A_k'[(\Lambda_k^{-1})' m_{k+1} + M_{k+1}\Lambda_k^{-1}c_k], \quad m_{K+1} = 0. \tag{6.13f}$$

Proof. Let us first note that, with the open-loop information structure and under condition (6.11a) (cf. Lemma 6.1), the game is a static strictly convex-concave quadratic zero-sum game which admits a unique saddle-point solution by Corollary 4.5. Second, it follows from Thm. 6.3 that this unique saddle-point solution should satisfy relations (6.9a)-(6.9d), which can be rewritten, for the affine-quadratic game, as follows:

$$x_{k+1}^* = A_k x_k^* + B_k^1 u_k^{1*} + B_k^2 u_k^{2*} + c_k,\ x_1^* = x_1, \tag{i}$$

$$H_k(p_{k+1}, u_k^{1*}, u_k^2, x_k^*) \leq H_k(p_{k+1}, u_k^{1*}, u_k^{2*}, x_k^*) \leq H_k(p_{k+1}, u_k^1, u_k^{2*}, x_k^*) \quad \forall u^1 \in \mathbf{R}^{Km_1}, u^2 \in \mathbf{R}^{Km_2}, \tag{ii}$$

$$p_k = A_k'[p_{k+1} + Q_{k+1}x_{k+1}^*], \quad p_{K+1} = 0, \tag{iii}$$

$$H_k(p_{k+1}, u_k^1, u_k^2, x_k) \triangleq \frac{1}{2}[|A_k x_k + B_k^1 u_k^1 + B_k^2 u_k^2 + c_k|_{Q_{k+1}}^2 + |u_k^1|^2 - |u_k^2|^2] + p_{k+1}'(A_k x_k + B_k^1 u_k^1 + B_k^2 u_k^2 + c_k), \quad k \in \mathbf{K}.$$

An inductive argument, as in the proof of Thm. 6.2, shows that this set of equations admits the solution:

$$u_k^{i*} = (-1)^i B_k^{i'}[M_{k+1}x_{k+1}^* + m_{k+1}],\ k \in \mathbf{K}, i = 1, 2, \tag{iv}$$

$$p_k = A_k'[M_{k+1}x_{k+1}^* + m_{k+1}],\ k \in \mathbf{K},$$

where the corresponding value of the state vector, x_k^*, satisfies

$$\Lambda_k x_{k+1}^* = A_k x_k^* + \eta_k.$$

Since there exists a unique saddle-point solution, there necessarily exists a unique relationship between x_k^* and x_{k+1}^*, implying that the matrix Λ_k should be invertible for each $k \in \mathbf{K}$. Hence,

$$x_{k+1}^* = \Lambda_k^{-1}[A_k x_k^* + \eta_k], \quad k \in \mathbf{K},$$

which verifies (6.13c). When this is used in (iv), it leads to (6.13a)-(6.13b), also in view of (6.13d)-(6.13f), thus completing the proof of the theorem. □

Remark 6.2 The statement of Thm. 6.4 is valid even if the "nonnegative definiteness" condition on Q_{k+1} does not hold, provided that some other appropriate conditions are imposed on the game parameters to ensure that L defined by (6.10b) is strictly convex in u^1. One such set of conditions, which are in fact tight, can be obtained directly from Lemma 6.1, by simply replacing Q_k by $-Q_k$, and B^2 by B^1, that is,

$$I + B_k^{1'} S_{k+1}^1 B_k^1 > 0 \quad (k \in \mathbf{K}), \tag{6.14a}$$

where S_k^1 is generated by

$$\begin{aligned} S_k^1 &= Q_k + A_k' S_{k+1}^1 A_k - A_k' S_{k+1}^1 B_k^1 [I + B_k^{1'} S_{k+1}^1 B_k^1]^{-1} B_k^{1'} S_{k+1}^1 A_k; \\ S_{K+1}^1 &= Q_{K+1}. \end{aligned} \tag{6.14b}$$

□

6.2.2 Closed-loop no-memory and feedback Nash equilibria

We now turn our attention to a discussion and derivation of Nash equilibria under the memoryless perfect state information pattern which provides the players with only the current value of the state at every stage of the game (and of course also the initial value of the state, which is known *a priori*). For this class of dynamic games we first obtain the following counterpart of Thm. 6.1, which provides a set of necessary conditions for any closed-loop no-memory Nash equilibrium solution to satisfy:

Theorem 6.5 *For an N-person discrete-time infinite dynamic game, let*

(i) $f_k(\cdot)$ be continuously differentiable on $\mathrm{R}^n \times U_k^1 \times \cdots \times U_k^N$, $(k \in \mathbf{K})$,

(ii) $g_k(\cdot)$ be continuously differentiable on $\mathrm{R}^n \times U_k^1 \times \cdots \times U_k^N \times \mathrm{R}^n$, $(k \in \mathbf{K}, i \in \mathbf{N})$.

Then, if $\{\gamma_k^{i*}(x_k, x_1) = u_k^{i*};\ k \in \mathbf{K}, i \in \mathbf{N}\}$ provides a closed-loop no-memory Nash equilibrium solution such that $\gamma_k^{i*}(\cdot, x_1)$ is continuously differentiable on R^n for all $k \in \mathbf{K}$, $i \in \mathbf{N}$, $f_k\ (x_k, u^1, \ldots, \gamma_k^{i*}(x_k, x_1), u^{i+1}, \ldots, u^N)$ is convex in $(x_k, u^1, \ldots, u^{i-1}, u^{i+1}, \ldots, u^N)$ for every $i \in \mathbf{N}$, $k \in \mathbf{K}$, and if $\{x_{k+1}^*; k \in \mathbf{K}\}$ is the corresponding state trajectory, there exists a finite sequence of n-dimensional (costate) vectors $\{p_2^i, \ldots, p_{K+1}^i\}$, for each $i \in \mathbf{N}$ such that the following relations are satisfied:

$$x_{k+1}^* = f_k(x_k^*, u_k^{1*}, \ldots, u_k^{N*}), \quad x_1^* = x_1, \tag{6.15a}$$

$$\begin{aligned}\gamma_k^{i*}(x_k^*, x_1) \ &\equiv \ u_k^{i*} = \arg \min_{u_k^i \in U_k^i} H_k^i(p_{k+1}^i, u_k^{1*}, \ldots, u_k^{i-1*}, u_k^i, \\ &\qquad\qquad u_k^{i+1*}, \ldots, u_k^{N*}, x_k^*),\end{aligned} \tag{6.15b}$$

$$\begin{aligned}p_k^i \ =\ & \frac{\partial}{\partial x_k} f_k(*)' \left[p_{k+1}^i + \left(\frac{\partial}{\partial x_{k+1}} g_k^i(*)\right)'\right] + \left[\frac{\partial}{\partial x_k} g_k^i(*)\right]' \\ & + \sum_{\substack{j \in \mathbf{N} \\ j \neq i}} \left[\frac{\partial}{\partial x_k}\gamma_k^j(x_k^*, x_1)\right]' \left[\frac{\partial}{\partial u_k^j} f_k(*)\right]' \left[p_{k+1}^i + \left(\frac{\partial}{\partial x_{k+1}} g_k^i(*)\right)'\right] \\ & + \sum_{\substack{j \in \mathbf{N} \\ j \neq i}} \left[\frac{\partial}{\partial x_k}\gamma_k^j(x_k^*, x_1)\right]' \left[\frac{\partial}{\partial u_k^j} g_k^i(*)\right]'; p_{K+1}^i = 0, i \in \mathbf{N}, k \in \mathbf{K},\end{aligned} \tag{6.15c}$$

where

$$\begin{aligned}H_k^i(p_{k+1}, u_k^1, \ldots, u_k^N, x_k) \ \triangleq\ & g_k^i(f_k(x_k, u_k^1, \ldots, u_k^N), u_k^1, \ldots, u_k^N, x_k) \\ & + p_{k+1}^{i'} f_k(x_k, u_k^1, \ldots, u_k^N),\ k \in \mathbf{K}, i \in \mathbf{N},\end{aligned} \tag{6.15d}$$

$$\frac{\partial}{\partial x_k} f_k(*) \triangleq \frac{\partial}{\partial x_k} f_k(x_k, u_k^{1*}, \ldots, u_k^{N*}) \mid_{x_k = x_k^*},$$

and a similar convention applies to

$$\frac{\partial}{\partial x_k} g_k^i(*) \text{ and } \frac{\partial}{\partial u_k^j} f_k(*).$$

Proof. The proof is similar to that of Thm. 6.1, but here we also have to take into account the possible dependence of u_k^{i*} on x_k ($k \in \mathbf{K}, i \in \mathbf{N}$). Accordingly, the ith inequality of (6.3) now says that $\{\gamma_k^{i*}(x_k, x_1); k \in \mathbf{K}\}$ minimizes the function

$$J^i(\gamma^{1*}, \ldots, \gamma^{i-1*}, \gamma^i, \gamma^{i+1*}, \ldots, \gamma^{N*}) = \sum_{k \in \mathbf{K}} g_k^i(x_{k+1}, \gamma_k^{1*}(x_k, x_1),$$
$$\ldots, \gamma_k^{i-1*}, (x_k, x_1), \gamma_k^i(x_k, x_1), \gamma_k^{i+1*}(x_k, x_1), \ldots, \gamma^{N*}(x_k, x_1), x_k)$$
$$\triangleq \sum_{k \in \mathbf{K}} \bar{g}_k^i(x_{k+1}, \gamma_k^i(x_k, x_1), x_k)$$

over Γ^i subject to the state equation constraint

$$x_{k+1} = f_k(x_k, \gamma_k^{1*}(x_k, x_1), \ldots, \gamma_k^{i-1*}(x_k, x_1), \gamma_k^i(x_k, x_1), \gamma_k^{i+1*}(x_k, x_1),$$
$$\ldots, \gamma_k^{N*}(x_k, x_1)) \triangleq \bar{f}_k(x_k, \gamma_k^i(x_k, x_1)), \quad k \in \mathbf{K}.$$

Then, the result follows directly from Thm. 5.5 by taking f_k as $\bar{f}_k$ and g_k as $\bar{g}_k^1$ which are continuously differentiable in their relevant arguments because of the hypothesis that the equilibrium strategies are continuously differentiable in their arguments. □

If the set of relations (6.15a)-(6.15d) is compared with (6.4a)-(6.4d), it will be seen that they are identical, except for the costate equations—the latter equation (6.15c) having two additional terms which are due to the dynamic nature of the information structure, allowing the players to utilize current values of the state. Furthermore, it is important to notice that every solution of the set (6.4a)-(6.4c) also satisfies relations (6.15a)-(6.15c), since every such solution is associated with static information, thereby causing the last two terms in (6.15c) to drop out. But, the "open-loop" solution is not the only one that the set (6.15a)-(6.15c) admits; surely it could have other solutions which explicitly depend on the current value of the state, thus leading to nonunique Nash equilibria, all of which are *weakly time consistent* (à la Def. 5.13). This phenomenon of multiplicity of Nash equilibria is closely related to the "informational nonuniqueness" feature of Nash equilibria under dynamic information, as introduced in Section 3.5 (see, in particular, Prop. 3.10 and Remark 3.15), whose counterpart in infinite games will be thoroughly discussed in the next section. Here, we deal with a more restrictive class of Nash equilibrium solutions under memoryless perfect state information pattern—the so-called *feedback* Nash equilibrium—which is devoid of any "informational nonuniqueness" (see also Def. 3.22) and is also strongly time consistent (cf. Def. 5.14).

Definition 6.2 *For an N-person K-stage discrete-time infinite dynamic game with memoryless perfect state information pattern,*[52] *let J^i denote the cost functional of* $\mathbf{P}i$ *($i \in \mathbf{N}$) defined on $\Gamma^1 \times \cdots \times \Gamma^N$. An N-tuple of strategies $\{\gamma^{i*} \in \Gamma^i; i \in \mathbf{N}\}$ constitutes a* feedback Nash equilibrium solution *if it satisfies the set of K N-tuple inequalities (3.28) for all $\gamma_k^i \in \Gamma_k^i$, $i \in \mathbf{N}, k \in \mathbf{K}$.*

Proposition 6.1 *Under the memoryless perfect state information pattern, every feedback Nash equilibrium solution of an N-person discrete-time infinite dynamic game is a closed-loop no-memory Nash equilibrium solution (but not vice versa).*

Proof. This result is the counterpart of Prop. 3.9 in the present framework and therefore its proof parallels that of Prop. 3.9. The result basically follows by showing that, for each $i \in \mathbf{N}$, the collection of all the ith inequalities of the K N-tuples imply the ith inequality of (6.3). □

Definition of the feedback Nash equilibrium solution directly leads to a recursive derivation which involves solutions of static N-person nonzero-sum games at every stage of the dynamic game. Again, a direct consequence of Def. 6.2 is that the feedback equilibrium solution depends only on x_k at stage k, and dependence on x_1 is only at stage $k = 1$.[53] By utilizing these properties, we readily arrive at the following theorem.

Theorem 6.6 *For an N-person discrete-time infinite dynamic game, the set of strategies $\{\gamma_k^{1*}(x_k); k \in \mathbf{K}, i \in \mathbf{N}\}$ provides a* feedback Nash equilibrium solution *if, and only if, there exist functions $V^i(k,\cdot): \mathbf{R}^n \to \mathbf{R}, k \in \mathbf{K}, i \in \mathbf{N}$, such that the following recursive relations are satisfied:*

$$\begin{aligned} V^i(k,x) &= \min_{u_k^i \in U_k^i} [g_k^i(\tilde{f}_k^{i*}(x,u_k^i), \gamma_k^{1*}(x), \ldots, \gamma_k^{i-1*}(x), u_k^i, \gamma_k^{i+1*}(x), \\ &\qquad \ldots, \gamma_k^{N*}(x), x) + V^i(k+1, \tilde{f}_k^{i*}(x,u_k^i))] \\ &= g_k^i(\tilde{f}_k^{i*}(x,\gamma_k^{i*}(x)), \gamma_k^{1*}(x), \ldots, \gamma_k^{N*}(x), x) \\ &\quad + V^i(k+1, \tilde{f}_k^{i*}(x,\gamma_k^{i*}(x))); \quad V^i(K+1,x) = 0, \quad i \in \mathbf{N}, \end{aligned} \tag{6.16}$$

where

$$\tilde{f}_k^{i*}(x,u_k^i) \triangleq f_k(x, \gamma_k^{1*}(x), \ldots, \gamma_k^{i-1*}(x), u_k^i, \gamma_k^{i+1*}(x), \ldots, \gamma_k^{N*}(x)).$$

Every such equilibrium solution is strongly time consistent, *and the corresponding Nash equilibrium cost for* $\mathbf{P}i$ *is $V^i(1,x_1)$.*

Proof. Let us start with the first set of N inequalities of (3.28). Since they have to hold true for all $\gamma_k^i \in \Gamma_k^i$, $i \in \mathbf{N}$, $k \leq K-1$, this necessarily implies

[52] The statement of this definition remains valid if the information pattern is instead "closed-loop perfect state".

[53] This statement is valid also under the "closed-loop perfect state" information pattern. Note that the feedback equilibrium solution retains its equilibrium property also under the feedback information pattern.

that they have to hold true for all values of state x_k which are reachable by utilization of some combination of these strategies. Let us denote that subset of $\mathbf{R}^n$ by X_K. Then, the first set of inequalities of (3.28) becomes equivalent to the problem of seeking Nash equilibria of an N-person static game with cost functionals

$$g_K^i(f_K(x_K, u_K^1, \ldots, u_K^N), u_K^1, \ldots, u_K^N, x_K), \quad i \in \mathbf{N}, \tag{i}$$

which should be valid for all $x_K \in X_K$. This is precisely what (6.16) says for $k = K$, with a set of associated Nash equilibrium controls denoted by $\{\gamma_K^{i*}(x_K); i \in \mathbf{N}\}$ since they depend explicitly on $x_K \in X_K$, but not on the past values of the state (including the initial state x_1). Now, with these strategies substituted into (i), a similar argument (as above) leads to the conclusion that the second set of inequalities of (3.28) defines a static Nash game with cost functionals

$$V^i(K, x_K) + g_{K-1}^i(x_K, u_{K-1}^1, \ldots, u_{K-1}^N, x_{K-1}), \quad i \in \mathbf{N},$$

where

$$x_K = f_{K-1}(x_{K-1}, u_{K-1}^1, \ldots, u_{K-1}^N),$$

and the Nash solution has to be valid for all $x_{K-1} \in X_{K-1}$ (where X_{K-1} is the counterpart of X_K at stage $k = K - 1$). Here again, we observe that the Nash equilibrium controls can only be functions of x_{K-1}, and (6.16) with $k = K - 1$ provides a set of necessary and sufficient conditions for $\{\gamma_{K-1}^{i*}(x_{K-1}): i \in \mathbf{N}\}$ to solve this static Nash game. The theorem then follows from a standard induction argument. Note that the "strong time consistency" property of the feedback Nash equilibrium, and the expression for the corresponding cost for each player, are direct consequences of the recursive nature of the construction of the solution. □

The following corollary, which is the counterpart of Thm. 6.2 in the case of feedback Nash equilibrium, now follows as a special case of Thm. 6.6.

Preliminary Notation for Corollary 6.1. Let P_k^i ($i \in \mathbf{N}, k \in \mathbf{K}$) be matrices of appropriate dimensions, satisfying the set of linear matrix equations

$$[R_k^{ii} + B_k^{i\prime} Z_{k+1}^i B_k^i] P_k^i + B_k^{i\prime} Z_{k+1}^i \sum_{\substack{j \in \mathbf{N} \\ j \neq i}} B_k^j P_k^j = B_k^{i\prime} Z_{k+1}^i A_k, \; i \in \mathbf{N}, \tag{6.17a}$$

where Z_k^i ($i \in \mathbf{N}$) are obtained recursively from

$$Z_k^i = F_k' Z_{k+1}^i F_k + \sum_{j \in \mathbf{N}} P_k^{j\prime} R_k^{ij} P_k^j + Q_k^i; \; Z_{K+1}^i = Q_{K+1}^i, \; i \in \mathbf{N}, \tag{6.17b}$$

and

$$F_k \triangleq A_k - \sum_{i \in \mathbf{N}} B_k^i P_k^i, \quad k \in \mathbf{K}. \tag{6.17c}$$

Furthermore, let $\alpha_k^i \in \mathbf{R}^{m_i}$ $(i \in \mathbf{N}, k \in \mathbf{K})$ be vectors satisfying the set of linear equations:

$$[R_k^{ii}+B_k^{i'}Z_{k+1}^iB_k^i]\alpha_k^i+B_k^{i'}Z_{k+1}^i\sum_{\substack{j\in\mathbf{N}\\ j\neq i}}B_k^j\alpha_k^j = B_k^{i'}(\zeta_{k+1}^i+Z_{k+1}^ic_k),\ i\in\mathbf{N},\quad(6.17\text{d})$$

where ζ_k^i $(i \in \mathbf{N})$ are obtained recursively from

$$\zeta_k^i = F_k'(\zeta_{k+1}^i + Z_{k+1}^i\beta_k) + \sum_{j\in\mathbf{N}} P_k^{j'} R_k^{ij}\alpha_k^j;\ \zeta_{K+1}^i = 0, i \in \mathbf{N}, \qquad (6.17\text{e})$$

and

$$\beta_k \triangleq c_k - \sum_{j\in\mathbf{N}} B_k^{j'}\alpha_k^j, \quad k \in \mathbf{K}. \qquad (6.17\text{f})$$

Finally, let $n_k^i \in \mathbf{R}$ $(i \in \mathbf{N},\ k \in \mathbf{K})$ be generated by

$$n_k^i = n_{k+1}^i + \frac{1}{2}|\beta_k|^2_{Z_{k+1}^i} + \zeta_{k+1}^{i'}\beta_k + \frac{1}{2}\sum_{j\in\mathbf{N}}|\alpha_k^j|^2_{R_k^{ij}},\ n_{K+1}^i = 0. \qquad (6.17\text{g})$$

Corollary 6.1 *An N-person affine-quadratic dynamic game (cf. Def. 6.1) with $Q_{k+1}^i \geq 0$ $(i \in \mathbf{N}, k \in \mathbf{K})$ and $R_k^{ij} \geq 0$ $(i,j \in \mathbf{N}, j \neq i, k \in \mathbf{K})$ admits a* unique feedback Nash equilibrium solution *if, and only if, (6.17a) and (6.17d) admit unique solution sets $\{P_k^{i*}; i \in \mathbf{N}, k \in \mathbf{K}\}$ and $\{\alpha_k^{i*}; i \in \mathbf{N}, k \in \mathbf{K}\}$, respectively, in which case the equilibrium strategies are given by*

$$\gamma_k^{i*}(x_k) = -P_k^{i*}x_k - \alpha_k^{i*} \qquad (k \in \mathbf{K}, i \in \mathbf{N}), \qquad (6.18\text{a})$$

and the corresponding feedback Nash equilibrium cost for each player is

$$J^i(\gamma^{1*},\ldots,\gamma^{N*}) = V^i(1,x_1) = \frac{1}{2}|x_1|^2_{Z_1^i} + \zeta_1^{i'}x_1 + n_1^i, \quad (i \in \mathbf{N}). \qquad (6.18\text{b})$$

Proof. Starting with $k = K$ in the recursive equation (6.16), we first note that the functional to be minimized (for each $i \in \mathbf{N}$) is strictly convex, since $R_K^{ii} + B_K^{i'}Q_{K+1}^iB_K^i > 0$. Then, the first-order necessary conditions for minimization are also sufficient and therefore we have (by differentiation) the unique set of equations

$$\begin{aligned}-[R_K^{ii} + B_K^{i'}Q_{K+1}^iB_K^i]\gamma_K^{i*}(x_K) - B_K^{i'}Q_{K+1}^i\sum_{\substack{j\in\mathbf{N}\\ j\neq i}}B_K^j\gamma_K^{j*}(x_k)&\\ = B_K^{i'}Q_{K+1}^i[A_Kx_K + c_K]; \qquad & i \in \mathbf{N},\end{aligned}$$

which readily leads to the conclusion that any set of Nash equilibrium strategies at stage $k = K$ has to be affine in x_K. Therefore, by substituting $\gamma_K^{i*} = -P_K^ix_K - \alpha_K^i$ $(i \in \mathbf{N})$ into the foregoing equation, and by requiring it to be satisfied for all possible x_K, we arrive at (6.17a) and (6.17d) for $k = K$. Further substitution of this solution into (6.16) for $k = K$ leads to $V^i(K,x) = \frac{1}{2}x'(Z^i_K - Q^i_K)x + \zeta^{i'}_K x + n_K$; that is, $V^i(K,\cdot)$ has a quadratic structure at stage $k = K$. Now, if this expression is substituted into (6.16) with $k = K-1$, and the outlined procedure is carried out for $k = K-1$, and this so (recursively) for all $k \le K-1$, one arrives at the conclusions that

(i) $V^i(k,x) = \frac{1}{2}x'(Z^i_k - Q^i_k)x + \zeta^{i'}_k x + n_k$ is the unique solution of the recursive equation (6.16) under the hypothesis of the corollary and by noting that $Z^i_k \ge 0$ $(i \in \mathbf{N}, k \in \mathbf{K})$, and

(ii) the minimization operation in (6.16) leads to the unique solution (6.18a) under the condition of unique solvability of (6.17a) and (6.17d).

The expression for the cost, (6.18b), follows directly from the expression derived for the "cost-to-go" $V^i(k,x)$. This, then, completes verification of Corollary 6.1. □

Remark 6.3 The result of Corollary 6.1 as well as the verification given above extends readily to more general affine-quadratic dynamic games where the cost functions of the players contain additional terms that are linear in x_k, that is, with g^i in (6.5b) replaced by

$$g^i_k(x_{k+1}, u^1_k, \ldots, u^N_k, x_k) = \frac{1}{2}(x'_{k+1}[Q^i_{k+1}x_{k+1} + 2l^i_{k+1}] + \sum_{j\in\mathbf{N}} u^{j'}_k R^{ij}_k u^j_k),$$

where l^i_{k+1} $(k \in \mathbf{K})$ is a known sequence of n-dimensional vectors for each $i \in \mathbf{N}$. Then, the statement of Corollary 6.1 remains intact, with only the equation (6.17e) that generates ζ^i_k now reading

$$\zeta^i_k = F'_k(\zeta^i_{k+1} + Z^i_{k+1}\beta_k) + \sum_{j\in\mathbf{N}} P^{j'}_k R^{ij}_k \alpha^j_k + l_k;\ \zeta^i_{K+1} = l_{K+1}, i \in \mathbf{N},$$

and the cost-to-go functions admitting the compatibly modified form

$$V^i(k,x) = \frac{1}{2}x'(Z^i_k - Q^i_k)x + (\zeta^i_k - l^i_k)'x + n_k, \quad i \in \mathbf{N}.$$

□

Remark 6.4 The "nonnegative-definiteness" requirements imposed on Q_{k+1} and R^{ij}_k $(i,j \in \mathbf{N}, j \ne i; k \in \mathbf{K})$ are sufficient for the functionals to be minimized in (6.16) to be strictly convex, but they are by no means necessary. A set of less stringent (but more indirect) conditions would be

$$R^{ii}_k + B^{i'}_k Z^i_{k+1} B^i_k > 0 \quad (i \in \mathbf{N}, k \in \mathbf{K}),$$

under which the statement of Corollary 6.1 still remains valid. Furthermore, it follows from the proof of Corollary 6.1 that, if (6.17a) admits more than one set of solutions, every such set constitutes a feedback Nash equilibrium solution, which is also strongly time consistent. □

Remark 6.5 It is possible to give a precise condition for the unique solvability of the sets of equations (6.17a) and (6.17d) for P_k^i and α_k^i ($i \in \mathbf{N}, k \in \mathbf{K}$), respectively. The said condition (which is the same for both) is the invertibility of matrices Φ_k, $k \in \mathbf{K}$, which are composed of block matrices, with the iith block given as $R_k^{ii} + B_k^{i'} Z_{k+1}^i B_k^i$ and the ijth block as $B_k^{i'} Z_{k+1}^i B_k^j$, where $i, j \in \mathbf{N}$, $j \neq i$. □

Zero-sum dynamic games

We now consider the general class of two-person discrete-time zero-sum dynamic games and determine the "cost-to-go" equation associated with the feedback saddle-point solution, as a special case of Thm. 6.6.

Corollary 6.2 *For a two-person discrete-time zero-sum dynamic game, the set of strategies* $\{\gamma_k^{i*}(x_k);\ k \in \mathbf{K}, i = 1, 2\}$ *provides a feedback saddle-point solution if, and only if, there exist functions* $V(k, \cdot) : \mathbf{R}^n \to \mathbf{R}$, $k \in \mathbf{K}$, *such that the following recursive relation is satisfied:*

$$
\begin{aligned}
V(k,x) &= \min_{u_k^1 \in U_k^1} \max_{u_k^2 \in U_k^2} [g_k(f_k(x,u_k^1,u_k^2),u_k^1,u_k^2,x) \\
&\qquad + V(k+1, f_k(x,u_k^1,u_k^2))] \\
&= \max_{u_k^2 \in U_k^2} \min_{u_k^1 \in U_k^1} [g_k(f_k(x,u_k^1,u_k^2),u_k^1,u_k^2,x) \\
&\qquad + V(k+1, f_k(x,u_k^1,u_k^2))] \\
&= g_k(f_k(x,\gamma_k^{1*}(x),\gamma_k^{2*}(x)),\gamma_k^{1*}(x),\gamma_k^{2*}(x),x) \\
&\qquad + V(k+1, f_k(x,\gamma_k^{1*}(x),\gamma_k^{2*}(x))); \\
V(K+1,x) &= 0.
\end{aligned} \tag{6.19}
$$

Every such saddle-point solution is strongly time consistent, and the unique saddle-point value of the game is $V(1, x_1)$.

Proof. The recursive equation (6.19) follows from (6.16) by taking $N = 2$, $g_k^1 \equiv -g_k^2 \triangleq g_k$ ($k \in \mathbf{K}$), since then $V^1 \equiv -V^2 \triangleq V$ and existence of a saddle point is equivalent to interchangeability of the min-max operations. □

For the further special case of an affine-quadratic zero-sum game, the solution of (6.19) as well as the feedback saddle-point solution can be explicitly determined in a simple structural form, but after a rather lengthy derivation. We accomplish this in two steps: first we obtain directly the special case of Corollary 6.1 for the affine-quadratic zero-sum game (see Corollary 6.3 below), and second we simplify these expressions further so as to bring them into a form compatible with the results of Thm. 6.4 (see Thm. 6.7 in the sequel).

Corollary 6.3 *For the two-person affine-quadratic zero-sum dynamic game described by (6.10a)-(6.10b), the unique solutions of (6.17a) and (6.17d) are given by*

$$P_k^1 = [I + K_k^1 Z_{k+1} B_k^1]^{-1} K_k^1 Z_{k+1} A_k, \tag{6.20a}$$

$$P_k^2 - -[I - K_k^2 Z_{k+1} B_k^2]^{-1} K_k^2 Z_{k+1} A_k, \tag{6.20b}$$

$$\alpha_k^1 = [I + K_k^1 Z_{k+1} B_k^1]^{-1} K_k^1 (\zeta_{k+1} + Z_{k+1} c_k), \tag{6.20c}$$

$$\alpha_k^2 = -[I - K_k^2 Z_{k+1} B_k^2]^{-1} K_k^2 (\zeta_{k+1} + Z_{k+1} c_k), \tag{6.20d}$$

where

$$K_k^1 \triangleq B_k^{1'} [I + Z_{k+1} B_k^2 (I - B_k^{2'} Z_{k+1} B_k^2)^{-1} B_k^{2'}], \tag{6.21a}$$

$$K_k^2 \triangleq B_k^{2'} [I - Z_{k+1} B_k^1 (I + B_k^{1'} Z_{k+1} B_k^1)^{-1} B_k^{1'}], \tag{6.21b}$$

and $Z_k^1 = -Z_k^2 \triangleq Z_k$ *and* $\zeta_k^1 = -\zeta_k^2 \triangleq \zeta_k$ *satisfy the recursive equations*

$$Z_k = F_k' Z_{k+1} F_k + P_k^{1'} P_k^1 - P_k^{2'} P_k^2 + Q_k; \; Z_{K+1} = Q_{K+1}, \tag{6.22a}$$

$$\zeta_k = F_k' (\zeta_{k+1} + Z_{k+1} c_k); \; \zeta_{K+1} = 0 \tag{6.22b}$$

and

$$F_k \triangleq A_k - B_k^1 P_k^1 - B_k^2 P_k^2,$$

$$\beta_k \triangleq c_k - B_k^1 \alpha_k^1 - B_k^2 \alpha_k^2.$$

Furthermore, the set of conditions of Remark 6.4 is equivalent to

$$I + B_k^{1'} Z_{k+1} B_k^1 > 0 \quad (k \in \mathbf{K}), \tag{6.23a}$$

$$I - B_k^{2'} Z_{k+1} B_k^2 > 0 \quad (k \in \mathbf{K}). \tag{6.23b}$$

Proof. By letting $N = 2$, $Q_{k+1}^1 = -Q_{k+1}^2 \triangleq Q_{k+1}$, $R_k^{11} = -R_k^{21} = I$, $R_k^{22} = -R_k^{12} = I$ in Corollary 6.1, we first observe that $Z_k^1 \equiv -Z_k^2$ and the equations for Z_k^1 and ζ_k^1 are the same as (6.22a) and (6.22b),[54] respectively, assuming of course that P_k^i's and α_k^i's in the these equations are correspondingly the same—a property which we now verify. Toward this end, start with (6.17a) when $i = 2$, solve for P_k^2 terms of P_k^1 (this solution is unique under (6.23b)), substitute this into (6.17a) when $i = 1$ and solve for P_k^1 from this linear matrix equation. The result is (6.20a), assuming this time that the matrix $[I + K_k^1 Z_{k+1} B_k^1]$ is invertible. Furthermore, by following the same procedure with indices 1 and 2 interchanged, we arrive at (6.20b), on account of the invertibility of $[I - K_k^2 Z_{k+1} B_k^2]$. Repeating the same steps for α_k^1 and α_k^2, this time by working with (6.17d), we arrive at (6.20c)-(6.20d), again under the same matrix invertibility

[54]Verification of (6.22b) requires some algebraic manipulations.

conditions. Therefore, for this special case, P_k^1 and P_k^2 given by (6.20a)-(6.20b), and α_k^1 and α_k^2 given by (6.20c)-(6.20d), are indeed the unique solutions of (6.17a) and (6.17d), respectively, provided that the two matrices in question are invertible. A direct manipulation on these matrices actually establishes nonsingularity under conditions (6.23a)-(6.23b); however, we choose here to accomplish this by employing an indirect method which is more illuminating.

First note that (6.23a) and (6.23b) correspond to the existence conditions given in Remark 6.4, and therefore they make the functionals to be minimized in (6.16) strictly convex in the relevant control variables. But, for the zero-sum game, (6.16) is equivalent to (6.19), and consequently (6.23a) and (6.23b) make the kernel in (6.19) strictly convex in u_k^1 and strictly concave in u_k^2 $(k \in \mathbf{K})$. Since every strictly convex-concave quadratic static game admits a (unique) saddle point (cf. Corollary 4.5), the sets (6.17a) and (6.17d) have to admit unique solutions for the specific problem (zero-sum game) under consideration. Hence, the required inverses in (6.21a) and (6.21b) should exist. This, then, completes the proof of Corollary 6.3. □

To further simplify (6.20a)-(6.22b), we now make use of the following matrix identity.

Lemma 6.2 *Let $Z = Z'$ and B be matrices of dimensions $(n \times n)$ and $(n \times m)$, respectively, and with the further property that $B'ZB$ does not have any unity eigenvalues. Then,*

$$[I_n + ZB(I_m - B'ZB)^{-1}B'] \equiv (I_n - ZBB')^{-1}. \tag{6.24}$$

Proof. First note that the matrix inverse on the RHS of this identity exists, under the hypothesis of the lemma, since nonzero eigenvalues of $B'ZB$ and ZBB' are the same (see Marcus and Minc, 1964, p. 24). Then, the result follows by multiplying both sides of the identity by $I_n - ZBB'$ from the right and by straightforward manipulations. □

Application of this identity to (6.21a) and (6.21b) readily yields (by identifying B_k^2 with B and Z_{k+1} with Z in the former, and B_k^1 with B and $-Z_{k+1}$ with Z in the latter)[55]

$$\begin{aligned} K_k^1 &= B_k^{1'}[I - Z_{k+1}B_k^2B_k^{2'}]^{-1}, \\ K_k^2 &= B_k^{2'}[I + Z_{k+1}B_k^1B_k^{1'}]^{-1}. \end{aligned}$$

Furthermore, if these are substituted into (6.20a)-(6.20d), some extensive, but straightforward, matrix manipulations which involve repeated application of (6.24) lead to the expressions

$$P_k^1 = B_k^{1'}Z_{k+1}[I + (B_k^1B_k^{1'} - B_k^2B_k^{2'})Z_{k+1}]^{-1}A_k, \tag{6.25a}$$

$$P_k^2 = -B_k^{2'}Z_{k+1}[I + (B_k^1B_k^{1'} - B_k^2B_k^{2'})Z_{k+1}]^{-1}A_k, \tag{6.25b}$$

[55]Here we suppress the dimensions of the identity matrices since they are clear from the context.

$$\alpha_k^1 = B_k^{1'}[I + Z_{k+1}(B_k^1 B_k^{1'} - B_k^2 B_k^{2'})]^{-1}(\zeta_{k+1} + Z_{k+1}c_k), \tag{6.25c}$$

$$\alpha_k^2 = -B_k^{2'}[I + Z_{k+1}(B_k^1 B_k^{1'} - B_k^2 B_k^{2'})]^{-1}(\zeta_{k+1} + Z_{k+1}c_k), \tag{6.25d}$$

and F_k, given in Corollary 6.3, is then expressed as

$$F_k = [I - (B_k^1 B_k^{1'} - B_k^2 B_k^{2'})Z_{k+1}]^{-1}A_k. \tag{6.25e}$$

Now, finally, if (6.25a)-(6.25b) are substituted into (6.22a), we arrive at the conclusion that Z_k satisfies the same equation as M_k, given by (6.12b), and hence $F_k = \Lambda_k^{-1}A_k$, and furthermore ζ_k generated by (6.22b) is identical with m_k generated by (6.13f). Therefore, we have the following.

Theorem 6.7 *The two-person affine-quadratic zero-sum dynamic game described by (6.10a)-(6.10b) admits a* unique feedback saddle-point solution *if, and only if,*

$$I + B_k^{1'} M_{k+1} B_k^1 > 0 \quad (k \in \mathbf{K}), \tag{6.26a}$$

$$I - B_k^{2'} M_{k+1} B_k^2 > 0 \quad (k \in \mathbf{K}), \tag{6.26b}$$

in which case the unique equilibrium strategies, which are also strongly time consistent, are given by

$$\gamma_k^{1*}(x_k) = -B_k^{1'} M_{k+1}\Lambda_k^{-1}[A_k x_k + c_k] - B_k^{1'}(\Lambda_k^{'})^{-1}\zeta_{k+1}, \tag{6.27a}$$

$$\gamma_k^{2*}(x_k) = B_k^{2'} M_{k+1}\Lambda_k^{-1}[A_k x_k + c_k] + B_k^{2'}(\Lambda_k^{'})^{-1}\zeta_{k+1}, \quad k \in \mathbf{K}, \tag{6.27b}$$

where

$$\zeta_k = (\Lambda_k^{-1}A_k)'(\zeta_{k+1} + M_{k+1}c_k), \quad \zeta_{K+1} = 0,$$

and M_{k+1}, Λ_k ($k \in \mathbf{K}$) are as defined in Thm. 6.4. The corresponding unique state trajectory $\{x_{k+1}^*; k \in \mathbf{K}\}$ satisfies the difference equation

$$x_{k+1}^* = \Lambda_k^{-1}[A_k x_k^* + \eta_k], \quad x_1^* = x_1, \tag{6.27c}$$

where η_k ($k \in \mathbf{K}$) is as defined in Thm. 6.4, with m_{k+1} replaced by the equivalent vector ζ_{k+1}. The corresponding saddle-point value is

$$J^* = J(\gamma^{1*}, \gamma^{2*}) = \frac{1}{2}x_1^{'} M_1 x_1 + \zeta_1^{'} x_1 + n_1.$$

Furthermore, if $Q_{k+1} \geq 0\ \forall k \in \mathbf{K}$, then $M_{k+1} \geq 0\ \forall k \in \mathbf{K}$, and hence condition (6.26a) becomes superfluous.

Proof. This result, with the exception of the last statement, follows from Corollary 6.1, Corollary 6.3 and the discussion given prior to the statement of the theorem. To prove the last statement, take $c_k \equiv 0$, without any loss of generality, and note that with $Q_{k+1} \geq 0$ the lower value of any stage-truncated (from below) version of the game is nonnegative (choose, e.g., $u_k^2 \equiv 0$), and hence the feedback saddle-point value of each such truncated game, which is $\frac{1}{2}x_k' M_k x_k$, is nonnegative for any arbitrary $x_k \in \mathbf{R}^n$. Hence, $M_k \geq 0$ for all $k \in \mathbf{K}$.[56] □

Remark 6.6 As in the case of Nash equilibria (cf. Remark 6.3), the result above extends naturally to the more general affine-quadratic zero-sum dynamic games which have in the cost function an additional linear term in x, to be denoted $x_{k+1}' l_{k+1}$. Then, the only modification will have to be made in the equation for ζ_k, which will now read

$$\zeta_k = (\Lambda_k^{-1} A_k)'(\zeta_{k+1} + M_{k+1} c_k) + l_k, \quad \zeta_{K+1} = l_{K+1}.$$

Otherwise, the statement of Thm. 6.7 remains intact. □

Remark 6.7 A comparison of Thms. 6.4 and 6.7 now readily reveals an important property of the saddle-point solution in such games; namely, *whenever they both exist, the unique open-loop saddle-point solution and the unique feedback saddle-point solution generate the same state trajectory* in affine-quadratic zero-sum games. Furthermore, the open-loop values (cf. Section 5.6) of the feedback saddle-point strategies (6.27a)-(6.27b) are correspondingly the same as the open-loop saddle-point strategies (6.13a)-(6.13b). These two features of the saddle-point solution under different information structures are in fact characteristic of not only affine-quadratic games, but of the most general zero-sum dynamic games treated in this section, as it will be verified in the next section (see, in particular, Thm. 6.9). In nonzero-sum dynamic games, however, the Nash equilibrium solutions obtained under different information structures do not exhibit such a feature, as it will also be clear from the discussion of the next section (see, in particular, subsection 6.3.1). □

Even though the two saddle-point solutions of Thms. 6.4 and 6.7 generate the same state trajectory, the existence conditions involved are not equivalent—a result that follows from a comparison of (6.11a) and (6.14b) with (6.26b) and (6.26a). This point is further elaborated on in the following proposition for the case $Q_{k+1} \geq 0$, under which (6.14b) and (6.26a) are automatically satisfied, which therefore leaves only (6.11a) and (6.26b) for comparison.

Proposition 6.2 *For the affine-quadratic two-person zero-sum dynamic game described by (6.10a)-(6.10b) and with $Q_{k+1} \geq 0 \ \forall k \in \mathbf{K}$, condition (6.11a) implies (6.26b), but not vice versa. In other words, every affine-quadratic two-person zero-sum dynamic game (with nonnegative cost on the state) that admits*

[56] It is possible to verify this property of M_k also by direct matrix manipulations; see Prop. 6.2 later for a hint in this direction.

a unique open-loop saddle point also admits a unique feedback saddle point, but existence of the latter does not necessarily imply existence of a saddle point in the open-loop strategies.

Proof. The proof of the implication from (6.11a) to (6.26b) is by induction (on k), where we have to show that for arbitrary $k \in \mathbf{K}$,

$$S_{k+1} \geq M_{k+1} \quad \Rightarrow \quad S_k \geq M_k.$$

To save from indices, let $S_{k+1} = S$, $M_{k+1} = M$, and likewise for B^1 and B^2, and start with the relation $S \geq M \geq 0$, where the nonnegative definiteness of M is a property that was already proven in Thm. 6.7. Consider the perturbed matrices

$$S_\epsilon \triangleq S + \epsilon I, \quad M_\epsilon \triangleq M + \epsilon I, \quad \epsilon > 0,$$

both of which are positive definite for each $\epsilon > 0$. The following sequence of implications ($\Rightarrow$) and equivalences ($\Leftrightarrow$) now follows from standard properties of matrices:[57]

$$\begin{array}{rrcl} & S_\epsilon & \geq & M_\epsilon \\ \Rightarrow & S_\epsilon^{-1} - B^2 B^{2'} & \geq & M_\epsilon^{-1} - B^2 B^{2'} + B^1 B^{1'} \\ \Leftrightarrow & S_\epsilon (I - B^2 B^{2'} S_\epsilon)^{-1} & \geq & M_\epsilon (I + (B^1 B^{1'} - B^2 B^{2'}) M_\epsilon)^{-1}. \end{array}$$

Since the last inequality holds $\forall \epsilon > 0$, and both sides are well defined as $\epsilon \downarrow 0$, we finally have (using the continuous dependence of eigenvalues of the matrices above on the parameter ϵ)

$$(I - B^2 B^{2'} S)^{-1} \geq M(I + (B^1 B^{1'} - B^2 B^{2'}) M)^{-1},$$

and pre- and post-multiplying both sides by A' and A and then adding Q yields the desired result $S_k \geq M_k$. This completes the proof by induction, since for $k = K + 1$ both matrices equal Q_{K+1}.

To verify that (6.26b) does not necessarily imply (6.11a), we simply produce a counter example. Consider the scalar two-stage game (i.e., $K = 2$) with $Q_{k+1} = B_k^1 = A_k = 1$, $B_k^2 = 1/\sqrt{2}$, $(k = 1, 2)$. Both (6.11a) and (6.26b) are satisfied for $k = 2$. For $k = 1$, however, $S_2 = 3$, $M_2 = 5/3$, and while (6.26b) is satisfied, the condition (6.11a) fails to hold true. □

Remark 6.8 The condition (6.26b) of Thm. 6.7 is quite tight for affine-quadratic dynamic games (with nonnegative cost on the state) under the CLPS information structure, in the sense that if the matrix in (6.26b) has any negative eigenvalues, then the upper value of the game becomes unbounded. This is

[57] For two positive definite matrices V and W, the inequality $V \geq W$ means that the matrix difference $V - W$ is nonnegative definite, which also implies that the difference of their inverses, $V^{-1} - W^{-1}$, is nonpositive definite. For a proof of this last inequality, see, for example, Bellman (1970).

easy to see when **P1** uses only the current value of the state, because then existence of a negative eigenvalue for the matrix in (6.26b) at some stage k would imply that the recursive derivation of the feedback saddle-point equilibrium solution encounters at stage k a static game that is not concave in the maximizing variable—making the upper value of that game unbounded (as in Remark 4.7). This argument can be extended also to the general CLPS information structure, implying that even if the minimizer is allowed to use also the past values of the state the upper value would still be unbounded when the matrix in (6.26b) has at least one negative eigenvalue.[58] The lower value of the game, however, could still be bounded, which means that if the minimizer is also allowed to have access to the maximizer's actions (controls), then condition (6.26b) can be further relaxed; this relaxed version of (6.26b) can in fact be obtained quite readily by allowing (in the recursive derivation of the feedback saddle-point equilibrium) the minimizer (**P1**) to choose u_k^1 dependent not only on x_k but also on u_k^2, and then requiring the resulting minimum value of the stagewise cost (in the truncated game) to be strictly concave in u_k^2.

If the information structure is CLPS, and the matrix in (6.26b) is nonnegative (but not positive) definite, then whether the upper value is bounded or not depends on the specific structures of the cost and system matrices, which we do not further discuss here. □

6.2.3 Linear-quadratic games with an infinite number of stages

We now consider stationary dynamic games with an infinite number of stages, and restrict attention to linear-quadratic structures, which means that in the formulation of Def. 6.1, all matrices are time invariant, $c \equiv 0$, and $K = \infty$ or $K \to \infty$. Feedback Nash equilibria of such games can be obtained in two different ways: as the limit of the feedback Nash solution of any time-truncated version (with, say, K stages) as the number of stages (K) goes to infinity, or from the outset as the Nash equilibrium of an infinite-horizon game under CLNM information pattern. Of course, a natural procedure to follow would be a combination of these two methods: find the limit of the finite-horizon solution as $K \to \infty$, and then verify that this limiting solution (if it exists) provides a Nash equilibrium solution for the infinite-horizon game. Toward this end, first note that a candidate solution can easily be obtained from Corollary 6.1, by simply taking $c \equiv 0$ and dropping the time indices from the various matrices. Let $Z_k^{i(K)}$ ($i \in \mathbf{N}$) denote the solution of (6.17b) when there are K stages (where we show explicit dependence on K, as in the one-player case discussed in subsection 5.5.1, Prop. 5.2), and let $\bar{Z}^i$ ($i \in \mathbf{N}$) be its limit as $K \to \infty$ for fixed k (assuming that this limit exists, and is independent of k). Likewise, let $\bar{P}^i$ ($i \in \mathbf{N}$) denote the limit of $P_k^{i(K)}$ ($i \in \mathbf{N}$) as $K \to \infty$. Then, these limiting

[58] See Başar (1991b) or Başar and Bernhard (1995) for details of this argument.

matrices necessarily satisfy the following two algebraic matrix equations

$$[R^{ii} + B^{i'}\bar{Z}^i B^i]\bar{P}^i + B^{i'}\bar{Z}^i \sum_{\substack{j\in\mathbf{N}\\ j\neq i}} B^j\bar{P}^j = B^{i'}\bar{Z}^i A,\ i \in \mathbf{N}, \tag{6.28a}$$

$$\bar{Z}^i = \bar{F}'\bar{Z}^i\bar{F} + \sum_{j\in\mathbf{N}} \bar{P}^{j'} R^{ij}\bar{P}^j + Q^i;\ i \in \mathbf{N}, \tag{6.28b}$$

where

$$\bar{F} \triangleq A - \sum_{i\in\mathbf{N}} B^i\bar{P}^i. \tag{6.28c}$$

The corresponding limiting feedback strategy for $\mathbf{P}i$ is then (from (6.18a))

$$\bar{\gamma}^{i*}(x_k) = -\bar{P}^i x_k \quad (i \in \mathbf{N},\ k = 1, 2, \ldots). \tag{6.29}$$

Conditions that will guarantee that such an N-tuple of stationary policies are in Nash equilibrium can be obtained readily from Prop. 5.2, by simply holding all but one (say, ith) players' strategies fixed at (6.29) and requiring that the resulting infinite-horizon optimal control problem be well defined, as explained in subsection 5.5.1. This result is given below in Prop. 6.3 after introducing some notation.

Let $\bar{F}_i$ and $\bar{Q}_i$ $(i \in \bar{N})$ be defined by

$$\bar{F}_i \triangleq A - \sum_{\substack{j\in\mathbf{N}\\ j\neq i}} B^j\bar{P}^j; \quad \bar{Q}_i \triangleq Q^i + \sum_{\substack{j\in\mathbf{N}\\ j\neq i}} \bar{P}^{j'} R^{ij}\bar{P}^j \quad (i \in \mathbf{N}).$$

Proposition 6.3 *Let there exist two N-tuples of matrices* $\{\bar{Z}^i, \bar{P}^i,\ i \in \mathbf{N}\}$ *satisfying (6.28a) and (6.28b), and further satisfying the conditions that for each* $i \in \mathbf{N}$ *the pair* $(\bar{F}_i, B^i)$ *is stabilizable and the pair* $(\bar{F}_i, \bar{Q}_i)$ *is detectable. Then,*

(i) the N-tuple of stationary feedback policies (6.29) provides a Nash equilibrium solution for the linear-quadratic nonzero-sum dynamic game of this subsection, leading to the finite infinite-horizon Nash equilibrium cost $\frac{1}{2}x_1'\bar{Z}^i x_1$ *for* $\mathbf{P}i$,

(ii) the resulting system dynamics, described by

$$x_{k+1} = \bar{F}x_k, \quad k = 1, 2, \ldots,$$

are stable.

A few remarks regarding the solution presented above are now in order. First, we have not given any conditions on the parameters of the game that will guarantee the existence of a solution set to (6.28a)-(6.28b); obtaining such conditions seems to be quite a challenging task. Second, even though the set of equations (6.28a)-(6.28b) was obtained by taking a limit on the solution for the time-truncated version of the game, it is quite possible that these equations will admit (other) solutions that are not necessarily related to the solution(s) of the finite-horizon game. But, these would also provide Nash equilibria, as long as the conditions on *stabilizability* and *detectability* are satisfied.

For the special class of linear-quadratic zero-sum games it is possible to obtain stronger results. First, let us rewrite the "stationary" counterparts of the feedback saddle-point policies (6.27a)-(6.27b), along with the "stationary" counterparts of the relevant matrices (Λ and M) which were introduced in Thm. 6.4, where we again use an "overbar" to denote the limiting values (as $K \to \infty$):

$$\bar{\gamma}^1(x_k) = -B^{1'}\bar{M}\bar{\Lambda}^{-1}Ax_k, \tag{6.30a}$$

$$\bar{\gamma}^2(x_k) = B^{2'}\bar{M}\bar{\Lambda}^{-1}Ax_k, \quad k = 1, 2, \ldots; \tag{6.30b}$$

$$\bar{\Lambda} \triangleq [I + (B^1B^{1'} - B^2B^{2'})\bar{M}], \tag{6.30c}$$

$$\bar{M} = Q + A'\bar{M}\bar{\Lambda}^{-1}A. \tag{6.30d}$$

Furthermore, the counterpart of the concavity condition (6.26b) is

$$I - B^{2'}\bar{M}B^2 > 0, \tag{6.30e}$$

and for convenience we also introduce the following relaxed version of (6.30e):

$$I - B^{2'}\bar{M}B^2 \geq 0. \tag{6.30f}$$

Now, the following lemma, whose proof can be found in (Başar and Bernhard, 1995), provides precise relationships between the limit of the sequence generated by (6.12b) and solutions of (6.30d) as well as the value of the underlying infinite-horizon game, when $Q \geq 0$. Before stating the lemma, let us introduce the notation $M_k^{(K)}$ to denote the unique nonnegative-definite matrix sequence generated by (6.12b) for a K-stage time-invariant linear-quadratic game, with the terminal condition (at $k = K+1$) taken as Q, and with condition (6.30f) satisfied.

Lemma 6.3 *For the linear-quadratic infinite-horizon zero-sum dynamic game with $Q \geq 0$, let the pair (A, Q) be observable (respectively, detectable). Then,*

(i) if there exists a nonnegative-definite solution to the generalized algebraic Riccati equation (GARE) (6.30d), $\bar{M}$, which also satisfies (6.30e), then $\bar{M}$ is necessarily positive definite,

(ii) if the GARE (6.30d) does not admit a positive (respectively, nonnegative) definite solution satisfying (6.30f), then the upper value of the game is unbounded,

(iii) if $\bar{M}$ denotes a positive (respectively, nonnegative) definite solution of (6.30d) satisfying (6.30e), then for all $K > 1$,

$$\bar{M} \geq M_k^{(K)} \geq M_{k+1}^{(K)}, \quad \textit{for all } k \leq K.$$

An important consequence of this lemma is that if the GARE (6.30d) admits multiple positive (respectively, nonnegative) definite solutions satisfying (6.30e), there is a minimal such solution (minimal in the sense of matrix partial ordering), denoted $\bar{M}^+$, to which the sequence $\{M_k^{(K)}\}$ converges as $K \to \infty$. In view of this observation, and the result of Lemma 6.3, the following theorem can be established. Its proof can be found in Başar (1991b) and Başar and Bernhard (1995).

Theorem 6.8 *Consider the infinite-horizon discrete-time linear-quadratic zero-sum dynamic game, with (A, Q) constituting an observable (respectively, detectable) pair. Then, we have the following.*

(i) The game has equal upper and lower values if, and only if, the GARE (6.30d) admits a positive (respectively, nonnegative) definite solution satisfying (6.30e), and only if it admits a positive (respectively, nonnegative) definite solution satisfying (6.30f).

(ii) If the GARE admits a positive (respectively, nonnegative) definite solution, satisfying (6.30e), then it admits a minimal such solution, to be denoted $\bar{M}^+$. Then, the finite value of the game is $\frac{1}{2}x_1'\bar{M}^+x_1$.

(iii) The upper (minimax) value of the game is finite if, and only if, the upper and lower values are equal.

(iv) If $\bar{M}^+ > 0$ (respectively, ≥ 0) exists, as given above, the controller $\bar{\gamma}^1$ given by (6.30a), with $\bar{M}$ replaced by $\bar{M}^+$, attains the finite upper value, in the sense that

$$\sup_{\gamma^2} J(\bar{\gamma}^1, \gamma^2) = x_1'\bar{M}^+x_1 \ ,$$

and a maximizing solution above is the stationary feedback policy given by (6.30b), again with $\bar{M}$ replaced by $\bar{M}^+$.

(v) If $\bar{M}^+ > 0$ (respectively, ≥ 0), exists, the closed-loop system under the policies (6.30a)-(6.30b), with $\bar{M}$ replaced by $\bar{M}^+$, is asymptotically stable, that is, the matrix $\bar{F}$ defined below is Hurwitz:

$$\bar{F} \triangleq (I - (BB' - DD')\bar{M}^+(\bar{\Lambda}^+)^{-1})A.$$

Note that the theorem above does not say that the feedback policies (6.30a)-(6.30b), with $\bar{M}$ replaced by $\bar{M}^+$, are in saddle-point equilibrium. It only says that **P1**'s feedback policy as given assures a finite upper value, which is also equal to the lower value. This finite value of the game may not, however, be assured by the corresponding feedback policy of **P2**; more will be said on this point in the context of continuous-time (differential) games, to be discussed in subsection 6.5.3.

6.3 Informational Properties of Nash Equilibria in Discrete-Time Dynamic Games

This section is devoted to an elaboration on the occurrence of "informationally nonunique" Nash equilibria in discrete-time dynamic games, and to a general discussion on the interplay between information patterns and existence-uniqueness properties of noncooperative equilibria in such games. First, we consider, in some detail, a scalar three-person dynamic game which admits uncountably many Nash equilibria, and which features several important properties of infinite dynamic games. Then, we discuss these properties in a general context.

6.3.1 A three-person dynamic game illustrating informational nonuniqueness

Consider a scalar three-person two-stage linear-quadratic dynamic game in which each player acts only once. The state equation is given by

$$x_3 = x_2 + u^1 + u^2; \quad x_2 = x_1 + u^3, \tag{6.31}$$

and the cost functionals are defined as

$$L^1 = (x_3)^2 + (u^1)^2; \quad L^2 = -L^3 = -(x_3)^2 + 2(u^2)^2 - (u^3)^2. \tag{6.32}$$

In this formulation, u^i is the scalar unconstrained control variable of **P**i ($i = 1, 2, 3$), and x_1 is the initial state whose value is known to all players. **P1** and **P2**, who act at stage 2, have also access to the value of x_2 (i.e., the underlying information pattern is CLPS (or, equivalently, MPS) for both **P1** and **P2**), and their permissible strategies are taken as twice continuously differentiable mappings from $\mathbf{R} \times \mathbf{R}$ into $\mathbf{R}$. A permissible strategy for **P3**, on the other hand, is any measurable mapping from $\mathbf{R}$ into $\mathbf{R}$. This, then, completes the description of the strategy spaces Γ^1, Γ^2 and Γ^3 (for **P1**, **P2** and **P3**, respectively), where we suppress the subindices denoting the corresponding stages since each player acts only once.

Now, let $\{\gamma^1 \in \Gamma^1, \gamma^2 \in \Gamma^2, \gamma^3 \in \Gamma^3\}$ denote any noncooperative (Nash) equilibrium solution for this three-person nonzero-sum dynamic game. Since L^i is strictly convex in u^i ($i = 1, 2$), a set of necessary and sufficient conditions for γ^1 and γ^2 to be in equilibrium (with $\gamma^3 \in \Gamma^3$ fixed) is obtained by differentiation of L^i with respect to u^i ($i = 1, 2$), thus leading to

$$\begin{aligned} \bar{x}_2(\gamma^3) + 2\gamma^1(\bar{x}_2, x_1) + \gamma^2(\bar{x}_2, x_1) &= 0, \\ -\bar{x}_2(\gamma^3) - \gamma^1(\bar{x}_2, x_1) + \gamma^2(\bar{x}_2, x_1) &= 0, \end{aligned}$$

where

$$\bar{x}_2 \triangleq \bar{x}_2(\gamma^3) = x_1 + \gamma^3(x_1).$$

Solving for $\gamma^1(\bar{x}_2, x_1)$ and $\gamma^2(\bar{x}_2, x_1)$ from the foregoing pair of equations, we obtain

$$\gamma^1(\bar{x}_2, x_1) = -\frac{2}{3}\bar{x}_2, \tag{6.33a}$$

$$\gamma^2(\bar{x}_2, x_1) = \frac{1}{3}\bar{x}_2, \tag{6.33b}$$

which are the side conditions on the equilibrium strategies γ^1 and γ^2, and which depend on the equilibrium strategy γ^3 of **P3**. Besides these side conditions, the Nash equilibrium strategies of **P1** and **P2** have no other natural constraints imposed on them. To put it in other words, *every Nash equilibrium strategy for* **P1** *will be a closed-loop representation (cf. Def. 5.12) of the open-loop value (6.33a), and every Nash equilibrium strategy for* **P2** *will be a closed-loop representation of (6.33b).*

To complete the solution of the problem, we now proceed to stage 1. Since $\{\gamma^1, \gamma^2, \gamma^3\}$ constitutes an equilibrium triple, with $\{u^1 = \gamma^1(x_2, x_1), u^2 = \gamma^2(x_2, x_1)\}$ substituted into L^3, the resulting cost functional of **P3** (denoted as $\tilde{L}^3$) should attain a minimum at $u^3 = \gamma^3(x_1)$; and since γ^1 and γ^2 are twice continuously differentiable in their arguments, this requirement can be expressed in terms of the relations

$$\frac{\mathrm{d}}{\mathrm{d}u^3}\tilde{L}^3(\gamma^3(x_1)) = x_3(1 + \gamma^1_{x_2} + \gamma^2_{x_2}) - 2\gamma^2_{x_2}\gamma^2 + \gamma^3(x_1) = 0,$$

$$\frac{\mathrm{d}^2}{(\mathrm{d}u^3)^2}\tilde{L}^3(\gamma^3(x_1)) = (1 + \gamma^1_{x_2} + \gamma^2_{x_2})^2 + x_3(\gamma^1_{x_2x_2} + \gamma^2_{x_2x_2}) - 2(\gamma^2_{x_2})^2 - 2\gamma^2_{x_2x_2}\gamma^2 + 1 > 0, \text{ [59]}$$

where we have suppressed the arguments of the strategies. Now, by utilizing the side conditions (6.33a)-(6.33b) in the above set of relations, we arrive at a simpler set of relations which are, respectively,

$$\frac{2}{3}[x_1 + \gamma^3(x_1)][1 + \gamma^1_{x_2}(\bar{x}_2, x_1)] + \gamma^3(x_1) = 0, \tag{6.34a}$$

$$[1 + \gamma^1_{x_2}(\bar{x}_2, x_1) + \gamma^2_{x_2}(\bar{x}_2, x_1)]^2 - \frac{2}{3}\bar{x}_2\gamma^1_{x_2x_2}(\bar{x}_2, x_1) - 2[\gamma^2_{x_2}(\bar{x}_2, x_1)]^2 + 1 > 0, \tag{6.34b}$$

[59]Here, we could of course also have nonstrict inequality (i.e., $\geq$) in which case we also have to look at higher-order derivatives of $\tilde{L}^3$. We avoid this by restricting our analysis at the outset only to those equilibrium triples $\{\gamma^1, \gamma^2, \gamma^3\}$ which lead to an $\tilde{L}^3$ that is locally strictly convex at the solution point.

where $\bar{x}_2$ is again defined as

$$\bar{x}_2 = x_1 + \gamma^3(x_1). \tag{6.34c}$$

These are the relations which should be satisfied by an equilibrium triple, in addition to (6.33a) and (6.33b). The following proposition summarizes the result.

Proposition 6.4 *Any triple $\{\gamma^{1*} \in \Gamma^1, \gamma^{2*} \in \Gamma^2, \gamma^{3*} \in \Gamma^3\}$ that satisfies (6.33a)-(6.33b) and (6.34a)-(6.34b), and also possesses the additional feature that (6.34a) with $\gamma^1 = \gamma^{1*}$ and $\gamma^2 = \gamma^{2*}$ admits a unique solution $\gamma^3 = \gamma^{3*}$, constitutes a Nash equilibrium solution for the nonzero-sum dynamic game described by (6.31)-(6.32).*

Proof. This result follows from the derivation outlined prior to the statement of the proposition. Uniqueness of the solution of (6.34a) for each pair $\{\gamma^{1*}, \gamma^{2*}\}$ is imposed in order to ensure that the resulting γ^{3*} is indeed a globally minimizing solution for $\tilde{L}^3$. □

We now claim that there exists an uncountable number of triplets that satisfy the requirements of Prop. 6.4. To justify this claim, and to obtain a set of explicit solutions, we consider the class of γ^1 and γ^2 described as

$$\begin{aligned} \gamma^1(x_2, x_1) &= -\tfrac{2}{3}x_2 + p[x_2 - \bar{x}_2(\gamma^3)], \\ \gamma^2(x_2, x_1) &= -\tfrac{1}{3}x_2 + q[x_2 - \bar{x}_2(\gamma^3)], \end{aligned}$$

where p and q are free parameters. These structural forms for γ^1 and γ^2 clearly satisfy the side conditions (6.33a) and (6.33b), respectively. With these choices, (6.34a) can be solved uniquely (for each p, q) to give

$$\gamma^3(x_1) = -[(2+6p)/(11+6p)]x_1, \quad p \neq -11/6,$$

with the existence condition (6.34b) reading

$$\left(\frac{2}{3} + p\right)^2 + 2pq - q^2 + \frac{7}{9} > 0. \tag{6.35}$$

The scalar $\bar{x}_2$ is then given by

$$\bar{x}_2 = [9/(11+6p)]x_1.$$

Hence, we have the following.

Proposition 6.5 *The set of strategies*

$$\begin{aligned} \gamma^{1*}(x_2, x_1) &= -\frac{2}{3}x_2 + p\{x_2 - [9/(11+6p)]x_1\}, \\ \gamma^{2*}(x_2, x_1) &= -\frac{1}{3}x_2 + q\{x_2 - [9/(11+6p)]x_1\}, \\ \gamma^{3*}(x_1) &= -[(2+6p)/(11+6p)]x_1 \end{aligned}$$

constitutes a Nash equilibrium solution for the dynamic game described by (6.31)-(6.32), for all values of the parameters p and q satisfying (6.35) and with $p \neq -11/6$. The corresponding equilibrium costs of the players are

$$\begin{aligned} J^{1*} &= 2[6/(11+6p)]^2(x_1)^2, \\ J^{2*} &= -J^{3*} = -[(22+24p+36p^2)/(11+6p)^2](x_1)^2. \end{aligned}$$

Several remarks and observations are in order here, concerning the Nash equilibrium solutions presented above.

(1) The nonzero-sum dynamic game of this subsection admits *uncountably many* Nash equilibrium solutions, each one leading to a different equilibrium cost triple.

(2) Within the class of linear strategies, Prop. 6.5 provides the complete solution to the problem, which is parameterized by p and q.

(3) The equilibrium strategy of **P3**, as well as the equilibrium cost values of all three players, depend only on p (not on q), whereas the existence condition (6.35) involves both p and q. There is indeed an explanation for this: the equilibrium strategies of **P1** and **P2** are in fact representations of the open-loop values (6.33a)-(6.33b) on appropriate trajectories. By choosing a specific representation of (6.33a), **P1** influences the cost functional of **P3** and thereby the optimization problem faced by him. Hence, for each different representation of (6.33a), **P3** ends up, in general, with a different solution to his minimization problem, which directly contributes to nonuniqueness of Nash equilibria. For **P2**, on the other hand, even though he may act analogously—i.e., choose different representations of (6.33b)—these different representations do not lead to different minimizing solutions for **P3** (but instead affect only the existence of a minimizing solution) since $L^2 \equiv -L^3$, i.e., **P2** and **P3** have completely conflicting goals (see Thm. 6.9, later in this section, and also the next remark for further clarification). Consequently, $\gamma^{3*}(x_1)$ is independent of q, but the existence condition explicitly depends upon q. (This is true also for nonlinear representations, as it can be seen from (6.34a) and (6.34b).)

(4) If **P1** has access to only x_2 (and not to x_1), then, necessarily, $p = 0$, and both **P1** and **P3** have unique equilibrium strategies which are $\{\gamma^{1*}(x_2) = -\frac{2}{3}x_2, \gamma^{3*}(x_1) = -(2/11)x_1\}$. (This is true also within the class of nonlinear strategies.) Furthermore, the equilibrium cost values are also unique (simply set $p = 0$ in $J^{i*}, i = 1,2,3$, in Prop. 6.5). However, the existence condition (6.35) still depends on q, since it now reduces to $q^2 < 11/9$. The reason for this is that **P2** still has the freedom of employing different representations of (6.33b), which affects existence of the equilibrium solution but not the actual equilibrium state trajectory, since **P2** and **P3** are basically playing a zero-sum game (in which case the equilibrium (i.e., saddle-point) solutions are interchangeable). (See Thm. 6.9, later in this section; also recall the feature discussed in Remark 6.7 earlier.)

(5) By setting $p = q = 0$ in Prop. 6.5, we obtain the *unique feedback* Nash equilibrium solution (cf. Corollary 6.1 and Remark 6.4) of the dynamic game under consideration (which exists since $p = 0 \neq 11/6$, and (6.35) is satisfied).

(6) Among the uncountable number of Nash equilibrium solutions presented in Prop. 6.5, there exists a subsequence of strategies which brings **P1**'s Nash cost arbitrarily close to zero which is the lowest possible value L^1 can attain. Note, however, that the corresponding cost for **P3** approaches $(x_1)^2$ which is unfavorable to him.

Before concluding this subsection, it is worthy to note that the linear equilibrium solutions presented in Prop. 6.5 are not the only ones that the dynamic game under consideration admits, since (6.34a) will also admit nonlinear solutions. To obtain an explicit nonlinear equilibrium solution, we may start with a nonlinear representation of (6.33a), for instance,

$$\gamma^1(x_2, x_1) = -\frac{2}{3}x_2 + p[x_2 - \bar{x}_2(\gamma^3)]^2,$$

substitute it into (6.34a) and solve for a corresponding $\gamma^3(x_1)$, checking at the same time satisfaction of the second-order condition (6.34b). For such a derivation (of nonlinear Nash solutions in a linear-quadratic game) the reader is referred to Başar (1974). See also Problem 5, Section 6.8.

6.3.2 General results on informationally nonunique equilibrium solutions

Section 3.5 has already displayed existence of "informationally nonunique" Nash equilibria in finite multi-act nonzero-sum dynamic games, which was mainly due to the fact that an increase in information to one or more players leaves the Nash equilibrium obtained under the original information pattern unchanged, but it also creates new equilibria (cf. Prop. 3.10). In infinite games, the underlying reason for occurrence of informationally nonunique Nash equilibria is essentially the same (though much more intricate), and a counterpart of Prop. 3.10 can be verified. Toward this end we first introduce the notion of "informational inferior" in such dynamic games.

Definition 6.3 *Let* I *and* II *be two N-person K-stage infinite dynamic games which admit precisely the same extensive form description except the underlying information structure (and, of course, also the strategy spaces whose descriptions depend on the information structure). Let η^i_{I} (respectively, η^i_{II}) denote the information pattern of* **P***i in the game* I *(respectively,* II*), and let the inclusion relation $\eta^i_{\mathrm{I}} \subseteq \eta^i_{\mathrm{II}}$ imply that whatever* **P***i knows at each stage of game* I *he also knows at the corresponding stages of game* II*, but not necessarily vice versa. Then,* I *is* informationally inferior *to* II *if $\eta^i_{\mathrm{I}} \subseteq \eta^i_{\mathrm{II}}$ for all $i \in \mathbf{N}$, with strict inclusion for at least one i.*

Proposition 6.6 *Let* I *and* II *be two N-person K-stage infinite dynamic games as introduced in Def. 6.3, so that* I *is informationally inferior to* II. *Furthermore, let the strategy spaces of the players in the two games be compatible with the given information patterns and the constraints (if any) imposed on the controls, so that $\eta^i_{\rm I} \subseteq \eta^i_{\rm II}$ implies $\Gamma^i_{\rm I} \subseteq \Gamma^i_{\rm II}$, $i \in \mathbf{N}$. Then,*

(i) any Nash equilibrium solution for I *is also a Nash equilibrium solution for* II,

(ii) if $\{\gamma^1, \ldots, \gamma^N\}$ is a Nash equilibrium solution for II *such that $\gamma^i \in \Gamma^i_{\rm I}$ for all $i \in \mathbf{N}$, then it is also a Nash equilibrium solution for* I.

Proof. Let $\{\gamma^{i*}; i \in \mathbf{N}\}$ constitute a Nash equilibrium solution for I. Then, by definition,

$$J^1(\gamma^{1*}, \gamma^{2*}, \ldots, \gamma^{N*}) \leq J^1(\gamma^1, \gamma^{2*}, \ldots, \gamma^{N*}), \quad \forall \gamma^1 \in \Gamma^1_{\rm I};$$

therefore, **P1** minimizes $J^1(\cdot, \gamma^{2*}, \ldots, \gamma^{N*})$ over $\Gamma^1_{\rm I}$, with the corresponding solution being $\gamma^{1*} \in \Gamma^1_{\rm I}$.

Now consider minimization of the same expression over $\Gamma^1_{\rm II}$ ($\supseteq \Gamma^1_{\rm I}$) which reflects an increase in deterministic information concerning the values of state. But, since we have a deterministic optimization problem, the minimum value of $J^1(\gamma^1, \gamma^{2*}, \ldots, \gamma^{N*})$ does not change with an increase in information (see Section 5.6). Hence,

$$\min_{\gamma^1 \in \Gamma^1_{\rm II}} J^1(\gamma^1, \gamma^{2*}, \gamma^{N*}) = J^1(\gamma^{1*}, \gamma^{2*}, \ldots, \gamma^{N*});$$

and furthermore, since $\gamma^{1*} \in \Gamma^1_{\rm II}$, we have the inequality

$$J^1(\gamma^{1*}, \gamma^{2*}, \ldots, \gamma^{N*}) \leq J^1(\gamma^1, \gamma^{2*}, \ldots, \gamma^{N*}), \quad \forall \gamma^1 \in \Gamma^1_{\rm II}.$$

Since **P1** was an arbitrary player in this discussion, it follows in general that

$$\begin{aligned} J^i(\gamma^{1*}, \ldots, \gamma^{i*}, \ldots, \gamma^{N*}) \quad &\leq \quad J^i(\gamma^{1*}, \ldots, \gamma^{i-1*}, \gamma^i, \gamma^{i+1*}, \ldots, \gamma^{N*}) \\ &\qquad \forall \gamma^i \in \Gamma^i_{\rm II} \quad (i \in \mathbf{N}) \end{aligned}$$

which verifies (i) of Prop. 6.6. Proof of (ii) is along similar lines. □

Since there corresponds at least one informationally inferior game (viz. a game with an open-loop information structure) to every multi-stage game with CLPS information, the foregoing result clearly provides one set of reasons for existence of "informationally nonunique" Nash equilibria in infinite dynamic games (as Prop. 3.10 did for finite dynamic games). However, this is not yet the whole story, as it does not explain occurrence of uncountably many equilibria in such games. What is really responsible for this is the existence of uncountably many *representations* of a strategy under dynamic information (cf. Section 5.6). To elucidate somewhat further, consider the scalar three-person dynamic game of subsection 6.3.1. We have already seen that, for each fixed equilibrium strategy γ^3 of **P3**, the equilibrium strategies of **P1** and **P2** have unique open-loop values given by (6.33a) and (6.33b), respectively, but they are otherwise free. We also know from Section 5.6 that there exist infinitely many closed-loop representations of such open-loop policies; and since each one has a different structure, this leads to infinitely many equilibrium strategies for **P3**, and consequently to a plethora of Nash equilibria.

The foregoing discussion is valid on a much broader scale (not only for the specific three-person game treated in subsection 6.3.1), it provides a general guideline for derivation of informationally nonunique Nash equilibria, thus leading to the algorithm given below after introducing the concept of "stagewise equilibrium".

Definition 6.4 *For an N-person K-stage infinite dynamic game, a set of strategies $\{\gamma^{i*}; i \in \mathbf{N}\}$ satisfying the following inequalities for all $\gamma_k^i \in \Gamma_k^i$, $i \in \mathbf{N}$, $k \in \mathbf{K}$, is a* stagewise (Nash) equilibrium solution*:*

$$J^i(\gamma^{1*}; \ldots; \gamma^{i*}; \ldots; \gamma^{N*}) \leq J^i(\gamma^{1*}; \ldots; \gamma_1^{i*}, \ldots, \gamma_{k-1}^{i*}, \gamma_k^i, \gamma_{k+1}^{i*}, \ldots, \gamma_K^{i*}; \ldots; \gamma^{N*}). \tag{6.36}$$

Remark 6.9 Every Nash equilibrium solution (cf. the set of inequalities (6.3)) is a stagewise Nash equilibrium solution, but not vice versa. □

An algorithm to obtain informationally nonunique Nash equilibria in infinite dynamic games

First determine the entire class of stagewise equilibrium solutions.

(1) Starting at the last stage of the game ($k = K$), fix the N-tuples of strategies at every other stage $k < K$, and solve basically an N-person static game at stage K (which is defined by (6.36) with $k = K$) to determine the open-loop values of the corresponding stagewise equilibrium strategies at $k = K$ as functions of the strategies applied previously. Furthermore, determine the equivalence class of representations of these N-tuples of open-loop values, again as functions of the previous strategies and so that they are compatible with the underlying information structure of the problem.

(2) Now consider inequalities (6.36) for $k = K - 1$ and adopt a specific member of the equivalence class determined at step 1 as a strategy N-tuple applied at stage K; furthermore fix the N-tuples of strategies at every stage $k < K - 1$, and solve basically an N-person static game to determine the open-loop values of the corresponding stagewise equilibrium strategies at stage $K - 1$ as functions of the strategies applied previously. Repeat this for every member of the equivalence class determined at step 1. Each will, in general, result in a different set of open-loop values. Now, for each set of open-loop values thus determined, find the corresponding equivalence class of representations which are also compatible with the underlying

information structure and which depend on the strategies applied at earlier stages.

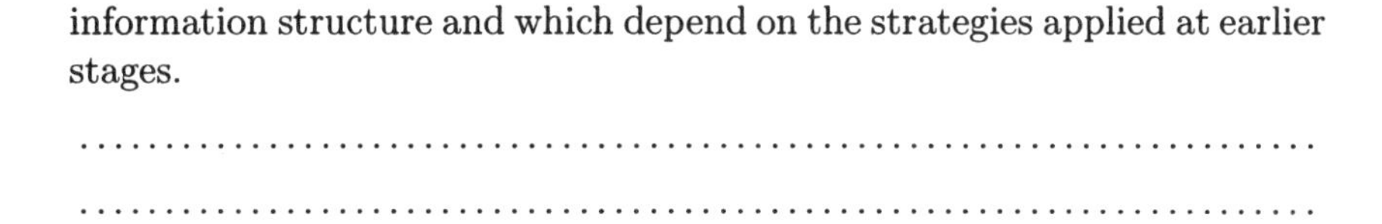

(K) Now, finally, consider the set of inequalities (6.36) for $k = 1$, and adopt specific (and compatible) members of the equivalence classes determined at steps $1, \dots, K-1$ as strategies applied at stages $K, \dots, 2$; solve the resulting N-person static game to determine the corresponding stagewise equilibrium strategies at stage 1, which will in general also require solution of some implicit equations. Repeat this for every compatible $(K-1)$-tuple of members of the equivalence classes constructed at steps $1, \dots, K-1$.

This then completes the construction of the entire class of stagewise equilibrium solutions of the given dynamic game. Finally, one has to check to see which of these stagewise equilibrium solutions also constitute Nash equilibria, by referring to the set of inequalities (6.3).

Remark 6.10 In obtaining a set of informationally nonunique equilibrium solutions for the scalar example of subsection 6.3.1, we have actually followed the steps of the foregoing algorithm; but there we did not have to check the last condition since every stagewise equilibrium solution was also a Nash equilibrium solution because every player acted only once. □

The foregoing algorithm leads to an uncountable number of informationally nonunique Nash equilibria in deterministic multi-stage infinite nonzero-sum games wherein at least one player has dynamic information. Furthermore, these informationally nonunique Nash equilibria are, in general, not interchangeable—thereby leading to infinitely many different equilibrium cost N-tuples. The reason why we use the words "in general" in the preceding sentence is because there exist some extreme cases of nonzero-sum games which do not exhibit such features. One such class is the so-called *team problems* for which $L^1 \equiv L^2 \equiv \cdots \equiv L^N$, i.e., there is a common objective function. The Nash equilibrium solution corresponds in this case to a person-by-person optimal solution and it becomes a team solution under some further restrictions (see the discussion given in Section 4.6). Deterministic team problems are no different from optimal control problems discussed in Section 5.5, and in particular the discussion of Section 5.6 leads to the conclusion that informational nonuniqueness of equilibria does not create any major difficulty in team problems, since all these equilibrium solutions are different representations of the same N-tuple of strategies which is associated with the global minimum of a single objective functional. Hence, for the special case of team problems, informational nonuniqueness does *not* lead to different equilibrium cost N-tuples.

Let us now consider the other extreme case—the class of two-person zero-sum infinite dynamic games—in which case $L^1 \equiv -L^2$. For such dynamic games, informational nonuniqueness does not lead to different (saddle-point) equilibrium cost pairs either, since multiple saddle points are interchangeable; however, as we have observed in subsection 6.3.1, existence of a saddle point will depend on the specific pair of representations adopted. Therefore, contrary to team solutions, every representation of a pair of saddle-point strategies is *not* necessarily a saddle-point strategy pair. The next theorem makes this statement precise, and it also provides a strengthened version of Prop. 6.6 for two-person zero-sum dynamic games. Before giving the theorem, it will be convenient first to introduce the notion of a *strongly unique saddle point* (say, $\{\gamma^{1*}, \gamma^{2*}\}$) on a given product strategy space $\Gamma^1 \times \Gamma^2$, as one where the pair $\{\gamma^{1*}, \gamma^{2*}\}$ is unique as a saddle-point solution, and in addition $\gamma^{1*} \in \Gamma^1$ is the unique response to γ^{2*} and $\gamma^{2*} \in \Gamma^2$ is the unique response to γ^{1*}. Then, we have the following theorem.

Theorem 6.9 *Let* I *be a two-person zero-sum K-stage dynamic game with closed-loop perfect state information pattern.*

(i) If I *admits a strongly unique feedback saddle-point solution, and also an open-loop saddle-point solution, then the latter is unique.*

(ii) If I *admits a strongly unique feedback saddle-point solution $\{\gamma^{1\mathrm{f}}, \gamma^{2\mathrm{f}}\}$ with a corresponding state trajectory $\{x^{\mathrm{f}}_{k+1}; k \in \mathbf{K}\}$, and $\{\gamma^{1*}, \gamma^{2*}\}$ denotes some other closed-loop saddle-point solution with corresponding state trajectory $\{x^*_{k+1}; k \in \mathbf{K}\}$, we have*

$$\begin{aligned} x^*_{k+1} &= x^{\mathrm{f}}_{k+1}, \quad k \in \mathbf{K}, \\ \gamma^{i*}_k(x^*_k, \ldots, x^*_2, x_1) &\equiv \gamma^{i\mathrm{f}}_k(x^{\mathrm{f}}_k), \quad k \in \mathbf{K}, \quad i = 1, 2. \end{aligned}$$

(iii) If I *admits a strongly unique open-loop saddle-point solution $\{\gamma^{1\mathrm{o}}, \gamma^{2\mathrm{o}}\}$ with a corresponding state trajectory $\{x^{\mathrm{o}}_{k+1}; k \in \mathbf{K}\}$, and if $\{\gamma^{1*}, \gamma^{2*}\}$ denotes some other closed-loop saddle-point solution with corresponding state trajectory $\{x^*_{k+1}; k \in \mathbf{K}\}$ we have*

$$\begin{aligned} x^*_{k+1} &= x^{\mathrm{o}}_{k+1}, \quad k \in \mathbf{K}, \\ \gamma^{i*}_k(x^*_k, \ldots, x^*_2, x_1) &\equiv \gamma^{i\mathrm{o}}_k(x_1), \quad k \in \mathbf{K}, \quad i = 1, 2. \end{aligned}$$

Proof. (i) Let $\{\gamma^{1\mathrm{f}}, \gamma^{2\mathrm{f}}\}$ be the strongly unique feedback saddle-point solution and $\{\gamma^{1\mathrm{o}}, \gamma^{2\mathrm{o}}\}$ be any open-loop saddle-point solution. It follows from Prop. 6.6 that $\{\gamma^{1\mathrm{f}}, \gamma^{2\mathrm{f}}\}$ and $\{\gamma^{1\mathrm{o}}, \gamma^{2\mathrm{o}}\}$ are also closed-loop saddle-point solutions. Furthermore, because of the ordered interchangeability property of saddle points, $\{\gamma^{1\mathrm{o}}, \gamma^{2\mathrm{f}}\}$ and $\{\gamma^{1\mathrm{f}}, \gamma^{2\mathrm{o}}\}$ are also closed-loop saddle-point strategy pairs, i.e.,

$$J(\gamma^{1\mathrm{o}}, \gamma^2) \le J(\gamma^{1\mathrm{o}}, \gamma^{2\mathrm{f}}) \le J(\gamma^1, \gamma^{2\mathrm{f}}), \ \forall \gamma^1 \in \Gamma^1, \gamma^2 \in \Gamma^2, \qquad (i)$$

and an analogous pair of inequalities for $\{\gamma^{1\mathrm{f}}, \gamma^{2\mathrm{o}}\}$, where Γ^i denotes the closed-loop strategy space of $\mathbf{P}i$. Now, let us consider the RHS inequality of (i), which is equivalent to

$$J(\gamma^{1\mathrm{o}}, \gamma^{2\mathrm{f}}) = \min_{\gamma^1 \in \Gamma^1} J(\gamma^1, \gamma^{2\mathrm{f}}).$$

This is a standard optimum control problem in discrete time with a cost functional

$$\hat{J}(\gamma^1) = \sum_{k\in\mathbf{K}} \hat{g}_k(x_{k+1}, u_k^1, x_k),$$
$$\hat{g}(x_{k+1}, u_k^1, x_k) \triangleq g_k(x_{k+1}, u_k^1, \gamma_k^{2\mathrm{f}}(x_k), x_k), \quad u_k^1 = \gamma_k^1(\cdot),$$

and a state equation

$$x_{k+1} = \hat{f}_k(x_k, u_k^1) \triangleq f_k(x_k, u_k^1, \gamma_k^{2\mathrm{f}}(x_k)).$$

Globally optimal solution of this problem over Γ^1 can be obtained (if it exists) via dynamic programming and as a feedback strategy. But by hypothesis (since $\{\gamma^{1\mathrm{f}}, \gamma^{2\mathrm{f}}\}$ is a strongly unique saddle-point pair), there exists a unique feedback strategy (namely, $\gamma^{1\mathrm{f}}$) that solves this problem. Since every closed-loop strategy admits a unique open-loop representation (cf. Section 5.6), it follows that there exists a unique open-loop solution to that optimum control problem, which is given by $\{\gamma_k^{1\circ}(x_1) = \gamma_k^{1\mathrm{f}}(x_k^\mathrm{f}),\ k \in \mathbf{K}\}$, where $\{x_{k+1}^\mathrm{f};\ k \in \mathbf{K}\}$ denotes the state trajectory corresponding to $\{\gamma^{1\mathrm{f}}, \gamma^{2\mathrm{f}}\}$. Therefore, $\gamma^{1\circ}$ is unique whenever $\gamma^{1\mathrm{f}}$ is. It can likewise be shown, by making use of the LHS inequality of (i), that $\gamma^{2\circ}(x_1) = \gamma^{2\mathrm{f}}(x_k^\mathrm{f})$, and hence $\gamma^{2\circ}$ is unique whenever $\gamma^{2\mathrm{f}}$ is unique. This then completes verification of part (i). We note that a converse statement cannot be made here (i.e., strong uniqueness of open-loop saddle point may not imply strong uniqueness of feedback saddle point), because there could be cost-irrelevant states on which the feedback solution can be chosen to be nonunique.

(ii) By Prop. 6.6 and the ordered interchangeability property of saddle points, $\{\gamma^{1*}, \gamma^{2\mathrm{f}}\}$ and $\{\gamma^{1\mathrm{f}}, \gamma^{2*}\}$ are also saddle-point strategy pairs; therefore

$$J(\gamma^{1*}, \gamma^{2\mathrm{f}}) = \min_{\gamma^1\in\Gamma^1} J(\gamma^1, \gamma^{2\mathrm{f}}).$$

This equality defines an optimum control problem in discrete time, and as in the proof of part (i), every globally optimal solution of this optimization problem can be obtained via dynamic programming; and by hypothesis, $\gamma^{1\mathrm{f}}$ is the unique feedback strategy that renders the global minimum. Hence, every solution in Γ^1 has the same open-loop representation as $\gamma^{1\mathrm{f}}$, that is, $\{\gamma_k^{1*}(x_k^\mathrm{f}, \ldots, x_2^\mathrm{f}, x_1) = \gamma_k^{1\mathrm{f}}(x_k^\mathrm{f}), k \in \mathbf{K}\}$. Similarly, this time by starting with $\{\gamma^{1\mathrm{f}}, \gamma^{2*}\}, \gamma_k^{2*}(x_k^\mathrm{f}, \ldots, x_2^\mathrm{f}, x_1) \equiv \gamma_k^{2\mathrm{f}}(x_k^\mathrm{f}), k \in \mathbf{K}$. Therefore, the pairs of strategies $\{\gamma^{1\mathrm{f}}, \gamma^{2\mathrm{f}}\}$ and $\{\gamma^{1*}, \gamma^{2*}\}$ have the same open-loop representations, and this necessarily implies that the corresponding state trajectories are the same.

(iii) The idea of the proof here is the same as that of part (ii). By Prop. 6.6 and the ordered interchangeability property of multiple saddle points, $\{\gamma^{1*}, \gamma^{2\circ}\}$ and $\{\gamma^{1\circ}, \gamma^{2*}\}$ are also saddle-point strategy pairs; the former of these implies that

$$J(\gamma^{1*}, \gamma^{2\circ}) = \min_{\gamma^1\in\Gamma^1} J(\gamma^1, \gamma^{2\circ})$$

which defines an optimization problem whose open-loop solution is known to be unique (namely, $\gamma^{1\circ}$). Therefore, every other solution of this optimization problem in Γ^1 has to be a closed-loop representation of $\gamma^{1\circ}$ on the trajectory $\{x_{k+1}^\circ; k \in \mathbf{K}\}$, i.e., $\gamma_k^{1*}(x_k^\circ, \ldots, x_2^\circ, x_1) \equiv \gamma_k^{1\circ}(x_1)$, $k \in \mathbf{K}$. Similarly, $\gamma_k^{2*}(x_k^\circ, \ldots, x_2^\circ, x_1) \equiv \gamma_k^{2\circ}(x_1), k \in \mathbf{K}$. Furthermore, because of this equivalence between the open-loop representations, we have $x_{k+1}^* = x_{k+1}^\circ$, $k \in \mathbf{K}$. □

Remark 6.11 The "uniqueness" assumptions of parts (ii) and (iii) of Thm. 6.9 serve to simplify the statement of the theorem, but do not bring in any loss of generality of conceptual nature. "Uniqueness" can be dispensed with provided that the optimal responses to each strategy in the set of saddle points is also a saddle-point strategy. To make this qualification more precise, and to indicate the modification to be incorporated in (ii) and (iii) of Thm. 6.9 in case of nonunique open-loop or feedback saddle points, suppose that there exist two open-loop saddle-point solutions, say $\{\gamma^{1\circ}, \gamma^{2\circ}\}$ and $\{\gamma^{1\circ\circ}, \gamma^{2\circ\circ}\}$. Then, because of the ordered interchangeability property, $\{\gamma^{1\circ}, \gamma^{2\circ\circ}\}$ and $\{\gamma^{1\circ\circ}, \gamma^{2\circ}\}$ will also be saddle-point solutions. Now suppose that $\gamma^{2\circ}$ and $\gamma^{2\circ\circ}$ are the only two optimal responses to $\gamma^{1\circ}$ as well as to $\gamma^{1\circ\circ}$, and likewise $\gamma^{1\circ}$ and $\gamma^{1\circ\circ}$ are the only two optimal responses to $\gamma^{2\circ}$ and $\gamma^{2\circ\circ}$. Then, if $\{\gamma^{1*}, \gamma^{2*}\}$ denotes some closed-loop saddle-point solution, the open-loop representation of this pair of strategies will be equivalent to one of the four open-loop saddle-point strategy pairs. The saddle-point value is, of course, the same for all these different saddle-point solutions. □

Theorem 6.9, together with Remark 6.11, provides a complete characterization of the saddle-point solutions of two-person zero-sum discrete-time dynamic games, if their open-loop and/or feedback saddle-point solutions are known; every other saddle point can be obtained as their representation under the given information pattern. Furthermore, whenever they both are strongly unique, the open-loop representation of the feedback saddle-point solution is necessarily the open-loop saddle-point solution. This does not imply, however, that if a feedback saddle point exists, an open-loop saddle point necessarily has to exist (and vice versa); we have already seen in subsection 6.2.2 (cf. Prop. 6.2) that the former statement is not true in affine-quadratic zero-sum games.

In conclusion, a nonzero-sum multi-stage infinite game with dynamic information will, in general, admit uncountably many (informationally nonunique) Nash equilibria which are not interchangeable, unless it is a team problem or a zero-sum game. In Başar and Selbuz (1976) it has actually been shown within the context of a class of dynamic two-stage nonzero-sum games that this is in fact a strict property, i.e., unless the nonzero-sum game can be converted into an equivalent team problem or a zero-sum game, it admits uncountably many Nash equilibria.

One possible way of removing informational nonuniqueness in the noncooperative equilibria of nonzero-sum dynamic games is to further restrict the equilibrium solution concept by requiring it to be a feedback Nash equilibrium (cf. Def. 6.2) which has been discussed thoroughly in subsection 6.2.2. However, it should be noted that this restriction (which is compatible with a *delayed commitment* mode of play or *strong time consistency*) makes sense only if all players have access to the current value of the state (e.g., the CLPS, MPS or FB information pattern). If at least one of the players has access to delayed or open-loop information, the feedback equilibrium solution cannot be defined, and one then has to resort to some other method for eliminating informational nonuniqueness. One such alternative is to formulate the nonzero-sum game problem in a *stochastic framework*, which is the subject of the next section.

6.4 Stochastic Nonzero-Sum Games with Deterministic Information Patterns

We have already seen in the previous sections that one way of eliminating "informational nonuniqueness" in the Nash solution of dynamic nonzero-sum games is to further restrict the equilibrium concept by requiring satisfaction of a "feedback" property (cf. Def. 6.2) which is, however, valid only under the CLPS and MPS information patterns. Yet another alternative to eliminate informational nonuniqueness is to formulate the dynamic game in a stochastic framework, in accordance with Def. 5.4; for, under an appropriate stochastic formulation, every strategy has a unique representation (cf. Prop. 5.5) and the statement of Prop. 6.6 is no longer valid. This section is, therefore, devoted to an investigation of existence, uniqueness and derivation of Nash equilibrium in stochastic dynamic games with deterministic information patterns.

The class of N-person K-stage discrete-time stochastic infinite dynamic games to be treated in this section will be as described in Def. 5.4, but with the state equation (5.11) replaced by

$$x_{k+1} = f_k(x_k, u_k^1, \ldots, u_k^N) + \theta_k, \quad k \in \mathbf{K}, \tag{6.37}$$

where $\{x_1, \theta_1, \ldots, \theta_K\}$ is a set of statistically independent Gaussian random vectors with values in R^n, and $\operatorname{cov}(\theta_k) > 0$, $\forall k \in \mathbf{K}$. Furthermore, the cost functionals of the players are taken to be stage-additive, i.e.,

$$L^i(u^1, \ldots, u^N) = \sum_{k=1}^{K} g_k^i(x_{k+1}, u_k^1, \ldots, u_k^N, x_k), \quad i \in \mathbf{N}. \tag{6.38}$$

Under the CLPS information pattern, the following theorem now provides a set of necessary and sufficient conditions for any Nash equilibrium solution of such a stochastic dynamic game to satisfy the following

Theorem 6.10 *Every Nash equilibrium solution of an N-person K-stage stochastic infinite dynamic game described by (6.37)-(6.38) and with the CLPS information pattern (for all players) comprises* only feedback strategies; *and for an N-tuple $\{\gamma_k^{i*}(x_k), k \in \mathbf{K}; i \in \mathbf{N}\}$ to constitute such a solution, it is* *necessary and sufficient that the following recursive relations are satisfied:*

$$\begin{aligned} V^i(k,x) &= \min_{u_k^i \in U_k^i} E_{\theta_k}[g_k^i(\tilde{f}_k^{i*}(x,u_k^i)+\theta_k, \gamma_k^{1*}(x),\ldots,\gamma_k^{i-1}(x), u_k^i, \\ &\quad \gamma_k^{i+1*}(x),\ldots,\gamma_k^{N*}(x),x) + V^i(k+1,\tilde{f}_k^{i*}(x,u_k^i)+\theta_k)] \\ &= E_{\theta_k}[g_k^i(\tilde{f}_k^{i*}(x,\gamma_k^{i*}(x))+\theta_k, \gamma_k^{1*}(x),\ldots,\gamma_k^{i*},\ldots, \\ &\quad \gamma_k^{N*}(x),x) + V^i(k+1,\tilde{f}_k^{i*}(x,\gamma_k^{i*}(x))+\theta_k)]; \\ &\quad V^i(K+1,x)=0, \quad i \in \mathbf{N}, \end{aligned} \tag{6.39}$$

where

$$\tilde{f}_k^{i*}(x,u_k^i) \triangleq f_k(x,\gamma_k^{1*}(x),\ldots,\gamma_k^{i-1*}(x),u_k^i,\gamma_k^{i+1*}(x),\ldots,\gamma^{N*}(x)) \tag{6.40}$$

and E_{θ_k} denotes the expectation operation with respect to the statistics of θ_k. Any solution obtained as above is strongly time consistent, *and the corresponding expected Nash equilibrium cost for* $\mathbf{P}i$ *is* $V^i(1,x_1)$.

Proof. The first step is to verify the statement of the theorem for stagewise equilibrium solution (cf. Def. 6.4); hence, start with inequalities (6.36) at stage $k=K$, which read

$$\begin{aligned} J^i(\gamma^{1*};\ldots;\gamma^{i*};\ldots;\gamma^{N*}) &\le J^i(\gamma^{1*};\ldots;\gamma_1^{i*},\ldots,\gamma_{K-1}^{i*},\gamma_K^i;\ldots;\gamma^{N*}), \\ &\quad \forall \gamma_K^i \in \Gamma_K^i, \quad i \in \mathbf{N}, \end{aligned}$$

where

$$J^i(\gamma^1;\ldots;\gamma^N) = E_{\theta^K,x_1}[L^i(u^1,\ldots,u^N) \mid u^j = \gamma^j(\cdot), j \in \mathbf{N}]$$

and

$$\theta^k \triangleq \{\theta_1,\ldots,\theta_k\}, \quad k \in \mathbf{K}.$$

Because of the stage-additive nature of L^i, these inequalities equivalently describe an N-person static game defined by

$$\begin{aligned} &\min_{u_K^i \in U_K^i} E_{\theta^K,x_1}[g_K^i(\tilde{f}_K^{i*}(x_K^*,u_K^i)+\theta_K, \gamma_K^{1*}(x_K^*),\ldots,\gamma_K^{i-1*}(x_K^*), \\ &\qquad u_K^i, \gamma_K^{i+1*}(x_K^*),\ldots,\gamma_K^{N*}(x_K^*),x_K^*)] \qquad (i) \\ &= E_{\theta^K,x_1}[g_K^i(\tilde{f}_K^{i*}(x_K^*,\gamma_K^{i*}(\cdot))+\theta_K,\gamma_K^{1*}(\cdot),\ldots,\gamma_K^{i*}(\cdot),\ldots,\gamma_K^{N*}(\cdot),x_K^*)], \end{aligned}$$

where x_k^* $(k \in \mathbf{K})$ is recursively defined by

$$x_{k+1}^* = f_k^i(x_k^*,\gamma_k^1(x_k^*),\ldots,\gamma_k^N(x_k^*)) + \theta_k, \quad x_1^* = x_1. \qquad (ii)$$

Now, since θ_K is statistically independent of $\{\theta^{K-1},x_1\}$, (i) can equivalently be written as

$$\begin{aligned} E_{\theta^{K-1},x_1}\Big\{ \min_{u_K^i \in U_K^i} E_{\theta_K}[g_K^i(\tilde{f}_K^{i*}(x_K^*,u_K^i)+\theta_K,\gamma_K^{1*}(\cdot),\ldots,\gamma_K^{i-1*}(\cdot), \\ u_K^i,\gamma_K^{i+1*}(\cdot),\ldots,\gamma_K^{N*}(\cdot),x_K^*)]\Big\} \end{aligned} \qquad (iii)$$

$$= E_{\theta^{K-1},x_1}\{E_{\theta_K}[g^i_K(\tilde{f}^{i*}_K(x^*_K,\gamma^{i*}_K(\cdot))+\theta_K,\gamma^{1*}_K(\cdot),\ldots,\gamma^{i*}_K(\cdot),\ldots,\gamma^{N*}_K(\cdot),x^*_K)]\},$$

which implies that the minimizing u^i_K will be a function of x^*_K; furthermore (in contrast with the deterministic version) x^*_K cannot be expressed in terms of $x^*_{K-1},\ldots,x^*_1$ since there is a noise term in (ii) which directly contributes to additional errors in case of such a substitution (mainly because every strategy has a unique representation when the dynamics are given by (ii)). Therefore, under the CLPS information pattern, any stagewise Nash equilibrium strategy for **P***i* at stage K will only be a function of x_K, i.e., a feedback strategy $u^{i*}_K = \gamma^{i*}_K(x_K)$, which further implies that *the choice of such a strategy is independent of all the optimal (or otherwise) strategies employed by the players at previous stages.* Finally note that, at stage $k = K$, the minimizing solutions of (iii) coincide with those of (6.39) since the minimization problems are equivalent.

Let us now consider inequalities (6.36) at stage $k = K-1$, i.e.,

$$J^i(\gamma^{1*};\ldots;\gamma^{N*}) \le J^i(\gamma^{1*};\ldots;\gamma^{i*}_1,\ldots,\gamma^i_{K-1},\gamma^{i*}_K;\cdot;\gamma^{N*}),$$
$$\forall \gamma^i_{K-1}\in\Gamma^i_{K-1}, i\in\mathbf{N},$$

and these can further be written as follows because of the stage-additive nature of L^i and since γ^{i*}_K ($i\in\mathbf{N}$) have already been determined at stage K as feedback strategies (independent of all the previous strategies):

$$\begin{aligned}
&E_{\theta^{K-2},x_1}\Big\{\min_{u^i_k\in U^i_K} E_{\theta_{K-1}}[g^i_{K-1}(\tilde{f}^{i*}_{K-1}(x^*_{K-1},u^i_{K-1})+\theta_{K-1},\\
&\qquad\gamma^{1*}_{K-1}(\cdot),\ldots,\gamma^{i-1*}_{K-1}(\cdot),u^i_{K-1},\gamma^{i+1*}_{K-1}(\cdot),\ldots,\gamma^{N*}_{K-1}(\cdot),x^*_{K-1})\\
&\qquad\qquad+V^i(K,\tilde{f}^{i*}_{K-1}(x^*_{K-1},u^i_{K-1}))+\theta_{K-1}]\Big\} \qquad (iv)\\
&= E_{\theta^{K-2},x_1}\Big\{E_{\theta^{K-1}}[g^i_{K-1}(\tilde{f}^{i*}_{K-1}(x^*_{K-1},\gamma^{i*}_{K-1}(\cdot))+\theta_{K-1},\\
&\qquad\gamma^{1*}_{K-1}(\cdot),\ldots,\gamma^{i*}_{K-1}(\cdot),\ldots,\gamma^{N*}_{K-1}(\cdot),x^*_{K-1})\\
&\qquad\qquad+V^i(K,\tilde{f}^{i*}_{K-1}(x^*_{K-1},\gamma^{i*}_{K-1}(\cdot))+\theta_{K-1}]\Big\}.
\end{aligned}$$

In writing down (iv), we have also made use of the statistical independence property of θ^{K-2}, x_1, θ_{K-1} and θ_K. Through a reasoning similar to the one employed at stage $k = K$, we readily conclude that any $\gamma^i_{K-1}(\cdot)$ that satisfies (iv) will have to be a feedback strategy, *independent of all the past strategies employed and the past values of the state*; therefore $u^{i*}_{K-1} = \gamma^{i*}_{K-1}(x_{K-1})$ ($i\in$ **N**). This is precisely what (6.39) says for $k = K-1$.

Proceeding in this manner, the theorem can readily be verified for stagewise equilibrium; that is, every set of strategies $\{\gamma^{i*}; i\in\mathbf{N}\}$ satisfying (6.39) constitutes a stagewise equilibrium solution for the stochastic game under consideration, and conversely every stagewise equilibrium solution of the stochastic dynamic game is comprised of feedback strategies which satisfy (6.39).

To complete the last phase of verification of the theorem, we first observe that every stagewise equilibrium solution determined in the foregoing derivation is also a feedback Nash equilibrium solution (satisfying (3.28)) since the construction of stagewise equilibrium strategies at stage $k = l$ did not depend on the strategies employed at stages $k < l$ (note the italicized statements above). Now, since every feedback Nash equilibrium under CLPS information pattern is also a Nash equilibrium (this is a trivial extension of Prop. 6.1 to stochastic dynamic games), and furthermore since every Nash equilibrium solution is a stagewise equilibrium solution (see Remark 6.9), it readily follows that the statement of Thm. 6.10 is valid also for the Nash equilibrium solution. The strong time consistency of the solution (cf. Def. 5.14) follows readily from the constructive nature of the derivation given above. Also, since $V^i(k,x)$ is the expected cost-to-go function for **P**i, $V^i(1,x_1)$ indeed follows as the expected Nash equilibrium cost for him. □

Remark 6.12 First, since every strategy admits a unique representation under the state dynamics description (6.37) (with $\operatorname{cov}(\theta_k) > 0$, $k \in \mathbf{K}$), the solution presented in Thm. 6.10 clearly does not exhibit any "informational nonuniqueness". Therefore if nonunique Nash equilibria exist in a stochastic dynamic game with CLPS information pattern, this is only due to the structure of f_k and g_k^i and not a consequence of the dynamic nature of the information pattern as in the deterministic case. Second, the Nash equilibrium solution presented in Thm. 6.10 will, in general, depend on the statistical properties (i.e., the mean and covariance) of the Gaussian noise term θ_k $(k \in \mathbf{K})$. For the special case of affine-quadratic stochastic dynamic games, however, no such dependence on the covariance exists as it is shown in the sequel. □

As an immediate application of Thm. 6.10, let us consider the class of affine-quadratic stochastic dynamic games wherein f_k and g_k^i are as defined by (6.5a)-(6.5b). Furthermore, assume that $E[\theta_k] = 0 \, \forall k \in \mathbf{K}$. Because of this latter assumption, and since f_k is affine and g_k^i is quadratic, the minimization problem in (6.39) becomes independent of the covariance of θ_k, as far as the optimum strategies are concerned, and a comparison with (6.16) readily leads to the conclusion that they should admit the same solution, which is affine in the current value of the state. Therefore, we arrive at the following corollary which we state here without any further verification.

Corollary 6.4 *An N-person affine-quadratic stochastic dynamic game as formulated above, and with $Q_{k+1}^i \geq 0$ $(i \in \mathbf{N}, k \in \mathbf{K})$, $R_k^{ij} \geq 0$ $(i,j \in \mathbf{N}, j \neq i, k \in \mathbf{K})$ and CLPS information pattern, admits a unique Nash equilibrium solution if, and only if, (6.17a) and (6.17d) admit unique solution sets $\{P_k^{i*}; i \in \mathbf{N}, k \in \mathbf{K}\}$ and $\{\alpha_k^{i*}; i \in \mathbf{N}, k \in \mathbf{K}\}$ in which case the equilibrium strategies are given by*

$$\gamma_k^{i*}(x_k) = -P_k^{i*} x_k - \alpha_k^{i*} \quad (k \in \mathbf{K}, i \in \mathbf{N}); \tag{6.41}$$

that is, the unique solution coincides with the unique strongly time-consistent feedback Nash equilibrium solution of the deterministic version of the game (cf. Corollary 6.1).

We therefore observe that an appropriate stochastic formulation in affine-quadratic dynamic games with CLPS information pattern serves to remove informational nonuniqueness of equilibria, since it leads to a unique equilibrium which is one of the uncountably many equilibria of the original deterministic game. This particular equilibrium solution (which is also a strongly time-consistent feedback equilibrium solution in this case) may be viewed as the unique *robust* solution of the deterministic game, which is insensitive to zero-mean random perturbations in the state equation. As an illustration, let us consider the three-person dynamic game example of subsection 6.3.1, and, out of the uncountably many Nash equilibrium solutions obtained there, let us attempt to select the one which features such a property. The new (stochastic) game is now the one which has zero-mean random perturbations in (6.31). More specifically, the state equation is

$$\left.\begin{array}{rcl} x_3 &=& x_2 + u^1 + u^2 + \theta_2, \\ x_2 &=& x_1 + u^3 + \theta_1, \end{array}\right\}$$

where $E[\theta_2] = E[\theta_1] = 0$; the variables θ_2, θ_1, x_1 are Gaussian, statistically independent, and $E[(\theta_2)^2] > 0$, $E[(\theta_1)^2] > 0$. The unique Nash equilibrium solution of this stochastic game is $\{\gamma^{1*}(x_2) = -(2/3)x_2,\ \gamma^{2*}(x_2) = (1/3)x_2,\ \gamma^{3*}(x_1) = -(2/11)x_1\}$ which is also the feedback Nash equilibrium solution of the associated deterministic game.

If the underlying information pattern of a stochastic dynamic game is not CLPS (or more precisely, if every player does not have access to the current value of the state), the statement of Thm. 6.10 is not valid. It is, in fact, not possible to obtain a general theorem which provides the Nash equilibrium solution under all different types of deterministic information patterns, and therefore one has to solve each such game separately. One common salient feature of all these solutions is that they are all free of any "informational nonuniqueness", and any nonuniqueness in equilibria (if it exists) is only due to the structure of the cost functionals and state equation. In particular, for affine-quadratic stochastic dynamic games, it can be shown that the Nash equilibrium solution will in general be unique under every deterministic information pattern. One such class of problems, viz. N-person linear-quadratic stochastic dynamic games wherein some players have CLPS information and others OL information, have been considered in Başar (1975) and a unique Nash equilibrium solution has been obtained in explicit form. This solution has the property that for the players with CLPS information the equilibrium strategies are dependent linearly not only on the current value of the state, but also on the initial state. The following example now illustrates the nature of this solution and also serves to provide a comparative study of Nash equilibria under four different information patterns.

Example 6.1 Consider the stochastic version of the three-person nonzero-sum game of subsection 6.3.1. Under four different information structures, namely, when (i) **P1** and **P2** have access to CLPS information and **P3** has OL information (this case is also covered by Corollary 6.4), (ii) **P1** has CLPS information and **P2**, **P3** have OL information, (iii) **P2** has CLPS information and **P1**, **P3** have OL information, (iv) all three players have OL information. In these stochastic games the state dynamics are described by

$$\left.\begin{aligned} x_3 &= x_2 + u^1 + u^2 + \theta_2, \\ x_2 &= x_1 + u^3 + \theta_1, \end{aligned}\right\}$$

where $\{\theta_1, \theta_2, x_1\}$ is a set of statistically independent Gaussian random variables with $E[\theta_k] = 0$, $E[(\theta_k)^2] > 0$, $k = 1, 2$.[60] The cost functionals are again as given by (6.32), but now their expected values determine the costs incurred to the players.

(i) Under the first information pattern (CLPS for **P1** and **P2**, OL for **P3**), the unique solution also follows from Corollary 6.4 and is given by (as discussed right after Corollary 6.4)

$$\gamma^{1*}(x_2) = -\frac{2}{3}x_2, \; \gamma^{2*}(x_2) = \frac{1}{3}x_2, \; \gamma^{3*}(x_1) = -\frac{2}{11}x_1.$$

The corresponding expected values of the cost functionals are

$$J^{1*} = \frac{72}{121}E[(x_1)^2] + \frac{8}{9}E[(\theta_1)^2] + E[(\theta_2)^2], \tag{6.42a}$$

$$J^{2*} = -J^{3*} = -\frac{2}{11}E[(x_1)^2] - \frac{2}{9}E[(\theta_1)^2] - E[(\theta_2)^2]. \tag{6.42b}$$

(ii) Under the second information pattern (CLPS for **P1**, OL for **P2** and **P3**), a direct derivation or utilization of the results of Başar (1975) leads to the unique Nash equilibrium solution

$$\gamma^{1*}(x_2, x_1) = -\frac{1}{2}x_2 - \frac{1}{8}x_1, \; \gamma^{2*}(x_1) = \frac{1}{4}x_1, \; \gamma^{3*}(x_1) = -\frac{1}{4}x_1.$$

Note that this unique solution of the stochastic game under consideration corresponds to one of the uncountably many equilibrium solutions presented in Prop. 6.5 for its deterministic version—simply set $p = 1/6$, $q = 1/3$. The corresponding expected values of the cost functionals are

$$J^{1*} = \frac{1}{2}E[(x_1)^2] + \frac{1}{2}E[(\theta_1)^2] + E[(\theta_2)^2], \tag{6.43a}$$

$$J^{2*} = -J^{3*} = -\frac{3}{16}E[(x_1)^2] - \frac{1}{4}E[(\theta_1)^2] - E[(\theta_2)^2]. \tag{6.43b}$$

(iii) Under the third information pattern (CLPS for **P2**, OL for **P1** and **P3**), direct derivation leads to a finite solution set

$$\gamma^{1*}(x_1) = -\frac{2}{5}x_1, \; \gamma^{2*}(x_2, x_1) = x_2 - \frac{2}{5}x_1, \; \gamma^{3*}(x_1) = -\frac{2}{5}x_1,$$

[60] This Gaussian distribution may be replaced by any probability distribution that assigns positive probability mass to every open subset of R^n (see also Remark 5.11), without affecting the results to be obtained in the sequel.

which also corresponds to one of the solution sets of Prop. 6.5, this time obtained by setting $p = 2/3$, $q = 2/3$. The corresponding expected values of the cost functionals are

$$J^{1*} = \frac{8}{25}E[(x_1)^2] + 4E[(\theta_1)^2] + E[(\theta_2)^2], \tag{6.44a}$$

$$J^{2*} = -J^{3*} = -\frac{6}{25}E[(x_1)^2] - 2E[(\theta_1)^2] - E[(\theta_2)^2]. \tag{6.44b}$$

(iv) Under the fourth information pattern (OL for all three players), the unique Nash equilibrium solution is

$$\gamma^{1*}(x_1) = -\frac{2}{5}x_1,\ \gamma^{2*}(x_1) = \frac{1}{5}x_1,\ \gamma^{3*}(x_1) = -\frac{2}{5}x_1,$$

which can also be obtained from Prop. 6.5 by letting $p = 2/3$, $q = -1/3$. The corresponding expected cost values are

$$J^{1*} = \frac{8}{25}E[(x_1)^2] + E[(\theta_1)^2] + E[(\theta_2)^2], \tag{6.45a}$$

$$J^{2*} = -J^{3*} = -\frac{6}{25}E[(x_1)^2] - E[(\theta_1)^2] - E[(\theta_2)^2]. \tag{6.45b}$$

A comparison of the expected equilibrium cost values of the players attained under the four different information patterns now reveals some interesting features of the Nash equilibrium solution.

(1) Comparing (6.43a) with (6.45a), we observe that when the other players' information patterns are OL, an increase in information to **P1** (that is, going from OL to CLPS) could improve his performance (if $E[(\theta_1)^2]$ is sufficiently large) or be detrimental (if $E[(\theta_1)^2]$ is relatively small).

(2) When **P2** has CLPS information, a comparison of (6.42a) and (6.44a) leads to a similar trend in the effect of an increase of **P1**'s information to his equilibrium performance.

(3) When **P1** has CLPS information, an increase in information to **P2** leads to degradation in J^{2*} (simply compare (6.42b) with (6.43b)).

(4) When **P1** has OL information, an increase in information to **P2** leads to improvement in his performance (compare (6.44b) with (6.45b)).

These results clearly indicate that, in nonzero-sum games and under the Nash equilibrium solution concept, an increase in information to one player does not necessarily lead to improvement in his equilibrium performance, and at times it could even be detrimental.

□

6.5 Open-Loop and Feedback Nash and Saddle-Point Equilibria of Differential Games

This section presents results which are counterparts of those given in Section 6.2, for differential games that fit the framework of Def. 5.5; that is, we consider N-person quantitative dynamic games defined in continuous time, and described by the state equation

$$\dot{x}(t) = f(t, x(t), u^1(t), \ldots, u^N(t)); \quad x(0) = x_0 \tag{6.46a}$$

and cost functionals

$$L^i(u^1, \ldots, u^N) = \int_0^T g^i(t, x(t), u^1(t), \ldots, u^N(t))\, \mathrm{d}t + q^i(x(T)); \quad i \in \mathbf{N}. \tag{6.46b}$$

Here, $[0, T]$ denotes the fixed prescribed duration of the game, x_0 is the initial state known by all players, $x(t) \in \mathbf{R}^n$ and $u^i(t) \in S^i \subseteq \mathbf{R}^{m_i}$ $(i \in \mathbf{N})$ $\forall t \in [0, T]$. Furthermore, f and $\gamma^i \in \Gamma^i$ $(i \in \mathbf{N})$ satisfy the conditions of Thm. 5.1 so that, for a given information structure (as one of (i), (ii) or (iv) of Def. 5.6), the state equation admits a unique solution for every corresponding N-tuple $\{\gamma^i \in \Gamma^i; i \in \mathbf{N}\}$. Nash equilibrium solution under a given information structure is again defined as one satisfying the set of inequalities (6.3) where J^i denotes the cost functional of $\mathbf{P}i$ for the differential game in strategic (normal) form.

6.5.1 Open-loop Nash equilibria

The results to be described in this subsection are counterparts of those given in Section 6.2.1; and in order to display this relationship explicitly, we shall present them here in the same order as their counterparts in subsection 6.2.1. We therefore first have the counterpart of Thm. 6.1 in the continuous time.

Theorem 6.11 *For an N-person differential game of prescribed fixed duration $[0, T]$, let*

(i) $f(t, \cdot, u^1, \ldots, u^N)$ be continuously differentiable on $\mathbf{R}^n$, $\forall t \in [0, T]$,

(ii) $g^i(t, \cdot, u^1, \ldots, u^N)$ and $q^i(\cdot)$ be continuously differentiable on $\mathbf{R}^n$, $\forall t \in [0, T], i \in \mathbf{N}$.

Then, if $\{\gamma^{i*}(t, x_0) = u^{i*}(t);\ i \in \mathbf{N}\}$ provides an open-loop Nash equilibrium solution, *and $\{x^*(t),\ 0 \le t \le T\}$ is the corresponding state trajectory, there exist N* costate functions *$p^i(\cdot) : [0, T] \to \mathbf{R}^n, i \in \mathbf{N}$, such that the following relations are satisfied:*

$$\dot{x}^*(t) = f(t, x^*(t), u^{1*}(t), \ldots, u^{N*}(t)), \quad x^*(0) = x_0, \tag{6.47a}$$

$$\begin{aligned} \gamma^{i*}(t, x_0) \ \equiv\ u^{i*}(t) = \arg \min_{u^i \in S^i} & H^i(t, p^i(t), x^*(t), u^{1*}(t), \\ & \ldots, u^{i-1*}(t), u^i, u^{i+1*}(t), \ldots, u^{N*}(t)), \end{aligned} \tag{6.47b}$$

$$\begin{aligned} \dot{p}^{i'}(t) &= -\frac{\partial}{\partial x} H^i(t, p^i(t), x^*, u^{1*}(t), \ldots, u^{N*}(t)), \\ p^{i'}(T) &= \frac{\partial}{\partial x} q^i(x^*(T)), \quad i \in \mathbf{N}, \end{aligned} \tag{6.47c}$$

where

$$H^i(t, p^i, x, u^1, \ldots, u^N) \triangleq g^i(t, x, u^1, \ldots, u^N) + p^{i'} f(t, x, u^1, \ldots, u^N),$$
$$t \in [0, T], \quad i \in \mathbf{N}.$$

Every such Nash equilibrium solution is weakly time consistent.

Proof. The proof follows the same lines as in the proof of Thm. 6.1, but now the minimum principle for continuous-time control systems (i.e., Thm. 5.4) is used instead of Thm. 5.5. □

Remark 6.13 One class of differential games for which the necessity condition of Thm. 6.11 is satisfied is that with *weakly coupled* players (à la Remark 4.1), that is one with the following state equation and cost functions (taking N=2, without any loss of generality):

$$\begin{aligned} \dot{x}_1(t) &= f_1(t, x_1(t), u^1(t)) + \epsilon f_{12}(t, x_2(t)); \quad x_1(0) = x_{10}, \\ \dot{x}_2(t) &= f_2(t, x_2(t), u^2(t)) + \epsilon f_{21}(t, x_1(t)); \quad x_2(0) = x_{20} \end{aligned} \tag{6.48a}$$

and cost functionals

$$\begin{aligned} L^i(u^1, u^2; \epsilon) = \int_0^T & [g^{ii}(t, x_i(t), u^i(t)) + \epsilon g^{ij}(t, x_j(t), u^j(t))] \, dt \\ & + q^{ii}(x_i(T)) + \epsilon q^{ij}(x_j(T)); \quad j \neq i, \quad i, j = 1, 2, \end{aligned} \tag{6.48b}$$

where ϵ is a sufficiently small scalar. Under some appropriate convexity (on g^{ij}'s) and differentiability conditions, it can be shown (see Srikant and Başar (1992)) that there exists an $\epsilon_0 > 0$ such that for all $\epsilon \in (-\epsilon_0, \epsilon_0)$, the differential game admits a unique open-loop Nash equilibrium solution that is stable with respect to Gauss–Seidel or Jacobi iterations (see Def. 4.6 for terminology). This solution can be obtained by expanding the state and control corresponding to the open-loop Nash equilibrium solution in power series in terms of ϵ,

$$x^*(t; \epsilon) = \sum_{k=0}^{\infty} x^{(k)}(t) \epsilon^k, \quad u^{i*}(t; \epsilon) = \sum_{k=0}^{\infty} u_i^{(k)}(t) \epsilon^k,$$

substituting these into (6.47a)-(6.47c), along with a similar expansion for $p^i(t)$, and solving for the different terms $x^{(k)}$ and $u_i^{(k)}$, $k = 0, 1, \ldots$, iteratively. It turns out that $u_i^{(0)}$ $(i = 1, 2)$ are the (open-loop) optimal controls associated with the decoupled optimal control problems:

$$\begin{aligned} \dot{x}_i &= f_i(t, x_i, u^i(t)), \quad x_i(0) = x_{i0}, \\ L^i(u^i) &= \int_0^T g^{ii}(t, x_i(t), u^i(t)) \, dt + q^{ii}(x_i(T)), \quad i = 1, 2, \end{aligned}$$

and $x^{(0)}$ is the corresponding state trajectory, with $x^{(0)} = (x_1^{(0)'}, x_2^{(0)'})'$. For $k \geq 1$, $u_1^{(k)}$ and $u_2^{(k)}$ are obtained by solving some appropriate linear-quadratic optimal control problems (see Srikant and Başar (1991)). Hence this approach decomposes the original two-player differential game into two nonlinear optimal control problems (the zeroth-order problems) and a sequence of iteratively constructed linear-quadratic control problems. Halting this iteration at the kth step yields an ϵ^k-approximate open-loop Nash equilibrium solution. □

As indicated earlier, Thm. 6.11 provides a set of necessary conditions for the open-loop Nash equilibrium solution to satisfy, and therefore it can be used to generate candidate solutions. For the special class of "affine-quadratic" differential games, however, a unique candidate solution can be obtained in explicit terms, which can further be shown to be an open-loop Nash equilibrium solution under certain convexity restrictions on the cost functionals of the players.

Definition 6.5 *An N-person differential game of fixed prescribed duration is of the* affine-quadratic *type if $S^i = \mathbf{R}^{m_i}$ $(i \in \mathbf{N})$ and*

$$\begin{aligned} f(t,x,u^1,\ldots,u^N) &= A(t)x + \sum_{i\in\mathbf{N}} B^i(t)u^i + c(t), \\ g^i(t,x,u^1,\ldots,u^N) &= \frac{1}{2}(x'Q^i(t)x + \sum_{j\in\mathbf{N}} u^{j'}R^{ij}(t)u^j), \\ q^i(x) &= \frac{1}{2}x'Q^i_f x, \end{aligned}$$

where $A(\cdot)$, $B^i(\cdot)$, $Q^i(\cdot)$, $R^{ij}(\cdot)$ are matrices of appropriate dimensions, $c(\cdot)$ is an n-dimensional vector, all defined on $[0,T]$, and with continuous entries $(i,j \in \mathbf{N})$. Furthermore $Q^i_f, Q^i(\cdot)$ are symmetric, and $R^{ii}(\cdot) > 0$ $(i \in \mathbf{N})$.
An affine-quadratic game is of the linear-quadratic *type if $c \equiv 0$.*

Theorem 6.12 *For an N-person affine-quadratic differential game with $Q^i(\cdot) \geq 0$, $Q^i_f \geq 0$ $(i \in \mathbf{N})$, let there exist a solution set $\{M^i; i \in \mathbf{N}\}$ to the coupled matrix Riccati differential equations*

$$\begin{aligned} \dot{M}^i + M^iA + A'M^i + Q^i - M^i \sum_{j\in\mathbf{N}} B^j(R^{jj})^{-1}B^{j'}M^j &= 0; \\ M^i(T) = Q^i_f \quad (i \in \mathbf{N}). \end{aligned} \tag{6.49}$$

Then, the differential game admits an open-loop Nash equilibrium solution *given by*

$$\gamma^{i*}(t,x_0) \equiv u^{i*}(t) = -R^{ii}(t)^{-1}B^{i'}(t)[M^i(t)x^*(t) + m^i(t)] \quad (i \in N), \tag{6.50}$$

where $\{m^i(\cdot),\ i \in \mathbf{N}\}$ solve uniquely the set of linear differential equations:

$$\begin{aligned} \dot{m}^i + A'm^i + M^ic - M^i \sum_{j\in\mathbf{N}} B^j(R^{jj})^{-1}B^{j'}m^j &= 0; \\ m^i(T) = 0 \quad (i \in \mathbf{N}), \end{aligned} \tag{6.51}$$

and $x^*(\cdot)$ denotes the (Nash) equilibrium state trajectory, generated by

$$\begin{aligned} x^*(t) &= \Phi(t,0)x_0 + \int_0^t \Phi(t,\sigma)\eta(\sigma)\,\mathrm{d}\sigma, \\ \frac{\mathrm{d}}{\mathrm{d}t}\Phi(t,\sigma) &= F(t)\Phi(t,\sigma); \quad \Phi(\sigma,\sigma) = I, \\ F(t) &\triangleq A - \sum_{i\in\mathbf{N}} B^i(R^{ii})^{-1}B^{i'}M^i(t), \\ \eta(t) &\triangleq c(t) - \sum_{i\in\mathbf{N}} B^i(R^{ii})^{-1}B^{i'}m^i(t). \end{aligned} \tag{6.52}$$

Proof. For the affine-quadratic differential game, and in view of the additional restrictions $Q^i(\cdot) \geq 0$, $Q^i_f \geq 0$, $L^i(u^1,\ldots,u^N)$ is a strictly convex function of $u^i(\cdot)$ for all permissible control functions $u^j(\cdot)$ $(j \neq i, j \in \mathbf{N})$ and for all $x_0 \in \mathbf{R}^n$. This then implies that Thm. 6.11 is also a sufficiency result and every solution set of the first-order conditions provides an open-loop Nash solution. Hence, we now show that the solution given in Thm. 6.12 indeed satisfies the first-order conditions.

First note that the Hamiltonian is

$$H^i(t,p,x,u^1,\ldots,u^N) = \frac{1}{2}(x'Q^ix + \sum_{j\in\mathbf{N}} u^{j'}R^{ij}u^j) + p^{i'}(Ax + c + \sum_{j\in\mathbf{N}} B^ju^j)$$

whose minimization with respect to $u^i(t) \in \mathbf{R}^{m_i}$ yields the unique relation

$$u^{i*}(t) = -(R^{ii}(t))^{-1}B^i(t)'p^i(t), \quad i \in \mathbf{N}. \tag{i}$$

Furthermore, the costate equations are

$$\dot{p}^i = -Q^ix^* - A'p^i, \quad p^i(T) = Q^i_fx^*(T) \quad (i \in \mathbf{N}),$$

and the optimal state trajectory is given by

$$\dot{x}^* = Ax^* + c - \sum_{i\in\mathbf{N}} B^i(R^{ii})^{-1}B^{i'}p^i, \quad x^*(0) = x_0. \tag{ii}$$

This set of differential equations constitutes a two-point boundary value problem. Now, substituting $p^i = M^ix^* + m^i$ $(i \in \mathbf{N})$ into the costate equations, we arrive at the conclusion that if M^i $(i \in \mathbf{N})$ and m^i $(i \in \mathbf{N})$ satisfy (6.49) and (6.51), respectively, this indeed solves the two-point boundary value problem, along with x^*. The expressions for the open-loop Nash strategies follow readily from (i) by substituting $p^i = M^ix^* + m^i$, and likewise the associated state trajectory (6.52) follows from (ii). □

Remark 6.14 It is possible to obtain the result of Thm. 6.12 in a somewhat different way, by directly utilizing Prop. 5.2; see Problem 10 in section 6.8. □

Theorem 6.12 provides a set of sufficiency conditions for the Nash equilibrium solution to exist. It will now be shown that a Nash equilibrium may exist even if (6.49) does not admit a solution. For simplicity of presentation assume now that $c(t) \equiv 0$. The differential equations for $p^i(t)$ and $x^*(t)$ can be written jointly as

$$\frac{\mathrm{d}}{\mathrm{d}t}\begin{pmatrix} x \\ p^1 \\ p^2 \\ \vdots \\ p^N \end{pmatrix} = -\begin{pmatrix} -A & \bar{S}^1 & \bar{S}^2 & \dots & \bar{S}^N \\ Q^1 & A' & 0 & \dots & 0 \\ Q^2 & 0 & A' & 0 & 0 \\ \vdots & \vdots & \ddots & \ddots & 0 \\ Q^N & 0 & \dots & 0 & A' \end{pmatrix}\begin{pmatrix} x \\ p^1 \\ p^2 \\ \vdots \\ p^N \end{pmatrix}, \tag{6.53}$$

where $\bar{S}^i \triangleq B^i(R^{ii})^{-1}B^{T'}$, $i = 1, 2, \dots, N$. The boundary conditions of (6.53) are $x(0) = x_0$, $p^i(T) = Q_f^i x(T)$, $i = 1, 2, \dots, N$. Symbolically (6.53) will be written as

$$\dot{z} = -Gz, \text{ with } G \triangleq \begin{pmatrix} -A & \bar{S}^1 & \bar{S}^2 & \dots & \bar{S}^N \\ Q^1 & A' & 0 & \dots & 0 \\ Q^2 & 0 & A' & 0 & 0 \\ \vdots & \vdots & \ddots & \ddots & 0 \\ Q^N & 0 & \dots & 0 & A' \end{pmatrix}, \tag{6.54}$$

and

$$Pz(0) + Qz(T) = (x_0' \, 0 \, \dots \, 0)', \tag{6.55}$$

with

$$P \triangleq \begin{pmatrix} I & 0 & \dots & \dots & 0 \\ 0 & \dots & \dots & 0 & \\ \vdots & & & & \vdots \\ \vdots & & & & \vdots \\ 0 & \dots & \dots & \dots & 0 \end{pmatrix}, \quad Q \triangleq \begin{pmatrix} 0 & 0 & \dots & \dots & 0 \\ -Q_f^1 & I & 0 & \dots & 0 \\ Q_f^2 & 0 & I & 0 & 0 \\ \vdots & \vdots & \ddots & \ddots & 0 \\ Q_f^N & 0 & \dots & 0 & I \end{pmatrix}. \tag{6.56}$$

From (6.55) we immediately get that a unique open-loop Nash equilibrium solution exists if and only if the equation

$$(Pe^{GT} + Q)e^{-GT}z(0) = (x_0' \, 0 \, \dots \, 0)' \tag{6.57}$$

is uniquely solvable in $z(0)$ for every x_0. The matrix G consists of $(N+1) \times (N+1)$ blocks. The same block structure will be applied to e^{Gt} and these blocks will be indicated by $W_{ij}(t)$, $i, j = 1, 2, \dots, N+1$. If we define

$$H(T) \triangleq W_{11}(T) + \sum_{k=1}^{N} W_{1,k+1}(T)Q_f^k, \tag{6.58}$$

then the solvability of (6.57) is equivalent to the invertibility of $H(T)$.

The following example now shows that an open-loop Nash equilibrium solution can exist even if the coupled set of Riccati differential equations (6.49) does not have a solution.

Example 6.2 Consider the two-person ($N = 2$) linear-quadratic game with

$$A = \begin{pmatrix} -1 & 0 \\ 0 & -5/22 \end{pmatrix}, \; B^1 = \begin{pmatrix} 1 & 0 \\ 0 & 1 \end{pmatrix}, \; B^2 = \begin{pmatrix} 1 \\ 0 \end{pmatrix},$$

$$Q^1 = \begin{pmatrix} 1 & 0 \\ 0 & 0 \end{pmatrix}, \; Q^2 = \begin{pmatrix} 1 & 1 \\ 1 & 2 \end{pmatrix},$$

$$R^{11} = \begin{pmatrix} 1 & 1 \\ 1 & 2 \end{pmatrix}^{-1}, \; R^{22} = 1, \; R^{12} = R^{21} = 0.$$

Choose $T = 0.1$. Then numerical calculation leads to

$$W_{11}(T) = \begin{pmatrix} 1.1155 & 0.0051 \\ 0.0051 & 1.0230 \end{pmatrix}, W_{12}(T) = \begin{pmatrix} 0.1007 & 0.1047 \\ 0.0964 & 0.2002 \end{pmatrix},$$

$$W_{13}(T) = \begin{pmatrix} 0.1005 & 0 \\ 0.0002 & 0 \end{pmatrix}.$$

Pick

$$Q_f^1 = \begin{pmatrix} 1 & h_1 \\ h_1 & h_1^2 + 1 \end{pmatrix}$$

and

$$Q_f^2 = \begin{pmatrix} 10 & h_2 \\ h_2 & (h_2^2 + 1)/10 \end{pmatrix},$$

where

$$h_1 = -\frac{(W_{11})_{12} + (W_{12})_{21} + 10(W_{13})_{21}}{(W_{12})_{22}},$$

$$h_2 = -\frac{(W_{11})_{22} + (W_{12})_{21}(Q_f^1)_{12} + (W_{12})_{22}(Q_f^1)_{22}}{(W_{13})_{21}}.$$

Here the notation $(W_{1i})_{jk}$ refers to the jkth element of the 2×2 matrix $W_{1i}(T)$. Clearly both Q_f^1 and Q_f^2 are positive definite, whereas

$$H(T) = W_{11}(T) + W_{12}(T)Q_f^1 + W_{13}(T)Q_f^2 = \begin{pmatrix} 2.1673 & -752.6945 \\ 0 & 0 \end{pmatrix}$$

is not invertible.

The conclusion therefore is that an open-loop Nash equilibrium does not exist and the corresponding set of Riccati differential equations does not have a solution. Now we modify the problem statement in one respect, viz. instead of $T = 0.1$ now take $T = 0.11$. Numerical calculation then shows that $H(T)$ is now invertible. However, the solution of the set of Riccati differential equations, when integrated backward and starting at $T = 0.11$, blows up (i.e., becomes ∞) at $t = 0.01$ (note that the solution of the Riccati equations is shift-invariant). Thus it has been shown that for $T = 0.11$ an open-loop Nash equilibrium solution exists despite the fact that the Riccati differential equations do not have a solution on the interval $[0, 0.11]$. □

What the preceding example has shown is that for a linear-quadratic game defined on a fixed interval $[0, T]$, existence of a Nash equilibrium does not imply existence of a solution to the set of Riccati equations (6.49). An equivalence between these two conditions exists, however, if T is a variable, as shown in (Engwerda, 1998). What can be shown is that the following three statements are equivalent, where T_f is a positive real number.

- For each $T \in (0, T_f]$ an open-loop Nash equilibrium solution exists for the linear-quadratic game defined on $[0, T]$.
- $H(T)$ is invertible for all $T \in [0, T_f]$.
- The set of Riccati differential equations (6.49) has a solution on $[0, T_f]$.

Open-loop saddle points of zero-sum differential games

For the class of two-person zero-sum differential games, we rewrite the cost functional (to be minimized by **P1** and maximized by **P2**) as

$$L(u^1, u^2) = \int_0^T g(t, x, u^1, u^2)\, \mathrm{d}t + q(x(T)),$$

and obtain the following set of necessary conditions for the saddle-point strategies to satisfy.

Theorem 6.13 *For a two-person zero-sum differential game of prescribed fixed duration* $[0, T]$, *let*

(i) $f(t, \cdot, u^1, u^2)$ *be continuously differentiable on* $\mathbf{R}^n$, $\forall t \in [0, T]$,

(ii) $g(t, \cdot, u^1, u^2)$ *and* $q(\cdot)$ *be continuously differentiable on* $\mathbf{R}^n, \forall t \in [0, T]$. *Then, if* $\{\gamma^{i*}(t, x_0) = u^{i*}(t);\ i = 1, 2\}$ *provides an open-loop saddle-point solution, and* $\{x^*(t), 0 \le t \le T\}$ *is the corresponding state trajectory, there exists a costate function* $p(\cdot) : [0, T] \to \mathbf{R}^n$ *such that the following relations are satisfied:*

$$\dot{x}^*(t) = f(t, x^*(t), u^{1*}(t), u^{2*}(t)); \quad x^*(0) = x_0, \tag{6.59a}$$

$$H(t, p, x^*, u^{1*}, u^2) \le H(t, p, x^*, u^{1*}, u^{2*}) \le H(t, p, x^*, u^1, u^{2*}), \quad \forall u^1(t) \in S^1, u^2(t) \in S^2,\ t \in [0, T], \tag{6.59b}$$

$$\dot{p}'(t) = -\frac{\partial}{\partial x} H(t, p(t), x^*, u^{1*}(t), u^{2*}(t)); \quad p'(T) = \frac{\partial}{\partial x} q(x^*(T)), \tag{6.59c}$$

where

$$H(t, p, x, u^1, u^2) \triangleq g(t, x, u^1, u^2) + p' f(t, x, u^1, u^2), \quad t \in [0, T]. \tag{6.59d}$$

Proof. This result follows directly from Thm. 6.11 by taking $N = 2$, $g^1 \equiv -g^2 \triangleq g, q^1 \equiv -q^2 \triangleq q$, and by noting that $p^1(\cdot) \equiv -p^2(\cdot) \triangleq p(\cdot)$ and $H^1(\cdot) \equiv -H^2(\cdot) \triangleq H(\cdot)$. □

For the special case of two-person zero-sum affine-quadratic differential games described by the state equation

$$\dot{x} = A(t)x + B^1(t)u^1 + B^2(t)u^2 + c(t); \quad x(0) = x_0 \tag{6.60a}$$

and the objective functional

$$\begin{aligned} L(u^1, u^2) &= \frac{1}{2}\int_0^T (x'(t)Q(t)x(t) + u^1(t)'u^1(t) - u^2(t)'u^2(t))\,\mathrm{d}t \\ &\quad + \frac{1}{2}x'(T)Q_f x(T), \end{aligned} \tag{6.60b}$$

we now first impose additional restrictions on the parameters so that $L(u^1, u^2)$ is strictly concave in u^2.

Lemma 6.4 *For the affine-quadratic two-person zero-sum differential game introduced above, the objective functional $L(u^1, u^2)$ is strictly concave in u^2 for all permissible $u^1(\cdot)$ if, and only if, there exists a unique bounded symmetric solution $S(\cdot)$ to the matrix Riccati equation*

$$\dot{S} + A'S + SA + Q + SB^2B^{2'}S = 0; \quad S(T) = Q_f, \tag{6.61}$$

on the interval $[0, T]$.

Proof. A reasoning similar to the one employed in the proof of Lemma 6.1 leads to the observation that the requirement of strict concavity is equivalent to the condition of existence of a unique solution to the optimal control problem

$$\min_{u(\cdot)} \left\{ \int_0^T [u(t)'u(t) - y(t)'Q(t)y(t)]\,\mathrm{d}t - y(T)'Q_f y(T) \right\}$$

subject to

$$\dot{y} = A(t)y + B^2(t)u; \quad y(0) = x_0.$$

The lemma then follows, in view of this observation, from Remark 5.8. □

Theorem 6.14 *For the two-person affine-quadratic zero-sum differential game described by (6.60a)-(6.60b) and with $Q_f \geq 0$, $Q(\cdot) \geq 0$, let there exist a unique bounded symmetric solution to the matrix Riccati differential equation (6.61) on $[0, T]$. Then,*

(i) there exists a unique bounded symmetric solution $M(\cdot)$ to the (generalized) matrix Riccati differential equation

$$\dot{M} + A'M + MA + Q - M(B^1B^{1'} - B^2B^{2'})M = 0; M(T) = Q_f, \tag{6.62a}$$

on $[0,T]$, *and again on* $[0,T]$ *a unique bounded solution* $m(\cdot)$ *to the linear vector differential equation*

$$\dot{m} + [A - (B^1 B^{1'} - B^2 B^{2'})M]'m + Mc = 0; \quad m(T) = 0, \tag{6.62b}$$

(ii) the game admits a unique open-loop saddle-point solution *given by*

$$\left.\begin{aligned} \gamma^{1*}(t,x_0) \equiv u^{1*}(t) &= -B^1(t)'[M(t)x^*(t) + m(t)], \\ \gamma^{2*}(t,x_0) \equiv u^{2*}(t) &= B^2(t)'[M(t)x^*(t) + m(t)], \end{aligned}\right\} \tag{6.63}$$

where $x^*(\cdot)$ *denotes the associated state trajectory, generated by*

$$\begin{aligned} x^*(t) &= \Phi(t,0)x_0 + \int_0^t \Phi(t,\sigma)[c(\sigma) - (B^1B^{1'} - B^2B^{2'})m(\sigma)]\,\mathrm{d}\sigma, \\ \frac{\mathrm{d}}{\mathrm{d}t}\Phi(t,\sigma) &= F(t)\Phi(t,\sigma); \quad \Phi(\sigma,\sigma) = I, \\ F &\triangleq A - (B^1B^{1'} - B^2B^{2'})M. \end{aligned} \tag{6.64}$$

Proof. The steps involved in the proof of this theorem are similar to those used in the proof of Thm. 6.4: we first establish existence of a unique saddle-point solution to the affine-quadratic differential game under consideration, and we then verify (i) and (ii) above in view of this property. For the former step we shall proceed rather casually, since a rigorous verification requires some prior knowledge of functional analysis which is beyond the scope of this book. The interested reader can, however, refer to Balakrishnan (1976) for a rigorous exposition of the underlying mathematics, or to Bensoussan (1971) and Lukes and Russell (1971) for a framework (Hilbert space setting) wherein similar mathematical techniques are used at a rigorous level but in a somewhat different context.[61]

Let U^i, the space of permissible control functions $u^i(\cdot)$, be a complete (Banach) space equipped with an inner product

$$\langle u, v\rangle_i \triangleq \int_0^T u(t)'v(t)\,\mathrm{d}t, \qquad \text{for } u, v \in U^i, \quad i = 1, 2.$$

Then, since $L(u^1, u^2)$ is quadratic in x, u^1,u^2, and x is linear in u^1 and u^2, the cost $L(u^1, u^2)$ can be written as

$$\begin{aligned} L(u^1,u^2) &= \frac{1}{2}\langle u^1, \mathbf{L}_1 u^1\rangle_1 - \langle u^2, \mathbf{L}_2 u^2\rangle_2 + \langle u^1, \mathbf{L}_{12}u^2\rangle_1 \\ &\qquad + \langle u^1, r_1\rangle_1 - \langle u^2, r_2\rangle_2 + r_3, \end{aligned} \tag{i}$$

where $\mathbf{L}_1 : U^1 \to U^1$, $\mathbf{L}^2 : U^2 \to U^2$, $\mathbf{L}_{12} : U^2 \to U^1$ are bounded linear operators, $r_1 \in U^1$, $r_2 \in U^2$ and $r_3 \in \mathbf{R}$. Moreover, since $Q(\cdot) \geq 0$, $Q_f \geq 0$, $\mathbf{L}_1$ is a strongly positive operator (written as $\mathbf{L}_1 > 0$), and furthermore under

[61] For an alternative rigorous verification of the result of this theorem, see Başar and Bernhard (1995).

the condition of unique solvability of (6.61) (cf. Lemma 6.4) $\mathbf{L}_2 > 0$. Now, if $L(u^1, u^2)$ admits a unique saddle point in open-loop policies, it is necessary and sufficient that it satisfy the following two relations (obtained by differentiation of (i))

$$\begin{aligned} \mathbf{L}_1 u^1 + \mathbf{L}_{12} u^2 + r_1 &= 0, \\ -\mathbf{L}_{12}^* u^1 + \mathbf{L}_2 u^2 + r_2 &= 0, \end{aligned}$$

where $\mathbf{L}_{12}^*$ denotes the adjoint of $\mathbf{L}_{12}$. Solving for $u^2 \in U^2$ from the second one (uniquely), and substituting it into the first one, we obtain the single relation

$$(\mathbf{L}_1 + \mathbf{L}_{12}\mathbf{L}_2^{-1}\mathbf{L}_{12}^*)u^1 + r_1 - \mathbf{L}_{12}\mathbf{L}_2^{-1}r_2 = 0 \tag{ii}$$

which admits a unique solution, since $\mathbf{L}_1 + \mathbf{L}_{12}\mathbf{L}_2^{-1}\mathbf{L}_{12}^* > 0$ and is thereby invertible. (This latter feature follows since $\mathbf{L}_1 > 0$ and $\mathbf{L}_{12}\mathbf{L}_2^{-1}\mathbf{L}_{12}^*$ is a nonnegative self-adjoint compact operator; see Balakrishnan (1976), for details.) We have thus established existence of a unique open-loop saddle-point strategy for **P1** under the condition of Lemma 6.4. This readily implies existence of a unique open-loop saddle-point strategy for **P2** since $\mathbf{L}_2$ is invertible.

Thus completing the proof of existence of a unique saddle point, we now turn to verification of (6.62a)-(6.62b) and (6.63). At the outset, let us first note that the unique saddle-point solution should necessarily satisfy relations (6.59a)-(6.59d) which can be rewritten for the affine-quadratic game as

$$\left.\begin{aligned} \dot{x}^* = A(t)x^* + B^1(t)u^{1*}(t) + B^2(t)u^{2*}(t) + c(t);\ x^*(0) = x_0, \\ u^{1*}(t) = -B^1(t)'p(t); \qquad u^{2*}(t) = B^2(t)'p(t), \\ \dot{p}(t) = -Q(t)x^*(t) - A(t)'p(t);\quad p(T) = Q_f x^*(T). \end{aligned}\right\} \tag{iii}$$

Since $r_1(\cdot)$ and $r_2(\cdot)$ introduced earlier are affine functions of x_0, the unique solution u^{1*} of (ii) is necessarily an affine function of x_0, and so is u^{2*}. Therefore, the dependence of $p(\cdot)$ on x_0 in (iii) should be affine. This readily implies that we can write, without any loss of generality, $p(t) = M(t)x^*(t) + m(t)$, where $M(\cdot)$ is some bounded matrix-valued function of dimension $(n \times n)$ and $m(\cdot)$ is some bounded vector-valued function of dimension n, both with continuously differentiable elements. Now, substitution of this relation into (iii), and requiring (iii) to hold true for all $x_0 \in \mathbf{R}^n$, leads to (6.62a)-(6.62b), which in turn yields (6.63). □

Remark 6.15 The statement of Thm. 6.14 is valid even if the "nonnegative-definiteness" condition on $Q(\cdot)$ does not hold, provided that some other appropriate conditions are imposed on the game parameters to ensure that L defined by (6.60b) is strictly convex in u^1. One such condition, which is in fact tight, can be obtained directly from Lemma 6.4, by simply replacing Q_f by $-Q_f$, $Q(\cdot)$ by $-Q(\cdot)$ and B^2 by B^1, which is the existence of a unique bounded symmetric solution $S^1(\cdot)$ to the matrix Riccati differential equation

$$\dot{S}^1 + A'S^1 + S^1A + Q - SB^1B^{1'}S = 0; \quad S^1(T) = Q_f, \tag{6.65}$$

on the interval $[0, T]$. □

6.5.2 Closed-loop no-memory and feedback Nash equilibria

Under the memoryless perfect state information pattern, the following theorem provides a set of necessary conditions for any closed-loop no-memory Nash equilibrium solution to satisfy.

Theorem 6.15 *For an N-person differential game of prescribed fixed duration $[0,T]$, let*

(i) $f(t,\cdot)$ be continuously differentiable on $\mathbf{R}^n \times S^1 \times \cdot \times S^N, \forall t \in [0,T]$,

(ii) $g^i(t,\cdot)$ and $q^i(\cdot)$ be continuously differentiable on $\mathbf{R}^n \times S^1 \times \cdots \times S^N$ and $\mathbf{R}^n$, respectively, $\forall t \in [0,T]$, $i \in \mathbf{N}$.

Then, if $\{\gamma^{i*}(t,x,x_0); i \in \mathbf{N}\}$ provides a closed-loop no-memory Nash equilibrium solution *such that $\gamma^{i*}(t,\cdot,x_0)$ is continuously differentiable on $\mathbf{R}^n$ ($\forall t \in [0,T], i \in \mathbf{N}$), and $\{x^*(t), 0 \le t \le T\}$ is the corresponding state trajectory, there exist N costate functions $p^i(\cdot) : [0,T] \to \mathbf{R}^n, i \in \mathbf{N}$, such that the following relations are satisfied:*

$$\dot{x}^*(t) = f(t, x^*(t), u^{1*}(t), \ldots, u^{N*}(t)); \quad x^*(0) = x_0, \tag{6.66a}$$

$$\begin{aligned}\gamma^{i*}(t,x^*,x_0) \;\; \equiv \;\; & u^{i*}(t) = \arg\min_{u^i \in S^i} H^i(t, p^i(t), x^*(t), u^{1*}(t), \ldots, \\ & u^{i-1*}(t), u^i, u^{i+1*}(t), \ldots, u^{N*}(t)),\end{aligned} \tag{6.66b}$$

$$\begin{aligned}\dot{p}^{i'}(t) &= -\frac{\partial}{\partial x} H^i(t, p^i(t), x^*, \gamma^{1*}(t,x^*,x_0), \ldots, \gamma^{i-1*}(t,x^*,x_0), u^{i*}, \\ &\qquad \gamma^{i+1*}(t,x^*,x_0), \ldots, \gamma^{N*}(t,x^*,x_0)), \\ p^{i'}(T) &= \frac{\partial}{\partial x} q^i(x^*(T)), \quad i \in \mathbf{N},\end{aligned} \tag{6.66c}$$

where H^i is as defined in Thm. 6.11.

Proof. Let us consider the ith inequality of (6.3), which fixes all players' strategies (except the ith one) at $\gamma^j = \gamma^{j*}$ ($j \neq i, j \in \mathbf{N}$) and constitutes an optimal control problem for $\mathbf{P}i$. Therefore, relations (6.66a)-(6.66c) directly follow from Thm. 5.4 (the minimum principle for continuous-time systems), as do relations (6.47a)-(6.47c) of Thm. 6.11. We should note, however, that the partial derivative (with respect to x) in the costate equations (6.66c) receives contribution also from dependence of the remaining $N-1$ players' strategies on the current value of x—a feature which was clearly absent in the costate equations of Thm. 6.11. □

The set of relations (6.66a)-(6.66c) in general admits an uncountable number of solutions, which correspond to "informationally nonunique" Nash equilibrium solutions of the differential game under memoryless perfect state information pattern. All of these are weakly time consistent, one of which is the open-loop solution given in Thm. 6.11. The analysis of subsection 6.3.1 can in fact be duplicated in the continuous time to produce explicitly a plethora of informationally nonunique Nash equilibria under dynamic information; and the reader is referred to Başar (1977b) for one such explicit derivation.

In order to eliminate informational nonuniqueness in the derivation of Nash equilibria under dynamic information (specifically under the MPS and CLPS patterns) we refine the Nash solution concept further, by requiring it to satisfy conditions similar to those of Def. 6.2, but now in the continuous time. Such a consideration leads to the concept of a "feedback Nash equilibrium solution" which is introduced below.

Definition 6.6 *For an N-person differential game of prescribed fixed duration $[0, T]$ and with memoryless perfect state (MPS) or closed-loop perfect state (CLPS) information pattern, an N-tuple of strategies $\{\gamma^{i*} \in \Gamma^i; i \in \mathbf{N}\}$*[62] *constitutes a* feedback Nash equilibrium solution *if there exist functionals $V^i(\cdot,\cdot)$ defined on $[0,T] \times \mathrm{R}^n$ and satisfying the following relations for each $i \in \mathbf{N}$:*

$$
\begin{aligned}
V^i(T,x) &= q^i(x),\\
V^i(t,x) &= \int_t^T g^i(s, x^*(s), \gamma^{1*}(s,\eta_s), \ldots, \gamma^{i*}(s,\eta_s), \ldots, \gamma^{N*}(s,\eta_s))\,\mathrm{d}s\\
&\qquad\qquad + q^i(x^*(T))\\
&\le \int_t^T g^i(s, x^i(s), \gamma^{1*}(s,\eta_s), \ldots, \gamma^{i-1*}(s,\eta_s), \gamma^i(s,\eta_s), \gamma^{i+1*}(s,\eta_s),\\
&\qquad \ldots, \gamma^{N*}(s,\eta_s))\,\mathrm{d}s + q^i(x^i(T)),\ \forall \gamma^i \in \Gamma^i, x \in \mathrm{R}^n,
\end{aligned}
\tag{6.67}
$$

where, on the interval $[t,T]$,

$$
\begin{aligned}
\dot{x}^i(s) &= f(s, x^i(s), \gamma^{1*}(s,\eta_s), \ldots, \gamma^{i-1*}(s,\eta_s), \gamma^i(s,\eta_s),\\
&\qquad \gamma^{i+1*}(s,\eta_s), \ldots, \gamma^{N*}(s,\eta_s));\ x^i(t) = x,\\
\dot{x}^*(s) &= f(s, x^*(s), \gamma^{1*}(s,\eta_s), \ldots, \gamma^{i*}(s,\eta_s), \ldots, \gamma^{N*}(s,\eta_s));\\
&\qquad x^*(t) = x,
\end{aligned}
\tag{6.68}
$$

and η_s, stands for either the data set $\{x(s), x_0\}$ or $\{x(\sigma), \sigma \le s\}$, depending on whether the information pattern is MPS or CLPS.

One salient feature of the concept of feedback Nash equilibrium solution introduced above is that if an N-tuple $\{\gamma^{i*}; i \in \mathbf{N}\}$ provides a feedback Nash equilibrium solution (FNES) to an N-person differential game with duration $[0,T]$, its restriction to the time interval $[t,T]$ provides an FNES to the same differential game defined on the shorter time interval $[t,T]$, with the initial state taken as $x(t)$, and this being so for all $0 \le t \le T$; hence, an FNES is strongly time consistent. An immediate consequence of this observation is that, under either MPS or CLPS information pattern, *feedback Nash equilibrium strategies*

[62] Here Γ^i is chosen to be compatible with the associated information pattern.

will depend only on the time variable and the current value of the state, but not on memory (including the initial state x_0).

The following proposition now verifies that an FNES is indeed a Nash equilibrium solution.

Proposition 6.7 *Under the memoryless (respectively, closed-loop) perfect state information pattern, every feedback Nash equilibrium solution of an N-person differential game is a closed-loop no-memory (respectively, closed-loop) Nash equilibrium solution (but not vice versa).*

Proof. Note that $V^i(t,x)$ is the value function (cf. subsection 5.5.2) associated with the optimal control problem of minimizing over $\gamma^i \in \Gamma^i$ the function $J^i(\gamma^{1*},\ldots,\gamma^{i-1*},\gamma^i,\gamma^{i+1*},\ldots,\gamma^{N*})$ (see in particular, equation (5.24a)). Therefore (6.67), together with the given terminal boundary condition, implies satisfaction of the ith inequality of (6.3). Since $i \in \mathbf{N}$ was arbitrary, and the foregoing argument is valid under both the MPS and CLPS information patterns, the result stands proven. □

If the value functions V^i $(i \in \mathbf{N})$ are continuously differentiable in both arguments, then N partial differential equations, related to the Hamilton–Jacobi–Bellman equation of optimal control (cf. subsection 5.5.2), replace (6.67).

Theorem 6.16 *For an N-person differential game of prescribed fixed duration $[0,T]$, and under either MPS or CLPS information pattern, an N-tuple of strategies $\{\gamma^{i*} \in \Gamma^i; i \in \mathbf{N}\}$ provides a* feedback Nash equilibrium solution *if there exist functions $V^i : [0,T] \times \mathbf{R}^n \to \mathbf{R}$, $i \in \mathbf{N}$, satisfying the partial differential equations*

$$\begin{aligned} -\frac{\partial V^i(t,x)}{\partial t} &= \min_{u^i \in S^i}\left[\frac{\partial V^i(t,x)}{\partial x}\tilde{f}^{i*}(t,x,u^i) + \tilde{g}^{i*}(t,x,u^i)\right] \\ &= \frac{\partial V^i(t,x)}{\partial x}\tilde{f}^{i*}(t,x,\gamma^{i*}(t,x)) + \tilde{g}^{i*}(t,x,\gamma^{i*}(t,x)), \\ V^i(T,x) &= q^i(x), \quad i \in \mathbf{N}, \end{aligned} \tag{6.69}$$

where

$$\begin{aligned} \tilde{f}^{i*}(t,x,u^i) &\triangleq f(t,x,\{\gamma^*_{-i}(t,x),u^i\}), \\ \tilde{g}^{i*}(t,x,u^i) &\triangleq g^i(t,x,\{\gamma^*_{-i}(t,x),u^i\}), \\ \{\gamma^*_{-i}(t,x),u^i\} &\triangleq \gamma^{1*}(t,x),\ldots,\gamma^{i-1*}(t,x),u^i,\gamma^{i+1*}(t,x),\ldots,\gamma^{N*}(t,x). \end{aligned}$$

Every such equilibrium solution is strongly time consistent, *and the corresponding Nash equilibrium cost for* $\mathbf{P}i$ *is* $V^i(0,x_0)$.

Proof. In view of the argument employed in the proof of Prop. 6.7, this result follows by basically tracing the standard technique of obtaining (5.25) from (5.24a). □

For the class of N-person affine-quadratic differential games (cf. Def. 6.5), it is possible to obtain an explicit solution for (6.69), which is quadratic in x. This also readily leads to a set of feedback Nash equilibrium strategies which are expressible in closed-form.

Corollary 6.5 *For an N-person affine-quadratic differential game with $Q^i(\cdot) \geq 0$, $Q^i_f \geq 0$, $R^{ij}(\cdot) \geq 0$ $(i, j \in \mathbf{N}, i \neq j)$, let there exist a set of matrix valued functions $Z^i(\cdot) \geq 0$, $i \in \mathbf{N}$, satisfying the following N coupled matrix Riccati differential equations:*

$$\dot{Z}^i + Z^i\tilde{F} + \tilde{F}'Z^i + \sum_{j\in\mathbf{N}} Z^j B^j R^{jj^{-1}} R^{ij} R^{jj^{-1}} B^{j'} Z^j + Q^i = 0; \qquad Z^i(T) = Q^i_f, \tag{6.70a}$$

where

$$\tilde{F}(t) \triangleq A(t) - \sum_{i\in\mathbf{N}} B^i(t)R^{ii}(t)^{-1}B^i(t)'Z^i(t). \tag{6.70b}$$

Then, under either the MPS or CLPS information pattern, the differential game admits a feedback Nash equilibrium solution, affine in the current value of the state, given by

$$\gamma^{i*}(t,x) = -R^{ii}(t)^{-1}B^i(t)'[Z^i(t)x(t) + \zeta^i(t)] \quad (i \in \mathbf{N}), \tag{6.71}$$

where ζ^i $(i \in \mathbf{N})$ are obtained as the unique solution of the coupled linear differential equations

$$\dot{\zeta}^i + \tilde{F}'\zeta^i + \sum_{i\in\mathbf{N}} Z^j B^j R^{jj^{-1}} R^{ij} R^{jj^{-1}} B^{j'} \zeta^j + Z^i\beta = 0; \; \zeta^i(T) = 0, \tag{6.72a}$$

with

$$\beta \triangleq c - \sum_{i\in\mathbf{N}} B^i R^{ii^{-1}} B^{i'} \zeta^i. \tag{6.72b}$$

The corresponding values of the cost functionals are

$$J^{i*} = V^i(0,x_0) = \frac{1}{2}x_0'Z^i(0)x_0 + x_0'\zeta^i(0) + n^i(0) \quad (i \in \mathbf{N}), \tag{6.73a}$$

where $n^i(\cdot)$ $(i \in \mathbf{N})$ are obtained from

$$\dot{n}^i + \beta'\zeta^i + \frac{1}{2}\sum_{j\in\mathbf{N}} \zeta^j B^j R^{jj^{-1}} R^{ij} R^{jj^{-1}} B^{j'} \zeta^j = 0; \; n^i(T) = 0. \tag{6.73b}$$

Proof. Simply note that, under the condition of solvability of the set of matrix Riccati differential equations (6.70a), the partial differential equation

(6.69) admits a solution in the form $V^i(t,x) = \frac{1}{2}x'Z^i(t)x + x'\zeta^i(t) + n^i(t)$ ($i \in \mathbf{N}$) with the corresponding minimizing controls being given by (6.71). The "nonnegative definiteness" requirement imposed on $Z^i(\cdot)$ is a consequence of the fact that $V^i(t,x) \geq 0 \; \forall x \in \mathbf{R}^n, t \in [0,T]$, this latter feature being due to the eigenvalue restrictions imposed *a priori* on $Q^i(\cdot)$, Q^i_f and $R^{ij}(\cdot)$. Finally, the corresponding "Nash" values for the cost functionals follow from the definition of $V(t,x)$ (see equation (6.67)). □

Remark 6.16 The foregoing corollary provides only one set of feedback Nash equilibrium strategies for the affine-quadratic game under consideration, and it does not attribute any uniqueness feature to this solution set. However, in view of the discrete-time counterpart of this result (cf. Corollary 6.1), one would expect the solution to be unique under the condition that (6.70a) admits a unique solution set; but, in order to verify this, one has to show that it is not possible to come up with other (possibly nonlinear) solutions that satisfy (6.67), and hitherto this has remained an unresolved problem. What has been accomplished, though, is verification of uniqueness of feedback Nash equilibrium when the players are restricted at the outset to affine memoryless strategies (cf. Lukes, 1971). □

Remark 6.17 As in the case of Remark 6.3, the result above extends readily to more general affine-quadratic dynamic games where the cost functions of the players contain additional terms that are linear in x, that is, with g^i and q^i in Def. 6.5 replaced, respectively, by

$$g^i = \frac{1}{2}\left(x'[Q^i(t)x + 2l^i(t)] + \sum_{j\in\mathbf{N}} u^{j'}_k R^{ij}_k u^j_k\right); \quad q^i = \frac{1}{2}x'(Q^i_f x + 2l^i_f),$$

where $l^i(\cdot)$ is a known n-dimensional vector-valued function, continuous on $[0,T]$, and l^i_f is a fixed n-dimensional vector, for each $i \in \mathbf{N}$. Then, the statement of Corollary 6.5 as well as its derivation remains intact, with only the differential equation (6.72a) that generates $\zeta^i(\cdot)$ now reading

$$\dot{\zeta}^i + \tilde{F}'\zeta^i + \sum_{j\in\mathbf{N}} Z^j B^j R^{jj^{-1}} R^{ij} R^{jj^{-1}} B^{j'} \zeta^j + Z^i\beta + l^i = 0; \;\; \zeta^i(T) = l^i_f.$$

□

Remark 6.18 For general nonlinear-nonquadratic differential games wherein the players are weakly coupled through the system equation as well as the cost functions, the features observed in Remark 6.13 for the open-loop Nash solution can also be derived for the feedback Nash solution, but now we have to use the sufficiency result of Thm. 6.16. Again confining ourselves to the two-player case, we take the system equation and the cost functions, as well as the expansion of the state, to be in the same structural form as in Remark 6.13, and only replace the expansion for $u^i(t;\epsilon)$ by a similar expansion for the feedback strategy:

$$\gamma^{i*}(t,x;\epsilon) = \sum_{k=0}^{\infty} \gamma_i^{(k)}(t,x)\epsilon^k.$$

Invoking a similar expansion on V^i,

$$V^{i*}(t,x;\epsilon) = \sum_{k=0}^{\infty} V_i^{(k)}(t,x)\epsilon^k,$$

it can be shown (see Srikant and Başar , 1991) that $V_i^{(0)}$ $(i = 1,2)$, the zeroth order terms, satisfy decoupled Hamilton–Jacobi–Bellman equations (associated with the optimal control problems obtained by setting $\epsilon = 0$), and the higher-order terms, $V_i^{(k)}$, $k \geq 1$, involve simple cost evaluations subject to some state equation constraints. Furthermore, the higher-order feedback strategies, $\gamma_i^{(k)}$, $k \geq 1$, are obtained from linear equations. More explicitly, for the zeroth order we have

$$\begin{aligned} -\frac{\partial}{\partial t}V_i^{(0)}(t,x) &= \min_{u^i \in S^i}\left[\frac{\partial}{\partial x_i}V_i^{(0)}(t,x_1,x_2)\cdot f_i(t,x_i,u^i) + g^{ii}(t,x_i,u^i)\right] \\ &= \frac{\partial}{\partial x_i}V_i^{(0)}(t,x_1,x_2)\cdot f_i(t,x_i,\gamma_i^{(0)}(t,x)) + g^{ii}(t,x_i,\gamma_i^{(0)}(t,x)), \end{aligned}$$

$$V_i^{(0)}(T,x) = q^{ii}(x_i), \quad i = 1,2, \tag{6.74a}$$

and to first-order $V_i^{(1)}$ satisfies

$$\begin{aligned} -\frac{\partial}{\partial t}V_i^{(1)}(t,x) &= \frac{\partial}{\partial x_i}V_i^{(1)}\cdot f_i(t,x_i,\gamma_i^{(0)}(t,x)) + \frac{\partial}{\partial x_i}V_i^{(0)}\cdot f_{ij}(x_j) \\ &+\frac{\partial}{\partial x_j}V_i^{(1)}\cdot f_j(t,x_j,\gamma_j^{(0)}(t,x)) + g^{ij}(t,x_j,\gamma_j^{(0)}(t,x)), \end{aligned}$$

$$V_i^{(1)}(T,x) = q^{ij}(x_j), \quad j \neq i,\ i,j = 1,2, \tag{6.74b}$$

and $\gamma_i^{(1)}$ is obtained from the linear equation

$$\begin{aligned} &\frac{\partial}{\partial u^i}f_i(t,x_i,\gamma_i^{(0)}(t,x))\left[\frac{\partial}{\partial x_i}V_i^{(1)}(t,x)\right]' + \frac{\partial}{\partial x_i}V_i^{(0)}\frac{\partial^2}{\partial u^{i2}}f_i(t,x_i,\gamma_i^{(0)}(t,x))\gamma_i^{(1)}(t,x) \\ &+\frac{\partial^2}{\partial u^{i2}}g^{ii}(t,x_i,\gamma_i^{(0)}(t,x))\gamma_i^{(1)}(t,x) = 0,\ i,j = 1,2. \end{aligned} \tag{6.74c}$$

Note that (6.74a) is the Hamilton–Jacobi–Bellman equation associated with the optimal control problem obtained by setting $\epsilon = 0$ in the the differential game with weakly coupled players, and hence $\gamma_i^{(0)}$ is the feedback representation of the open-loop policy $u_i^{(0)}$ obtained in Remark 6.13, on a common zeroth-order trajectory $x_i^{(0)}$. Hence, to zeroth order, there exists a complete equivalence between open-loop and feedback Nash equilibrium solutions in weakly coupled differential games—as in the case of optimal control (see, for example, Bensoussan, 1988). For higher-order terms, however, no such correspondence exists in nonzero-sum differential games, which is one explanation for the nonequivalence of the equilibrium trajectories under different information structures—a feature we have observed throughout this chapter. □

Feedback saddle-point solution of zero-sum differential games

We now first specialize the result of Thm. 6.16 to the class of two-person zero-sum differential games of prescribed fixed duration, and obtain a single partial differential equation to replace (6.69).

Corollary 6.6 *For a two-person zero-sum differential game of prescribed fixed duration $[0,T]$, and under either MPS or CLPS information pattern, a pair of strategies $\{\gamma^{i*} \in \Gamma^i; i = 1,2\}$ provides a feedback saddle-point solution if there exists a function $V : [0,T] \times \mathbf{R}^n \to \mathbf{R}$ satisfying the partial differential equation*

$$\begin{aligned} -\frac{\partial V(t,x)}{\partial t} &= \min_{u^1\in S^1}\max_{u^2\in S^2}\left[\frac{\partial V(t,x)}{\partial x}f(t,x,u^1,u^2)+g(t,x,u^1,u^2)\right] \\ &= \max_{u^2\in S^2}\min_{u^1\in S^1}\left[\frac{\partial V(t,x)}{\partial x}f(t,x,u^1,u^2)+g(t,x,u^1,u^2)\right] \\ &= \frac{\partial V(t,x)}{\partial x}f(t,x,\gamma^{1*}(t,x),\gamma^{2*}(t,x)) \\ &\qquad +g(t,x,\gamma^{1*}(t,x),\gamma^{2*}(t,x)). \end{aligned} \tag{6.75}$$

Every such saddle-point solution is strongly time consistent, and the unique saddle-point value of the game is $V(0,x_0)$.

Proof. This result follows as a special case of Thm. 6.16 by taking $N=2$, $g^1(\cdot) \equiv -g^2(\cdot) \triangleq g(\cdot)$ and $q^1(\cdot) \equiv -q^2(\cdot) \triangleq q$, in which case $V^1 \equiv -V^2 \triangleq V$ and existence of a saddle point is equivalent to interchangeability of the min-max operations. □

Remark 6.19 The partial differential equation (6.75) was first obtained by Isaacs in the early 1950s (see Isaacs, 1954-1957; see also Isaacs, 1975) and is therefore generally referred to as *Isaacs equation*, and also as *Hamilton–Jacobi–Isaacs equation.* Furthermore, the requirement for interchangeability of the "min" and "max" operations in (6.75) is also known as the "Isaacs condition". For more details on this, the reader is referred to Chapter 8. □

We now obtain the solution of the Isaacs equation (6.75) for the case of a two-person zero-sum affine-quadratic differential game described by (6.60a)-(6.60b). Since the boundary condition is quadratic in x (i.e., $V(T,x) = q(x) = \frac{1}{2}x'Q_f x$), we try out a solution in the form $V(t,x) = \frac{1}{2}x'Z(t)x + x'\zeta(t) + n(t)$, where $Z(\cdot)$

is symmetric, with $Z(T) = Q_f$, and $\zeta(T) = 0$, $n(T) = 0$. Substitution of this general quadratic structure in (6.75) yields the following equation:

$$\begin{aligned} -\frac{1}{2}x'\dot{Z}(t)x - x'\dot{\zeta}(t) - \dot{m}(t) \\ = \min_{u^1 \in \mathbf{R}^{m_1}} \max_{u^2 \in \mathbf{R}^{m_2}} \Big\{ [x'Z(t) + \zeta(t)'] \left[A + B^1u^1 + B^2u^2 + c\right] \qquad (i) \\ + \frac{1}{2}x'Q(t)x + \frac{1}{2}u^{1'}u^1 - \frac{1}{2}u^{2'}u^2 \Big\}. \end{aligned}$$

Since the kernel on the RHS is strictly convex (respectively, strictly concave) in u^1 (respectively, u^2), and is further separable, it admits a unique saddle-point solution given by

$$u^i = \gamma^i(t,x) = (-1)^i B^i(t)'[Z(t)x + \zeta(t)], \quad i = 1,2.$$

If these unique values for u^1 and u^2 are substituted back into the RHS of the Isaacs equation, we finally obtain the following differential equations for $Z(\cdot)$, $\zeta(\cdot)$ and $n(\cdot)$, by also taking into account the end-point conditions, the starting assumption that Z is symmetric (which is no loss of generality) and the requirement that equation (i) should be satisfied for all $x \in \mathbf{R}^n$:

$$\dot{Z} + A'Z + ZA + Q - Z(B^1B^{1'} - B^2B^{2'})Z = 0; \quad Z(T) = Q_f, \qquad (6.76a)$$

$$\dot{\zeta} + A'\zeta + Zc - Z(B^1B^{1'} - B^2B^{2'})\zeta = 0; \quad \zeta(T) = 0, \qquad (6.76b)$$

$$n + c'\zeta - \frac{1}{2}\zeta'(B^1B^{1'} - B^2B^{2'})\zeta = 0; \quad n(T) = 0. \qquad (6.76c)$$

The following theorem now even strengthens this result.

Theorem 6.17 *Let there exist a unique symmetric bounded solution to the generalized matrix Riccati differential equation (6.76a) on the interval* $[0,T]$. *Then, the two-person affine-quadratic zero-sum differential game described by (6.60a)-(6.60b), and under the MPS or CLPS information pattern, admits a* unique feedback saddle-point solution *given by*

$$\gamma^{i*}(t,x) = (-1)^i B^i(t)'[Z(t)x(t) + \zeta(t)], \quad i = 1,2, \qquad (6.77)$$

where $\zeta(\cdot)$ *is obtained as the unique solution of (6.76b). The state trajectory associated with this unique pair of feedback saddle-point strategies is given by the solution of*

$$\dot{x}(t) = [A - (B^1B^{1'} - B^2B^{2'})Z]x - (B^1B^{1'} - B^2B^{2'})\zeta(t) + c(t); \quad (6.78a)$$

subject to the initial condition $x(0) = x_0$, *and the saddle-point value of the game is*

$$J(\gamma^{1*}, \gamma^{2*}) = \frac{1}{2}x_0'Z(0)x_0 + x_0'\zeta(0) + n(0). \qquad (6.78b)$$

If, furthermore, $Q(\cdot) \geq 0$ *and* $Q_f \geq 0$, *then the solution to (6.76a),* $Z(t)$, *is nonnegative definite for all* $t \in [0,T]$.

Proof. With the exception of the "uniqueness" of the feedback saddle-point solution, and the nonnegative definiteness of $Z(\cdot)$, the result follows as a special case of Corollary 6.6, as explained prior to the statement of the theorem. To verify uniqueness of the solution, one approach is to go back to the Isaacs equation and investigate existence and uniqueness properties of solutions of this partial differential equation. This, however, requires some advance knowledge on the theory of partial differential equations, which is beyond the scope of our coverage in this book. In its stead, we shall choose here an alternative approach which makes direct use of the affine-quadratic structure of the differential game. This approach will provide us with the uniqueness result, as well as the equilibrium strategies (6.77).

Toward this end, let us first note that, under the condition of unique solvability of (6.76a), the linear differential equations (6.76b) and (6.76c) admit unique solutions. Furthermore, $x(T)'Q_f x(T)$ can be written as

$$\begin{aligned} x(T)'Q_f x(T) &= x(T)'Z(T)x(T) + 2x(T)'\zeta(T) + 2m(T) \\ &\equiv x_0'Z(0)x_0 + 2x_0'\zeta(0) + 2m(0) \\ &\quad + \int_0^T \frac{\mathrm{d}}{\mathrm{d}t}[x(t)'Z(t)x(t) + 2x(t)'\zeta(t) + 2m(t)]\,\mathrm{d}t \\ &= x_0'Z(0)x_0 + 2x_0'\zeta(0) + 2m(0) \\ &\quad + \int_0^T \{x(t)'\dot{Z}(t)x(t) + 2[x(t)'Z(t) + \zeta(t)'] \\ &\qquad + \dot{x}(t) + 2x(t)'\dot{\zeta}(t) + 2\dot{m}(t)\}\,\mathrm{d}t, \end{aligned}$$

and using (6.76a)-(6.76c), the integrand can equivalently be written as

$$\begin{aligned} &x'[-A'Z - ZA - Q + Z(B^1B^{1'} - B^2B^{2'})Z]x + 2[x(t)'Z(t) + \zeta(t)']\dot{x}(t) \\ &+2x'[-A'\zeta - Zc + Z(B^1B^{1'} - B^2B^{2'})\zeta] - c'\zeta + \frac{1}{2}\zeta'(B^1B^{1'} - B^2B^{2'})\zeta, \end{aligned}$$

where we have suppressed the t-dependencies of various terms. If we now substitute for $\dot{x}(\cdot)$ its expression from (6.60a), we obtain (after canceling out some common terms)

$$\begin{aligned} x(T)'Q_f x(T) &\equiv x_0'Z(0)x_0 + \int_0^T [2u^{1'}B^{1'}[Zx + \zeta] + 2u^{2'}B^{2'}[Zx + \zeta] \\ &\quad -x'Qx + [Zx + \zeta]'(B^1B^{1'} - B^2B^{2'})[Zx + \zeta]]\,\mathrm{d}t \\ &\quad +2x_0'\zeta(0) + 2m(0). \end{aligned}$$

If this identity is used in (6.60b), the result is the following equivalent expression for $L(u^1, u^2)$:

$$\begin{aligned} L(u^1, u^2) &= \frac{1}{2}x_0'Z(0)x_0 + x_0'\zeta(0) + m(0) + \frac{1}{2}\int_0^T |u^1 + B^{1'}[Zx + \zeta]|^2\,\mathrm{d}t \\ &\quad -\frac{1}{2}\int_0^T |u^2 - B^{2'}[Zx + \zeta]|^2\,\mathrm{d}t. \end{aligned}$$

Since the first term above is independent of u^1 and u^2, $L(u^1, u^2)$ is separable in u^1 and u^2, and it clearly admits the saddle-point solution given by (6.77). The saddle-point value of $L(u^1, u^2)$ also readily follows as $\frac{1}{2}x_0'Z(0)x_0 + x_0'\zeta(0) + m(0)$.

To prove that $Z(\cdot) \geq 0$ whenever $Q(\cdot) \geq 0$ and $Q_f \geq 0$, follow the line of reasoning used in verifying the last statement of Thm. 6.7 (for the discrete-time counterpart). With $c(\cdot) \equiv 0$, the value function is $V(t, x) = \frac{1}{2}x'Z(t)x$, which is necessarily nonnegative since the maximizer can assure such a lower bound by simply choosing $u^2(\cdot) \equiv 0$. □

Remark 6.20 As in the case of Nash equilibria (cf. Remark 6.17), the result above extends naturally to the more general affine-quadratic zero-sum dynamic games which have in the cost function additional linear terms in x, to be denoted $x(T)'l_f$ for the terminal cost term, and $x(t)'l(t)$ for the integrand. Then, the only modification will have to be made in equation (6.76b), for $\zeta(\cdot)$, which will now read:

$$\zeta + A'\zeta + Zc + l - Z(B^1B^{1'} - B^2B^{2'})\zeta = 0; \quad \zeta(T) = l_f.$$

Otherwise, the statement of Thm. 6.17 remains intact. □

Remark 6.21 A comparison of Thms. 6.14 and 6.17 reveals one common feature of the open-loop and feedback saddle-point solutions in affine-quadratic differential games with $Q(\cdot) \geq 0$, $Q_f \geq 0$, which is akin to the one observed for games described in discrete time (cf. Remark 6.7). The two matrix Riccati differential equations (6.62a) and (6.76a) are identical, and so are the two linear differential equations for m and ζ ((6.62b) and (6.76b), respectively) and by this token the state trajectories generated by the open-loop saddle-point solution and feedback saddle-point solution are identical; and furthermore the open-loop values of the feedback saddle-point strategies are correspondingly the same as the open-loop saddle-point strategies. A comparison of the existence conditions of Thms. 6.14 and 6.17, on the other hand, readily leads to the conclusion that *existence of an open-loop saddle-point solution necessarily implies existence of a feedback saddle-point solution* (since then (6.62a) (equivalently (6.76a)) admits a unique bounded symmetric solution), but not *vice versa*. This latter feature is a consequence of the following observation: for the class of affine-quadratic differential games satisfying the restriction

$$B^1(t)'B^1(t) - B^2(t)'B^2(t) \geq 0 \quad \forall t \in [0, T], \tag{6.79}$$

the matrix Riccati differential equation (6.76a) admits a unique nonnegative definite solution, whenever $Q_f \geq 0$, $Q(\cdot) \geq 0$ (since then (6.76a) becomes comparable with (5.35a)); however, (6.79) is not sufficient for (6.61) to admit a bounded solution on $[0, T]$. This situation is now exemplified in the sequel. □

Example 6.3 Consider a scalar linear-quadratic two-person zero-sum differential game described by

$$\dot{x} \;=\; \sqrt{2}u^1 - u^2; \; x(0) = x_0, \quad t \in [0, 2];$$

$$L(u^1, u^2) = \frac{1}{2}[x(2)]^2 + \frac{1}{2}\int_0^2 \{[u^1(t)]^2 - [u^2(t)]^2\}\,dt.$$

For these parametric values, the Riccati differential equation (6.76a) can be written as

$$\dot{Z} - Z^2 = 0; \quad Z(2) = 1$$

which admits the unique solution

$$Z(t) = \frac{1}{3-t}, \quad t \in [0, 2].$$

Hence, the differential game under consideration admits the unique feedback saddle-point solution

$$\gamma^{1*}(t, x) = -[2/(3-t)]x(t), \quad \gamma^{2*}(t, x) = -[1/(3-t)]x(t).$$

Let us now investigate the existence of an open-loop saddle-point solution for the same scalar differential game. The Riccati differential equation (6.61) can be written, for this example, as

$$\dot{S} + S^2 = 0; \quad S(T) = 1 \quad (t \in [0, T], T = 2)$$

which admits a bounded solution only if $T < 1$, in which case

$$S(t) = \frac{1}{t - T + 1}.$$

For $T > 1$, the Riccati differential equation does not admit a solution, and it displays (in retrograde time) *finite escape* at $t = T - 1$; i.e., it has a *conjugate point* at $t = T - 1$. In our case, since $T = 2 > 1$, the existence condition of Thm. 6.14 is thus not satisfied, and therefore an open-loop saddle-point solution does not exist. This implies, in view of Lemma 6.4, that under the open-loop information structure $L(u^1, u^2)$ is not concave in u^2, and hence the upper value of the game is unbounded (since it is scalar and quadratic). □

Even though feedback saddle-point equilibrium requires less stringent existence conditions than open-loop saddle-point equilibrium in affine-quadratic differential games (as also exemplified in the preceding example), this feature does not necessarily imply (at the outset) that the least stringent conditions for existence of saddle-point equilibrium under closed-loop information structure are required by the feedback saddle-point solution. This also brings up the question of whether the affine-quadratic differential game still admits a saddle-point solution in spite of existence of a conjugate point to the matrix Riccati equation (6.76a) on the interval $[0, T]$. In other words, can a saddle point *survive* a conjugate point? Intuitively, this could happen if at a conjugate point t^*, while Z goes off to infinity the corresponding state trajectory x^* goes to zero at least at the same rate, so that the control actions dictated by the feedback strategies remain (square-) integrable, and that Z does not change sign in any small open neighborhood of t^*. To permit for such behavior, the allowable strategies can no longer be continuously differentiable, or even continuous, but their realizations will have to be square-integrable for every square-integrable open-loop policy of the opponent player. Bernhard (1979) has shown, by working in this more general class of strategies, that a saddle point may survive a conjugate point, if the conjugate point is of an "even" type.

To give a brief description of this result, let us first note that in the absence of a conjugate point, the solution of (6.76a) can be decomposed as

$$Z(t) = Y(t)X^{-1}(t), \tag{6.80}$$

where X and Y are $n \times n$ matrices satisfying the linear matrix differential equations

$$\dot{X} = AX - (B'B^{1\prime} - B^2B^{2\prime})Y, \quad X(T) = I, \tag{6.81a}$$

$$\dot{Y} = -QX - A'Y, \qquad Y(T) = Q_f. \tag{6.81b}$$

Clearly, being linear, these equations admit unique solutions regardless of the length of the time interval, and hence the conjugate point of Z is the first instant (in retrograde time) when X becomes singular. Let us adopt the convention that at points where X is singular, X^{-1} in (6.80) is replaced by its pseudo-inverse $X^{\dagger}$. This then extends the solution of (6.76a) uniquely beyond a conjugate point. The following result, which is from (Bernhard, 1979), now provides the necessity part of Thm. 6.17.

Theorem 6.18 *Consider the affine-quadratic zero-sum differential game described by (6.60a)-(6.60b), with $Q(\cdot) \geq 0$, $Q_f \geq 0$, and under the CLPS information structure. The game admits a saddle point (with finite value) if, and only if, the following four conditions are satisfied:*

(i) $x_0 \in$ Range $(X(0))$,

(ii) rank of $X(t)$ is piecewise constant,

(iii) Range $(B^2(t)) \subset$ Range $(X(t))$, $\forall t \in [0,T]$, *except, perhaps, at isolated points,*

(iv) $Z(t) \geq 0$ $\forall t \in [0,T]$, *under the extended definition of a solution provided by (6.80) and (6.81a)-(6.81b).*

Then, the pair of feedback strategies (6.77) provides a saddle-point solution, and the saddle-point value is given by (6.78b). If the conditions above are not satisfied, then the upper value of the game becomes unbounded.

The following example (again due to Bernhard (1979)) illustrates the foregoing result, and demonstrates the occurrence of an "even" conjugate point.

Example 6.4 Consider the time-varying scalar linear-quadratic differential game described by

$$\begin{aligned}\dot{x} &= (2-t)u^1 + tu^2, \qquad x(0) = 1,\\ L(u^1,u^2) &= \frac{1}{2}[x(2)]^2 + \int_0^2 ([u^1(t)]^2 - [u^2(t)]^2)\,dt.\end{aligned}$$

The Riccati equation (6.76a) in this case is

$$\dot{Z} - 4(1-t)Z^2 = 0, \qquad Z(2) = \frac{1}{2},$$

which has a conjugate point at $t = 1$. Using the construction (6.80), however, we can obtain an extended solution as

$$Z(t) = \frac{1}{2(1-t)^2}, \qquad t \in [0,2], t \neq 1.$$

This shows that the conjugate point $t^* = 1$ is of the even type (i.e., $Z(t)$ does not change sign in an open neighborhood of t^*). Using this solution, the strategies (6.77) can be written as

$$\gamma^{1^*}(t,x) = -\frac{(2-t)}{2(1-t)^2}\,x, \qquad \gamma^{2^*}(t,x) = \frac{t}{2(1-t)^2}\,x$$

with the corresponding state trajectory satisfying

$$\dot{x}^* = -\frac{2}{1-t}x^*, \qquad x^*(0) = 1.$$

Note that, for $t < 1$, $x^*(t) = (1-t)^2$, and hence open-loop representations of both γ^{1^*} and γ^{2^*} remain bounded as $t \uparrow 1$, and so does the value function. Note also that all four conditions of Thm. 6.18 are satisfied in this case, with $X(t) = (1-t)^2$. □

Another perspective on the relationship between existence of conjugate points and saddle points in zero-sum affine-quadratic games can be obtained by studying the dependence of the solution on a scalar parameter. Of particular importance (as we shall later see in Section 6.6) is the case when the parameter is chosen as the weighting on the control of the maximizer. Consider, therefore, the modified cost (in place of (6.60b))

$$\begin{aligned}L_\lambda(u^1,u^2) \;=\; &\frac{1}{2}\int_0^T (x'(t)Qx(t) + u^1(t)'u^1(t) - \lambda u^2(t)'u^2(t))\,dt\\ &+\frac{1}{2}x'(T)Q_f x(T),\end{aligned} \tag{6.82}$$

where λ is a scalar positive parameter, $Q(\cdot) \geq 0$ and $Q_f \geq 0$. Let the state dynamics be again described by (6.60a). Then, the counterpart of the generalized Riccati differential equation (6.76a) is

$$\dot{Z} + A'Z + ZA + Q - Z\left(B^1B^{1'} - \frac{1}{\lambda}B^2B^{2'}\right)Z = 0; \quad Z(T) = Q_f, \tag{6.83}$$

whose solution over the interval $[0, T]$ we will denote by $Z_\lambda(\cdot)$. Note that the limiting case $\lambda = \infty$ yields the Riccati equation (5.35a) associated with the one-player game (with the maximizer being absent), which admits a nonnegative definite solution. Hence, by continuity, for every fixed T there exists some range of values of λ for which (6.83) does not have any conjugate point on $[0, T]$. What can in fact be shown (see Başar and Bernhard (1995)) is that there exists some threshold value of λ, say λ^*, such that (i) for all $\lambda > \lambda^*$, (6.83) has no conjugate points (and hence the game has a finite value under CLPS information), (ii) for all $\lambda \leq \lambda^*$, (6.83) has a conjugate point, and (iii) for all $\lambda < \lambda^*$, the upper value of the differential game is unbounded (regardless of the underlying information structure). We summarize this result in the following theorem.

Theorem 6.19 *Consider the parameterized affine-quadratic zero-sum differential game described by (6.60a) and (6.82), with $Q(\cdot) \geq 0$, $Q_f \geq 0$. There exists a critical value $\lambda^* > 0$ for λ such that*

(i) For each $\lambda > \lambda^*$, (6.83) admits a unique nonnegative definite solution $Z_\lambda(\cdot)$ (which is positive definite if either $Q_f > 0$ or $Q(\cdot) > 0$), and the game has a finite saddle-point value, given by (6.78b) with Z replaced by Z_λ, and $B^2B^{2\prime}$ in (6.76b) and (6.76c) replaced by $\frac{1}{\lambda}B^2B^{2\prime}$.

(ii) For $\lambda \leq \lambda^*$, (6.83) has a conjugate point.

(iii) For $\lambda < \lambda^*$, the upper value of the game is unbounded, even if the minimizer has access to full information (on the control action of the maximizer).

(iv) For $\lambda > \lambda^*$, the unique feedback saddle-point strategies of the players are

$$\gamma^{1^*}(t,x) = -B^{1\prime}[Z_\lambda(t)x + \zeta_\lambda(t)]; \qquad \gamma^{2^*}(t,x) = \frac{1}{\lambda}B^{2\prime}[Z_\lambda(t)x + \zeta_\lambda(t)];$$

where $\zeta_\lambda(\cdot)$ is given by (6.76b) with Z_λ replacing Z and $\frac{1}{\lambda}B^2B^{2\prime}$ replacing $B^2B^{2\prime}$.

Proof. A proof of this result for the linear case (with $c = 0$) can be found in Chapter 9 of Başar and Bernhard (1995), which readily extends to the affine case in view of Thm. 6.17. □

6.5.3 Linear-quadratic differential games on an infinite time horizon

In this subsection, we will study the limiting cases of the Nash equilibrium solution of the linear-quadratic differential game as the time horizon becomes unbounded—first for the open-loop case and then for the closed-loop feedback case.

The open-loop case

Here, we will study the limiting case of Thm. 6.12 as the time horizon becomes unbounded. The starting point will be Def. 6.5, with the assumptions that $N = 2$ and all matrices involved are time-invariant, and $Q_f^1 = Q_f^2 = 0$. Moreover, we will restrict ourselves to the linear-quadratic case, i.e., $c \equiv 0$.

The so-called algebraic Riccati equations (compare with (6.49))

$$M^1 A + A' M^1 + Q^1 - M^1 \bar{S}^1 M^1 - M^1 \bar{S}^2 M^2 = 0, \quad (6.84)$$

$$M^2 A + A' M^2 + Q^2 - M^2 \bar{S}^2 M^2 - M^2 \bar{S}^1 M^1 = 0, \quad (6.85)$$

where, as before, $\bar{S}^i \triangleq B^i (R^{ii})^{-1} B^{T'}$, $i = 1, 2$, play an important role here. Under certain technical conditions solutions of (6.49), with $N = 2$, will approach the solutions of (6.84) and (6.85) as $T \to \infty$. Since the dependence of the solutions of (6.49) on the final time T will be emphasized, these solutions will be written as $M^i(t, T)$. We will elaborate on the relationship between the solutions of these equations and the matrix G:

$$G \triangleq \begin{pmatrix} -A & \bar{S}^1 & \bar{S}^2 \\ Q^1 & A' & 0 \\ Q^2 & 0 & A' \end{pmatrix}. \quad (6.86)$$

Define

$$\mathcal{G}^{\text{inv}} \triangleq \{\mathcal{K} | G\mathcal{K} \subset \mathcal{K}\},$$

i.e., $\mathcal{G}^{\text{inv}}$ is the set of all G-invariant subspaces of $\mathbf{R}^{3n}$. It is well known that this set consists of a finite number of (different) elements if and only if all eigenvalues of G have geometric multiplicity one (otherwise there is an infinite number of different elements). Next define the subset $\mathcal{G}^{\text{sub}}$ of $\mathcal{G}^{\text{inv}}$ such that

$$\mathcal{G}^{\text{sub}} \triangleq \{\mathcal{K} \in \mathcal{G}^{\text{inv}} | \mathcal{K} \oplus \Im \begin{pmatrix} 0 & 0 \\ I & 0 \\ 0 & I \end{pmatrix} = \mathbf{R}^{3n}\},$$

where $\Im$ refers to "image of". Note that the elements in $\mathcal{G}^{\text{sub}}$ are the images of suitably chosen matrices of size $3n \times n$. In Engwerda (1998) the following theorem has been proved.

Theorem 6.20 *The set of coupled algebraic Riccati equations (6.84) and (6.85) have a real solution (M^1, M^2) if, and only if, $M^i = X_i X^{-1}$, $i = 1, 2$ for matrices X, X_1 and X_2 such that*

$$\mathcal{K} \triangleq \Im \begin{pmatrix} X \\ X_1 \\ X_2 \end{pmatrix} \in \mathcal{G}^{\text{sub}}.$$

Moreover, if the control functions $u^{i*}(t) = -(R^{ii})^{-1} B^{i'} M^i \Phi(t, 0) x_0$ are used in the system equation $\dot{x} = Ax + \sum_{j=1}^{2} B^j u^j$, the eigenvalues of the closed-loop matrix $A - \sum_{j=1}^{2} \bar{S}^j M^j$ coincide with the eigenvalues of $-G|_{\mathcal{K}}$, which is the restriction of $-G$ to the subspace $\mathcal{K}$.

It follows that every element of $\mathcal{G}^{\text{sub}}$ defines one solution of (6.84) and (6.85). If the geometric multiplicity of all eigenvalues of G equals one, then the set $\mathcal{G}^{\text{sub}}$ has a finite number of different elements and consequently (6.84) and (6.85) have at most a finite number of real solutions. Equations (6.84) and (6.85) do not have a real solution if and only if $\mathcal{G}^{\text{sub}}$ is empty. An immediate consequence of Thm. 6.20 is the following corollary.

Corollary 6.7 *The algebraic Riccati equations (6.84) and (6.85) have a set of stabilizing solutions (i.e., the closed-loop matrix $A - \sum_{j=1}^{2} \bar{S}^j M^j$ has all its eigenvalues in the left-half plane) if, and only if a G-invariant subspace $\mathcal{K} \in \mathcal{G}^{\text{sub}}$ exists such that $\Re\lambda > 0$ for all $\lambda \in \sigma(G|_{\mathcal{K}})$ (the symbol σ refers to the spectrum of its argument).*

We now introduce a useful concept, which will lead to a technical condition needed in the main result.

Definition 6.7 *The $3n \times 3n$ matrix G is called dichotomically separable if subspaces $\mathcal{V}_1$ and $\mathcal{V}_2$, with $\dim \mathcal{V}_1 = n$ and $\dim \mathcal{V}_2 = 2n$, exist such that $G\mathcal{V}_i \in \mathcal{V}_i$, $i = 1, 2$, $\mathcal{V}_1 \oplus \mathcal{V}_2 = \mathrm{R}^{3n}$ and $\Re\lambda > \Re\mu$ for all $\lambda \in \sigma(G|_{\mathcal{V}_1})$, $\mu \in \sigma(G|_{\mathcal{V}_2})$.*

It can be shown that dichotomic separability implies that $\mathcal{G}^{\text{sub}}$ is nonempty, but the reverse implication is not necessarily true.

Theorem 6.21 *Assume that $H(T)$ is invertible for all finite T, G is dichotomically separable and*

$$\text{span} \begin{pmatrix} I \\ Q_f^1 \\ Q_f^2 \end{pmatrix} \oplus \mathcal{V}_2 = \mathbf{R}^{3n}.$$

Then, $M^i(0,T) \to \bar{X}_i \bar{X}^{-1}$, $i = 1, 2$ if $T \to \infty$, where $\bar{X}$ and $\bar{X}_i$ are defined as $\mathcal{V}_1 \triangleq \text{span}(\bar{X}'\ \bar{X}_2'\ \bar{X}_2')'$.

Corollary 6.7 and Thm. 6.21 immediately lead to the following corollary.

Corollary 6.8 *If the time horizon of the linear quadratic differential game tends to infinity, then the unique open-loop Nash equilibrium converges to $u^{i*}(t) = -(R^{ii})^{-1} B^{i'} M^i \Phi(0,t) x_0$, $i = 1, 2$, which leads to a stable system, if the following conditions are satisfied:*

- *The conditions of Thm. 6.21 are fulfilled;*
- $\Re\lambda > 0$ *for all* $\lambda \in \sigma(G|_{\mathcal{V}_1})$.

For the cost function of each player to remain bounded, we assume that $Q^i > 0$, $i = 1, 2$, and under the equilibrium solution the state of the system converges to zero as t tends to infinity.

Theorem 6.22 *Let M^i, $i = 1, 2$, be solutions of (6.84) and (6.85) satisfying the additional constraint that the eigenvalues of the resulting system matrix $A - \sum_{i=1}^{2} \bar{S}^i M^i$ are all in the left-half plane. Then the controls $u^i(t) = -(R^{ii})^{-1} B^{i'} M^i \Phi(0,t) x_0$, $i = 1, 2$, form an open-loop Nash equilibrium solution.*

Proof. Let $L^i(u^1, u^2; T)$ denote the cost function of $\mathbf{P}^i$, on the time interval $[0, T]$. It can then be shown that the optimization problem

$$\min_{u_1} \lim_{T \to \infty} L^1(u^1, u^{2*}; T),$$

where $u^{2*}(t) = -(R^{22})^{-1} B^{2'} M^2 e^{(A - \bar{S}^1 M^1 - \bar{S}^2 M^2)t} x_0$, admits a solution given by $u^{1*}(t) = -(R^{11})^{-1} B^{1'} M^1 e^{(A - \bar{S}^1 M^1 - \bar{S}^2 M^2)t} x_0$. A similar reasoning holds for the optimization problem

$$\min_{u_2} \lim_{T \to \infty} L^1(u^{1*}, u^2; t),$$

leading to the conclusion that (u^{1*}, u^{2*}) forms an open-loop Nash equilibrium solution. □

The following example now demonstrates the somewhat surprising fact that an infinite-horizon differential game may admit a Nash equilibrium solution in open-loop policies, while its finite-horizon version (with a particular terminal state weighting) may not.

Example 6.5 Reconsider Example 6.2 and replace the A-matrix by

$$A = \begin{pmatrix} 1 & 0 \\ 0 & \frac{5}{22} \end{pmatrix}.$$

The eigenvalues of the related G-matrix are $\{\frac{5}{22}, 1.8810, 0.1883, \frac{1}{2}, -1.7966\}$. Further numerical calculation shows that the algebraic Riccati equations have three stabilizing solutions. Hence, according to Thm. 6.22, the infinite-horizon problem has at least three open-loop Nash equilibrium solutions.

On the other hand, the matrices Q_f^i, $i = 1, 2$, can suitably be chosen such that the matrix $G(t)$ fails to be invertible for certain t, e.g., $t = 0.1$ as in Example 6.2. Thus the infinite-horizon problem can have solutions whereas the finite-horizon problem fails to have a solution. □

One might conjecture that if the finite-horizon game has a unique solution for all finite T, then this solution converges to a solution of the infinite-horizon game. This is not true, as shown in Engwerda (1998), which presents an example in which the limiting solution does lead to a solution of the coupled algebraic Riccati equations, but fails to be a stabilizing solution.

The closed-loop feedback case

We now study the limiting cases of Corollary 6.5 and Thm. 6.17 (as well as Thm. 6.19) as the time horizon becomes unbounded (that is, $T \to \infty$), when

$c \equiv 0$, $Q_f = 0$, and all matrices are time-invariant. This will constitute the counterparts of the discrete-time results of subsection 6.2.3 in the continuous time.

First we consider the convex-quadratic Nash differential game covered by Corollary 6.5, and note that any asymptotically convergent (as $T \to \infty$) limit of the solution of (6.70a) will have to satisfy the coupled algebraic Riccati equations:

$$\bar{Z}^i \bar{F} + \bar{F}' \bar{Z}^i + \sum_{j \in \mathbf{N}} \bar{Z}^j B^j R^{jj^{-1}} R^{ij} R^{jj^{-1}} B^{j\prime} \bar{Z}^j + Q^i = 0, \qquad (6.87a)$$

where $\bar{F}$ is given by (6.70b), with all matrices replaced by their time-invariant counterparts.[63] The limiting stationary feedback (Nash) strategies are (from (6.71))

$$\bar{\gamma}^{i*}(x) = -R^{ii^{-1}} B^{i\prime} \bar{Z}^i x, \qquad i \in \mathbf{N}. \qquad (6.87b)$$

Furthermore, introduce for each $i \in \mathbf{N}$ the matrices

$$\bar{F}_i \triangleq A - \sum_{\substack{j \in \mathbf{N} \\ j \neq i}} B^j R^{jj^{-1}} B^{j\prime} \bar{Z}^j; \quad \bar{Q}_i \triangleq Q^i + \sum_{\substack{j \in \mathbf{N} \\ j \neq i}} \bar{Z}^i B^j R^{jj^{-1}} R^{ij} R^{jj^{-1}} B^{j\prime} \bar{Z}^j.$$

Then the following result is the natural counterpart of Prop. 6.3 in the continuous time.

Proposition 6.8 *Let there exist an N-tuple of matrices $\{\bar{Z}^i, i \in \mathbf{N}\}$ satisfying (6.87a), and further satisfying the conditions that for each $i \in \mathbf{N}$ the pair $(\bar{F}_i, B^i)$ is stabilizable and the pair $(\bar{F}_i, \bar{Q}_i)$ is detectable.*[64] *Then,*

(i) the N-tuple of stationary feedback policies (6.87b) provides a Nash equilibrium solution for the linear-quadratic nonzero-sum differential game of this subsection, leading to the infinite-horizon Nash equilibrium cost $\frac{1}{2} x_0' \bar{Z}^i x_0$ for **P***i,*

(ii) the resulting system dynamics, described by

$$\dot{x} = \bar{F} x, \qquad x(0) = x_0$$

are asymptotically stable.

Again, as in the discrete-time case, there could exist a Nash equilibrium for the infinite-horizon game (as covered by Prop. 6.8) which is not obtainable as the limit of any Nash equilibrium of the time-truncated finite-horizon version. Furthermore, we do not have any direct conditions (on the system parameters) that would guarantee satisfaction of the indirect conditions of Prop. 6.8 for any

[63] Note that here we are using the same notation as in the discrete-time case, but this should not create any confusion.

[64] See subsection 5.5.2 for the terminology.

sufficiently general class of nonzero-sum differential games. For zero-sum games, however, more explicit results can be obtained as will be discussed next.

Consider the class of zero-sum differential games described by (6.60a)-(6.60b), with $c \equiv 0$, $Q_f = 0$, all matrices time-invariant, $Q \geq 0$, and $T = \infty$. As in subsection 6.2.3, we would expect the feedback saddle-point solution for this game to be in the form

$$\bar{\gamma}^1(x) = -B^{1'}\bar{Z}x; \tag{6.88a}$$

$$\bar{\gamma}^2(x) = B^{2'}\bar{Z}x, \tag{6.88b}$$

where $\bar{Z}$ is the limiting solution of (6.76a) which (if it exists) should satisfy the continuous-time generalized algebraic Riccati equation (GARE)

$$A'\bar{Z} + \bar{Z}A + Q - \bar{Z}(B^1B^{1'} - B^2B^{2'})\bar{Z} = 0. \tag{6.89}$$

Denote the solution of the generalized Riccati differential equation (6.76a) by $Z(t;T)$, where dependence on the terminal time is explicitly recognized. Now, the first question that would be of interest is: If $Z(t;T)$ is bounded for all $T > 0$ (i.e., (6.76a) does not have any conjugate points on $[0,T]$ for any T) and $\lim_{T\to\infty} Z(t,T) = \bar{Z}$ exists, does this necessarily imply that the pair (6.88a)-(6.88b) is in saddle-point equilibrium? The answer is "no", as the following (counter) example, taken from (Mageirou, 1976) clearly illustrates.

Example 6.6 Consider the scalar state dynamics

$$\dot{x} = x + u^1 + u^2, \quad x(0) = x_0$$

and the objective function

$$L(u^1,u^2) = \int_0^\infty (x^2 + (u^1)^2 - 2(u^2)^2)\, dt.$$

The generalized Riccati differential equation (6.76a) associated with this game is

$$\dot{Z} + 2Z + 1 - \frac{1}{2}Z^2 = 0; \quad Z(T;T) = 0$$

which admits the unique solution

$$Z(t;T) = \sqrt{6}\tanh\left[\sqrt{6}(T-t) + \tanh^{-1}(-2\sqrt{6})\right] + 2.$$

This function has a well-defined limit, as $T \to \infty$, which is t-independent:

$$\bar{Z} = 2 + \sqrt{6}.$$

Hence the policies (6.88a)-(6.88b) are

$$\bar{\gamma}^1(x) = -(2+\sqrt{6})x; \quad \bar{\gamma}^2(x) = \frac{2+\sqrt{6}}{2}x.$$

Note that the closed-loop system under these policies is

$$\dot{x} = -\frac{\sqrt{6}}{2}x$$

which is asymptotically stable. Also, the system only under $\bar{\gamma}^1$ is

$$\dot{x} = -(1 - \sqrt{6})x + u^2$$

which is bounded input-bounded state stable.

Now it is easy to see (using Prop. 5.4) that if $u^1 = \bar{\gamma}^1(\cdot)$, the policy for **P2** that maximizes L is $\bar{\gamma}^2$. However, when $u^2 = \bar{\gamma}^2(\cdot)$, $\bar{\gamma}^1$ is not a minimizing solution for **P1**. The underlying minimization problem is

$$\int_0^\infty \left[-\left(4 + 2\sqrt{6}\right) x^2 + (u^1)^2 \right] dt \to \text{ minimum}$$

subject to

$$\dot{x} = \frac{4 + \sqrt{6}}{2}x + u^1.$$

Since the open-loop system is unstable, and there is a negative cost on state in the cost function, the cost can be driven to $-\infty$ (pick, for example, $u^1 \equiv 0$).

Even though there is no continuity in the minimizer's policy at $T = \infty$, nevertheless the value of the game (even in the absence of a saddle point) is continuous. It can be shown that given an $\epsilon > 0$, there exists a $t_\epsilon > 0$ such that by choosing the time-varying feedback policy

$$\gamma_\epsilon^2(t, x(t)) = \begin{cases} (1/2)Z(t; t_\epsilon)x(t), & 0 \leq t < t_\epsilon, \\ 0, & t \geq t_\epsilon, \end{cases}$$

P2 can assure that

$$\min_{u^1} L(u^1, \gamma_\epsilon^2) \geq (2 + \sqrt{6})(x_0)^2 - \epsilon(x_0)^2,$$

where

$$(2 + \sqrt{6})(x_0)^2 = \lim_{T \to \infty} Z(t; T)(x_0)^2$$

which is the limit of the value of the finite-horizon game (as $T \to \infty$).[65] Since $\epsilon > 0$ can be chosen arbitrarily small, this shows that the infinite-horizon game of this example indeed has a value. □

What has been demonstrated in the last part of the example above actually turns out to be the case for the general linear-quadratic differential game, under some general conditions, as stated in the following theorem, which is the counterpart of Thm. 6.8, as well as of Lemma 6.3.

[65] t_ϵ above is chosen such that $Z(0; t_\epsilon) + \epsilon \geq 2 + \sqrt{6}$.

Theorem 6.23 *Consider the infinite-horizon time-invariant linear-quadratic differential game, where $Q_f = 0$, $Q \geq 0$ and the pair (A, Q) is observable. Then,*

(i) for each fixed t, the solution to (6.76a), $Z(t;T)$, is nondecreasing in T; that is, if (6.76a) has no conjugate point in a given interval $[0,T]$, $Z(t;T') - Z(t;T'') \geq 0$, $T > T' > T'' \geq 0$;

(ii) every nonnegative definite solution of (6.89) is in fact positive definite, and if there exists a positive definite solution there is a minimal such solution, denoted $\bar{Z}^+$. This matrix has the property that $\bar{Z}^+ - Z(t;T) \geq 0$ for all $T \geq 0$;

(iii) the game has equal upper and lower values if, and only if, the GARE (6.89) admits a positive definite solution, in which case the common value is $J^{\infty *} = x_0' \bar{Z}^+ x_0$;

(iv) the upper value is finite if, and only if, the upper and lower values are equal;

(v) if $\bar{Z}^+ > 0$ exists, as given above, the controller $\bar{\gamma}^1$ given by (6.88a), with $\bar{Z}$ replaced by $\bar{Z}^+$, attains the finite upper value, in the sense that

$$\sup_{\bar{\gamma}^2} J(\bar{\gamma}^1, \gamma^2) = x_0' \bar{Z}^+ x_0,$$

and the maximizing feedback solution above is given by (6.88b), again with $\bar{Z} = \bar{Z}^+$;

(vi) the minimal positive definite solution of (6.89), $\bar{Z}^+$, is feedback stabilizing, i.e., the closed-loop system

$$\dot{x} = [A - (B^1 B^{1'} - B^2 B^{2'}) \bar{Z}^+] x$$

has no eigenvalues in the right-half plane. If $Q > 0$, then $\bar{Z}^+$ is the only feedback stabilizing solution of (6.89).

Proof. See Mageirou (1976) or Başar and Bernhard (1995). □

The result of Thm. 6.23 above is not the strongest possible, since the given policies (6.88a)-(6.88b) are not *strictly feedback stabilizing* (i.e., the closed-loop feedback matrix could have some eigenvalues on the imaginary axis of the complex plane) and furthermore the policies are not necessarily in saddle-point equilibrium (as also observed in Example 6.6). In view of this one would like to obtain conditions under which the policies (6.88a)-(6.88b) would be in saddle-point equilibrium and/or they would be strictly feedback stabilizing.

If we assume that there exists a solution to GARE (6.89) that is strictly feedback stabilizing, then it can be shown (see Jacobson (1977)) that the strategy pair (6.88a)-(6.88b) is in saddle-point equilibrium in the restricted class of feedback policies that are strictly feedback stabilizing and under which $x(t) \to 0$ as $t \to \infty$ for all $x_0 \in \mathrm{R}^n$. This result follows essentially from a completion of squares argument, using the GARE (6.89). An alternative condition, again under the assumption of existence of a strictly feedback stabilizing solution to (6.89), is to require that under $\bar{\gamma}^2$, the minimization problem faced by **P1** is well defined; this is equivalent to requiring (using some results of Willems (1971)) that an associated ARE has a negative definite solution (see Mageirou (1976)). But, this is a rather complicated condition, and admits no clean interpretation in terms of the parameters of the game.

Yet another set of conditions can be obtained, by parameterizing the differential game, as in the finite-horizon case discussed earlier. Consider the infinite-horizon version of the cost function (6.82), where λ is again a scalar parameter (to be varied). For $\lambda = \infty$, the underlying problem is essentially a single player game, which is the same as the linear regulator problem discussed in subsection 5.5.2. We know from Prop. 5.4 that if (A, B) is stabilizable and (A, Q) is observable (respectively, detectable), then the associated ARE admits a positive (respectively, nonnegative) definite solution, which is further strictly feedback stabilizing. Since eigenvalues of the feedback matrix in the game case are continuous functions of λ, the result will also hold for some range of finite values for λ. This result is now made precise under also a weaker condition of detectability in the following theorem, which is the infinite-horizon counterpart of Thm. 6.19. Its proof can be found in Başar and Bernhard (1995).

Theorem 6.24 *For the parameterized infinite-horizon linear-quadratic differential game formulated above, let (A, B) be stabilizable and (A, Q) be observable (respectively, detectable). Then, there exists a threshold value $\lambda_\infty^* > 0$ for λ such that the following three properties hold.*

(i) For every $\lambda > \lambda_\infty^*$, the GARE

$$A'\bar{Z} + \bar{Z}A + Q - \bar{Z}(B^1B^{1'} - {\lambda_\infty^*}^{-1}B^2B^{2'})\bar{Z} = 0 \tag{6.90}$$

admits a minimal positive (respectively, nonnegative) definite solution, $\bar{Z}_\lambda$.

(ii) For every $\lambda > \lambda_\infty^*$, the game has a finite upper value, which is achieved by the pair of strategies

$$\bar{\gamma}_\lambda^1(x) = -B^{1'}\bar{Z}_\lambda^+ x, \qquad \bar{\gamma}_\lambda^2(x) = \frac{1}{\lambda}B^{2'}\bar{Z}_\lambda^+ x. \tag{6.91}$$

Furthermore, the closed-loop matrix

$$\bar{F}_\lambda = A - (B^1B^{1'} - \lambda^{-1}B^2B^{2'})\bar{Z}_\lambda^+$$

is Hurwitz (has all its eigenvalues in the open left-half plane).

(iii) For $\lambda < \lambda_\infty^*$, the upper value of the game is unbounded.

6.6 Applications in Robust Controller Designs: H$^\infty$-Optimal Control

The results presented in the previous section on zero-sum differential games, and especially on linear-quadratic games, as well as their counterparts in Section 6.3, have important applications in a special class of worst-case controller design problems, known as "H^∞-optimal control" problems (see Zames (1981); Francis (1987); Doyle et al. (1989); Başar and Bernhard (1995); Stoorvogel (1992), Green and Limebeer (1995) to cite just a few representative papers and books from this voluminous literature). In these design problems, the (linear) plant has two types of inputs (*controlled* and *disturbance*) and two types of outputs (*regulated* and *measured*), and the objective is to obtain a controller (that uses the measured output) such that the effect of the disturbance on the regulated output is minimized. Even though both frequency and time domain formulations are possible, the latter is preferable since it allows for transient analysis and can handle time varying as well as nonlinear plants.

In this section we first present a general formulation of this worst-case design problem, and then show how the theory of the previous section can directly be used to construct minimax (H^∞-optimal) controllers. The presentation will be brief, and be restricted to the continuous-time problem only. Further details, and counterparts of these results in the discrete-time, can be found in the book by Başar and Bernhard (1995).

To introduce the H^∞-optimal control problem (in the continuous-time), let us adopt the plant dynamics

$$\dot{x} = Ax + Bu + Dw, \qquad x(0) = 0 \tag{6.92a}$$

to replace the linear state equation (6.60a). Here u is the controlled input, w is the disturbance, and A, B, D are matrices which could have time-varying entries if the time horizon is not infinite. Let the regulated and measured outputs be given, respectively, by

$$z = Hx + Gu, \tag{6.92b}$$

$$y = Cx + Ew, \tag{6.92c}$$

where H, G, C, E are again appropriate dimensional matrices, with possibly time-variant entries for the finite-horizon problem. The controlled input is chosen as a function of the present and past values of the measured output y, a relationship that we will write as

$$u(t) = \mu(t, y(s), s \leq t), \qquad t \geq 0, \tag{6.92d}$$

or simply as $u = \mu(y)$. Here μ is the (causal) control policy. If in (6.92c) $C = I$ and $E = 0$, then we have the CLPS information pattern, where the control has access to all components of the state.

Now for any given $\mu \in \mathcal{M}$, where $\mathcal{M}$ is chosen so that some measurability and smoothness conditions are satisfied, substitution of u into (6.92a) and (6.92c) leads to a unique mapping from w to z, obtained from,

$$z = Hx + G\mu(Cx + Ew), \tag{6.93a}$$
$$\dot{x} = Ax + B\mu(Cx + Ew) + Dw, \quad x(0) = 0. \tag{6.93b}$$

Let us denote this mapping by $\mathcal{G}_\mu$, i.e.,

$$z = \mathcal{G}_\mu(w). \tag{6.93c}$$

Then, the H^∞-optimal control problem is to minimize the induced (operator) norm of $\mathcal{G}_\mu$, that is, to solve the optimization problem

$$\inf_{\mu\in\mathcal{M}} \sup_{w\in\mathcal{H}} \{\|\mathcal{G}_\mu(w)\|_z / \|w\|_w\}, \tag{6.94}$$

where $\|\cdot\|_z$ and $\|\cdot\|_w$ denote Hilbert space norms on the regulated output and disturbance spaces, respectively, where the latter space is denoted by $\mathcal{H}$. More specifically, for the finite-horizon problem, defined on the interval $[0, T]$,

$$\|z\|_z^2 \triangleq z'(T)\tilde{Q}_f z(T) + \int_0^T z(t)'z(t)\,\mathrm{d}t, \qquad \|w\|_w^2 \triangleq \int_0^T w(t)'w(t)\,\mathrm{d}t$$

and for the infinite-horizon problem $T = \infty$, and $\tilde{Q}_f = 0$. Note that (6.94) defines the *upper value* of a differential game where the minimizer (**P1**) is the controller, and the maximizer (**P2**) is the disturbance. Let the square of this upper value be denoted by $\hat{\lambda}$, and assume for the moment that there exists a control policy, μ^*, that achieves it. Then, (6.94) can equivalently be expressed as (where we suppress the subindices on different norms)

(i)

$$\|\mathcal{G}_{\mu^*}(w)\|^2 \leq \hat{\lambda}\|w\|^2, \qquad \forall w \in \mathcal{H}, \tag{6.95a}$$

and

(ii) there exists no other $\mu \in \mathcal{M}$ (say, $\tilde{\mu}$), and a corresponding $\lambda = \tilde{\lambda} < \hat{\lambda}$, such that

$$\|\mathcal{G}_{\tilde{\mu}}(w)\|^2 \leq \tilde{\lambda}\|w\|^2, \qquad \forall w \in \mathcal{H}. \tag{6.95b}$$

Now, introducing the parameterized (in $\lambda \geq 0$) family of cost functions

$$L_\lambda(\mu, w) \triangleq \|\mathcal{G}_\mu(w)\|^2 - \lambda\|w\|^2, \tag{6.96}$$

(i) and (ii) above become equivalent to the problem of finding the "smallest" value of $\lambda \geq 0$ under which the upper value of the associated game with objective function $L_\lambda(\mu, w)$ is *zero*, and finding the corresponding controller that achieves this upper value.

We now relate this problem to the ones covered by Thms. 6.17, 6.19, 6.23 and 6.24, when the information pattern is CLPS. Toward this end let us first make the simplifying assumptions that $H'G = 0$, $G'G = I$, $H'(T)\tilde{Q}_f G(T) = 0$ and $G'(T)\tilde{Q}_f G(T) = 0$. Then, L_λ can be written as (6.82), with

$$Q_f \triangleq H(T)'\tilde{Q}_f H(T), \qquad Q \triangleq H'H. \tag{6.97}$$

In the finite horizon, we then have precisely the linear-quadratic game covered in Thm. 6.19, with B^1 replaced by B, B^2 replaced by D, and $x_0 = 0$. We know from Thm. 6.19, together with Thm. 6.18, that the upper and lower values are equal whenever the upper value is bounded, and that this finite value is *zero* (since $x_0 = 0$). Furthermore, where this breaks down (that is, when the upper value becomes unbounded) is at the largest value of λ for which the generalized Riccati differential equation (6.83), that is,

$$\dot{Z} + A'Z + ZA + Q - Z\left(BB' - \frac{1}{\lambda}DD'\right)Z = 0, \quad Z(T) = Q_f, \tag{6.98}$$

has a conjugate point. This value of λ (which we had denoted λ^*) is precisely the $\hat{\lambda}$ introduced earlier as the square of (6.94). Hence the following result follows as a direct corollary to Thm. 6.19.

Corollary 6.9 *For the finite-horizon H^∞-optimal control problem formulated above, and with CLPS information, let λ^* be defined as in Thm. 6.19, in connection with the Riccati differential equation (6.98). Then,*

(i) $\hat{\lambda} = \lambda^*$,

(ii) for all $\lambda > \lambda^*$, the control policy

$$\mu_\lambda(t, x) = -B(t)'Z_\lambda(t)x, \qquad t \geq 0$$

delivers a performance level of at least $\sqrt{\lambda}$, that is, for all $w \in \mathcal{H}$,

$$\{\|\mathcal{G}_{\mu_\lambda}(w)\| / \|w\|\} \leq \sqrt{\lambda}.$$

For the infinite-horizon problem, the counterpart of Corollary 6.9 is the following corollary to Thm. 6.24.

Corollary 6.10 *For the infinite-horizon H^∞-optimal control problem, and with CLPS information, let (A, B) be stabilizable and (A, Q) be observable (respectively, detectable). Let λ^* be the smallest positive scalar with the property that for all $\lambda > \lambda^*$, the GARE*

$$A\bar{Z} + \bar{Z}A + Q - \bar{Z}\left(BB' - \frac{1}{\lambda}DD'\right)\bar{Z} \;=\; 0 \tag{6.99}$$

admits a minimal positive (respectively, nonnegative) definite solution $\bar{Z}_\lambda$. Then, *(i)* $\hat{\lambda} = \lambda^*$,

(ii) for all $\lambda > \lambda^*$, *the feedback control policy*

$$\bar{\mu}_\lambda(x) = -B'\bar{Z}_\lambda x, \quad t \geq 0$$

delivers a performance level of at least $\sqrt{\lambda}$, *and under it the linear system*

$$\dot{x} = (A - BB'\bar{Z}_\lambda)x + Dw$$

is bounded-input bounded-state stable.

The following scalar example (taken from Başar and Bernhard (1995)) illustrates the two corollaries above.

Example 6.7 Consider the scalar plant with dynamics

$$\dot{x} = u + w,$$

and H^∞-performance index:

$$\left[\int_0^T ([x(t)]^2 + [u(t)]^2)\,\mathrm{d}t\right]^{\frac{1}{2}} \bigg/ \left[\int_0^T [w(t)]^2\,\mathrm{d}t\right]^{\frac{1}{2}}. \qquad (i)$$

The associated GRDE is

$$\dot{Z} + 1 - \left(1 - \frac{1}{\lambda}\right) Z^2 = 0, \qquad Z(T) = 0, \qquad (ii)$$

whose unique solution takes two different forms, depending on whether $\lambda > 1$ or not. For the former case,

$$\lambda > 1: \quad Z_\lambda(t) = \tfrac{1}{\sigma}\tanh[\sigma(T-t)], \quad \sigma \triangleq \sqrt{(\lambda-1)/\lambda},$$

and for the latter,

$$\lambda < 1: \quad Z_\lambda(t) = \tfrac{1}{m}\tan[m(T-t)], \quad m \triangleq \sqrt{(1-\lambda)/\lambda},$$

which exists provided that

$$mT < \frac{\pi}{2} \Longleftrightarrow \lambda > \lambda^*(T) \triangleq \frac{4T^2}{4T^2 + \pi^2}.$$

Note that the given solution family is continuous at $\lambda = 1$. Now, for fixed T, the GRDE has a conjugate point in the interval $[0,T]$ if $\lambda \leq \lambda^*(T)$, and for all $\lambda < \lambda^*(T)$ the upper value of the related game is unbounded. Hence, the optimum H^∞-performance level (that is, the upper value of the game with performance index (i)) is $\sqrt{\lambda^*(T)}$. For $\lambda > \lambda^*(T)$, the controller that guarantees this level of performance is

$$\mu_\lambda^*(t,x) = -Z_\lambda(t)x(t), \qquad 0 \leq t \leq T. \qquad (iii)$$

Note that as $\lambda \downarrow \lambda^*(T)$, $Z_\lambda(0) \uparrow \infty$, and hence the "optimum" controller is a high gain controller, which may not be desirable. For any $\lambda > \lambda^*(T)$, however, the control gain is finite, and one can achieve ϵ-optimal performance (that is, one that is within an ϵ-neighborhood of $\lambda^*(T)$) using finite gain controllers of the type (iii).

For the infinite-horizon case, we simply let $T \uparrow \infty$, leading to $\lambda^* = 1$, and the unique stationary feedback controller

$$\bar{\mu}_\lambda(x) = -\frac{\sqrt{\lambda}}{\sqrt{\lambda - 1}} x, \qquad \text{for } \lambda > 1.$$

Again note that the solution ceases to exist as $\lambda \downarrow \lambda^* = 1$, and hence an "optimal" controller does not exist, but an ϵ-optimal controller does. □

We now consider the imperfect state measurements case, as originally formulated. For technical reasons we will treat the initial state x_0 as also unknown and as part of the disturbance, and attach weight to it in the cost function. Furthermore, we take the system and measurement disturbances to be independent (which is tantamount to taking $ED' = 0$), and take without any loss of generality $EE' = I$ and let $Ew = v$, which is viewed as a separate disturbance (in addition to w and x_0). Then, the associated parameterized game is one with state dynamics

$$\dot{x} = Ax + Bu + Dw, \qquad x(0) = x_0, \tag{6.100a}$$

measurement equation

$$y(t) = Cx(t) + v(t), \qquad t \geq 0, \tag{6.100b}$$

and objective function (for finite horizon)

$$\begin{aligned} L_\lambda(u; w, v, x_0) = |x(T)|^2_{Q_f} + \int_0^T (|x(t)|^2_Q + |u(t)|^2)\, dt \\ -\lambda \left[|x_0|^2_{Q_0} + \int_0^T (|w(t)|^2 + |v(t)|^2)\, dt \right]. \end{aligned} \tag{6.100c}$$

For the infinite horizon, we simply let $T = \infty$, $Q_f = 0$. In view of our earlier discussion in this section, we seek the smallest value of λ, say λ^*, such that for all $\lambda > \lambda^*$, and with $\omega \triangleq \{w, v, x_0\}$,

$$\inf_{\mu \in \mathcal{M}} \sup_{\omega} L_\lambda(\mu(y); \omega) = 0, \tag{6.101}$$

where $\mathcal{M}$ is defined here as the class of all (measurable, smooth) control policies that depend causally on y.

The theory we have developed in this chapter does not directly apply to this problem, since the information structure is neither CLPS nor OL. Similar game theoretic techniques can be used, however, to obtain a complete, very appealing solution to the problem. The result is given below in Thm. 6.25 for the finite horizon, and Thm. 6.26 for the infinite-horizon case. Verifications of these results will be sketchy, since the details are lengthy and fairly technical. A complete treatment can be found in Başar and Bernhard (1995).

Before presenting Thm. 6.25, we introduce a second GARE, dual to (6.98), again defined on $[0,T]$:

$$-\dot{\Sigma} + A\Sigma + \Sigma A' + DD' - \Sigma\left(C'C - \frac{1}{\lambda}Q\right)\Sigma = 0; \quad \Sigma(0) = Q_0^{-1}. \quad (6.102)$$

We further define

$$G(t) \triangleq Z\left[I - \lambda^{-1}\Sigma Z\right]^{-1} \quad (6.103)$$

whenever it exists, and introduce two different equations

$$\dot{\hat{x}} = [A-(BB'-\lambda^{-1}DD')Z]\hat{x}+[I-\lambda^{-1}\Sigma Z]^{-1}\Sigma C(y-C\hat{x}); \quad \hat{x}(0) = 0 \quad (6.104a)$$

and

$$\dot{\tilde{x}} = (A - \Sigma CC' + \lambda^{-1}\Sigma Q)\tilde{x} + Bu + \Sigma C'y; \quad \tilde{x}(0) = 0. \quad (6.104b)$$

Theorem 6.25 *Let Λ be the set of all $\lambda > 0$ such that*

(a) the GRDE (6.98) does not have a conjugate point on $[0,T]$,

(b) the GRDE (6.102) does not have a conjugate point on $[0,T]$,

and

(c) the matrix $\lambda I - \Sigma(t)Z(t)$ has only positive eigenvalues for all $t \in [0,T]$, or equivalently

$$\rho(\Sigma(t)Z(t)) < \lambda, \qquad t \in [0,T], \quad (6.105)$$

where $\rho(\cdot)$ is the spectral radius.

For each $\lambda \in \Lambda$, let a controller $\mu_\lambda \in \mathcal{M}$ be defined by

$$\mu_\lambda(t,y) = -B(t)'Z_\lambda(t)\hat{x}_\lambda(t), \qquad t \geq 0, \quad (6.106a)$$

where Z_λ is the solution of (6.98), and $\hat{x}_\lambda$ is generated by (6.102). Then,

(i) $\inf\{\lambda \geq 0 : \lambda \in \Lambda\} = \lambda^*$, where λ^* is the smallest positive scalar λ such that (6.101) holds;

(ii) for each $\lambda > \lambda^*$, the controller given by (6.106a) delivers a performance level of at least $\sqrt{\lambda}$;

(iii) for each $\lambda > \lambda^*$, the controller (6.106a) can equivalently be written as

$$\mu_\lambda(t,y) = -B(t)'G_\lambda(t)\tilde{x}_\lambda(t), \qquad t \geq 0, \tag{6.106b}$$

where G_λ *is given by (6.103) and* $\tilde{x}$ *is generated by (6.104b).*

Proof. We will sketch here a proof of sufficiency, using completion of squares. First note that since $Q_0^{-1} > 0$, the solution of (6.102) is positive definite, thus admitting an inverse. Define the function

$$V(t) = |\tilde{x}(t)|^2_{G(t)} + \lambda|x(t) - \tilde{x}(t)|^2_{\Sigma^{-1}(t)}$$

and add the identically *zero* function

$$V(0) - V(T) + \int_0^T \dot{V}(t)\,\mathrm{d}t$$

to L_λ given by (6.100c). The derivative $\dot{V}$ involves the derivatives of G, Σ^{-1} and $\epsilon \triangleq x - \tilde{x}$, which can be shown to be given by (through elementary but extensive manipulations)

$$\dot{G} + G(A + \lambda^{-1}\Sigma Q) + (A + \lambda^{-1}\Sigma Q)'G + Q - G(BB' - \lambda^{-1}\Sigma C'C\Sigma)G = 0,$$
$$\dot{\Sigma}^{-1} + \Sigma^{-1}A + A'\Sigma^{-1} + \Sigma^{-1}DD'\Sigma^{-1} - C'C + \lambda^{-1}Q = 0; \quad \Sigma^{-1}(0) = Q_0,$$
$$\dot{\epsilon} = (A - \Sigma C'C)\epsilon - \Sigma C'v + Dw - \lambda^{-1}\Sigma Q\tilde{x}; \quad \epsilon(0) = x_0.$$

If all these relationships are used in L_λ, it simplifies to

$$\begin{aligned} L_\lambda(u; w, v, x_0) = &-|\epsilon(T) + [Q_f - \lambda\Sigma^{-1}(T)]^{-1}Q_f\tilde{x}(T)|^2_{\lambda\Sigma^{-1}(T)-Q_f} \\ &-2|\tilde{x}(T)|^2_{Q_f(\lambda I - \Sigma(T)Q_f)^{-1}\Sigma(T)Q_f} + \|u + B'G\tilde{x}\|^2 \qquad (\diamond) \\ &-\lambda\|v + C(\epsilon - \lambda^{-1}\Sigma G\tilde{x})\|^2 - \lambda\|w - D'\Sigma^{-1}\epsilon\|^2, \end{aligned}$$

where $\|\cdot\|$ denotes the appropriate Hilbert space norm. Now, invoking condition (c) of the theorem we conclude that

$$\lambda\Sigma^{-1}(T) - Q_f > 0, \quad Q_f(\lambda I - \Sigma(T)Q_f)^{-1}\Sigma(T)Q_f \geq 0,$$

and hence the first two terms in ($\diamond$) are nonpositive. Furthermore, since the control given by (6.106b) makes the third (nonnegative) term in ($\diamond$) zero, we have the bound

$$L_\lambda(\mu_\lambda; w, v, x_0) \leq 0, \qquad \forall w, v, x_0.$$

Its maximum is actually equal to zero, which can be seen by picking in ($\diamond$),

$$w = D'\Sigma^{-1}\epsilon, \quad v = -C(\epsilon - \lambda^{-1}\Sigma G\tilde{x}), \quad x_0 = 0.$$

This completes the proof of (ii). To prove (iii), we start with the transformation

$$\hat{x}(t) = [I - \lambda^{-1}\Sigma(t)Z(t)]^{-1}\tilde{x}(t)$$

and differentiate the RHS with respect to t. This leads (after some algebra) to the conclusion that $\hat{x}$ satisfies (6.104a) whenever $\tilde{x}$ satisfies (6.104b). □

Remark 6.22 The proof given above for sufficiency (that is, parts (ii) and (iii)) is an indirect one, and follows the arguments of Uchida and Fujita (1989). Başar and Bernhard (1995) provide a more constructive proof, for all three parts of the theorem. Other proofs can be found in Limebeer et al. (1989), Khargonekar (1991) and Khargonekar, Nagpal and Poolla (1991) to cite just a few. □

An important observation to be made in the context of Thm. 6.25 is the duality between the two GRDEs (6.98) and (6.102). This duality enables us to deduce the behavior of the minimax controller μ_λ given by (6.106a) for the time-invariant problem, as $T \to \infty$, by making direct use of the results of subsection 6.5.3, as well as that of Corollary 6.10. This leads to the following counterpart of Corollary 6.10.

Theorem 6.26 *For the infinite-horizon H^∞-optimal control problem with imperfect state measurements, as defined above, let (A, B) be stabilizable, (A, Q) be detectable, (A, C) be detectable, and (A, D) be stabilizable. Let Λ^∞ be the set of all $\lambda > 0$ such that (6.99) admits a minimal nonnegative definite solution (denoted $\bar{Z}_\lambda^+$), the following dual GARE admits a minimal nonnegative definite solution (denoted $\bar{\Sigma}_\lambda^+$):*

$$A\bar{\Sigma} + \bar{\Sigma}A' + DD' - \bar{\Sigma}\left(C'C - \frac{1}{\lambda}Q\right)\bar{\Sigma} = 0, \tag{6.107}$$

and the following spectral radius condition is satisfied:

$$\rho(\bar{\Sigma}_\lambda^+ \bar{Z}_\lambda^+) < \lambda. \tag{6.108}$$

Then,

(i) $\inf\{\lambda \geq 0 : \lambda \in \Lambda^\infty\} = \lambda_\infty^*$, *where λ_∞^* is the smallest positive scalar λ such that the infinite-horizon version of (6.101) holds;*

(ii) for each $\lambda > \lambda_\infty^*$, the controller

$$\bar{\mu}_\lambda(y) = -B'\bar{Z}^+(I - \lambda^{-1}\Sigma^+ Z^+)^{-1}\tilde{x}, \tag{6.109a}$$

where

$$\dot{\tilde{x}} = (A + \lambda^{-1}\bar{\Sigma}^+ Q)\tilde{x} + Bu + \bar{\Sigma}^+ C'(y - C\tilde{x});\ \tilde{x}(0) = 0, \tag{6.109b}$$

delivers a performance level of at least $\sqrt{\lambda}$.

We close this section by revisiting Example 6.7 for the imperfect state measurements case.

Example 6.8 Consider Example 6.7 now with a measurement equation

$$y = x + v,$$

and the H^∞-performance index compatibly replaced by

$$(\|x\|^2 + \|u\|^2)^{1/2} \Big/ (\|w\|^2 + \|v\|^2)^{1/2},$$

where we have abused the notation and have taken $Dw = w$. Note that since $x_0 = 0$, we have taken in (6.100c) $Q_0^{-1} = 0$. In view of this, the second GRDE (6.102) reads

$$\dot{\Sigma} = \left(\frac{1}{\lambda} - 1\right)\Sigma^2 + 1, \qquad \Sigma(0) = 0,$$

which is the dual of (ii) in Example 6.7 (simply the time is reversed), and hence requires the same conjugate-point condition ($\lambda < 4T^2/(4T^2 + \pi^2)$) as the Z equation. Using the notation of Example 6.7, for $\lambda > \lambda^*(T)$ we have

$$\Sigma_\lambda = \frac{1}{m}\tan[mt],$$

and hence the spectral radius condition (6.105) is

$$\frac{1}{m^2}\tan[mt]\tan[m(T-t)] < \lambda \Longleftrightarrow \frac{1}{m}\tan\left[\frac{mT}{2}\right] < \sqrt{\lambda}$$

which is more restrictive than the other two conditions (which are identical): $\lambda < \lambda^*(T)$. Hence, in this case the spectral radius condition is binding.

For the infinite-horizon case, again from duality existence of $\bar{Z}^+$ and $\bar{\Sigma}^+$ requires the same condition, which is $\lambda > 1$. The spectral radius condition is

$$\bar{Z}^+\bar{\Sigma}^+ = \frac{\lambda}{\lambda - 1} < \lambda \Leftrightarrow \lambda > 2.$$

Hence $\lambda^*_\infty = 2$, and again the spectral radius condition is binding. □

6.7 Stochastic Differential Games with Deterministic Information Patterns

As it has been discussed in subsection 6.5.2, one possible way of eliminating "informational nonuniqueness" of Nash equilibria in nonzero-sum differential games is to require the equilibrium solution also to satisfy a feedback property (cf. Def. 6.2), which necessarily leads to strongly time consistent solutions; but this is a valid approach only if the information structure of the problem permits the players to have access to the current value of the state. Yet another approach to eliminate informational nonuniqueness is to formulate the problem in a stochastic framework, and this has been done in Section 6.4 for dynamic games defined in discrete time.

A stochastic formulation for quantitative differential games (i.e., dynamic games defined in continuous time) of prescribed duration involves (cf. Section 5.3) a stochastic differential equation

$$\mathrm{d}x_t = F(t, x_t, u^1(t), \ldots, u^N(t))\,\mathrm{d}t + \tilde{\sigma}(t, x_t)\,\mathrm{d}w_t,\ x_t\,|_{t=0} = x_0, \qquad (6.110)$$

which describes the evolution of the state (and replaces (6.37)), and N cost functionals

$$L^i(u^1,\ldots,u^N) = \int_0^T g^i(t,x_1,u^1(t),\ldots,u^N(t))\,\mathrm{d}t + q^i(x_T);\ i\in\mathbf{N}. \quad (6.111)$$

If Γ^i $(i \in \mathbf{N})$ denotes the strategy-space of $\mathbf{P}i$ (compatible with the deterministic information pattern of the game), his expected cost functional, when the differential game is brought to normal form, is given by

$$J^i(\gamma^1,\ldots,\gamma^N) = E[L^i(u^1,\ldots,u^N) \mid u^j(\cdot) = \gamma^j(\cdot,\eta^j), j\in\mathbf{N}] \quad (6.112)$$

with $\gamma^j \in \Gamma^j (j \in \mathbf{N})$ and $E[\cdot]$ denoting the expectation operation taken with respect to the statistics of the standard Wiener process $\{w_t, t\geq 0\}$. Furthermore, F and σ are assumed to satisfy the requirements of Thm. 5.2, and in particular σ is assumed to be nonsingular.

This section presents counterparts of some of the results of Section 6.4 for stochastic nonzero-sum differential games of the type described above; the analysis, however, will be rather informal since the underlying mathematics (on stochastic differential equations, stochastic calculus, etc.) is beyond the scope of this volume. Appropriate references will be given throughout for the interested reader who requires a more rigorous exposition of our treatment here.

First, by way of introduction, let us consider the one-person version of the N-person stochastic differential game, which is a stochastic optimal control problem of the type described in Chapter 5, for which a sufficiency condition for existence of an optimal control has been given in Prop. 5.7. This condition involves the existence of a solution to a particular partial differential equation (specifically, (5.71)), which corresponds to the Hamilton–Jacobi–Bellman equation (5.25) of dynamic programming, but now we have the extra second term which is due to the stochastic nature of the problem. We should note that if the control was allowed to depend also on the past values of the state, the optimal control would still be a feedback strategy since the underlying system is Markovian (i.e., F and σ depend only on x_t and not on $\{x_s, s<t\}$ at time t).

We now observe an immediate application of Prop. 5.7 to the N-person nonzero-sum stochastic differential game of this section.

Theorem 6.27 *For an N-person nonzero-sum stochastic differential game of prescribed fixed duration $[0,T]$, as described by (6.110)-(6.111), and under FB, MPS or CLPS information pattern, an N-tuple of feedback strategies $\{\gamma^{i*}\in\Gamma^i; i\in\mathbf{N}\}$ provides a Nash equilibrium solution if there exist suitably smooth functions $W^i : [0,T]\times\mathbf{R}^n \to \mathbf{R}$, $i\in\mathbf{N}$, satisfying the coupled semilinear parabolic partial differential equations*

$$\begin{aligned}
-\frac{\partial W^i}{\partial t} &- \frac{1}{2}\sum_{k,j}\sigma^{kj}(t,x)\frac{\partial^2 W^i}{\partial x_k \partial x_j} \\
&= \min_{u^i\in S^i}\,[\nabla_x W^i \cdot \tilde{F}^{i*}(t,x,u^i) + \tilde{g}^{i*}(t,x,u^i)] \qquad (6.113)\\
&= \nabla_x W^i \cdot \tilde{F}^{i*}(t,x,\gamma^{i*}(t,x)) + \tilde{g}^{i*}(t,x);\\
W^i(T,x) &= q^i(x) \quad (i\in\mathbf{N}),
\end{aligned}$$

where σ^{kj} *is the* kj*th element of the symmetric matrix* $\tilde{\sigma}\tilde{\sigma}'$,

$$\begin{aligned}
\tilde{F}^{i*}(t,x,u^i) &\triangleq F(t,x,\{\gamma^*_{-i},u^i\}),\\
\tilde{g}^{i*}(t,x,u^i) &\triangleq g^i(t,x,\{\gamma^*_{-i},u^i\}),\\
\{\gamma^*_{-i},u^i\} &\triangleq (\gamma^{1*}(t,x),\ldots,\gamma^{i-1*}(t,x),u^i,\gamma^{i+1*}(t,x),\ldots,\gamma^{N*}(t,x)).
\end{aligned}$$

Proof. This result follows readily from the definition of Nash equilibrium and from Prop. 5.7, since by fixing all players' strategies, except the ith ones, at their equilibrium choices (which are known to be feedback by hypothesis), we arrive at a stochastic optimal control problem of the type covered by Prop. 5.7 and whose optimal solution (if it exists) is a feedback strategy. □

Theorem 6.27 actually involves two sets of conditions: (i) existence of minima to the RHS of (6.113), and (ii) existence of "sufficiently smooth" solutions to the set of partial differential equations (6.113). A sufficient condition for the former is the existence of functions $u^{i\circ}:[0,T]\times\mathbf{R}^n\times\mathbf{R}^n\to S^i$ which satisfy the inequalities

$$\begin{aligned}
H^i(t,x,p^i;u^{1\circ},\ldots,u^{i\circ},\ldots,u^{N\circ}) &\leq H^i(t,x,p^i;u^{1\circ},\ldots,\\
&u^{i-1\circ},u^i,u^{i+1\circ},\ldots,u^{N\circ}),\quad \forall u^i\in S^i\quad (i\in\mathbf{N}),
\end{aligned}\tag{6.114a}$$

where $t\in[0,T]$, $p^i\in\mathbf{R}^n$, $x\in\mathbf{R}^n$ and

$$H^i(t,x,p^i;u)\triangleq p^{i'}F(t,x,u^1,\ldots,u^N)+g^i(t,x,u^1,\ldots,u^N).\tag{6.114b}$$

If such functions $u^{i\circ}(t,x,p^i)$ $(i\in\mathbf{N})$ can be found, then we say that the *Nash condition* holds for the associated differential game.[66] In fact, Uchida (1978) has shown, by utilizing some of the results of Davis and Varaiya (1973) and Davis (1973) obtained for stochastic optimal control problems, that under the Nash condition and certain technical restrictions on F, $\tilde{\sigma}$, S^i, g^i and q^i $(i\in\mathbf{N})$ there indeed exists a set of sufficiently smooth solutions to the partial differential equations (6.113). Therefore, the Nash condition is sufficient for a suitably well-defined class of stochastic differential games to admit a Nash equilibrium solution;[67] and a set of conditions sufficient for the Nash condition to hold are given in Uchida (1979).

For the special case of zero-sum stochastic differential games, i.e., when $L^l\equiv -L^2\triangleq L$ and $N=2$, every solution set of (6.113) is given by $W^1\equiv -W^2\triangleq W$,

[66] Note that such a condition could also have a counterpart in Thm. 6.16, for the deterministic problem.

[67] Uchida (1978) actually proves this result for a more general class of problems for which the state equation is not necessarily Markovian, in which case a more general version replaces (6.113).

where W satisfies

$$-\frac{\partial W}{\partial t} - \frac{1}{2}\sum_{i,j} \sigma^{ij}(t,x)\frac{\partial^2 W}{\partial x_i \partial x_j} = \min_{u^1\in S^1}\max_{u^2\in S^2} [\nabla_x W \cdot F(t,x,u^1,u^2)$$
$$+g(t,x,u^1,u^2)] = \max_{u^2\in S^2}\min_{u^1\in S^1} [\nabla_x W \cdot F(t,x,u^1,u^2) + g(t,x,u^1,u^2)];$$
$$W(T,x) = q(x). \tag{6.115}$$

Furthermore, the Nash condition (6.114a) can be re-expressed for the stochastic zero-sum differential game as existence of functions u^{io}: $[0,T]\times \mathbf{R}^n \times \mathbf{R}^n \to S^i$, $i = 1,2$, which satisfy

$$\begin{aligned} H(t,x,p;u^{1o},u^{2o}) &= \min_{u^1\in S^1}\max_{u^2\in S^2} H(t,x,p;u^1,u^2) \\ &= \max_{u^2\in S^2}\min_{u^1\in S^1} H(t,x,p;u^1,u^2), \end{aligned} \tag{6.116a}$$

where

$$H(t,x,p;u^1,u^2) \triangleq p'F(t,x,p;u^1,u^2) + g(t,x,u^1,u^2). \tag{6.116b}$$

Because of the similarity with the deterministic problem, condition (6.116a) is known as the *Isaacs condition.* Now, it can actually be shown (see Elliott, 1976) that, if the Isaacs condition holds and if S^1 and S^2 are compact, there exists a suitably smooth solution to the semilinear partial differential equation (6.115) and consequently a saddle-point solution to the stochastic zero-sum differential game under consideration. Therefore, we have the following.[68]

Corollary 6.11 *For the two-person zero-sum stochastic differential game described as an appropriate special case of (6.110)-(6.111), and satisfying certain technical restrictions, let S^1 and S^2 be compact and let the Isaacs condition (6.116a) hold. Then, there exists a saddle-point solution in feedback strategies, which is*

$$\left.\begin{aligned} \gamma^{1*}(t,x_1) &= u^{1o}(t,x_t,\nabla_x W(t,x_t)), \\ \gamma^{2*}(t,x_1) &= u^{2o}(t,x_t,\nabla_x W(t,x_t)), \end{aligned}\right\}$$

where $W(t,x)$ satisfies (6.115) and $u^{io}(t,x,p)$ $(i \in \mathbf{N})$ are determined from (6.116a). This solution is strongly time consistent.

For the special case of affine-quadratic stochastic differential games, both Thm. 6.27 and Corollary 6.11 permit explicit expressions for the equilibrium strategies. Toward this end, let F, g^i and q^i $(i \in \mathbf{N})$ be as given in Def. 6.5 (with F replacing f), and let $\tilde{\sigma}(t,x_t) = \tilde{\sigma}(t)$ which is nonsingular. Furthermore, assume that $S^i = \mathbf{R}^{m_i}$ $(i \in \mathbf{N})$.[69] Then, u^{io} from (6.114a) is uniquely given by

$$u^{io} = -R^{ii}(t)^{-1}B^i(t)'p^i, \quad (i \in \mathbf{N}),$$

[68] Elliott (1976) has proven this result for a non–Markovian problem in which case (6.115) is replaced by a more complex relation; here we provide a version of his result which is applicable to Markovian systems.

[69] In order to apply Corollary 6.11 directly, we in fact have to restrict ourselves to an appropriate (sufficiently large) compact subset of $\mathbf{R}^{m_j}$.

in view of which (6.113) can be written as

$$\begin{aligned}
-\frac{\partial W^i}{\partial t} &- \frac{1}{2}\sum_{k,j} \sigma^{kj}(t)\frac{\partial^2 W^i}{\partial x_k \partial x_j} \\
&= \nabla_x W^i \cdot \left(Ax + c - \sum_{j\in \mathbf{N}} B^j R^{jj^{-1}} B^{j'} \nabla_x W^{j'}\right) \\
&\quad + \frac{1}{2}x' Q^i x + \frac{1}{2}\sum_{j\in \mathbf{N}} \nabla_x W^j \, B^{j'} R^{jj^{-1}} R^{ij} R^{jj^{-1}} B^j \nabla_x W^{j'}; \\
W^i(T,x) &= \frac{1}{2}x' Q^i_f x,
\end{aligned}$$

which admits the unique solution set

$$W^i(t,x) = \frac{1}{2}x' Z^i(t) x + x'\zeta^i(t) + m^i(t) + \xi^i(t) \quad (i \in \mathbf{N}),$$

where Z^i, $i \in \mathbf{N}$, satisfy the N coupled matrix Riccati differential equations (6.70a), ζ^i and m^i satisfy (6.72a) and (6.73b), respectively, and ξ^i satisfies the differential equation

$$\dot{\xi}^i = -\frac{1}{2}\mathrm{Tr}[\tilde{\sigma}(\mathrm{t})\tilde{\sigma}(\mathrm{t})' \mathrm{Z}^i(\mathrm{t})].$$

Now, since γ^{i*} is given by

$$\gamma^{i*}(t, x_t) = -R^{ii}(t)^{-1} B^i(t)' \nabla_x W(t, x_t)',$$

it readily follows that the Nash equilibrium solution of the stochastic affine-quadratic differential game coincides with that of the deterministic game, as provided in Corollary 6.5. Therefore, we have the following.

Corollary 6.12 *The set of Nash equilibrium strategies (6.71) for the deterministic affine-quadratic differential game with MPS or CLPS information pattern also provides a Nash equilibrium solution for its stochastic counterpart, with σ not depending on the state x_t.*

A similar analysis and reasoning also leads to the following.

Corollary 6.13 *Under the conditions of Thm. 6.17, the set of saddle-point strategies (6.77) for the deterministic affine-quadratic two-person zero-sum differential game with MPS or CLPS information pattern also provides a saddle-point solution for its stochastic counterpart, provided that σ does not depend on the state x_t.*

Remark 6.23 Under appropriate convexity restrictions, which hold for the affine-quadratic nonzero-sum game treated here, existence of a solution set to (6.113) is also sufficient for existence of a Nash equilibrium (see Varaiya, 1976);

therefore the solution referred to in Corollary 6.12 is the unique Nash equilibrium solution for the affine-quadratic stochastic differential game within the class of feedback strategies. It should be noted that we do not rule out existence of some other Nash equilibrium solution that depends also on the past values of the state (though we strongly suspect that this should not be the case), since Thm. 6.27 provides a sufficiency condition only within the class of feedback strategies. The reader should note, however, that we cannot have informationally nonunique equilibrium solutions here, since $\{w_t, t \geq 0\}$ is an independent-increment process and $\tilde{\sigma}$ is nonsingular. For the zero-sum differential game, on the other hand, an analogous uniqueness feature may be attributed to the result of Corollary 6.13, by making use of a sufficiency result of Elliott (1976), which is obtained for compact S^1 and S^2 but can easily be generalized to noncompact action spaces for the affine-quadratic problem. □

6.8 Problems

1. An alternative derivation of the result of Thm. 6.2 makes use of Prop. 5.1. First show that the ith inequality of (6.3) dictates an optimization problem to $\mathbf{P}i$ which is similar to the one covered by Prop. 5.1, with c_k in Prop. 5.1 replaced by $c_k + \sum_{j \in \mathbf{N}, j \neq i} B_k^j u_k^{j*}$ which is only a function of x_1 (and hence can be considered as a constant). Then the open-loop Nash equilibrium solution is given as the solution of the set of relations

$$\begin{aligned}
u_k^{i*} &= -P_k^i S_{k+1}^i A_k x_k^* - P_k^i \Big(s_{k+1}^i + S_{k+1}^i c_k + S_{k+1}^i \sum_{\substack{j \in \mathbf{N} \\ j \neq i}} B_k^j u_k^{j*} \Big), \\
P_k^i &= [R_k^{ii} + B_k^{i'} S_{k+1}^i B_k^i]^{-1} B_k^{i'}, \\
S_k^i &= Q_k^i + A_k' S_{k+1}^i [I - B_k^i P_k^i S_{k+1}^i] A_k; \; S_{k+1}^i = Q_{k+1}^i, \\
s_k^i &= A_k' [I - B_k^i p_k^i S_{k+1}^i]' [s_{k+1}^i + S_{k+1}^i c_k + S_{k+1}^i \sum_{\substack{j \in \mathbf{N} \\ j \neq i}} B_k^j u_k^{j*}];
\end{aligned}$$

$$s_{K+1}^i = 0, \quad k \in \mathbf{K}, \; i \in \mathbf{N},$$

where x_{k+1}^* ($k \in \mathbf{K}$) denotes the corresponding state trajectory. Now finally prove that the preceding set of relations admits a unique solution as given in Thm. 6.2, under precisely the same existence condition.

2. Consider the general formulation of an N-person nonzero-sum dynamic game, as described by (6.1) and (6.2), where the players have access to the state with a delay of one time step, i.e., the information available to each player at stage k is

$$\eta_k = \{x_{k-1}, x_{k-2}, \ldots, x_1\}, \quad k \geq 2.$$

We wish to obtain a strongly time consistent Nash equilibrium solution for this class of dynamic games.

(i) Obtain the counterpart of Thm. 6.6 under this information structure.

(ii) Specialize the result above to affine-quadratic dynamic games, so as to obtain the counterpart of Corollary 6.1 for this one-step delay CLPS pattern.

3. For the class of affine-quadratic zero-sum dynamic games covered by Thm. 6.7, modify the information structure so that now **P1** is allowed to have access to the current and past actions of **P2**, i.e., **P1**'s information structure is

$$\eta_k^1 = \{x_k, u_k^2, x_{k-1}, u_{k-1}^2, \ldots, x_1, u_1^2\}, \quad k \geq 1.$$

Develop the counterpart of the result of Thm. 6.7 for this class of games. In particular, show that

(i) there exists a saddle point under conditions that are less stringent (for **P1**) than the conditions of Thm. 6.7;

(ii) the saddle-point policies of the players that are also strongly time consistent are affine functions of the current information (which is $\{x_k, u_k^2\}$ for **P1**, and x_k for **P2**, at stage k);

(iii) the corresponding state trajectory is the same as the one generated by the feedback saddle-point solution of Thm. 6.7, whenever the latter exists.

4. Consider the infinite-horizon two-player linear-quadratic dynamic game whose two-dimensional state $x_k \triangleq (x_k^1, x_k^2)'$ is described by

$$\begin{aligned} x_{k+1}^1 &= 2x_k^1 + u_k^1 + \epsilon x_k^2; \quad x_1^1 = 1, \\ x_{k+1}^2 &= -2x_k^2 + u_k^2 - \epsilon x_k^1; \quad x_1^2 = 1, \end{aligned}$$

where $\epsilon > 0$ is a scalar parameter. Let the cost functions be given by

$$L^i = \sum_{k=1}^{\infty} (x_{k+1}^i)^2 + (u_k^i)^2, \quad i = 1, 2.$$

(i) Find the largest value of $\epsilon > 0$ (say, ϵ^*), such that for all $\epsilon < \epsilon^*$, the game admits a stationary feedback Nash equilibrium solution.

(ii) Obtain the feedback Nash equilibrium solution when $\epsilon = \epsilon^*/2$, by directly using (6.28a)-(6.28c), and also by iterating on the solution of the finite-horizon game.

5. Derive a set of nonlinear Nash equilibrium solutions for the dynamic game of subsection 6.3.1. Specifically, obtain a counterpart of the result of Prop. 6.5 by starting with the representations

$$\begin{aligned} \gamma^1(x_2, x_1) &= -\frac{2}{3}x_2 + p[x_2 - \bar{x}_2(\gamma^3)]^2, \\ \gamma^2(x_2, x_1) &= -\frac{1}{3}x_2 + q[x_2 - \bar{x}_2(\gamma^3)]^2, \end{aligned}$$

and by mimicking the analysis that precedes Prop. 6.5. How does the underlying existence condition compare with (6.35)?

6. Consider the scalar two-person three-stage stochastic dynamic game described by the state equation

$$\begin{aligned} x_4 &= x_3 + u_3^1 + \theta, \\ x_3 &= x_2 + u_2^1 + u_2^2 + \theta, \\ x_2 &= x_1 + u_1^2 + \theta, \end{aligned}$$

and cost functionals

$$\begin{aligned} L^1 &= (x_4)^2 + (u_3^1)^2 + (u_2^1)^2, \\ L^2 &= (x_4)^2 + 2(u_2^2)^2 + (u_1^2)^2. \end{aligned}$$

Here θ denotes a Gaussian random variable with mean zero and variance 1, and both players have access to CLPS information.

Show that this dynamic game admits an uncountable number of Nash equilibria, and obtain one such set which is linear in the state information available to both players. Will there be additional Nash equilibria if θ has instead a two-point distribution with mean zero and variance 1?

7. For the stochastic dynamic game of the previous problem, prove that the Nash equilibrium solution is unique if the underlying information pattern is, instead, one-step delay CLPS pattern. Obtain the corresponding solution, and identify it as a member of the solution set determined in Problem 6.

8. Consider the class of two-person K-stage linear-quadratic stochastic dynamic games with cost functional (6.10b) and state equation

$$x_{k+1} = A_k x_k + B_k^1 u_k^1 + B_k^2 u_k^2 + \theta_k, k \in \mathbf{K},$$

where $\{\theta_k; k \in \mathbf{K}\}$ denotes a set of independent Gaussian vectors with mean zero and covariance I_n. Obtain the saddle-point solution when (i) both players have access to one-step delayed CLPS information, (ii) **P1** has access to CLPS information and **P2** has access to one-step delayed CLPS information, and (iii) **P1** has access to CLPS information whereas **P2** has only access to the value of x_1 (i.e., OL information). How do the existence conditions compare in the three cases, among themselves and with (6.26a)-(6.26b)? (Hint: Make use of Thm. 6.9 in order to simplify the derivation of the saddle-point strategies.)

9. Execute the procedure outlined in Remark 6.13 (for two-person games with weakly coupled players) when the state dynamics in (6.48a) are linear and the cost functions in (6.48b) are quadratic, and obtain expressions for the first three terms in the expansions for the open-loop Nash strategies of the

10. Show that Thm. 6.12 can also be obtained by directly utilizing Prop. 5.2 in a manner which is parallel to the one outlined in Problem 1 for the discrete-time counterpart.

11. This problem addresses a generalization of Problem 1 in Section 5.8. Assume that we now have two companies, **P1** and **P2**, with **P***i* having $x_i(t)$ customers at time t and making a total profit of

$$f^i(t) = \int_0^t [c^i x_i(s) - u^i(s)]\, ds$$

up to time t, where c^i is a constant. The function $u^i(t)$, restricted by $u^i(t) \geq 0$, $f^i(t) \geq 0$, represents the money put into advertisement by **P***i*, which helps to increase the number of customers attracted to that firm, according to

$$\dot{x}_i = u^i - u^j, \quad x_i(0) = x_{i0} > 0,\ i, j = 1, 2, \quad j \neq i.$$

The total number of customers shared between the two companies is a constant, so that we may write, without any loss of generality, the constraint relation

$$x_1(t) + x_2(t) = 1, \quad \forall t \in [0, T],$$

where T denotes the end of the time period when each firm wishes to maximize its total profit (i.e., $f^i(T)$).

Obtain the open-loop Nash equilibrium solution of this two-person nonzero-sum differential game wherein the decision variable for **P***i* is u^i. (For details see Olsder, 1976).

12. Consider a variation on the previous problem so that now the state equation is described by

$$\begin{aligned} \dot{x}_1 &= x_2 u^1 - x_1 u^2, \\ \dot{x}_2 &= x_1 u^2 - x_2 u^1, \end{aligned}$$

and the decision variables satisfy the constraints

$$|u^i(t)| \leq 1, \quad i = 1, 2.$$

The objective for **P***i* is still maximization of $f^i(T)$.

(i) Obtain the open-loop Nash equilibrium solution.

(ii) Obtain the feedback Nash equilibrium solution when both players have access to the current values of x_1 and x_2.

13. Obtain the complete set of informationally nonunique linear Nash equilibrium solutions of the scalar linear-quadratic two-person nonzero-sum differential game described by the state dynamics

$$\dot{x} = 2u^1 - u^2, \quad x(0) = 1, \quad t \in [0, 2]$$

and cost functionals

$$L^i = [x(2)]^2 + \int_0^2 [u^i(t)]^2 \, dt, \quad i = 1, 2,$$

and with the underlying information structure being MPS for both players. Which of these would still constitute a Nash equilibrium when (i) **P2** has, instead, access to OL information, (ii) both players have access to OL information? (Note that, in this case, OL information is in fact "no information" since the initial state is a known number.)

14. Prove that the statement of Thm. 6.9 is also valid for two-person zero-sum differential games with prescribed fixed duration.

15. Repeat Problem 9 for the CLNM information structure and using the feedback Nash equilibrium solution. Here Remark 6.18 replaces Remark 6.13, and Corollary 6.5 replaces Thm. 6.12.

16. Generalize the result of Thm. 6.17 to the case when the cost function has, under the integral, the additional term $u^1(t)'K(t)u^2(t)$, where $K(\cdot)$ is a matrix of appropriate dimensions and with continuous entries. Is it possible for the differential game to have a bounded lower value even if the conditions for existence of a saddle point fail?

17. This problem uses the notation and terminology introduced in Section 6.6. We wish to design optimal controllers for the following two-dimensional system, under different information patterns, using the H^∞ criterion:

$$\begin{aligned} \dot{x}_1 &= x_2 + w; \\ \dot{x}_2 &= x_1 + x_2 + u; \\ z &= x_1 + x_2; \quad y = x_1 + v. \end{aligned}$$

Here, x_1 and x_2 denote the two states, whose initial values are completely unknown, u is the scalar control variable, y is the measured output, z is the regulated output, w is the scalar system disturbance and v is the measurement disturbance. With this system, we associate the performance index

$$\left\{ \int_0^T [|z(t)|^2 + |u(t)|^2] \, dt \right\} / \left\{ |x_1(0)|^2 + |x_2(0)|^2 + \int_0^T [|w(t)|^2 + |v(t)|^2] \, dt \right\}$$

where T is the terminal time.

(i) Let $T = \infty$, and the controller have access to perfect state information. Denote the optimum (minimax) performance level for this problem by λ^*. Obtain the value of λ^*, and a controller that will ensure a performance level no worse than $\lambda = \lambda^* + 0.01$.

(ii) Repeat the above now under the original imperfect state information.

(iii) Now take the time horizon to be finite, and $T = 1$. Furthermore take the initial state to be known to be zero. Obtain (approximately) the λ^* for this problem under perfect state measurements by numerically solving the corresponding generalized Riccati differential equation for different values of λ.

18. The following two-dimensional system description depends on a parameter ϵ:

$$\dot{x}_1 = x_1 + 2x_2 + 2u + w; \quad \epsilon\dot{x}_2 = 2x_1 + 2x_2 + u + 3w.$$

Given the cost function:

$$\left\{\int_0^\infty (2x_1^2 + 2x_1x_2 + x_2^2 + u^2)\,dt\right\} / \left\{\int_0^\infty |w(t)|^2\,dt\right\},$$

obtain the H^∞-optimal performance level $(\lambda^*(\epsilon))$, under perfect state measurements, for $\epsilon = 1, 0.1, 0.01, 0.001, 0.0001$, to the nearest four decimal places.

Solution: 2.4332, 2.7843, 2.9724, 2.9972, 2.9997.

19. Repeat Problem 18 for the following system and cost function:

$$\dot{x}_1 = 2x_1 + x_2 + 2u + w; \quad \epsilon\dot{x}_2 = -x_1 - 2x_2 + 2u + 3w,$$

$$\left\{\int_0^\infty (2x_1^2 + 2x_1x_2 + 3x_2^2 + u^2)\,dt\right\} / \left\{\int_0^\infty |w(t)|^2\,dt\right\}.$$

Solution: 1.3306, 1.3393, 1.4212, 1.4237, 1.4240.

20. Consider the following two-dimensional system which again depends on a parameter ϵ:

$$\dot{x}_1 = x_1 + x_2 + 3u + 2w; \quad \epsilon\dot{x}_2 = -x_1 + 2x_2 + 2u + w.$$

We now have a measurement equation

$$y = 2x_1 + x_2 + 3v,$$

where v is the measurement noise. Given the cost function,

$$\left\{\int_0^\infty (2x_1^2 + 2x_1x_2 + 3x_2^2 + u^2)\,dt\right\} / \left\{\int_0^T [|w(t)|^2 + |v(t)|^2]\,dt\right\},$$

obtain the H^∞-optimal performance level $(\lambda^*(\epsilon))$, under imperfect state measurements, for $\epsilon = 1, 0.1, 0.01, 0.001, 0.0001$, to the nearest four decimal places, and determine a controller (for each case) that assures a performance no worse than $\lambda^*(\epsilon) + 0.01$.

Partial Solution: $\lambda^* = 55.1993, 17.3221, 9.9331, 9.1197, 9.0308$.

21. Consider the differential game of Problem 13, but with the state equation replaced by

$$\dot{x} = 2u^1 - u^2 + \sigma,$$

where σ is a random variable taking the values $+1$ and -1 with equal probability 1/2. Under the CLPS information pattern for both players, does this differential game, which also incorporates a chance move, admit a unique Nash equilibrium solution? In case the reply is "yes", obtain the unique solution; in case it is "no", obtain the complete set of Nash equilibria that are linear in the current value of the state.

22. Obtain the saddle-point solution of the class of two-person zero-sum stochastic differential games described by the state equation

$$dx_t = [A(t)x_t + B^1(t)u^1 + B^2(t)u^2]dt + \tilde{\sigma}(t)dw_t$$

and cost functional (6.60b), where **P1** has access to CLPS information and **P2** has OL information. Here, $\tilde{\sigma}(\cdot)$ is a nonsingular matrix, $\{w_t, t \geq 0\}$ is the standard Wiener process, and the initial state vector x_0 is a known quantity. (For details see Başar (1977b)).

6.9 Notes

Section 6.2. Results pertaining to Nash equilibria in infinite dynamic games first appeared in the works of Case (1967, 1969) and Starr and Ho (1969a, b) but in the continuous time (i.e., for differential games). Here we present counterparts of these results in discrete time, and with special attention paid to affine-quadratic (nonzero-sum and zero-sum) games, and existence conditions for equilibria under open-loop and memoryless perfect state information patterns. A possible extension of these results would be to the class of game problems in which the players also have the option of whether they should make a perfect state measurement or not, if it is costly to do so; for a discussion of such problems see Olsder (1977b). For applications of some of these results in macroeconomics, see Pindyck (1977), Kydland (1975) and Başar, Turnovsky and d'Orey (1986). The last reference also studies, within the context of a two-country macroeconomic model, the computation and stabilizability of stationary feedback Nash equilibria using the solutions of a sequence of time-truncated dynamic games.

Section 6.3. Existence of uncountably many informationally nonunique Nash equilibria in dynamic games under closed-loop information was first displayed in Başar (1974), Başar (1976a) for discrete-time games, and in Başar (1977b) for differential games. The underlying idea in this section therefore follows these references. The algorithm of subsection 6.3.2 is from Başar (1977d), and Thm. 6.9 was first presented in Başar (1977a); some further discussion on the relation between existence of a saddle point and the underlying information pattern in zero-sum dynamic games can be found in Başar (1976b) and Witsenhausen (1971b), where the latter reference also discusses the impact of information on the upper and lower values of a game when a saddle point does not exist.

Section 6.4. Elimination of informational nonuniqueness in Nash equilibria through a stochastic formulation was first discussed in Başar (1976a), and further elaborated on in Başar (1975, 1979b, 1989b); this section follows basically the lines of those references. The fact that an increase in information could be detrimental for a player under Nash equilibria was first pointed out in Başar and Ho (1974), but within the context of stochastic games with noisy (imperfect) observation. For some results on the equilibrium solution of linear-quadratic stochastic games with noisy observation, see Başar (1978), Başar and Mintz (1972, 1973).

Section 6.5. Derivation of open-loop and feedback Nash equilibria in nonzero-sum differential games was first discussed in Case (1967, 1969) and Starr and Ho (1969a, b), where the latter two references also display the differences (in value) between these two types of equilibria. Derivation of open-loop saddle-point equilibria, however, predates this development, and was first discussed in Berkovitz (1964) and Ho et al. (1965), with the latter devoted to a particular class of linear-quadratic differential games of the pursuit-evasion type, whose results were later put into rigorous framework in Schmitendorf (1970); see also Halanay (1968) which extends the results of Ho et al. (1965) to differential games with delays in the state equation. For an excellent exposition on the status of nonzero-sum and zero-sum differential game theory in the early 1970s, see the survey paper by Ho (1970). Some selective references which display the advances on Nash and saddle-point equilibria of deterministic differential games since the early 1970s are (in addition to those already cited in the text) Bensoussan (1974), Ho, Luh and Olsder (1980), Leitmann (1974, 1977), Scalzo (1974), Vidyasagar (1976) and related papers in the edited volumes Blaquière (1973), Hagedorn et al. (1977), Ho and Mitter (1976), Kuhn and Szegő (1971), Başar (1986b), Başar and Bernhard (1989), Hämäläinen and Ehtamo (1991), Başar and Haurie (1994). For applications of this theory to problems in economics, see Case (1971, 1979), Clemhout et al. (1973), Pohjola (1986), Jorgensen (1986), Dockner and Feichtinger (1986) and Leitmann and Wan, Jr. (1979), where the last one discusses a worst-case approach toward stabilization of uncertain systems, that utilizes the zero-sum differential game theory. Tolwinski (1978b) provides a rather complete account of the results on existence of open-loop Nash equilibria in nonzero-sum differential games, and Tolwinski (1978a) discusses a possible computational scheme for the numerical evaluation of the Nash equilibrium solution; see also Varaiya (1967, 1970), Rosen (1965) and Friedman (1971) for existence results concerning noncooperative equilibria. For the special class of linear-quadratic games, Papavassilopoulos and Cruz, Jr. (1979b) and Papavassilopoulos et al. (1979) discuss existence of a unique solution to the coupled set of Riccati equations (6.70a), which plays an important role in the characterization of feedback Nash equilibria; for computational algorithms for some specially structured games, see Başar (1991a), Srikant and Başar (1991), Gajic, Petkovski and Shen (1990). Papavassilopoulos and Olsder (1984) demonstrate that an infinite-horizon linear-quadratic differential game might have multiple Nash equilibria even though every time-truncated version of it has a unique feedback Nash equilibrium. Eisele (1982) was one of the first to point out that linear-quadratic differential games might admit multiple open-loop Nash equilibria, whose results were later extended by Engwerda (1998) who showed that an open-loop Nash equilibrium solution may exist for a linear quadratic game even if the corresponding differential Riccati equations may not have a solution. The latter reference also studied the limiting behavior of finite time horizon open-loop solutions of linear-quadratic games as the final time approaches infinite. For a discussion of non-cooperative equilibria in differential games with state dynamics described by partial differential equations, see Roxin (1977).

Section 6.6. The theory briefly presented here has extensions to other types of information structures, such as sampled-data (both perfect state and imperfect state), and delayed measurements, as well as to nonlinear systems; see Başar and Bernhard (1995). It is also possible to study the robustness (or sensitivity) of the H^∞-optimal controllers to unmodeled fast plant dynamics, within the framework of singular perturbations; see Pan and Başar (1993, 1994a,b). These papers also develop, as a byproduct, a complete theory for the saddle-point equilibria of singularly perturbed linear-quadratic differential games under perfect and imperfect state measurements. It is also possible to combine Nash and saddle-point equilibria to develop a counterpart of H^∞-optimal control theory for multiple controller systems under multiple worst-case design criteria (Başar 1992).

Section 6.7. For some explicit results in the case of stochastic quadratic differential games with mixed information patterns, see Başar (1977b,c, 1979a, 1980a). Extensions of these results to stochastic differential games with noisy observation meet with formidable difficulties, some of which have been resolved in the case of linear-quadratic zero-sum games (see Willman (1969), Bagchi and Olsder (1981) and Başar (1981c), where the latter reference deals with the case of identical noisy measurements for the players). For a distributed algorithm for the computation of Nash equilibria in linear-quadratic stochastic differential games, see Başar and Li (1989).

Chapter 7

Stackelberg Equilibria of Infinite Dynamic Games

7.1 Introduction

This chapter is devoted to derivation of the Stackelberg solution in infinite dynamic games with fixed prescribed duration. The chapter starts with a treatment of dynamic games defined in discrete time and with the number of players restricted to two. Continuous-time counterparts of most of these results and possible extensions to many-player games are treated in the latter part.

The next two sections, i.e., Sections 7.2 and 7.3, deal, respectively, with the Stackelberg solution under open-loop information and the feedback Stackelberg solution under CLPS information. These solutions are obtained using two standard techniques of optimal control theory, viz. the minimum principle and dynamic programming, respectively. Derivation of the (global) Stackelberg solution under the CLPS information pattern, however, requires a much more subtle analysis since all standard techniques and approaches of optimal control theory fail to provide the solution. Therefore, Section 7.4 is devoted exclusively to this topic and to elucidation of an indirect approach toward derivation of the closed-loop Stackelberg solution. This indirect approach is first introduced within the context of two scalar examples (cf. subsections 7.4.1 and 7.4.2), and then its application is extended to the class of two-person linear-quadratic dynamic games in subsection 7.4.3. There is a clear relationship with the theory of "incentives", and this is the subject of subsection 7.4.4.

Section 7.5 discusses derivation of the Stackelberg solution in stochastic dynamic games with deterministic information patterns, and primarily for the open-loop information (cf. subsection 7.5.1) and for the CLPS information structure under the "feedback Stackelberg" solution (cf. subsection 7.5.2). Derivation of the (global) Stackelberg solution under the CLPS information pattern is a challenging topic, which is briefly discussed also in subsection 7.5.1. Stochastic incentive problems are discussed in subsection 7.5.2, as the counterpart of subsection 7.4.4 in the stochastic case.

Section 7.6 treats continuous-time games, first presenting the Stackelberg solution under open-loop information for the leader (subsection 7.6.1), and then deriving the feedback Stackelberg solution (defined as a natural extension of the discrete-time FB Stackelberg solution concept) under CLPS information (subsection 7.6.2).

7.2 Open-Loop Stackelberg Solution of Two-Person Dynamic Games in Discrete Time

In this section we discuss the Stackelberg solution for the class of two-person deterministic discrete-time infinite dynamic games of prescribed fixed duration (cf. Def. 5.1) when the underlying information structure is open-loop (for both players) and **P1** acts as the leader. Hence, abiding by our standard terminology and notation, the state evolves according to

$$x_{k+1} = f_k(x_k, u_k^1, u_k^2), \qquad k \in \mathbf{K}, \tag{7.1}$$

where $x_k \in \mathbf{X} = \mathrm{R}^n$, x_1 is specified *a priori*, and $u_k^i \in U_k^i \subseteq \mathrm{R}^{m_i}$, $i = 1, 2$; and the stage-additive cost functional for $\mathbf{P}i$ is introduced as

$$J^i(\gamma^1, \gamma^2) \triangleq L^i(u^1, u^2) = \sum_{k=1}^{K} g_k^i(x_{k+1}, u_k^1, u_k^2, x_k), \quad i = 1, 2. \tag{7.2}$$

Because of the open-loop information structure, the game is already in normal form, and therefore Defs 3.26, 3.27 and 3.28 readily introduce the Stackelberg equilibrium solution with **P1** acting as the leader (since finiteness of the game was not essential in those definitions).

One method of obtaining the Stackelberg solution for this class of games is to make use of the theory of Section 4.5, since these games can be viewed as static ones under the open-loop information structure; such an approach also readily leads to conditions for existence of a Stackelberg solution. Toward this end, we first note that, by recursive substitution of (7.1) into (7.2), the cost functional $J^i(u^1, u^2)$ can be made to depend only on the control vectors u^1, u^2 (which are of dimensions $m_1 K$ and $m_2 K$, respectively) and the initial state x_1, which is known *a priori*. Now, if

(i) J^i is continuous on $U^1 \times U^2 \quad (i = 1, 2)$,

(ii) $J^2(u^1, \cdot)$ is strictly convex on U^2 for all $u_1 \in U^1$,

(iii) U^i is a closed and bounded (thereby compact) subset of $\mathbf{R}^{m_i} (i = 1, 2)$,

then it follows from Corollary 4.4 (more generally from Thm. 4.8) that a Stackelberg equilibrium solution exists for the open-loop infinite game. A brute-force method to obtain the corresponding solution would be first to determine the unique reaction curve of the follower by minimizing $J^2(u^1, u^2)$ over $u^2 \in U^2$ for every fixed $u^1 \in U^1$, which is a meaningful optimization problem because of assumptions (i)-(iii) above. Denoting this unique mapping by $T^2 : U^1 \to U^2$, the optimization problem faced by **P1** (the leader) is then minimization of $J^1(u^1, T^2u^1)$ over U^1, thus yielding a Stackelberg strategy for the leader in this open-loop game.

Despite its straightforwardness, such a derivation is not practical, especially if the number of stages in the game is large—which directly contributes to the dimensions of the control vectors u^1, u^2. An alternative derivation, which parallels that of Thm. 6.1 in the case of open-loop Nash equilibria, utilizes techniques of optimal control theory. Toward this end, we first determine the unique optimal response of the follower to every announced strategy $\gamma^1 \in \Gamma^1$ of the leader. Since the underlying information pattern is open-loop, the optimization problem faced by the follower is (for each fixed $u^1 \in U^1$)

$$\min_{u^2 \in U^2} L^2(u^1, u^2)$$

subject to

$$x_{k+1} = f_k(x_k, u_k^1, u_k^2), \quad x_1 \text{ given.}$$

This is a standard optimal control problem whose solution exists and is unique under conditions *(i)-(iii)*, and which can further be obtained by utilizing Thm. 5.5.

Lemma 7.1 *In addition to conditions (i)-(iii) assume that*[70]

(iv) $f_k(\cdot, u_k^1, u_k^2)$ *is continuously differentiable on* $\mathbf{R}^n$, *and* $f_k(\cdot, u_k^1, \cdot)$ *is convex on* $\mathbf{R}^n \times U^2$, $(k \in \mathbf{K})$,

(v) $g_k^2(\cdot, u_k^1, u_k^2, \cdot)$ *is continuously differentiable on* $\mathbf{R}^n \times \mathbf{R}^n$, $(k \in \mathbf{K})$.

Then, to any announced strategy $u^1 = \gamma^1 \in \Gamma^1$ *of the leader, there exists a unique optimal response of the follower (to be denoted by* $\bar{\gamma}^1(x_1) = \bar{u}^2$*) satisfying the following relations:*

$$\begin{aligned} \bar{x}_{k+1} &= f_k(\bar{x}_k, u_k^1, \bar{u}_k^2), \quad \bar{x}_1 = x_1, && (7.3a)\\ \bar{u}_k^2 &= \arg \min_{u_k^2 \in U_k^2} H_k^2(p_{k+1}, u_k^1, u_k^2, \bar{x}_k), && (7.3b) \end{aligned}$$

$$\begin{aligned} p_k = &\frac{\partial}{\partial x_k} f_k(\bar{x}_k, u_k^1, \bar{u}_k^2)' \left[p_{k+1} + \left(\frac{\partial}{\partial x_{k+1}} g_k^2(\bar{x}_{k+1}, u_k^1, \bar{u}_k^2, \bar{x}_k) \right)' \right] \\ &+ \left[\frac{\partial}{\partial x_k} g_k^2(\bar{x}_{k+1}, u_k^1, \bar{u}_k^2, \bar{x}_k) \right]'; \quad p_{K+1} = 0, \end{aligned} \quad (7.3c)$$

[70]Except for very special cases, conditions (iv) and (v) given below, and conditions (vi), (vii), (ix) and (x) given later, are required for satisfaction of condition (i) (i.e., they are, in general, implicit in condition (i)); but we nevertheless rewrite them here since they will be needed explicitly in some of the results obtained in the sequel.

where

$$H_k^2(p_{k+1}, u_k^1, u_k^2, x_k) \triangleq g_k^2(f_k(x_k, u_k^1, u_k^2), u_k^1, u_k^2, x_k) + p'_{k+1} f_k(x_k, u_k^1, u_k^2), \quad k \in \mathbf{K}. \tag{7.3d}$$

Here, $\{p_1, \ldots, p_{K+1}\}$ *is a sequence of n-dimensional costate vectors associated with this optimal control problem.*

If we further assume

(vi) $f_k(x_k, u_k^1, \cdot)$ is continuously differentiable on U_k^2 $(k \in \mathbf{K})$,

(vii) $g_k^2(x_{k+1}, u_k^1, \cdot, x_k)$ is continuously differentiable on U_k^2 $(k \in \mathbf{K})$,

(viii) $\bar{u}^2$ in Lemma 7.1 is an inner-point solution for every $u^1 \in U^1$,

then (7.3b) can equivalently be written as

$$\nabla_{u_k^2} H_k^2(p_{k+1}, u_k^1, \bar{u}_k^2, \bar{x}_k) = 0 \quad (k \in \mathbf{K}). \tag{7.4}$$

Now, to obtain the Stackelberg strategy of the leader, we have to minimize $L^1(u^1, u^2)$, in view of the unique optimal response of the follower to be determined from (7.4) in conjunction with (7.3a) and (7.3c). Therefore, **P1** is faced with the optimal control problem

$$\min_{u^1 \in U^1} L^1(u^1, u^2) \tag{7.5a}$$

subject to

$$x_{k+1} = f_k(x_k, u_k^1, u_k^2), \qquad x_1 \text{ given}, \tag{7.5b}$$

$$p_k = F_k(x_k, u_k^1, u_k^2, p_{k+1}), \qquad p_{K+1} = 0, \tag{7.5c}$$

$$\nabla_{u_k^2} H_k^2(p_{k+1}, u_k^1, u_k^2, x_k) = 0 \qquad (k \in \mathbf{K}), \tag{7.5d}$$

where H_k^2 is defined by (7.3d) and

$$F_k \triangleq \frac{\partial}{\partial x_k} f_k(x_k, u_k^1, u_k^2)' p_{k+1} + \left[\frac{\partial}{\partial x_k} g_k^2(f_k(x_k, u_k^1, u_k^2), u_k^1, u_k^2, x_k)\right]'. \tag{7.5e}$$

Note that, in this optimal control problem the constraint equations involve boundary conditions at both end points, and hence Thm. 5.5 is not applicable; but a standard result of nonlinear programming (cf. Canon et al., 1970) can be utilized to yield the following conclusion.

Theorem 7.1 *In addition to conditions (i)-(viii) stated in this section, assume that*

(ix) $f_k(x_k, \cdot, u_k^2)$, $g_k^2(x_{k+1}, \cdot, u_k^2, x_k)$ *are continuously differentiable on* U_k^1, $k \in \mathbf{K}$,

(x) $g_k^1(\cdot,\cdot,\cdot,\cdot)$ *is continuously differentiable on* $\mathbf{R}^n \times U_k^1 \times U_k^2 \times \mathbf{R}^n$, $k \in \mathbf{K}$,

(xi) $f_k(\cdot, u_k^1, \cdot)$ *is twice continuously differentiable on* $\mathbf{R}^n \times U_k^2$, *and* $g_k^2(\cdot, u_k^1, \cdot, \cdot)$ *is twice continuously differentiable on* $\mathbf{R}^n \times U_k^2 \times \mathbf{R}^n$, $k \in \mathbf{K}$.

Then, if $\{\gamma_k^{1*}(x_1) = u_k^{1*} \in \mathring{U}_k^1, k \in \mathbf{K}\}$ *denotes an* open-loop Stackelberg equilibrium *strategy for the leader in the dynamic game formulated in this section, there exist finite vector sequences* $\{\lambda_1, \ldots, \lambda_K\}$, $\{\mu_1, \ldots, \mu_K\}$, $\{\nu_1, \ldots, \nu_K\}$ *that satisfy the following relations:*

$$x_{k+1}^* = f_k(x_k^*, u_k^{1*}, u_k^{2*}), \quad x_1^* = x_1, \tag{7.6a}$$

$$\nabla_{u_k^1} H_k^1(\lambda_k, \mu_k, \nu_k, p_{k+1}^*, u_k^{1*}, u_k^{2*}, x_k^*) = 0, \tag{7.6b}$$

$$\nabla_{u_k^2} H_k^1(\lambda_k, \mu_k, \nu_k, p_{k+1}^*, u_k^{1*}, u_k^{2*}, x_k^*) = 0, \tag{7.6c}$$

$$\lambda_{k-1}' = \frac{\partial}{\partial x_k} H_k^1(\lambda_k, \mu_k, \nu_k, p_{k+1}^*, u_k^{1*}, u_k^{2*}, x_k^*), \quad \lambda_K = 0, \tag{7.6d}$$

$$\mu_{k+1}' = \frac{\partial}{\partial p_{k+1}} H_k^1(\lambda_k, \mu_k, \nu_k, p_{k+1}^*, u_k^{1*}, u_k^{2*}, x_k^*), \quad \mu_1 = 0, \tag{7.6e}$$

$$\nabla_{u_k^2} H_k^2(p_{k+1}^*, u_k^{1*}, u_k^{2*}, x_k^*) = 0, \tag{7.6f}$$

$$p_k^* = F_k(x_k^*, u_k^{1*}, u_k^{2*}, p_{k+1}^*), \quad p_{K+1}^* = 0, \tag{7.6g}$$

where

$$\begin{aligned} H_k^1 \;=\; & g_k^1(f_k(x_k, u_k^1, u_k^2), u_k^1, u_k^2, x_k) + \lambda_k' f_k(x_k, u_k^1, u_k^2) \\ & + \mu_k' F_k(x_k, u_k^1, u_k^2, p_{k+1}) + \nu_k' (\nabla_{u_k^2} H_k^2(p_{k+1}, u_k^1, u_k^2, x_k))', \end{aligned} \tag{7.6h}$$

H_k^2 *is defined by (7.3d),* F_k *by (7.5e), and* $\mathring{U}_k^1$ *denotes the interior of* U_k^1. *Furthermore,* $\{u_k^{2*}, k \in \mathbf{K}\}$ *is the corresponding unique open-loop Stackelberg strategy of the follower* (**P2**), *and* $\{x_{k+1}^*, k \in \mathbf{K}\}$ *is the state trajectory associated with the Stackelberg solution.*

Proof. As discussed prior to the statement of the theorem, the leader (**P1**) is faced with the optimal control problem of minimizing $L^1(u^1, u^2)$ over U^1 and subject to (7.5b)-(7.5d). Since the problem is formulated in discrete time, it is equivalent to a finite dimensional nonlinear programming problem the "Lagrangian" of which is

$$\begin{aligned} L \;=\; & \sum_{k \in \mathbf{K}} \{ g_k^1(x_{k+1}, u_k^1, u_k^2, x_k) + \lambda_k' [f_k(x_k, u_k^1, u_k^2) - x_{k+1}] \\ & + \mu_k' [F_k(x_k, u_k^1, u_k^2, p_{k+1}) - p_k] + \nu_k' [\frac{\partial}{\partial u_k^2} H_k^2(p_{k+1}, u_k^1, u_k^2, x_k)]' \}, \end{aligned}$$

where λ_k, μ_k, ν_k ($k \in \mathbf{K}$) denote appropriate Lagrange-multiplier vectors. Now, if $\{u_k^{1*}, k \in \mathbf{K}\}$ is a minimizing solution and $\{x_{k+1}^*, p_{k+1}^*, u_k^{2*}; k \in \mathbf{K}\}$ are the corresponding values of the other variables so that the constraints (7.5b)-(7.5d) are satisfied, then it is necessary that (see, e.g., Canon et al., 1970, p. 51)

$$\nabla_{u_k^1} L = 0, \quad \nabla_{u_k^2} L = 0, \quad \nabla_{x_k} L = 0, \quad \nabla_{p_{k+1}} L = 0, \quad (k \in \mathbf{K})$$

wherefrom (7.6b)-(7.6e) follow. □

Remark 7.1 If the follower (**P**2) has, instead, access to closed-loop perfect state (CLPS) information, his optimal response (cf. Lemma 7.1) will be any closed-loop representation of the open-loop policy $\{\bar{u}_k^2, k \in \mathbf{K}\}$; however, since the constraints (7.5b)-(7.5d) are basically open-loop relations, these different representations do not lead to different optimization problems for the leader (**P**1). Therefore, the solution presented in Thm. 7.1 also constitutes a Stackelberg equilibrium for the case in which the follower has access to CLPS information (with the leader still having OL information). Under this latter information pattern, we may still talk about a unique open-loop Stackelberg strategy for the leader; whereas for the follower the corresponding optimal response strategy will definitely not be unique, since any closed-loop representation of u^{2*} given in Thm. 7.1 will constitute an optimal strategy for **P**2. □

Remark 7.2 The open-loop Stackelberg (OLS) solution is not weakly time consistent (à la Def. 5.13), since the OLS solution of any subgame does not necessarily coincide with the OLS solution of the full game, not even on the equilibrium trajectory. □

Linear-quadratic games

We now specialize the result of Thm. 7.1 to the class of linear-quadratic dynamic games (cf. Def. 6.1) wherein

$$f_k(x_k, u_k^1, u_k^2) = A_k x_k + B_k^1 u_k^1 + B_k^2 u_k^2, \tag{7.7a}$$

$$g_k^i(x_{k+1}, u_k^1, u_k^2, x_k) = \frac{1}{2}(x_{k+1}' Q_{k+1}^i x_{k+1} + u_k^{i'} u_k^i + u_k^{j'} R_k^{ij} u_k^j), \quad j \neq i, \; i, j = 1, 2. \tag{7.7b}$$

We further assume that $Q_{k+1}^i \geq 0$, $R_k^{12} > 0$, and $U_k^i = \mathbf{R}^{m_i}$, $\forall k \in \mathbf{K}$, $i = 1, 2$. Under these assumptions, it readily follows that the optimal response of the follower to any announced control sequence of the leader is unique and affine, and moreover the minimum of L^1 under the constraint imposed by this affine response curve is attained uniquely by an open-loop control sequence (for **P**1) that is linear in x_1.[71] This unique solution can be obtained explicitly by utilizing Thm. 7.1 and in terms of the current (open-loop) value of the state; it should be noted that Thm. 7.1 provides in this case both necessary and

[71]This follows since $L^1(u^1, u^2)$ is quadratic and strictly convex on $\mathbf{R}^{m_1} \times \mathbf{R}^{m_2}$. It should also be noted that the assumptions made on Q_{k+1}^i and R_k^{12} are sufficient, but not necessary, for the open-loop Stackelberg solution to exist and be unique; they are made here for convenience in the derivation to follow.

sufficient conditions because of strict convexity.[72] The result is given below in Corollary 7.1, which is obtained, however, under a condition of invertibility of a matrix related to the parameters of the game.

Corollary 7.1 *The two-person linear-quadratic dynamic game described by (7.7a)-(7.7b) under the parametric restrictions* $Q^i_{k+1} \geq 0$, $R^{12}_k > 0$ $(k \in \mathbf{K}, i = 1,2)$ *admits a unique open-loop Stackelberg solution with* **P1** *acting as the leader, which is given, under the condition of invertibility of the matrix appearing in the braces in (7.9c), by*

$$u^{1*}_k = \gamma^{1*}_k(x_1) = -B^{1'}_k K^1_k x^*_k, \tag{7.8a}$$

$$u^{2*}_k = \gamma^{2*}_k(x_1) = -B^{2'}_k K^2_k x^*_k \quad (k \in \mathbf{K}), \tag{7.8b}$$

where

$$\begin{aligned} K^1_k \triangleq\ & [I + Q^2_{k+1}B^2_k B^{2'}_k]^{-1}\Lambda_{k+1}\Phi_k \\ & + Q^2_{k+1}B^2_k[I + B^{2'}_k Q^2_{k+1}B^2_k]^{-1}R^{12}_k B^{2'}_k P_{k+1}\Phi_k \\ & + Q^2_{k+1}[I + B^2_k B^{2'}_k Q^2_{k+1}]^{-1}A_k M_k, \end{aligned} \tag{7.9a}$$

$$K^2_k \triangleq P_{k+1}\Phi_k \tag{7.9b}$$

$$\begin{aligned} \Phi_k \triangleq\ & \Big\{I + B^2_k B^{2'}_k P_{k+1} + B^1_k B^{1'}_k Q^2_{k+1} B^2_k\,(I + B^{2'}_k Q^1_{k+1}B^2_k)^{-1}R^{12}_k B^{2'}_k P_{k+1} \\ & + B^1_k B^{1'}_k(I + Q^2_{k+1}B^2_k B^{2'}_k)^{-1}\Lambda_{k+1}\Big\}^{-1} \\ & \cdot[A_k - B^1_k B^{1'}_k Q^2_{k+1}(I + B^2_k B^{2'}_k Q^2_{k+1})^{-1}A_k M_k], \end{aligned} \tag{7.9c}$$

and P_k, Λ_k, M_k *are appropriately dimensioned matrices generated by the difference equations*

$$P_k = A'_k P_{k+1} + A'_k Q^2_{k+1}\Phi_k; \quad P_{K+1} = Q^2_{K+1}, \tag{7.10a}$$

$$\Lambda_k = A'_k[Q^1_{k+1}\Phi_k + \Lambda_{k+1} + Q^2_{k+1}A_k M_k - Q^2_{k+1}B^2_k N_k]; \ \Lambda_{K+1} = Q^1_{K+1} \tag{7.10b}$$

$$M_{k+1} = A_k M_k - B^2_k N_k; \quad M_1 = 0 \tag{7.10c}$$

with

$$N_k \triangleq (I + B^{2'}_k Q^2_{k+1} B^2_k)^{-1}[B^{2'}_k(\Lambda_{k+1}\Phi_k + Q^2_{k+1}A_k M_k) - R^{12}_k B^{2'}_k K^2_k]. \tag{7.10d}$$

Furthermore, the unique state trajectory associated with this pair of strategies satisfies

$$x^*_{k+1} = \Phi_k x^*_k, \quad x^*_1 = x_1. \tag{7.10e}$$

---

[72]Note that condition (iii) in the general formulation is also met here since we already know that a unique solution exists with U^i_k taken as R^{m_i} and therefore it can be replaced by an appropriate compact subset of R^{m_i} $(i = 1, 2)$.

Proof. Existence of a unique solution to the linear-quadratic open-loop Stackelberg game has already been verified in the discussion prior to the statement of the corollary. We now apply Thm. 7.1 to obtain the corresponding equilibrium strategies of the players, in the structural form (7.8a)-(7.8b). Toward this end, we first rewrite (7.6b)-(7.6g), under the "linear-quadratic" assumption (7.7a)-(7.7b), respectively, as follows:

$$u_k^{1*} + B_k^{1'} Q_{k+1}^1 x_{k+1}^* + B_k^{1'}\lambda_k + B_k^{1'} Q_{k+1}^2 A_k \mu_k + B_k^{1'} Q_{k+1}^2 B_k^2 \nu_k = 0, \qquad (i)$$

$$R_k^{12} u_k^{2*} + B_k^{2'} Q_{k+1}^1 x_{k+1}^* + B_k^{2'}\lambda_k + B_k^{2'} Q_{k+1}^2 A_k \mu_k + (I + B_k^{2'} Q_{k+1}^2 B_k^2)\nu_k = 0, \quad (ii)$$

$$\lambda_{k-1} = A_k' Q_{k+1}^1 x_{k+1}^* + A_k'\lambda_k + A_k' Q_{k+1}^2 (A_k\mu_k + B_k^2\nu_k); \quad \lambda_K = 0, \qquad (iii)$$

$$\mu_{k+1} = A_k\mu_k + B_k^2\nu_k; \quad \mu_1 = 0, \qquad (iv)$$

$$u_k^{2*} + B_k^{2'}(Q_{k+1}^2 x_{k+1}^* + p_{k+1}^*) = 0, \qquad (v)$$

$$p_k^* = A_k'[p_{k+1}^* + Q_{k+1}^2 x_{k+1}^*]; \quad p_{K+1}^* = 0. \qquad (vi)$$

The coefficient of ν_k in (ii) is clearly positive definite; therefore, solving for ν_k from (ii) and substituting this in (i), (iii) and (iv), we obtain, respectively,

$$\begin{aligned} u_k^{1*} \;=\; & -B_k^{1'}[Q_{k+1}^1 - Q_{k+1}^2 B_k^2 (I + B_k^{2'} Q_{k+1}^2 B_k^2)^{-1} B_k^{2'} Q_{k+1}^1] x_{k+1}^* \\ & -B_k^{1'}[I - Q_{k+1}^2 B_k^2 (I + B_k^{2'} Q_{k+1}^2 B_k^2)^{-1} B_k^{2'}]\lambda_k \\ & -B_k^{1'} Q_{k+1}^2 [I - B_k^2 (I + B_k^{2'} Q_{k+1}^2 B_k^2)^{-1} B_k^{2'} Q_{k+1}^2] A_k \mu_k \\ & +B_k^{1'} Q_{k+1}^2 B_k^2 (I + B_k^{2'} Q_{k+1}^2 B_k^2)^{-1} R_k^{12} u_k^{2*}, \end{aligned} \qquad (vii)$$

$$\begin{aligned} \lambda_{k-1} \;=\; & A_k'[Q_{k+1}^1 - Q_{k+1}^2 B_k^2 (I + B_k^{2'} Q_{k+1}^2 B_k^2)^{-1} B_k^{2'} Q_{k+1}^1] x_{k+1}^* \\ & +A_k'[I - Q_{k+1}^2 B_k^2 (I + B_k^{2'} Q_{k+1}^2 B_k^2)^{-1} B_k^{2'}]\lambda_k \\ & +A_k' Q_{k+1}^2 [I - B_k^2 (I + B_k^{2'} Q_{k+1}^2 B_k^2)^{-1} B_k^{2'} Q_{k+1}^2] A_k \mu_k \\ & -A_k' Q_{k+1}^2 (I + B_k^{2'} Q_{k+1}^2 B_k^2)^{-1} R_k^{12} u_k^{2*}; \quad \lambda_K = 0, \end{aligned} \qquad (viii)$$

$$\begin{aligned} \mu_{k+1} \;=\; & [I - B_k^2 (I + B_k^{2'} Q_{k+1}^2 B_k^2)^{-1} B_k^{2'} Q_{k+1}^2] A_k \mu_k \\ & -B_k^{2'} (I + B_k^{2'} Q_{k+1}^2 B_k^2)^{-1} \\ & \cdot [R_k^{12} u_k^{2*} + B_k^{2'} (Q_{k+1}^1 x_{k+1}^* + \lambda_k)]; \quad \mu_1 = 0. \end{aligned} \qquad (ix)$$

Now, letting $p_{k+1}^* = (P_{k+1} - Q_{k+1}^2) x_{k+1}^*$, $\lambda_k = (\Lambda_{k+1} - Q_{k+1}^1) x_{k+1}^*$, $\mu_k = M_k x_k^*$, and substituting (vii) and (v) into the state equation, we obtain the relation

$$x_{k+1}^* = \Phi_k x_k^* \qquad (x)$$

assuming that the required inverse in the expression for Φ_k (defined by (7.9c)) exists. In determining the expression for Φ_k, as given by (7.9c), we have also utilized the matrix identity of Lemma 6.2 in simplifying some of the terms involved. Now, finally, by further making use of (x) in (vii), (v), (vi), (viii) and (ix), we arrive at (7.9a), (7.9b), (7.10a), (7.10b) and (7.10c), respectively. □

Remark 7.3 The coupled matrix difference equations (7.10a)-(7.10c) constitute a two-point boundary value problem, since the starting conditions are specified at both end points. If the problem is initially formulated as a (high dimensional) static quadratic game problem, however, such a two-point boundary value problem does not show up; but then one is faced with the task of inverting high dimensional matrices (i.e., the difficulty again shows up in a different form). □

Remark 7.4 Corollary 7.1 extends naturally (without any conceptual difficulties) to affine-quadratic games where the state equation has an additional control-independent driving term and the cost functions have cross terms between the current values of the state vector and the players' control vectors. For this more general case, the follower's optimal response will again be an affine function of the leader's control, and the leader's open-loop Stackelberg control will be affine in the initial state x_1, obtained from the solution of a generalized quadratic optimization problem. □

7.3 Feedback Stackelberg Solution Under CLPS Information Pattern

The feedback Stackelberg solution concept introduced in Def. 3.29 for finite games is equally valid for infinite dynamic games defined in discrete time and with additive cost functionals, provided that the underlying information structure allows the players to have access to the current value of the state (such as in the case of CLPS or MPS information pattern). Furthermore, in view of Prop. 3.15 (which is also valid for infinite games defined in discrete time), it is possible to obtain the solution recursively (in retrograde time), by employing a dynamic programming argument and by solving static Stackelberg games at every stage. Toward this end, we consider, in this section, the class of two-person dynamic games described by the state equation (7.1), and the cost functionals $L^i(u^1, u^2)$ given by (7.2). Under the delayed-commitment mode of play, we restrict attention to feedback strategies, i.e., $u_k^i = \gamma_k^i(x_k)$, $k \in \mathbf{K}$, $i = 1, 2$, and seek to obtain the feedback Stackelberg solution that is valid for all possible initial states $x_1 \in \mathbf{R}^n$. To complete the description of the dynamic Stackelberg game, let Γ_k^i denote the class of all permissible strategies of $\mathbf{P}i$ at stage k (i.e., measurable mappings from $\mathbf{R}^n$ into U_k^i). Then, the cost functional for $\mathbf{P}i$, when the game is in normal form, is (as opposed to (7.2))

$$J^i(\gamma^1, \gamma^2) = \sum_{k=1}^{K} g_k^i(x_{k+1}, \gamma_k^1(x_k), \gamma_k^2(x_k), x_k); \quad \gamma_k^j \in \Gamma_k^j, \ k \in \mathbf{K}, \ j = 1, 2.$$

Because of the additive nature of this cost functional, the following theorem now readily follows from Def. 3.29 and Prop. 3.15 (interpreted appropriately for the infinite dynamic game under consideration), and it provides a sufficient condition for a pair of strategies to constitute a feedback Stackelberg solution. Such a solution is clearly also a *strongly time consistent* one, using the terminology introduced in Def. 5.14 with the *sol* operator taken as the feedback Stackelberg equilibrium solution.

Theorem 7.2 *For the two-person discrete-time infinite dynamic game formulated in this section, a pair of strategies* $\{\gamma^{1*} \in \Gamma^1, \gamma^{2*} \in \Gamma^2\}$ *constitutes a feedback Stackelberg solution with* **P1** *as the leader if*

$$\min_{\gamma_k^1 \in \Gamma_k^1, \gamma_k^2 \in R_k^2(\gamma_k^1)} \tilde{G}_k^1(\gamma_k^1, \gamma_k^2, x_k) = \tilde{G}_k^1(\gamma_k^{1*}, \gamma_k^{2*}, x_k), \quad \forall x_k \in \mathbf{R}^n \quad (k \in \mathbf{K}), \tag{7.11}$$

where $R_k^2(\gamma_k^1)$ *is a* singleton *set defined by*

$$R_k^2(\gamma_k^1) = \{\beta_k^2 \in \Gamma_k^2 \colon \tilde{G}_k^2(\gamma_k^1, \beta_k^2, x_k) = \min_{\gamma_k^2 \in \Gamma_k^2} \tilde{G}_k^2(\gamma_k^1, \gamma_k^2, x_k) \ \ \forall x_k \in \mathbf{R}^n\}, \tag{7.12a}$$

$$\tilde{G}_k^i(\gamma_k^1, \gamma_k^2, x_k) \triangleq G_k^i(f_k(x_k, \gamma_k^1(x_k), \gamma_k^2(x_k)), \gamma_k^1(x_k), \gamma_k^2(x_k), x_k), \quad i = 1, 2, \quad k \in \mathbf{K}, \tag{7.12b}$$

and G_k^i *is recursively defined by*

$$\begin{aligned} G_k^i(x_{k+1}, \gamma_k^1(x_k), \gamma_k^2(x_k), x_k) = G_{k+1}^i(f_{k+1}(x_{k+1}, \gamma_{k+1}^{1*}(x_{k+1}), \\ \gamma_{k+1}^{2*}(x_{k+1})), \gamma_{k+1}^{1*}(x_{k+1}), \gamma_{k+1}^{2*}(x_{k+1}), x_{k+1}) + \\ g_k^i(x_{k+1}, \gamma_k^1(x_k), \gamma_k^2(x_k), x_k); \\ G_{K+1}^i = 0, \ i = 1, 2. \end{aligned} \tag{7.13}$$

Furthermore, every such solution is strongly time consistent.

Remark 7.5 If the optimal response set $R_k^2(\cdot)$ of **P2** (the follower) is not a singleton, then (in accordance with Def. 3.29) we have to take the maximum of the LHS of (7.11) over all these nonunique response functions. However, if $\tilde{G}_k^2(\gamma_k^1, \cdot, x_k)$ is strictly convex on $U_k^2 \quad (k \in \mathbf{K})$, the underlying assumption of "unique follower response" is readily met, and the theorem provides an easily implementable algorithm for computation of feedback Stackelberg strategies. □

Linear-quadratic games

For the special case of linear-quadratic games with strictly convex cost functionals, the feedback Stackelberg solution of Thm. 7.2 can be determined in closed form. Toward this end, let f_k and g_k^i be as defined by (7.7a) and (7.7b), respectively, with $Q_{k+1}^i \geq 0$, $R_k^{ij} \geq 0 \ \forall k \in \mathbf{K}$, $i, j = 1, 2$, $i \neq j$, and take $U_k^i = \mathbf{R}^{m_j}$. Then, solving recursively for G_k^i, in view of (7.11) and (7.12a), we obtain the quadratic structure

$$G_k^i(x_{k+1}, u_k^1, u_k^2) = \frac{1}{2}(x_{k+1}' L_{k+1}^i x_{k+1} + u_k^{i'} u_k^i + u_k^{j'} R_k^{ij} u_k^j) \ i, j = 1, 2; \ j \neq i,$$

where L_k^i satisfies the backward recurrence relation

$$L_k^i = (A_k - B_k^1 S_k^1 - B_k^2 S_k^2)' L_{k+1}^i (A_k - B_k^1 S_k^1 - B_k^2 S_k^2) + S_k^{i'} S_k^i + S_k^{j'} R_k^{ij} S_k^j + Q_k^i; \quad L_{K+1}^i = Q_{K+1}^i \quad (i,j = 1,2;\ j \neq i), \tag{7.14}$$

$$\begin{aligned} S_k^1 \triangleq\ & [B_k^{1'}(I + L_{k+1}^2 B_k^2 B_k^{2'})^{-1} L_{k+1}^1 (I + B_k^2 B_k^{2'} L_{k+1}^2)^{-1} B_k^1 B_k^{1'} L_{k+1}^{2'} B_k^2 \\ & \cdot (I + B_k^{2'} L_{k+1}^2 B_k^2)^{-1} R_k^{12} (I + B_k^{2'} L_{k+1}^2 B_k^2)^{-1} B_k^{2'} L_{k+1}^2 B_k^1 + I]^{-1} \\ & \cdot B_k^{1'} [(I + L_{k+1}^2 B_k^2 B_k^{2'})^{-1} L_{k+1}^1 (I + B_k^2 B_k^{2'} L_{k+1}^2) + L_{k+1}^{2'} B_k^2 \\ & \cdot (I + B_k^2 L_{k+1}^2 B_k^2)^{-1} R_k^{12} (I + B_k^{2'} L_{k+1}^2 B_k^2)^{-1} B_k^{2'} L_{k+1}^2] A_k, \end{aligned} \tag{7.15a}$$

$$S_k^2 \triangleq (I + B_k^{2'} L_{k+1}^2 B_k^2)^{-1} B_k^{2'} L_{k+1}^2 (A_k - B_k^1 S_k^1). \tag{7.15b}$$

It should be noted that $L_{k+1}^i \geq 0$, $\forall k \in \mathbf{K}$, $i = 1,2$ (since $Q_{k+1}^i \geq 0$, $R_k^{ij} \geq 0$), and therefore the required inverses in (7.15a) and (7.15b) exist. By the same token, $\tilde{G}_k^2(\gamma_k^1, \cdot, x_k)$ is strictly convex on $\mathbf{R}^{m_2}$ and hence R_k^2 is a singleton for every $k \in \mathbf{K}$; furthermore, the minimization problem (7.11) admits a unique solution, again because of the strict convexity (on $\mathbf{R}^{m_1}$). The implication, then, is the following corollary to Thm. 7.2, which summarizes the feedback Stackelberg solution of the linear-quadratic game.

Corollary 7.2 *The two-person linear-quadratic discrete-time dynamic game, as defined by (7.7a) and (7.7b) and with strictly convex cost functionals and FB, CLPS or MPS information pattern, admits a unique feedback Stackelberg solution with* **P1** *as the leader (under the delayed-commitment mode of play), which is strongly time consistent and linear in the current value of the state:*

$$\gamma_k^{i*}(x_k) = -S_k^i x_k, \quad k \in \mathbf{K};\ i = 1,2.$$

Here S_k^1 and S_k^2 are defined by (7.15a) and (7.15b), respectively, via the recursive relations (7.14).

Remark 7.6 The counterpart of Remark 7.4 holds here, in the sense that Corollary 7.2 extends naturally to affine-quadratic games where the state equation has an additional control-independent driving term and the cost functions have cross terms between the current values of the state vector and the players' control vectors. In this case, the unique feedback Stackelberg strategies of the players will be in the form:

$$\gamma_k^{i*}(x_k) = -\hat{S}_k^i x_k + s_k^i, \quad k \in \mathbf{K}; \qquad i = 1,2,$$

where $\hat{S}_k^1$ and $\hat{S}_k^2$ are some matrix sequences generated as in (7.15a) and (7.15b), and s_k^1 and s_k^2 are state-independent vector sequences, again generated in retrograde time. The precise expressions for these can readily be obtained by observing that in this more general case the counterpart of G_k^i (cost-to-go for **P**i) will also contain a state-independent term, a term that is linear in x_{k+1}, and some (quadratic) cross terms between x_k, u_k^1 and u_k^2. □

7.4 (Global) Stackelberg Solution Under CLPS Information Pattern

We consider, in this section, the derivation of (global) Stackelberg solutions when the leader has access to dynamic information (e.g., CLPS information). Such decision problems cannot be solved by utilizing standard techniques of optimal control theory (as in Sections 7.2 and 7.3), because the reaction set of the follower cannot, in general, be determined in closed form, for all possible strategies of the leader, and hence the optimization problem faced by the leader on this reaction set becomes quite an implausible one (at least, using the standard available techniques of optimization). In finite games (cf. Section 3.6) this difficulty can definitely be circumvented by converting the original game in extensive form into normal form (which is basically a static (matrix) game) and then by performing only a finite number of comparisons. In infinite dynamic games, however, such an approach is apt to fail because of the "infinite" nature of the strategy space of the leader (i.e., he has uncountably many alternatives).

In the sequel we first consider a specific example to elucidate the difficulties to be encountered in a brute-force derivation of the Stackelberg solution when the leader has access to dynamic information, and introduce (within the context of that example which involves two stages—each player acting only once, first the follower and then the leader) an indirect method to obtain the solution (see subsection 7.4.1). Then, we extend this indirect approach to a modified version of the example of subsection 7.4.1, in which the follower also acts at the second stage (see subsection 7.4.2) and we obtain in subsection 7.4.3 closed-form expressions for linear Stackelberg strategies in general linear-quadratic dynamic games, by utilizing the methods introduced in subsections 7.4.1 and 7.4.2. Finally, in subsection 7.4.4 we provide a brief discussion on the closely related topic of incentives.

7.4.1 An illustrative example (Example 7.1)

Example 7.1 Consider the two-stage scalar dynamic game described by the state equations

$$\left.\begin{aligned} x_2 &= x_1 - u^2, \\ x_3 &= x_2 - u^1 \end{aligned}\right\} \tag{7.16}$$

and cost functionals

$$L^1 = (x_3)^2 + 2(u^1)^2 + \beta(u^2)^2, \quad \beta \geq 0, \tag{7.17a}$$
$$L^2 = (x_3)^2 + (u^2)^2. \tag{7.17b}$$

(Note that we have suppressed the subindices on the controls, designating the stages, since every player acts only once in this game.) The information structure of the problem is CLPS, i.e., **P1** (the leader), acting at stage 2, has access to both x_1 and x_2, while **P2** (the follower) has only access to the value of x_1; therefore, a permissible strategy γ^1 for **P1** is a measurable mapping from $\mathbf{R} \times \mathbf{R}$

into R, and a permissible strategy γ^2 for **P2** is a measurable mapping from R into R. At this point, we impose no further restrictions (such as continuity, differentiability) on these mappings, and denote the strategy spaces associated with them by Γ^1 and Γ^2, respectively.

Now, to any announced strategy $\gamma^1 \in \Gamma^1$ of the leader, the follower's optimal reaction (for fixed x_1) will be the solution of the minimization problem

$$\min_{u^2 \in \mathbf{R}} \{[x_1 - u^2 - \gamma^1(x_1 - u^2, x_1)]^2 + [u^2]^2\} \tag{7.18}$$

which determines $\mathbf{R}^2(\gamma^1)$. For the sake of simplicity in the discussion to follow, let us now confine our analysis to only those strategies in Γ^1 which lead to a singleton $\mathbf{R}^2(\gamma^1)$. Denote the class of such strategies for the leader by $\tilde{\Gamma}^1$, and note that the follower's response is now a mapping $\tilde{\gamma}^2 : \mathbf{R} \times \tilde{\Gamma}^1 \to \mathbf{R}$, which we symbolically write as $\tilde{\gamma}^2[x_1; \gamma^1]$. There is, in general, no simple characterization of $\tilde{\gamma}^2$ since the minimization problem (7.18) depends "structurally" on the choice of $\gamma^1 \in \tilde{\Gamma}^1$. Now, the optimization problem faced by the leader is

$$\min_{\gamma^1 \in \tilde{\Gamma}^1} \{[x_2^{\gamma^1} - \gamma^1(x_2^{\gamma^1}, x_1)]^2 + 2[\gamma^1(x_2^{\gamma^1}, x_1)]^2 + \beta[\tilde{\gamma}^2[x_1; \gamma^1]]^2\},$$

where

$$x_2^{\gamma^1} \triangleq x_1 - \tilde{\gamma}^2[x_1; \gamma^1].$$

This is a constrained optimization problem which is, however, not of the standard type, since the constraint is in the form of the minimum of a functional. If we further restrict the permissible strategies of the leader to be twice continuously differentiable in the first argument (in which case the underlying Γ^1 becomes a smaller set), the first- and second-order conditions associated with the minimization problem (7.18) can be obtained, which implicitly determine $\tilde{\gamma}^2[x_1; \gamma^1]$. The problem faced by the leader then becomes one of calculus of variations, with equality and inequality constraints, which is quite intractable even for the "simple" problem formulated in this subsection. Besides, even if the solution of this constrained calculus of variations problem is obtained (say, through numerical techniques), we cannot be sure that it constitutes a Stackelberg strategy for the leader in the game problem under consideration (i.e., it might not be the best choice for the leader) since the *a priori* assumption of twice differentiability might be overly restrictive. We shall, in fact, see in the sequel that, depending on the value of β, the Stackelberg strategy of the leader could be nondifferentiable.

These considerations then force us to abandon the direct approach toward the solution of the Stackelberg game under CLPS information pattern, and to seek indirect methods. One such indirect method involves determination of a (tight) lower bound on the (Stackelberg) cost of the leader, and then finding an appropriate strategy for the leader that realizes that bound in view of possible rational responses of the follower.

In an attempt to explore this indirect method somewhat further, let us first observe that the lowest possible cost the leader can expect to attain is

$$\min_{\gamma^1 \in \Gamma^1} \min_{\gamma^2 \in \Gamma^2} J^1(\gamma^1, \gamma^2) \tag{7.19}$$

which is the one obtained when the follower cooperates with him to (globally) minimize his cost functional. This quantity clearly provides a lower bound to the leader's Stackelberg cost; but it is not as yet known whether it can be realized,[73] and this is what we investigate in the sequel.

First, solving the optimization problem (7.19) in view of (7.16) and (7.17a), we obtain (by utilizing dynamic programming or directly Prop. 5.1) the pair of feedback strategies

$$\gamma^{1^t}(x_2) = \frac{1}{3} x_2, \tag{7.20a}$$

$$\gamma^{2^t}(x_1) = [2/(3\beta + 2)] x_1 \tag{7.20b}$$

as providing the *unique* globally minimizing solution within the class of *feedback strategies.* The corresponding optimum state trajectory, in this two-stage team problem, is described by

$$\left.\begin{aligned} x_2^t &= [3\beta/(3\beta+2)] x_1, \\ x_3^t &= (2/3) x_2^t, \end{aligned}\right\} \tag{7.20c}$$

and the minimum cost is

$$J^{1^t} \triangleq J^1(\gamma^{1^t}, \gamma^{2^t}) = [2\beta/(3\beta+2)] x_1^2. \tag{7.20d}$$

The superscripts in (7.20a)-(7.20d) stand for "team", since (7.19) actually describes a two-member team problem with the common objective being the minimization of (7.17a), and (7.20a)-(7.20b) is the unique team-optimal solution in feedback strategies. We should note, however, that if **P1** is also allowed to know (and utilize) the value of the initial state x_1, (7.20a) is *no longer* a *unique* team-optimal strategy for **P1**, but any *representation* of it on the state trajectory (7.20c) also constitutes an optimal strategy (cf. Section 5.6). Denoting the class of all such strategies by Γ^{1^t}, we now have the more general result that any pair

$$\gamma^{1^t} \in \Gamma^{1^t}, \gamma^{2^t}(x_1) = [2/(3\beta + 2)] x_1$$

constitutes a team-optimal solution to (7.19). For each such pair, the state trajectory and the corresponding cost value are still as given by (7.20c) and (7.20d), respectively.

Now, if (7.19), or equivalently (7.20d), is going to be realized as the Stackelberg cost of the leader, there should be an element of Γ^{1^t}, say γ^{1*}, which forces

[73] Note that the follower in fact attempts to minimize his own cost functional which is different from J^1.

the follower to play γ^{2^t} even though he is minimizing his own cost functional (which is J^2—the counterpart of L^2 in the extensive form description). This, inevitably, asks for a relation of the form

$$\gamma^{2^t} = \arg \min_{\gamma^2 \in \Gamma^2} J^2(\gamma^{1*}, \gamma^2),$$

and this being so uniquely, i.e., the minimum of $J^2(\gamma^{1*}, \cdot)$ on Γ^2 should be attained uniquely at γ^{2^t}.[74] Intuitively, an appropriate strategy for the leader is the one that dictates (7.20a) if the follower plays (7.20b), and penalizes the follower rather heavily otherwise. Since the leader acts, in this game, after the follower does, and since the state equation is scalar and linear, he (the leader) is in a position to implement such a strategy (by utilizing solely the state information available to him). In particular, the strategy

$$\gamma^{1*}(x_2, x_1) = \begin{cases} \frac{1}{3}x_2 & \text{if } x_2 = [3\beta(3\beta+2)]x_1, \\ -kx_1 & \text{otherwise,} \end{cases} \tag{7.21}$$

where k is any positive number, is one such candidate. Such a strategy is clearly in Γ^{1^t}. Moreover, one may readily check validity of the inequality

$$\min_{\gamma^2 \in \Gamma^2} J^2\left(\gamma^1 = \tfrac{1}{3}x_2, \gamma^2\right) < \min_{\gamma^2 \in \Gamma^2} J^2(\gamma^1 = -kx_1, \gamma^2)$$

for $k > 0$,[75] which implies that the strategy (7.21) does indeed force the follower to the team strategy (7.20b), since otherwise he incurs a cost which is higher than the one obtained under the strategy (7.20a). The conclusion, therefore, is that the strategy (7.21) constitutes, for $k > 0$, a Stackelberg strategy for the leader, with the unique optimal response of the follower being (7.20b). □

To recapitulate:

(1) For the dynamic game described by (7.16), (7.17a) and (7.17b) and under the CLPS information pattern, a Stackelberg solution exists for all nonnegative values of β.

(2) The Stackelberg cost of the leader (**P1**) is equal to the global minimum value of his cost function (obtained by cooperative action of the follower), even though the follower (**P2**) seeks to minimize his own (different) cost functional and may not know the cost functional of **P1**.

(3) The leader's Stackelberg strategy is a representation of his optimal feedback strategy in the related team problem described by (7.19), and it necessarily involves *memory*. (A feedback strategy for **P1** *cannot* be a Stackelberg strategy.)

[74] In case of nonunique solutions, we have to take the supremum of J^1 over all those solutions, which should equal to J^{1^t}.

[75] This inequality, in fact, holds for $k > -1 + \sqrt{8/13}$, but we consider only positive values of k for the sake of simplicity in presentation.

(4) The Stackelberg strategy of **P1** is *nonunique* (parameterized, in this case, by a positive scalar k), but the optimal response of **P2** to all those strategies of **P1** is unique (and independent of k); it is the strategy that minimizes the leader's cost function, even though the follower's objective is quite different.

(5) The Stackelberg strategy of the leader, as given by (7.21), is *not* even continuous; this, however, does not rule out the possibility for existence of continuous (and differentiable) Stackelberg strategies.

(6) The strategy given by (7.21) may be viewed as a *threat strategy* on part of the leader, since he essentially threatens the follower to make his cost worse if he does not play the optimal team strategy (7.20b).

One natural question that arises at this point is whether there exists a Stackelberg strategy for the leader which is structurally more appealing (such as continuous, differentiable, etc.). In our investigation in this direction, we necessarily have to consider only the strategies in Γ^{1^t}, since the quantity (7.19) constitutes a tight lower bound on the cost of the leader. In particular, we now consider only those strategies in Γ^{1^t} which are linear in the available information, which can be written as

$$\gamma^1(x_2, x_1) = \frac{1}{3}x_2 + p\left(x_2 - \frac{3\beta}{3\beta+2}x_1\right), \tag{7.22}$$

where p is a scalar parameter. To determine the value of p for which (7.22) constitutes a Stackelberg strategy for the leader, we have to substitute (7.22) into J^2, minimize the resulting expression over $\gamma^2 \in \Gamma^2$ by also utilizing (7.16), and compare the argument of this minimization problem with the strategy (7.20b). Such an analysis readily leads to the unique value

$$p = \frac{2}{3} - \frac{1}{\beta}$$

provided, of course, that $\beta \neq 0$. Hence, we have the following.

Proposition 7.1 *Provided that $\beta \neq 0$, the linear strategy*

$$\gamma^{1*}(x_2, x_1) = \frac{1}{3}x_2 + \frac{2\beta-3}{3\beta}\left(x_2 - \frac{3\beta}{3\beta+2}x_1\right)$$

constitutes a Stackelberg strategy for the leader in the dynamic game described by (7.16), (7.17a) and (7.17b), and the unique optimal response of the follower is as given by (7.20b). The corresponding state trajectory is described by (7.20c) and the leader's Stackelberg cost is equal to the team cost (7.20d).

Therefore, with the exception of the singular case $\beta = 0$, the leader can force the follower to the team strategy (7.20b) by announcing a linear (continuously differentiable) strategy, which is definitely more appealing than (7.21) which was discontinuous on the equilibrium trajectory. For the case $\beta = 0$, however, a continuously differentiable Stackelberg strategy does not exist for the leader (see Başar (1979b), pp. 23-25) for a verification), and he has to employ a nondifferentiable strategy such as (7.21).

Remark 7.7 The linear strategy presented in Prop. 7.1 and the discontinuous strategies given by (7.21) do not constitute the entire class of Stackelberg strategies of the leader in the dynamic game of this subsection; there exist other, nonlinear (continuous and discontinuous) strategies for the leader which force the follower to the optimal team strategy (7.20b). Each such strategy can be determined by basically following the analysis that led to Prop. 7.1 under a different nonlinear representation of (7.20a) (instead of the linear representation (7.22). □

Remark 7.8 The results of this example can also be visualized graphically for each fixed x_1 and β. Let us take $x_1 = 1$ and leave β as a parameter, in which case the iso-cost curves of **P2** look as depicted in Fig. 7.1,and the (open-loop) team solution for **P1** can be written as (from (7.20a)-(7.20d))

$$u^{1^t} = \beta/(2+3\beta), \quad u^{2^t} = 2/(2+3\beta).$$

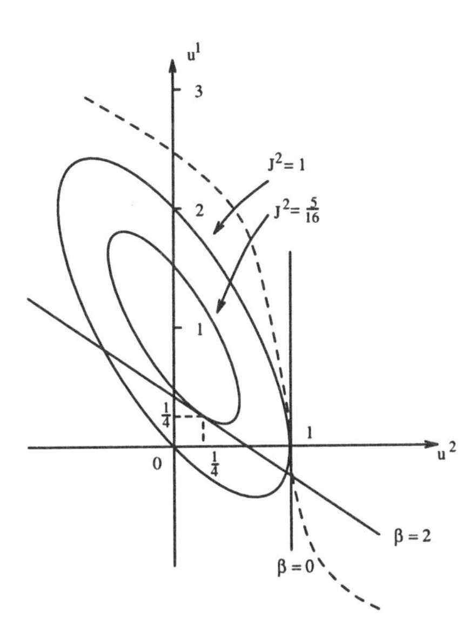

Figure 7.1: Graphical illustration of Example 7.1.

Since there is a one-to-one correspondence between x_2 and u^2, any permissible strategy of **P1** can also be written as a function of u^2, i.e., $u^1 = \gamma^1(u^2)$. Now, if we can find a strategy γ^{1*} with the property that its graph has only the point (u^1, u^2) in common with the set

$$D \triangleq \{(u^1, u^2) \colon J^2(u^1, u^2) \leq J^2(u^{1^t}, u^{2^t})\},$$

then the follower (**P2**) will definitely choose u^{2^t}. Thus γ^{1*} constitutes a Stackelberg strategy for **P1**, leading to the global minimum of his cost function.

It easily follows from Fig. 7.1 that, for $\beta > 0$, there exists a unique linear γ^{1*} which satisfies the above-mentioned requirement; it is the tangent at (u^{1^t}, u^{2^t}) to the corresponding iso-cost curve of **P2**. For $\beta = 0$ this tangent line turns out to be vertical (parallel to the u^1-axis), and hence cannot be described by a linear strategy; however, a Stackelberg strategy exists in the class of continuous strategies for the leader. (In Fig. 7.1 we have drawn the tangent lines for $\beta = 2$ and $\beta = 0$, and for the latter case the dashed curve provides a continuous nonlinear Stackelberg strategy for the leader.) □

Remark 7.9 A common property of the Stackelberg solutions obtained for the dynamic game described by (7.16), (7.17a) and (7.17b) and under CLPS information, is that they are also Nash equilibrium solutions. To see this, we first note the obvious inequality

$$J^2(\gamma^{1*}, \gamma^{2*}) \leq J^2(\gamma^{1*}, \gamma^2) \quad \forall \gamma^2 \in \Gamma^2$$

for such a Stackelberg equilibrium solution pair. Furthermore, the inequality

$$J^1(\gamma^{1*}, \gamma^{2*}) \leq J^1(\gamma^1, \gamma^{2*}) \quad \forall \gamma^1 \in \Gamma^1$$

also holds, since the Stackelberg solution is also team-optimal (under J^1) in this case. These two inequalities then imply that every such pair also constitutes a Nash equilibrium solution. We already know that the dynamic game under consideration admits an uncountable number of Nash equilibrium solutions (cf. Section 6.3); the above discussion then shows that, since the leader can announce his strategy ahead of time in a Stackelberg game, he can choose those particular Nash equilibrium strategies (and only those) which lead to an equilibrium cost that is equal to the global minimum of his cost functional. □

7.4.2 A second example (Example 7.2): Follower acts twice in the game

Example 7.2 The indirect method introduced in the foregoing subsection for the derivation of the Stackelberg solution of dynamic games under CLPS information basically involves two steps: (1) determination of a tight lower bound on the cost function of the leader, which coincides with the global minimum value of a particular team problem; (2) adoption of a particular representation of the optimal team strategy of the leader in this team problem, which forces the follower to minimize (in effect) the cost function of the team problem while he is in fact minimizing his own cost function. Even though this general approach is applicable in a much more general context (i.e., applicable to dynamic games with several stages and more complex dynamics), the related team problem is not always the one determined solely by the cost function of the leader (as in the previous subsection), in particular if the follower also acts at the last stage of the game. The following scalar dynamic game now exemplifies derivation of the related team problem in such a situation.

As a modified version of the scalar dynamic game of subsection 7.4.1, consider the one with dynamics

$$\left.\begin{aligned} x_2 &= x_1 - u_1^2, \\ x_3 &= x_2 - u^1 + u_2^2 \end{aligned}\right\} \tag{7.23}$$

and cost functionals

$$L^1 = (x_3)^2 + 2(u^1)^2 + \beta(u_1^2)^2, \quad \beta > 0, \tag{7.24a}$$

$$L^2 = (x_3)^2 + (u_1^2)^2 + (u_2^2)^2. \tag{7.24b}$$

The information structure of the problem is CLPS, and **P1** (the leader) acting at stage 2 has a single control variable u^1, while **P2** (the follower) acting at both stages 1 and 2 has control variables u_1^2 and u_2^2, respectively.

Let us introduce the strategy spaces Γ^1 (for **P1**), Γ_1^2 (for **P2** at stage 1) and Γ_2^2 (for **P2** at stage 2) in a way compatible with the governing information structure. Then, to any announced strategy $\gamma^1 \in \Gamma^1$ of the leader, the follower's optimal reaction at stage 2 will be

$$\gamma_2^2[x_2;\gamma^1] = -\frac{1}{2}[x_2 - \gamma^1(x_2,x_1)]$$

which is obtained by minimization of J^2 over $\gamma_2^2 \in \Gamma_2^2$. Hence, regardless of what strategy $\gamma^1 \in \Gamma^1$ the leader announces, and regardless of what strategy the follower employs at stage 1, his (the follower's) optimal strategy at stage 2 is a unique linear function of x_2 and γ^1, as given above. Substituting this structural form into J^1 and J^2, derived from (7.24a) and (7.24b), respectively, we obtain the reduced cost functionals

$$\tilde{J}^1 = \frac{1}{4}[x_2 - \gamma^1(x_2,x_1)]^2 + 2[\gamma^1(x_2,x_1)]^2 + \beta[\gamma_1^2(x_1)]^2,$$

$$\tilde{J}^2 = \frac{1}{2}[x_2 - \gamma^1(x_2,x_1)]^2 + [\gamma_1^2(x_1)]^2,$$

and therefore the lowest possible cost value the leader can achieve is

$$\min_{\gamma^1\in\Gamma^1}\ \min_{\gamma_1^2\in\Gamma_1^2} \tilde{J}^1(\gamma^1,\gamma_1^2) \tag{7.25}$$

which is the quantity that replaces (7.19) for this game.

The original Stackelberg game has thus been converted into one wherein the follower does not act at the last stage (his only strategy now being γ_1^2), and the related team problem is described by (7.25), whose optimal team solution in feedback strategies is

$$\gamma^{1^t}(x_2) = \frac{1}{9}x_2, \tag{7.26a}$$

$$\gamma_1^{2^t}(x_1) = [2/(9\beta+2)]x_1 \tag{7.26b}$$

with the corresponding unique team trajectory being

$$\left.\begin{aligned} x_2^t &= [9\beta/(9\beta+2)]x_1, \\ x_3^t &= (4/9)x_2^t. \end{aligned}\right\} \tag{7.26c}$$

The associated minimum team cost is

$$\bar{J}^{1^t} = [2\beta/(9\beta+2)](x_1)^2 \tag{7.26d}$$

which forms a lower bound on the leader's cost in this game, and is definitely higher than the quantity

$$\min_{\gamma^1\in\Gamma^1}\ \min_{\gamma_2^2\in\Gamma_2^2}\ \min_{\gamma_1^2\in\Gamma_1^2} J^1(\gamma^1,\gamma_2^2,\gamma_1^2).$$

Now, paralleling the arguments of subsection 7.4.1, we obtain the following representation of (7.26a), on the trajectory (7.26c), to be a Stackelberg strategy for the leader, forcing the follower at the first stage to the team strategy (7.26b):[76]

$$\gamma^{1*}(x_2,x_1) = \begin{cases} \frac{1}{9}x_2 & \text{if } x_2 = [9\beta/(9\beta+2)]x_1, \\ -kx_1 & \text{otherwise,} \end{cases} \tag{7.27}$$

where $k > -1 + \sqrt{(96/113)}$. The optimal response of the follower to any such strategy is

$$\gamma_1^{2*}(x_1) = [2/(9\beta+2)]x_1, \tag{7.28a}$$

$$\gamma_2^{2*}(x_2,x_1) = -\frac{1}{2}[x_2 - \gamma^{1*}(x_2,x_1)] \tag{7.28b}$$

leading to (7.26d) as the Stackelberg cost of the leader. □

To obtain the counterpart of Prop. 7.1 for the dynamic game of this subsection, we again start with a general linear representation of (7.26a), and by following similar arguments we arrive at the following proposition.

Proposition 7.2 *Provided that $\beta \neq 0$, the linear strategy*

$$\gamma^{1*}(x_2,x_1) = \frac{1}{9}x_2 + \frac{16\beta-9}{18\beta}\left(x_2 - \frac{9\beta}{9\beta+2}x_1\right)$$

constitutes a Stackelberg strategy for the leader in the dynamic game described by (7.23), (7.24a) and (7.24b), and the unique optimal response of the follower is given by (7.28a), (7.28b). The corresponding state trajectory is described by (7.26c), and the leader's Stackelberg cost is as given by (7.26d).

Remark 7.10 It can again be shown that, for $\beta = 0$, no continuously differentiable Stackelberg solution exists for the leader; therefore, he has to announce a strategy which is discontinuous in the derivative, such as (7.27), or the one

[76]The reader is asked to fill in the details of this derivation.

analogous to that discussed in Remark 7.7. Furthermore, the reader should note that, for the dynamic game under consideration, there does not exist a Stackelberg solution which also constitutes a Nash equilibrium solution; i.e., a counterpart of Remark 7.8 does not apply here. The main reason for this is the fact that the follower also acts at the last stage of the game. □

7.4.3 Linear Stackelberg solution of linear-quadratic dynamic games

We now utilize and extend the indirect methods introduced in subsections 7.4.1 and 7.4.2 to solve for the Stackelberg solution of two-person linear-quadratic dynamic games within the class of linear-memory strategies, so as to obtain the counterpart of Props 7.1 and 7.2 in this more general context. The class of dynamic games under consideration is described by the state dynamics

$$x_{k+1} = A_k x_k + B_k^1 u_k^1 + B_k^2 u_k^2 \tag{7.29}$$

and cost functionals (for **P1** and **P2**, respectively)

$$L^1 = \frac{1}{2}\sum_{k=1}^{K}(x'_{k+1}Q^1_{k+1}x_{k+1} + u_k^{1'}u_k^1 + u_k^{2'}R_k^{12}u_k^2) \tag{7.30a}$$

$$L^2 = \frac{1}{2}\sum_{k=1}^{K}(x'_{k+1}Q^2_{k+1}x_{k+1} + u_k^{2'}u_k^2 + u_k^{1'}R_k^{21}u_k^1) \tag{7.30b}$$

with $Q^1_{k+1} \geq 0$, $R_k^{12} > 0$, $Q^2_{k+1} \geq 0$, $R_k^{21} \geq 0$, $\forall k \in \mathbf{K} \triangleq \{1, \ldots, K\}$, $\dim(x) = n$, $\dim(u_k^i) = m_i$, $i = 1, 2$. The information structure of the problem is CLPS, under which the closed-loop strategy spaces for **P1** (the leader) and **P2** (the follower) are denoted by Γ_k^1 and Γ_k^2, respectively, at stage $k \in \mathbf{K}$. Furthermore, let J^1 and J^2 denote the cost functionals derived from (7.30a) and (7.30b), respectively, for the normal (strategic) form of the game, under these strategy spaces.

Since B_K^2 does not necessarily vanish in this formulation, the follower also acts at the last stage of the game, and this fact has to be taken into account in the derivation of a tight lower bound on the leader's cost functional. Proceeding as in subsection 7.4.2, we first note that, to any announced strategy K-tuple $\{\gamma_k^1 \in \Gamma_k^1; k \in \mathbf{K}\}$ by the leader, there corresponds a "robust" optimal reaction by the follower at stage K, which is determined by minimizing $J^2(\gamma^1, \gamma^2)$ over $\gamma_K^2 \in \Gamma_K^2$, with the remaining strategies held fixed. This quadratic strictly convex minimization problem readily leads to the relation

$$\gamma_K^2[x_K; \gamma_K^1] = -[I + B_K^{2'}Q^2_{K+1}B_K^2]^{-1}B_K^{2'}Q^2_{K+1}[A_K x_K + B_K^1\gamma_K^1(\eta_K)], \tag{7.31}$$

where $\eta_K \triangleq \{x_1, x_2, \ldots, x_K\}$, and $\gamma_K^2[x_K; \gamma_K^1]$ stands for $\gamma_K^2(\eta_k)$—the strategy of **P2** at stage K—and displays the explicit dependence of the choice of γ_K^2 on γ_K^1. It should be noted, however, that the *structure* of this optimal response of the follower *does not depend* on the choice of γ_K^1: regardless of what γ_K^1 and the previously applied strategies are, γ_K^2 depends *linearly* on x_K and $\gamma_K^1(\eta_K)$. The unique relation (7.31) can therefore be used in (7.30a) and (7.30b) without any loss of generality, in order to obtain an equivalent Stackelberg game wherein the follower does not act at the last stage. The cost functionals of this new dynamic game in extensive form are

$$\begin{aligned}\tilde{L}^1 &= \frac{1}{2}(\tilde{x}'_{K+1}\tilde{Q}^1_{K+1}\tilde{x}_{K+1} + u^{1'}_K u^1_K) \\ &+\frac{1}{2}\sum_{k=1}^{K-1}(x'_{k+1}Q^1_{k+1}x_{k+1} + u^{1'}_k u^1_k + u^{2'}_k R^{12}_k u^2_k),\end{aligned} \tag{7.32a}$$

$$\begin{aligned}\tilde{L}^2 &= \frac{1}{2}(\tilde{x}'_{K+1}\tilde{Q}^2_{K+1}\tilde{x}_{K+1} + u^{1'}_K R^{21}_K u^1_K) \\ &+\frac{1}{2}\sum_{k=1}^{K-1}(x'_{k+1}Q^2_{k+1}x_{k+1} + u^{2'}_k u^2_k + u^{1'}_k R^{21}_k u^1_k),\end{aligned} \tag{7.32b}$$

where

$$\begin{aligned}\tilde{x}_{K+1} &= A_K x_K + B^1_K u^1_K, \\ \tilde{Q}^1_{K+1} &= [I - B^2_K T]' Q^1_{K+1}[I - B^2_K T] + T' R^{12}_K T, \\ \tilde{Q}^2_{K+1} &= Q^2_{K+1}[I - B^2_K T], \\ T &\triangleq (B^{2'}_K Q^2_{K+1} B^2_K + I)^{-1} B^{2'}_K Q^2_{K+1}.\end{aligned}$$

Let us further introduce the notation

$$\tilde{\gamma}^2 \in \tilde{\Gamma}^2 \to \{\gamma^2_k \in \Gamma^2_k; k = 1, \ldots, K-1\},$$

$$\tilde{J}^i(\gamma^1, \tilde{\gamma}^2) \triangleq \tilde{L}^i \mid \{u^1_k = \gamma^1_k(\eta_k), k \in \mathbf{K}; u^2_k = \gamma^2_k(\eta_k), k = 0, \ldots, K-1\}.$$

Then, a lower bound on the leader's Stackelberg cost is, clearly, the quantity

$$\min_{\gamma^1 \in \Gamma^1} \min_{\tilde{\gamma}^2 \in \tilde{\Gamma}^2} \tilde{J}^1(\gamma^1, \tilde{\gamma}^2) \tag{7.33}$$

which is the minimum cost of a team problem in which both **P1** and **P2** strive to minimize the single objective functional $\tilde{J}^1$. This team problem admits a unique optimal solution within the class of feedback strategies, which is given in the following lemma whose proof readily follows from Prop. 5.1 by an appropriate decomposition.

Lemma 7.2 *In feedback strategies, the joint optimization (team) problem defined by (7.33) admits a unique solution given by*

$$\begin{aligned}\gamma^{1^t}_k(x_k) &= -L^1_k x_k, \quad k \in \mathbf{K}, \\ \gamma^{2^t}_k(x_k) &= -L^2_k x_k, \quad k \in \mathbf{K} - \{K\}\end{aligned}$$

and with minimum cost

$$\tilde{J}^{1^t} = \frac{1}{2}x_1' M_1 x_1,$$

where

$$\begin{aligned}
L_k^1 &= [I + S_k^1 B_k^1]^{-1} S_k^1 A_k, \quad k \in \mathbf{K}, \\
L_k^2 &= [R_k^{12} + S_k^2 B_k^2]^{-1} S_k^2 A_k, \quad k \le K-1, \\
S_K^1 &= B_K^{1'} \tilde{Q}_{K+1}^1, \\
S_k^1 &= B_k^{1'} M_{k+1}\{I - B_k^2 [R_k^{12} + B_k^{2'} M_{k+1} B_k^2]^{-1} B_k^{2'} M_{k+1}\}, \\
&\quad k \le K-1, \\
S_k^2 &= B_k^{2'} M_{k+1}\{I - B_k^1 [I + B_k^{1'} M_{k+1} B_k^1]^{-1} B_k^{1'} M_{k+1}\}, \\
&\quad k \le K-1.
\end{aligned}$$

$M_{(\cdot)}$ *is defined recursively by*

$$\begin{aligned}
M_K &= Q_K^1 + F_K' \tilde{Q}_{K+1}^1 F_K + L_K^{1'} L_K^1, \\
M_k &= Q_k^1 + F_k' M_{k+1} F_k + L_k^{1'} L_k^1 + L_k^{2'} R_k^{12} L_k^2, \quad k \le K-1,
\end{aligned}$$

where

$$\begin{aligned}
F_K &\triangleq A_K - B_K^1 L_K^1, \\
F_k &\triangleq A_k - B_k^1 L_k^1 - B_k^2 L_k^2, \quad k \le K-1.
\end{aligned}$$

The optimum team trajectory is then described by

$$x_{k+1}^t = F_k x_k^t, \quad k \in \mathbf{K}, \quad x_1^t = x_1. \tag{7.34}$$

In view of the analysis of subsection 7.4.2, we now attempt to find a representation of the team strategy $\{\gamma_k^{1^t}; k \in \mathbf{K}\}$ on the optimum team trajectory (7.34), which provides a Stackelberg strategy for the leader. The following proposition justifies this indirect approach for the problem under consideration.

Proposition 7.3 *Let* $\{\gamma_k^{1*}; k \in \mathbf{K}\}$ *be a representation of the team strategy* $\{\gamma_k^{1^t}; k \in \mathbf{K}\}$ *on the optimum team trajectory (7.34), such that every solution of the minimization problem*

$$\min_{\tilde{\gamma}^2 \in \tilde{\Gamma}^2} \tilde{J}^2(\gamma^{1*}, \tilde{\gamma}^2)$$

is a representation of $\{\gamma_k^{2^t}; k \in \mathbf{K} - \{K\}\}$ *on the same team trajectory. Then,* $\{\gamma_k^{1*}; k \in \mathbf{K}\}$ *provides a Stackelberg strategy for the leader.*

Proof. We have already shown, prior to the statement of Lemma 7.2, that the quantity (7.33) provides a lower bound on the cost function of the leader, for the dynamic game under consideration. But, under the hypothesis of Prop. 7.3, this lower bound is tight, and is realized if the leader announces the strategy $\{\gamma_k^{1*}; k \in \mathbf{K}\}$ which, thereby, manifests itself as a Stackelberg strategy for the leader. □

To obtain some explicit results, we now confine our attention to linear one-step memory strategies for the leader (more precisely, to those representations of $\{\gamma_k^{1^t}; k \in \mathbf{K}\}$ which are linear in the current and most recent past values of the state), viz.

$$\gamma_k^1(x_k, x_{k-1}) = -L_k^1 x_k + P_k[x_k - F_{k-1}x_{k-1}], \quad k \in \mathbf{K} - \{1\}, \tag{7.35}$$

where $\{P_k; k \in \mathbf{K} - \{1\}\}$ is a matrix sequence which is yet to be determined. This matrix sequence is independent of x_k and x_{k-1}, but it may, in general, depend on the initial state x_1 which is known *a priori*. Our objective for the remaining portion of this subsection may now be succinctly stated as (i) determination of conditions under which a Stackelberg strategy for the leader exists in the structural form (7.35) and also satisfies the hypothesis of Prop. 7.3, (ii) derivation of the corresponding strategy of the leader (more precisely, the matrices P_k, $k \in \mathbf{K} - \{1\}$) recursively, whenever those conditions are fulfilled. Toward this end, we substitute (7.35) into $\tilde{J}^2(\gamma^1, \tilde{\gamma}^2)$, minimize the resulting functional over $\tilde{\gamma}^2 \in \tilde{\Gamma}^2$ and compare the minimizing solutions with $\{\gamma_k^{2t}; k \in \mathbf{K}\}$ and their representations on the optimum team trajectory. Such an analysis leads to Thm. 7.3 given in the sequel.

Preliminary notation for Theorem 7.3

$\Sigma_{(\cdot)}$: an $(m_2 \times m_2)$-matrix defined recursively by

$$\Sigma_k = Q_k^2 + F_k'\Sigma_{k+1}F_k + L_k^{1'}R_k^{21}L_k^1 + L_k^{2'}L_k^2; \quad \Sigma_{K+1} = \tilde{Q}_{K+1}^2$$

Λ_k: an $(m_1 \times n)$-matrix defined recursively by (as a function of $\{P_{k+1}, \ldots, P_K\}$)

$$\Lambda_k = B_k^{1'}P_{k+1}'\Lambda_{k+1}F_k - R_k^{21}L_k^1 + B_k^{1'}\Sigma_{k+1}F_k; \quad \Lambda_{K+1} = 0. \tag{7.36}$$

Condition 7.1 *For a given $x_1 \in \mathbf{R}^n$, let there exist at least one matrix sequence $\{P_K, P_{K-1}, \ldots, P_2\}$ that satisfies recursively the vector equation*

$$[B_k^{2'}P_{k+1}'\Lambda_{k+1}F_k + B_k^{2'}\Sigma_{k+1}F_k - L_k^2]x_k^t = 0, \quad k+1 \in \mathbf{K} \tag{7.37}$$

where Λ_k is related to $\{P_K, \ldots, P_{k+1}\}$ through (7.36), and x_k^t is a known linear function of x_1, as determined through (7.34).

Theorem 7.3 *Let Condition 7.1 be satisfied and let $\{P_K^*(x_1), \ldots, P_2^*(x_1)\}$ denote one such sequence. Then, there exists a Stackelberg solution for the dynamic game described by (7.29)-(7.30b) and under the CLPS information, which* *is given by*

$$\left.\begin{aligned}\gamma_k^{1*}(x_k, x_{k-1}, x_1) &= [P_k^*(x_1) - L_k^1]x_k - P_k^*(x_1)F_{k-1}x_{k-1}, \\ &\qquad\qquad k \in \mathbf{K} - \{1\} \\ \gamma_1^{1*}(x_1) &= -L_1^1 x_1,\end{aligned}\right\} \tag{7.38a}$$

$$\gamma_k^{2*}(x_k) = -L_k^2 x_k, \quad k \in \mathbf{K} - \{K\}, \tag{7.38b}$$

where L_k^i $(i = 1, 2;\ k \in \mathbf{K})$ *were defined in Lemma 7.2. The corresponding Stackelberg costs of the leader and the follower are given, respectively, by*

$$J^{1*} = \frac{1}{2} x_1' M_1 x_1, \tag{7.39a}$$

$$J^{2*} = \frac{1}{2} x_1' \Sigma_1 x_1. \tag{7.39b}$$

Proof. Substituting $u_k^1 = \gamma_k^{1*}(\cdot)$, $k \in \mathbf{K}$, as given by (7.38a), into $\tilde{L}^2$ defined by (7.32b), we obtain a functional which is quadratic and strictly convex in $\{u_k^2; k \in \mathbf{K} - \{K\}\}$. Let us denote this functional by $L(u_1^2, \ldots, u_{K-1}^2)$, which is, in fact, what the follower will minimize in order to determine his optimal response. Strict convexity of L readily implies that the optimal response of the follower to the announced strategy K-tuple (7.38a) of the leader is unique in open-loop policies, and that every other optimal response in $\tilde{\Gamma}^2$ is a representation of this unique open-loop strategy on the associated state trajectory. Consequently, in view of Prop. 7.3, verification of the theorem amounts to showing that the minimum of $L(u_1^2, \ldots, u_{K-1}^2)$ is attained by the open-loop representation of the feedback strategy $(K-1)$-tuple (7.38b) on the state trajectory described by (7.34); in other words, we should verify that the set of controls

$$u_k^{2*} = -L_k^2 x_k^t, \quad k \in \mathbf{K} - \{K\}$$

minimizes $L(u_1^2, \ldots, u_{K-1}^2)$.

Now, since L is quadratic and strictly convex, it is both necessary and sufficient that $\{u_k^{2*}\}$ be a stage-by-stage optimal solution; i.e., the set of inequalities

$$L(u_k^{2*}, \{u^{2*}\}_{-k}) \leq L(u_k^2, \{u^{2*}\}_{-k}) \tag{i}$$

should be satisfied for all $u_k^2 \in \mathbf{R}^{m_2}$ and all $k = 1, \ldots, K-1$, where $\{u^{2*}\}_{-k}$ denotes the entire sequence $u_1^{2*}, \ldots, u_{K-1}^{2*}$ with only u_k^{2*} missing. Because of the additive nature of the cost function L, *(i)* can equivalently be written as

$$L_k(u_{K-1}^{2*}, \ldots, u_k^{2*}) \leq L_{k-1}(u_{K-1}^{2*}, \ldots, u_{k+1}^{2*}, u_k^2) \tag{ii}$$

for all $u_k^2 \in \mathrm{R}^{m_2}$ and all $k = 1, \ldots, K-1$, where, for fixed k,

$$\begin{aligned} L_k(u_{K-1}^{2*}, \ldots, u_{k+1}^{2*}, u_k^2) &\triangleq L_k^*(u_k^2) = \frac{1}{2}(y_{K+1}' \tilde{Q}_{K+1}^2 y_{K+1} + \mu_K' R_K^{21} \mu_k) \\ &+ \frac{1}{2}\sum_{j=k}^{K-1}(y_{j+1}' Q_{j+1}^2 y_{j+1} + \mu_j' R_j^{21} \mu_j) + \frac{1}{2}\sum_{j=k+1}^{K-1} u_j^{2*'} u_j^{2*} + \frac{1}{2} u_k^{2'} u_k^2, \end{aligned} \tag{iii}$$

with

$$\left.\begin{aligned} y_{K+1} &= F_K y_K + B_K^1 P_K^* [y_K - F_{K-1} y_{K-1}], \\ &\vdots \\ y_{k+2} &= F_{k+1} y_{k+1} + B_{k+1}^1 P_{k+1}^* [y_{k+1} - F_k x_k^t], \\ y_{k+1} &= (A_k - B_k^1 L_k^1) x_k^t + B_k^2 u_k^2, \end{aligned}\right\} \qquad (iv)$$

and

$$\mu_j = \begin{cases} -L_j^1 y_j + P_k^* [y_j - F_{j-1} y_{j-1}], & j \geq k+1, \\ -L_k^1 x_k^t, & j = k. \end{cases} \qquad (v)$$

It is important to note that, at the RHS of *(ii)*, L_k is not only a function of u_k^2 but also of x_k, which we take at its equilibrium value x_k^t because of the RHS of *(i)* (i.e., the sequence $\{u_1^2, \ldots, u_{k-1}^2\}$ that determines x_k is already at equilibrium while treating *(ii)*). It is for the same reason that y_k, on the last row of *(iv)*, is taken to be equal to x_k^t.

Now, if the relation in the last line of *(iv)* is used iteratively in the remaining relations of *(iv)*, we can express y_j, $j \geq k+2$, in terms of y_{k+1} and x_k^t, and if this is used in *(v)*, μ_j, $j \geq k+1$, can be expressed in terms of the same variables. The resulting expressions are

$$\begin{aligned} y_j &= \mathbf{F}(j,k+1) y_{k+1} + \mathbf{H}(j,k+1) B_{k+1}^1 P_{k+1}^* [y_{k+1} - F_k x_k^t], \quad j \geq k+2, \\ \mu_j &= -L_j^1 \mathbf{F}(j,k+1) y_{k+1} + [\mathbf{Y}(j,k+1) - L_j^1 \mathbf{H}(j,k+1) B_{k+1}^1] P_{k+1}^* \\ &\qquad \cdot [y_{k+1} - F_k x_k^t], \quad j \geq k+1, \end{aligned}$$

where

$$\begin{aligned} \mathbf{F}(j,k) &= F_{j-1} \ldots F_k, \quad \mathbf{F}(j,j) = I, \\ \mathbf{H}(j,k) &= \mathbf{H}(j,k+1) B_{k+1}^1 P_{k+1}^* + \mathbf{F}(j,k+1), \\ \mathbf{H}(j,j-1) &= I, \quad \mathbf{H}(j,i) = 0, \quad j \leq i, \\ \mathbf{Y}(j,k) &= P_j^* B_{j-1}^1 \ldots P_{k+1}^* B_k^1, \quad \mathbf{Y}(j,j) = I. \end{aligned}$$

Utilizing these expressions, we can next show that

$$\nabla_{u_k^2} L^*(u_{K-1}^{2*}, \ldots, u_{k+1}^{2*}, u_k^2) \mid_{u_k^2 = u_k^{2*}} = 0, \quad \forall k = 1, \ldots, K-1. \qquad (vi)$$

Toward this end, we first obtain

$$\begin{aligned} \nabla_{u_k^2} L^* \mid_{u_k^2 = u_k^{2*}} = [& B_k^{2'} P_{k+1}^{*'} \{ B_{k+1}^{1'} \mathbf{H}'(K+1,k+1) \tilde{Q}_{K+1}^2 \mathbf{F}(K+1,k) \\ &+ \sum_{j=k+1}^{K} [B_{k+1}^{1'} \mathbf{H}'(j,k+1) Z_j \mathbf{F}(j,k) - \mathbf{Y}'(j,k+1) R_j^{21} L_j^1 \mathbf{F}(j,k)] \} \qquad (vii) \\ &+ B_k^{2'} \Sigma_{k+1} F_k - L_k^2] x_k^t, \end{aligned}$$

where

$$Z_j = L_j^{1'} R_j^{21} L_j^1 + L_j^{2'} L_j^2 + Q_j^2.$$

Some further manipulations prove that *(vii)* can equivalently be written as

$$\nabla_{u_k^2} L^* \mid_{u_k^2 = u_k^{2*}} = [B_k^{2'} P_{k+1}^{*'} \Lambda_{k+1}^* F_k + B_k^{2'} \Sigma_{k+1} F_k - L_k^2] x_k^t,$$

where Λ_{k+1}^* satisfies the recursive equation (7.36) with P_{k+1} replaced by P_{k+1}^*. But, by Condition 7.1 [(7.37)], the foregoing expression identically vanishes; this proves (vi), and in turn establishes the desired result that u_k^{2*} minimizes L at stage k and is in equilibrium with $\{u^{2*}\}_{-k}$. (Note that, since L is a strictly convex functional, (vi) is a sufficient condition for minimization.) Since k was an arbitrary integer, the proof of the main part of the theorem follows. Expression (7.39a) is readily obtained from Lemma 7.2 under the stipulation that the optimum team trajectory is realized, since then J^{1*} is identical with $\tilde{J}^{1^t}$. Expression (7.39b), on the other hand, follows from substitution of the team solution given in Lemma 7.2 into $\tilde{J}^2$. □

Remark 7.11 Condition 7.1, and thereby the Stackelberg solution presented in Thm. 7.3, depends, in general, on the value of the initial state x_1; that is, the Stackelberg strategy of the leader at stage k is not only a function of x_k and x_{k-1}, but also of x_1. This latter dependence may be removed under a more stringent condition (than Condition 7.1) on the parameters of the game, in which case the vector equation (7.37) is replaced by the matrix equation

$$B_k^{2'} P_{k+1}' \Lambda_{k+1} F_k + B_k^{2'} \Sigma_{k+1} F_k - L_k^2 = 0 \quad (k+1 \in \mathbf{K}). \tag{7.40}$$

Such a restriction is, in fact, inevitable if we seek a solution to the infinite horizon problem (i.e., with $K \to \infty$), in which case our attention is solely confined to stationary controls (see Başar and Selbuz (1979b), for an illustrative example). □

Remark 7.12 The analyses of subsections 7.4.1 and 7.4.2 have already displayed the validity of the possibility that a linear-quadratic dynamic game might not admit a Stackelberg solution in linear strategies. Therefore, depending on the parameters of the dynamic game, Condition 7.1 might not be fulfilled. In such a case, we have to adopt, instead of (7.35), a parameterized nonlinear representation of $\{\gamma_k^{1^t}; k \in \mathbf{K}\}$ as a candidate Stackelberg strategy of the leader, and then determine those parameter values in view of Prop. 7.3. For a derivation of such a nonlinear Stackelberg solution, the reader is referred to Tolwinski (1981b).

Another important point worth noting is that, for the linear-quadratic game of this section, and for general parameter values, $\tilde{J}_1^t$ given in Lemma 7.2 is not always the Stackelberg cost of the leader, i.e., there may not exist a strategy $\tilde{\gamma}^{1*}$ in $\tilde{\Gamma}^1$ (linear or otherwise) under which the follower is forced to choose a representation of $\tilde{\gamma}^{2^t}$ on the optimal team trajectory. This would be the case if, for example, there exist certain control variables of the follower, at intermediate stages, which do not affect the state variable, and thereby cannot be "detected" by the leader. In such a case, the leader cannot influence the cost function of the follower through these control variables and hence cannot enforce the team solution. This then necessitates derivation of a new team cost (which is realizable as the Stackelberg cost of the leader), by taking into account the robust optimal responses of the follower in terms of these controls (as we did at the final stage). The resulting Stackelberg cost will be higher than $\tilde{J}^{1^t}$.

Another such case occurs if the leader can detect (through his state observation) a linear combination of the control variables of the follower, but does not observe them separately. (This would be the case if, for example, the dimension of the state is lower than that of the control vector of the follower.) Here again, we have to determine a new minimum team cost (different from $\tilde{J}^{1^t}$) to be realized as the Stackelberg cost of the leader, by taking into account the freedom allotted to the follower in the nondetectable (by the leader) region of his control space. We do not pursue this point any further here, and refer the reader to Tolwinski (1981b) for such a derivation, and also to Başar (1980a) and Başar (1982) for indirect derivations of the Stackelberg cost value of the leader in general dynamic games when the leader has access to closed-loop imperfect state information (cf. Def. 5.2). See also Problem 5 in Section 7.7. □

7.4.4 Incentives (deterministic)

The idea of declaring a reward (or punishment) for a decision maker **P1** according to his particular choice of action in order to induce a certain "desired" behavior on the part of another decision maker **P2** is known as an *incentive* (or in case of the punishment, as a threat). Mathematical formulation and analysis of such decision problems bear strong connections with the theory of Stackelberg games presented in the previous subsections, which is what we will be discussing next, for deterministic scenarios. Counterparts of these results in the stochastic case will be presented later in Section 7.5 (particularly, subsection 7.5.3).

Following the convention of the previous subsections, we call, in the above scenario, **P1** the leader and **P2** the follower. Then, the action outcome desired by the leader is:

$$(u^{1^t}, u^{2^t}) = \arg \min_{u^1 \in S^1, u_2 \in S^2} L^1(u^1, u^2). \tag{7.41}$$

The incentive problem can now be stated as: Find a $\gamma^1 \in \Gamma^1$, where Γ^1 is an admissible subclass of all mappings from S^2 into S^1, such that

$$\arg \min_{u^2} L^2(\gamma^1(u^2), u^2) = u^{2^t}, \tag{7.42a}$$

$$\gamma^1(u^{2^t}) = u^{1^t}. \tag{7.42b}$$

Note that (7.42a) and (7.42b) require choosing a set of m_1 scalar functions which together map S^2 into S^1 so as to satisfy $m_1 + m_2$ equations. If this set of m_1 functions has $m_1 + m_2$ or more parameters, then we might in general accomplish this by choosing the parameters appropriately.

Incentive problems do arise in real life decision making. Think of **P1** as a government and of **P2** as a citizen. The income tax which **P2** has to pay is a fraction (say k) of his income (before taxation) u^2. The amount of money that the government receives is $u^1 = ku^2$. It is up to **P2** how hard to work and thus how much money to earn. The incentive here is $u^1 = \gamma^1(u^2) = ku^2$. The government will choose k so as to achieve certain goals, but it cannot choose its own income, u^1, directly. In reality, the γ^1-functions will often be nonlinear, but that does not take away the incentive phenomenon.

Example 7.3 Consider $L^1 = (u^1)^2+(u^2)^2$ and $L^2 = (u^1-1)^2+(u^2-1)^2$, where the u^i are scalars. By inspection, $u^{1^t} = u^{2^t} = 0$. Consider the choice $u^1 = ku^2$, with k approaching ∞ if necessary, as a possible incentive mechanism for **P1**. The idea is that any choice of $u^2 \neq 0$ will make L^2 approach ∞ if $k \to \infty$ and thus force **P2** to choose u^2 arbitrarily close to u^{2^t} in his own interest. However, by substituting $u^1 = ku^2$ into L^2, it is easily shown that the minimizing action is $u^2 = (k+1)/(k^2+1)$ and consequently $u^1 = (k^2+k)/(k^2+1)$. Thus u^2 approaches $u^{2^t} = 0$ and u^1 approaches 1 (which is different from u^{1^t}) as k approaches ∞ and hence (7.42b) is violated. Consequently "infinite threat" as just described is not generally feasible. Besides, such a threat may not be credible in practice.

Let us now consider an incentive γ^1 of the form

$$u^1 = \gamma^1(u^2) = u^{1^t} + g(u^2, u^{2^t}), \tag{7.43}$$

where g is a function which satisfies $g(u^{2^t}, u^{2^t}) = 0$. With this restriction on g, equation (7.42b) is automatically satisfied. Let us try the linear function $g = k(u^2 - u^{2^t})$. Equation (7.42a) reduces to $(k+1)/(k^2+1) = 0$ and hence k must be equal to -1. Graphically, the incentive $u^1 = ku^2 = -u^2$ is a line through the team solution (u^{1^t}, u^{2^t}) and has only this point in common with the set of points (u^1, u^2) defined by $L^2(u^1, u^2) \leq L^2(u^{1^t}, u^{2^t})$. By announcing the γ^1-function, **P1** ensures that the solution (u^1, u^2) will lie on the line $u^1 = -u^2$ in the (u^1, u^2) plane, independent of the action of **P2**. Being rational, **P2** will choose that point on this line which minimizes his cost function; such a choice is $u^2 = 0$. The relationship of this with the derivation of the Stackelberg solution discussed in subsection 7.4.1 and particularly with the depiction of Fig. 7.1 should be clear now. From this graphical interpretation, it should also be clear that any (nonlinear, continuous or discontinuous) incentive policy $u^1 = \gamma^1(u^2)$ which passes through the point characterized by the team solution and has only this point in common with the set just described, will lead to the team solution for **P1**. If we restrict ourselves to linear incentives, then the solution is unique in this example.

This example exhibits yet another feature. Note that with $k = -1$,

$$L^2(ku^2, u^2) = 2(u^2)^2 + 2 = L^1(u^{1^t}, u^2) + 2\,.$$

In other words, by this choice of incentive, the objectives of both players are identical (apart from a constant), thus fulfilling the old adage "if you wish other people to behave in your own interest, then make them see things your way".

In general making the cost functions identical by the choice of an appropriate γ^1-function will be too strong a requirement. A weaker form, however, which also leads to the team solution, is

$$\arg\min_{u^2} L^2(\gamma^1(u^2), u^2) = \arg\min_{u^2} L^1(u^{1^t}, u^2), \tag{7.44}$$

where it is assumed that γ^1 is of the form as described by (7.43). □

Definition 7.1 *The incentive problem, as defined in this subsection, is* (linearly) incentive controllable, *if there exists a (linear) γ^1-function such that (7.42a) and (7.42b) are satisfied.*

Of course, not all problems are incentive controllable. What can **P1** achieve in problems that are not incentive controllable? The method to be employed to answer this question will be described in the context of an example, given below. It should then be clear how the method would apply to more elaborate problems.

Example 7.4 Consider $L^1 = (u^1 - 4)^2 + (u^2 - 4)^2$ and $L^2 = (u^1)^2 + (u^2 - 1)^2$, where the u^i are scalars; $u^1 \in S^1 = [0, 3]$ and $u^2 \in S^2 = [0, 6]$. The team solution in this case is: $u^{1^t} = 3, u^{2^t} = 4$ and $L^1(u^{1^t}, u^{2^t}) = 1$. This is depicted in Fig. 7.2, where some contours of L^1 have been drawn; here the horizontal axis stands for u^2 (which is the independent variable) and the vertical one corresponds to u^1. The worst possible outcome for **P2**, even if he minimizes his cost function with respect to his own decision variable, is

$$\min_{u^2 \in S^2} [\max_{u^1 \in S^1} L^2(u^1, u^2)].$$

This occurs for $u^2 = 1, u^1 = 3$ (point A in the figure), and $L^2(3, 1) = 9$. Whatever choice **P1** makes for γ^1, the cost for **P2** will never be higher than 9. If **P1** chooses $u^1 = \gamma^1(u^2) = 3$ on the interval $[0, 6]$, then the outcome becomes $u^2 = 1, u^1 = 3$, (point A), and the costs for **P1** and **P2** become 10 and 9, respectively. This, however, is not optimal for **P1**. He should instead consider $\min_{u^1} L^1$ subject to $L^2 \leq 9$. This latter region has been shaded in the figure. The solution of this minimization problem is $u^1 = 12/5, u^2 = 14/5$ (point B in the figure). Now, any γ^1-curve, in the rectangle $0 \leq u^1 \leq 3, 0 \leq u^2 \leq 6$, which has only the points A and B with the shaded region in common, would lead to a nonunique choice for **P2**; he might either choose $u^2 = 1$ or $u^2 = 14/5$. Both choices lead to $L^2 = 9$. The costs for **P1** are, respectively, 10 and 4 for these choices. Therefore, **P1** will choose a γ^1 function as just described, with one exception; it will have a little "dip" in the shaded area near point B, such that the choice of **P2** will be unique again. (A possible choice is: $\gamma^1(u^2) = 3$ for $0 \leq u^2 < 14/5 - \varepsilon$, where $\varepsilon > 0$ and $\gamma^1(u^2) = 12/5$ for $14/5 - \varepsilon \leq u^2 \leq 6$.) The outcome will now be a point u^1, u^2 near point B, just within the shaded area. **P1** can keep his costs arbitrarily close to 4 (but not equal to 4). □

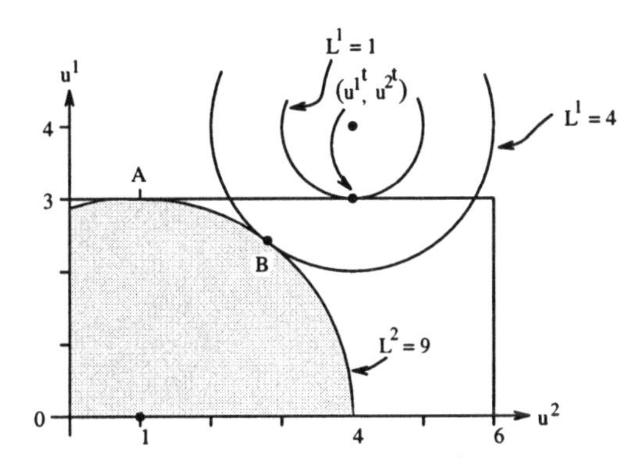

Figure 7.2: Graphical illustration of Example 7.4.

Extensions of the foregoing analysis are possible in different directions, such as the multi-stage problems (as in the case of the example of subsection 7.4.2) or problems with multiple hierarchies; see the *Notes* section 7.7 for selected references that cover these extensions. Another possibility is the many-followers case which we now briefly discuss. If there are two or more followers in the decision problem, the relationship (that is, the solution concept to be adopted) between the followers must be specified.[77] We shall illustrate a few of these, barring the formation of coalitions between followers. Let **P1** be the leader, **P**$i, i = 2, \ldots, M$, be the $M-1$ followers. An *incentive policy* (equivalently, *incentive mechanism*) for the leader is a mapping $\gamma^1 : S^2 \times \cdots \times S^m \to S^1$. Suppose that $u^{i^t}, i = 1, \ldots, M$, is the M-tuple of actions desired by the leader. An incentive mechanism γ^1 is said to induce a *dominant strategy* solution if

$$\arg\min_{u^i} L^i(\gamma^1(u^2, \ldots, u^M), u^2, \ldots, u^M) = u^{i^t},$$
$$\text{with arbitrary} \quad u^j, \forall j \neq i, i = 2, \ldots, M.$$

For an illustration of a three decision maker problem, let $u^1 = (u_1^1, u_2^1)$, and

$$L^1 = (u^2 + u^3)^2, \; L^i = (u^i - 1)^2 + u_{i-1}^1, \;\; i = 2, 3.$$

Then, the incentive mechanism $u_1^1 = \gamma_1^1(u^2) = 2u^2$ will induce $u_1^1 = 0$ regardless of the value of u^3, and similarly $u_2^1 = \gamma_2^1(u^3) = 2u^3$ will induce $u_2^1 = 0$ for all values of u^2, and hence the concatenation of γ_1^1 and γ_2^1 constitutes a dominant strategy. Such a policy is the most desirable one (for the leader), since it effectively decouples the followers from each other, and the leader can control each one's cost function separately. However, such a solution is generally difficult to realize, since the cost functions of the followers may not have the required structure. An alternative is the *Nash equilibrium* concept (among the followers),

[77] Recall the discussion in Section 3.6 in the context of finite games.

where we only require

$$\begin{aligned} &\arg\min_{u^i} L^i(\gamma^1(u^2,\dots,u^M),u^2,\dots,u^M) = u^{i^t}, \\ &\quad\text{with}\quad u^j = u^{j^t}\ \ \forall j\neq i,\ i,j=2,\dots,M, \end{aligned} \tag{7.45}$$

i.e., each follower will behave desirably conditioned on the fact (expectation) that the others will do the same. A particular subcase of this equilibrium occurs if the incentive mechanism can be chosen so that under it the cost functions of all followers become identical. The followers then face a team problem which has only one "reasonable" solution; see Problem 10 in Section 7.7 for an illustration of such a situation. The same problem also deals with the case of two followers facing a zero-sum game after implementation of the correct incentive mechanism. The saddle-point solution of this zero-sum game leads to the team solution of the leader who thus practices the adage "divide and rule".

7.5 Stochastic Dynamic Games with Deterministic Information Patterns

This section is devoted to a brief discussion on possible extensions of the results of the previous sections to stochastic dynamic games (cf. Def. 5.4) wherein the state evolves according to

$$x_{k+1} = f_k(x_k, u_k^1, u_k^2) + \theta_k, \quad k\in \mathbf{K}, \tag{7.46}$$

where $\{\theta_1,\dots,\theta_K\}$ is a set of statistically independent Gaussian vectors with values in $\mathbf{R}^n$ and with $\operatorname{cov}(\theta_k) > 0$, $k\in\mathbf{K}$. In a general context, the cost functional of **P***i*, for the game in extensive form, is again taken to be stage-additive, viz.

$$L^i(u^1,u^2) = \sum_{k=1}^{K} g_k^i(x_{k+1}, u_k^1, u_k^2, x_k), \quad i=1,2, \tag{7.47}$$

and, abiding by our standard convention, **P1** is taken to act as the leader and **P2** as the follower. In the discussion to follow, we shall consider three different information structures; namely, (A) open-loop for both players, (B) open-loop for the leader and closed-loop perfect state (CLPS) for the follower, (C) CLPS for both players. In addition to derivation of the (global) Stackelberg solution (cf. subsection 7.5.1), we shall also discuss derivation of the feedback Stackelberg solution under the third information pattern listed above (cf. subsection 7.5.2), and stochastic incentive problems (as the counterpart of the material presented in subsection 7.4.4).

7.5.1 (Global) Stackelberg solution

A. Open-loop information for both players

If the underlying deterministic information structure is open-loop for both players (in which case the controls depend on the initial state x_1 which is assumed to be known *a priori*), then the Stackelberg solution can be obtained by basically converting the game into equivalent static normal form and utilizing the first method outlined in Section 7.2 for the deterministic open-loop Stackelberg solution. Toward this end, we first recursively substitute (7.46) into (7.47), and take expected values of L^i over the statistics of the random variables $\{\theta_1, \ldots, \theta_K\}$, to obtain a cost functional $(J^i(u^1, u^2))$ which depends only on the control vectors, u^1, u^2 (and also on the initial state vector x_1 which is already known by the players). Since such static games have already been treated in Section 4.5, we do not discuss them here any further, with the exception of one special case (see the next paragraph and Prop. 7.4). It should be noted that the Stackelberg solution will, in general, depend on the statistical moments of the random disturbances in the state equation—unless the system equation is linear and the cost functionals are quadratic.

For the special case of linear-quadratic dynamic games (i.e., when f_k in (7.46) is given as in (7.7a), and g_k^i in (7.47) is structured as in (7.7b), and under the open-loop information structure, the cost functional $J^i(u^1, u^2)$ can be written as

$$J^i(u^1, u^2) = \sum_{k=1}^{K} \frac{1}{2}(y'_{k+1} Q^i_{k+1} y_{k+1} + u_k^{i'} u_k^i + u_k^{j'} R_k^{ij} u_k^j) + \xi^i; j \neq i, j = 1, 2,$$

where

$$y_{k+1} = A_k y_k + B_k^1 u_k^1 + B_k^2 u_k^2 + E[\theta_k]; \quad y_1 = x_1,$$

and ξ^i is independent of the controls and x_1, and depends only on the covariances of $\{\theta_1, \ldots, \theta_K\}$. This result is, of course, valid as long as the set $\{\theta_1, \ldots, \theta_K\}$ is statistically independent of x_1, which was one of our underlying assumptions at the beginning. Hence, for the linear-quadratic stochastic game, the stochastic contribution completely separates out; and, furthermore, if $E[\theta_k] = 0 \ \forall k \in \mathbf{K}$, the open-loop Stackelberg solution matches (completely) with the one given in Corollary 7.1 for the deterministic problem. This result is now summarized below in Prop. 7.4.

Proposition 7.4 *Consider the two-person linear-quadratic stochastic dynamic game described by (7.46)-(7.47) together with the structural assumptions (7.7a)-(7.7b), and under the parametric restrictions* $Q^i_{k+1} \geq 0$, $R_k^{12} > 0$, $E[\theta_k] = 0$ $(k \in \mathbf{K}; i = 1, 2)$. *It admits a unique open-loop Stackelberg solution with* **P1** *acting as the leader, which is given, under the condition of invertibility of the matrix appearing in the braces in (7.9c), by (7.8a)-(7.8b).*

B. OL information for P1 and CLPS information for P2

For the deterministic Stackelberg game, we have already seen in Section 7.2 that an additional state information for the follower does not lead to any difference in the open-loop Stackelberg strategy of the leader, provided, of course, that he (the leader) still has only open-loop information (cf. Remark 7.1). The only difference between the OL-OL and OL-CLPS Stackelberg solutions then lies in the optimum response strategy of the follower, which, in the latter case, is any closed-loop representation of the open-loop response strategy on the equilibrium trajectory associated with the OL Stackelberg solution; hence, the optimum strategy of the follower under the OL-CLPS information pattern is definitely not unique, but it can, nevertheless, be obtained from the OL Stackelberg solution.

For the stochastic Stackelberg game, however, the issue is more subtle, especially if the state dynamics is not linear and the cost functional is not quadratic. In general, the OL-OL Stackelberg solution does not coincide with the OL-CLPS Stackelberg solution, and the latter has to be obtained independently of the former. The steps involved in this derivation are as follows:

(1) For each fixed $\{u_k^1 \in U_k^1; k \in \mathbf{K}\}$, minimize

$$J^2(u^1, \gamma^2) = E\left[\sum_{k=1}^{K} g_k^2(x_{k+1}, u_k^1, u_k^2, x_k) \mid u_k^2 = \gamma_k^2(x_\ell, \ell \leq k), \quad k \in \mathbf{K}\right]$$

subject to (7.46) and over the permissible class of strategies, Γ^2. Any solution $\gamma^2 \in \Gamma^2$ of this stochastic control problem satisfies the dynamic programming equation (see Section 5.6, equation (5.61b))

$$\begin{aligned} V(k, x) &= \min_{u_k^2 \in U_k^2} E_{\theta_k}[g_k^2(f_k(x_k, u_k^1, u_k^2) + \theta_k, u_k^1, u_k^2, x_k) \\ &\qquad + V(k+1, f_k(x_k, u_k^1, u_k^2) + \theta_k)] \\ &= E_{\theta_k}[g_k^2(f_k(x_k, u_k^1, \gamma_k^{2\circ}(\eta_k^2)) + \theta_k, u_k^1, \gamma_k^{2\circ}(\eta_k^2), x_k) \\ &\qquad + V(k+1, f_k(x_k, u_k^1, \gamma_k^{2\circ}(\eta_k^2)) + \theta_k)], \end{aligned} \tag{7.48}$$

where

$$\eta_k^2 \triangleq \{x_\ell, \ell \leq k\},$$

and E_{θ_k} denotes the expectation operation with respect to the statistics of θ_k.

(2) Now, minimize the cost functional

$$J^1(\gamma^1, \gamma^{2\circ}) = E[L^1(u^1, u^2) \mid u_k^1 = \gamma_k^1(x_1), u_k^2 = \gamma_k^{2\circ}(\eta_k^2), k \in \mathbf{K}] \tag{7.49}$$

over the permissible class of strategies, Γ^1, and subject to the constraints imposed by the dynamic programming equation (7.48) and the state equation (7.46) with u_k^2 replaced by $\gamma_k^{2\circ}(\eta_k^2)$.[78] The solution of this optimization problem constitutes the Stackelberg strategy of the leader in the stochastic dynamic game under consideration.[79]

---

[78] Note that $\gamma_k^{2\circ}(\cdot)$ is, in fact, dependent on γ^1, but this dependence cannot be written explicitly (in closed-form), unless the cost functional of **P2** and the state equation have a specific simple structure (such as quadratic and linear, respectively).

[79] The underlying assumption here is that step (i) provides a unique solution $\{\gamma_k^{2\circ}; k \in \mathbf{K}\}$ for every $\{u_k^1 \in U_k^1; k \in \mathbf{K}\}$.

For the class of linear-quadratic games, these two steps can readily be carried out, to lead to a unique Stackelberg solution in closed-form. In this case, the leader's equilibrium strategy, in fact, coincides with its counterpart in the OL-OL game (presented in Prop. 7.4)—a property that can be verified without actually going explicitly through the two steps of the foregoing procedure. This result is given below in Prop. 7.5, whose proof, as provided here, is an indirect one.

Proposition 7.5 *Consider the two-person linear-quadratic stochastic Stackelberg game of Prop. 7.4, but under the OL-CLPS information pattern. Provided that the condition of matrix invertibility of Prop. 7.4 is satisfied, this stochastic game admits a unique Stackelberg solution with* **P1** *as the leader, which is given for* $k \in \mathbf{K}$ *by*

$$\gamma_k^{1*}(x_1) = -B_k^{1'} K_k^1 x_k^*, \tag{7.50a}$$

$$\gamma_k^{2*}(x_k, x_1) = -P_k^2 S_{k+1} A_k x_k - P_k^2 (s_{k+1} - S_{k+1} B_k^1 B_k^{1'} K_k^1 x_k^*), \tag{7.50b}$$

where

$$P_k^2 = [I + B_k^{2'} S_{k+1} B_k^2]^{-1} B_k^{2'}, \tag{7.51a}$$

$$S_k = Q_k^2 + A_k' S_{k+1} [I - B_k^2 P_k^2 S_{k+1}] A_k; \; S_{K+1} = Q_{K+1}^2, \tag{7.51b}$$

$$s_k = A_k' [I - B_k^2 P_k^2 S_{k+1}]' [s_{k+1} - S_{k+1} B_k^1 B_k^{1'} K_k^1 x_k^*]; \; s_{K+1} = 0, \tag{7.51c}$$

and K_k^1 *is defined by (7.9a), and* x_k^* *by (7.10e).*

Proof. For each fixed control strategy $\{u_k^1 = \gamma_k^1(x_1); k \in \mathbf{K}\}$ of the leader, the dynamic programming equation (7.48) can explicitly be solved to lead to the unique solution

$$\gamma_k^{2\circ}(x_k, x_1) = -P_k^2 S_{k+1} A_k x_k - P_k^2 (\bar{s}_{k+1} + S_{k+1} B_k^1 u_k^1), \quad k \in \mathbf{K}, \tag{i}$$

where P_k^2 and S_{k+1} are defined by (7.51a) and (7.51b), respectively, and $\bar{s}_k$ is defined by

$$\bar{s}_k = A_k' [I - B_k^2 P_k^2 S_{k+1}]' [\bar{s}_{k+1} + S_{k+1} B_k^1 u_k^1]; \; \bar{s}_{K+1} = 0. \tag{ii}$$

(This result follows from Prop. 5.4 by taking $E[\theta_k] = B_k^1 u_k^1$.)

Now, at step (2) of the derivation, (i) and (ii) above will have to be used, together with the state equation, as constraints in the minimization of (7.49) over Γ^1. By recursive substitution of (ii) and the state equation into (i), we obtain the following equivalent expression for (i), to be utilized as a constraint in the optimization problem faced by the leader:

$$\gamma_k^{2\circ} = \sum_{\ell=1}^{K} T_k(\ell) u_\ell^1 + M_k x_1 + \sum_{\ell=1}^{k-1} N_k(\ell) \theta_\ell, \tag{iii}$$

where $T_k(\ell)$, M_k, $N_k(\ell)$ are the coefficient matrices associated with this representation, whose exact expressions will not be needed in the sequel. Now, (iii) being a linear relation, and L^1 being a quadratic expression in $\{x_{k+1}, u_k^1, u_k^2; k \in \mathbf{K}\}$, it follows that, with $u^1 \triangleq (u_1^{1'}, \ldots, u_K^{1'})'$, $J^1(\gamma^1, \gamma^{2\circ})$ can be written as

$$J^1(\gamma^1, \gamma^{2\circ}) = E[u^{1'}T^1u^1 + u^{1'}T^2x_1 + x_1'T^3x_1 \mid u_k^1 = \gamma_k^1(x_1), k \in \mathbf{K}] + J_{\mathrm{m}},$$

where T^1, T^2, T^3 are deterministic matrices, and J_{m} is independent of u^1 and is determined solely by the covariance matrices of $\{\theta_k; k \in \mathbf{K}\}$—this decomposition being due to the fact that the leader's information is open-loop and hence his controls are independent of the random variables $\{\theta_k; k \in \mathbf{K}\}$ which were taken to have zero mean. Since the Stackelberg strategy of the leader is determined as the global minimum of $J^1(\gamma^1, \gamma^{2\circ})$, it readily follows that it should be independent of the covariance of $\{\theta_k; k \in \mathbf{K}\}$. Now, if there were no noise in the state equation, the follower's response (though not unique) would still be expressible in the form (i), and therefore the leader's Stackelberg strategy would still be determined as the control that minimizes $[u^{1'}T^1u^1 + u^{1'}T^2x_1 + x_1'T^3x_1]$. Hence the leader's Stackelberg strategy in this stochastic game is the same as the one in its deterministic counterpart—thus leading to (7.50a) (see Corollary 7.1 and Remark 7.1). The follower's unique optimal response (7.50b) then follows from (i) by substituting (7.50a) into (ii). Note that this strategy is, in fact, a particular closed-loop representation of (7.8b) on the open-loop equilibrium trajectory of the deterministic problem (which is described by (7.10e)). □

C. CLPS information for both players

When the leader has access to dynamic information, derivation of the Stackelberg solution in stochastic dynamic games meets with insurmountable difficulties.[80] First, any direct approach readily fails at the outset (as in its deterministic counterpart, discussed in Section 7.4) since the optimal response of the follower to any announced strategy of the leader (from a general strategy space) cannot be expressed analytically (in terms of that strategy), even for linear-quadratic games. Second, the indirect method of Section 7.4, which is developed for deterministic dynamic games, cannot be extended to stochastic games, since, in the latter case, every strategy has a *unique* representation (see Section 5.6) as opposed to the existence of infinitely many closed-loop representations of a given strategy in a deterministic system. Consequently, derivation of the closed-loop Stackelberg solution of stochastic dynamic games remains, today, as a challenge for the researchers.

If, however, we make some structural assumptions on the possible strategies of the leader—which is tantamount to seeking sub-optimal solutions—then the problem may become tractable. In particular, if, under the stipulated structural assumptions, the class of permissible strategies of the leader can be described by a finite number of parameters, and if the follower's optimal response can be determined analytically as a function of these parameters, then the original game

[80]Consistent with the CLPS information pattern, we are assuming here that the leader does not have direct access to the follower's control. If this is not the case, and the leader's information exhibits redundancy, then the problem might be more tractable. This case will be discussed later in subsection 7.5.3 in the context of stochastic incentive problems.

may be viewed as a static one in which the leader selects his strategy from a Euclidean space of appropriate dimension; such a static Stackelberg game is, in general, solvable—but more often numerically than analytically. The following example now illustrates this approach and demonstrates that even for simple stochastic dynamic games, and under the crudest type of structural assumptions, the corresponding (sub-optimal) Stackelberg solution cannot be obtained analytically, but only through some numerical minimization techniques.

Example 7.5 Consider the two-stage scalar stochastic dynamic game described by the state equations

$$\left.\begin{aligned} x_2 &= x_1 - u^2 + \theta_1, \\ x_3 &= x_2 - u^1 + \theta_2 \end{aligned}\right\} \tag{7.52}$$

and cost functionals

$$\begin{aligned} L^1 &= (x_3)^2 + 2(u^1)^2 + (u^2)^2, \\ L^2 &= (x_3)^2 + (u^2)^2. \end{aligned}$$

Here, θ_1 and θ_2 are taken as independent random variables with mean zero and variances σ_1 and σ_2, respectively. The leader (**P1**) acts at stage 2 and has access to both x_1 and x_2, while the follower (**P2**) acts at stage 1 and has only access to x_1.

If $\gamma^i \in \Gamma^i$ denotes a general strategy of **P**i ($i = 1, 2$), the expected (average) cost functional of **P2** can be written as

$$\begin{aligned} J^2(\gamma^1, \gamma^2) &= E\{[x_2 - \gamma^1(x_2, x_1) + \theta_2]^2 + [\gamma^2(x_1)]^2\} \\ &= E\{[x_2 - \gamma^1(x_2, x_1)]^2 + [\gamma^2(x_1)]^2\} + \sigma_2 \end{aligned}$$

which has to be minimized over $\gamma^2 \in \Gamma^2$, to determine the optimal response of **P2** to $\gamma^1 \in \Gamma^1$. Barring the stochastic aspects, this is similar to the problem treated in subsection 7.4.1, where the difficulties involved in working with a general γ^1 have been delineated. We therefore now restrict our investigation (also in line with the discussion preceding the example) to a subclass of strategies in Γ^1 which are affine in x_2, that is, to strategies of the form[81]

$$\gamma^1(x_2, x_1) = \alpha x_2 + \beta x_1, \tag{7.53}$$

where α and β are free parameters which are yet to be determined. They will, in general, be dependent on x_1 which is, though, known *a priori* by both players.

Under the structural restriction (7.53), J^2 admits a unique minimum, thus leading to the optimal response strategy (for the follower)

$$\gamma^{2o}(x_1) = \frac{(1-\alpha)(1-\alpha-\beta)}{1+(1-\alpha)^2} x_1 \tag{7.54}$$

[81]The apparent linear structure of the second term below is adopted for the sake of convenience in the analysis to follow; it could also have been taken as a single function β.

which explicitly depends on the parameters α and β that characterize the leader's strategy. To determine their optimal values, we now substitute (7.53)-(7.54) into $J^1(\gamma^1, \gamma^2)$, together with the corresponding values of x_3 and x_2 from (7.52), to obtain the function F given below, which has to be minimized over $\alpha \in \mathbf{R}$, $\beta \in \mathbf{R}$ for fixed x_1:

$$F(\alpha, \beta) = \left\{ \frac{[1-\alpha-\beta]^2}{1+(1-\alpha)^2} + \frac{2[\alpha+2\beta-\alpha\beta]^2}{[1+(1-\alpha)^2]^2} \right\} x_1^2 + [(1-\alpha)^2 + \alpha^2]\sigma_1 + \sigma_2.$$

Let us now note that

(i) F is jointly continuous in (α, β), $F(\alpha, \beta) \geq 0 \ \forall (\alpha, \beta) \in \mathbf{R} \times \mathbf{R}$, and $F(\alpha, \beta) \to \infty$ as $|\alpha|, |\beta| \to \infty$. Therefore, we can restrict our search for a minimum on $\mathbf{R}^2$ to a closed and bounded subset of $\mathbf{R}^2$, and consequently there exists (by the Weierstrass theorem[82]) at least one pair (α^*, β^*) that minimizes F for any given pair (x_1, σ_1).

(ii) The optimum pair (α^*, β^*) depends on (x_1, σ_1), but not on σ_2, and it cannot be expressed analytically as a function of (x_1, σ_1). Hence, for each fixed (x_1, σ_1), $F(\alpha, \beta)$ has to be minimized numerically, by utilizing one of the available optimization algorithms (Luenberger, 1973).

(iii) With (α^*, β^*) determined as above, the linear-in-x_2 strategy $\gamma^{1*}(x_2, x_1) = \alpha^*(x_1)x_2 + \beta^*(x_1)x_1$ is only a sub-optimal Stackelberg strategy for the leader, since he may possibly achieve a better performance by announcing a strategy outside the "linear-in-x_2" class. □

7.5.2 Feedback Stackelberg solution

In this subsection, we extend the results of Section 7.3 to stochastic dynamic games described by (7.46)-(7.47), and, in particular, we obtain the counterpart of Thm. 7.2 in the present context (see Thm. 7.4 below). In this analysis, we do not restrict our attention at the outset to feedback strategies (as it was done in Section 7.3), but rather start with general closed-loop strategies $u_k^i = \gamma_k^i(x_\ell, \ell \leq k)$, $k \in \mathbf{K}$, $i = 1, 2$. The conclusion, however, is that the feedback Stackelberg solution can be realized only in feedback strategies. To see this, we start (in view of Def. 3.29) at the last stage $k = K$ and solve basically a static Stackelberg game between **P1** and **P2**, with their strategies denoted by γ_K^1 and γ_K^2, respectively. Because of the additive nature of the cost functional of each player, and since $\text{cov}\,(\theta_K) > 0$, θ_K is Gaussian and statistically independent of the remaining random vectors, every Stackelberg solution at this stage will depend only on x_K and not on the past values of the state vector. Proceeding to the next stage $k = K-1$, after substitution of the Stackelberg solution at stage $k = K$ into the state equation, we conclude, by the same reasoning, that the Stackelberg strategies at stage $k = K-1$ are only functions of x_{K-1}. An inductive argument (as in the proof of Thm. 6.10) then verifies

[82]See Section 5 of Appendix A.

the "feedback" property of the feedback Stackelberg strategies, and it also simultaneously leads to the equations that yield the corresponding solution for the stochastic dynamic game. Theorem 7.4, below, now provides the complete solution under the assumption of singleton reaction sets for the follower.

Theorem 7.4 *Every feedback Stackelberg solution of a two-person K-stage stochastic infinite dynamic game described by (7.46)-(7.47) and with the CLPS information pattern (for both players) comprises only feedback strategies, and is strongly time consistent. A pair of strategies $\{\gamma^{1*} \in \Gamma^1, \gamma^{2*} \in \Gamma^2\}$ constitutes a feedback Stackelberg solution with* **P1** *as the leader if*

$$\min_{\gamma_k^1 \in \Gamma_k^1, \gamma_k^2 \in \bar{R}_k^2(\gamma_k^2)} \bar{G}_k^1(\gamma_k^1, \gamma_k^2, x_k) = G_k^1(\gamma_k^{1*}, \gamma_k^{2*}, x_k), \forall x_k \in \mathbf{R}^n \ (k \in \mathbf{K}),$$

where $\bar{R}_k^2(\cdot)$ is a singleton set defined by

$$\bar{R}_k^2(\gamma_k^1) = \left\{ \beta_k^2 \in \Gamma_k^2; \bar{G}_k^2(\gamma_k^1, \beta_k^2, x_k) = \min_{\gamma_k^2 \in \Gamma_k^2} \bar{G}_k^2(\gamma_k^1, \gamma_k^2, x_k) \ \forall x_k \in \mathbf{R}^n \right\},$$

$$\bar{G}_k^i(\gamma_k^1, \gamma_k^2, x_k) \triangleq E_{\theta_k}[\hat{G}_k^i(f_k(x_k, \gamma_k^1(x_k), \gamma_k^2(x_k) + \theta_k), \gamma_k^1(x_k), \gamma_k^2(x_k), x_k)], \quad i = 1, 2, \quad k \in \mathbf{K}$$

and $\hat{G}_k^i$ is recursively defined by

$$\begin{aligned}\hat{G}_k^i(x_{k+1}, \gamma_k^1(x_k), \gamma_k^2(x_k), x_k) = {} & E_{\theta_{k+1}}[\hat{G}_{k+1}^i(f_{k+1}(x_{k+1}, \gamma_{k+1}^{1*}(x_{k+1}), \\ & \gamma_{k+1}^{2*}(x_{k+1})) + \theta_{k+1}, \gamma_{k+1}^{1*}(x_{k+1}), \gamma_{k+1}^{2*}(x_{k+1}), x_{k+1})] \\ & + g_k^i(x_{k+1}, \gamma_k^1(x_k), \gamma_k^2(x_k), x_k); \quad \hat{G}_{K+1}^i = 0, \quad i = 1, 2,\end{aligned}$$

Proof. This result follows from a direct application of Def. 3.29 and Prop. 3.15 (interpreted appropriately for the infinite stochastic dynamic game under consideration), by employing the techniques and lines of thought used in the proof of Thm. 6.10 for the feedback Nash solution. □

Remark 7.3 has a natural counterpart here for the stochastic game; and for the special case of the linear-quadratic stochastic dynamic game in which $E[\theta_k] = 0 \ \forall k \in \mathbf{K}$, it may readily be verified that the feedback Stackelberg solution of Thm. 7.4 is precisely the one given in Corollary 7.2.

7.5.3 Stochastic incentive problems

We have seen in subsection 7.5.1 that for stochastic dynamic games, and under CLPS information, it is generally very difficult, if not impossible, to obtain the global Stackelberg solution—with the indirect method developed in the deterministic case (cf. Section 7.4) not applicable here because of uniqueness of strategy representations. What if, however, the leader has also access to the follower's past control actions, in addition to the state information based on which these actions were obtained? In this enlarged information structure, a given strategy of the leader will have multiple representations, thus opening the possibility of enforcement of a team solution (to the leader's advantage) by selecting an appropriate representation of the team-optimal strategy (of the leader). To illustrate this line of thought now, let us revisit Example 7.5, but with the following enlarged information structure.

Example 7.6 Consider the two-stage scalar stochastic dynamic game of Example 7.5, but with the enlarged information structure that allows **P1** to have access to u^2, in addition to x_1 and x_2. Note that since u^2 depends only on x_1, this enlarged information structure carries the same statistical information as the earlier one for each fixed (pure) policy of the follower; however, as we will see shortly, the informational redundancy that it generates will bring a substantial advantage to the leader.

In the spirit of the analysis of Example 7.1 for the deterministic case, let us first determine the best performance the leader would achieve if the follower were cooperating with him (in the minimization of the leader's expected cost function). The associated team problem is

$$J^t \triangleq J^1(\gamma^{1^t}, \gamma^{2^t}) = \min_{\gamma^1 \in \Gamma^1} \min_{\gamma^2 \in \Gamma^2} J^1(\gamma^1, \gamma^2),$$

where

$$J^1(\gamma^1, \gamma^2) = E\{[x_2 - \gamma^1(x_2, x_1) + \theta_2]^2 + 2[\gamma^1(x_2, x_1)]^2 + [\gamma^2(x_1)]^2\}.$$

Here, the cost shows dependence on $u^2 = \gamma^2(x_1)$ not only directly, but also through x_2 as given by (7.52), which has to be taken into account in the minimization. Furthermore, the strategy spaces Γ^1 and Γ^2 are taken as in Example 7.1, since the additional knowledge of u^2 for **P1** does not help in further reducing the minimum team cost J^t. Now, this team problem is in fact a standard LQ stochastic control problem of the type covered by Prop. 5.4, and its solution can readily be obtained as:

$$\gamma^{1^t}(x_2, x_1) = \frac{1}{3}x_2, \quad \gamma^{2^t}(x_1) = \frac{5}{14}x_1 \tag{7.55}$$

which is the *unique* minimizing pair in $\Gamma^1 \times \Gamma^2$. It is not, however, unique in the enlarged strategy space for the leader, as (for example) the following parameterized strategy also constitutes an optimal solution, along with γ^{2^t} given above, for every $\alpha \in \mathbf{R}$:

$$\gamma^1_\alpha(x_2, x_1, u^2) = \frac{1}{3}x_2 + \alpha(u^2 - \frac{5}{14}x_1). \tag{7.56}$$

This in fact characterizes the complete class of linear (in x_2, x_1, u^2) optimal strategies, but of course there are also nonlinear ones—all leading to the same (minimum) expected value for the leader. By a slight generalization of the terminology introduced in Def. 5.11, we will refer to all these "minimum expected cost achieving" strategies as *representations* of γ^{1^t} under the team-optimal solution $(\gamma^{1^t}, \gamma^{2^t})$. This is a rich family of strategies, among which we seek one with the additional property that if the follower instead minimizes his own expected cost function, then the strategy in Γ^2 that achieves this minimum is still γ^{2^t}. The corresponding strategy (representation) for the leader would then clearly constitute a global Stackelberg solution, leading to the best possible performance for him.

Let us now conduct the search in the family of linear representations (7.56), which leads to the quadratic optimization problem:

$$\min_{\gamma^2 \in \Gamma^2} E\{[x_2 - \gamma^1_\alpha(x_2, x_1, u^2) + \theta_2]^2 + [u^2]^2\},$$

where

$$x_2 = x_1 - u^2 + \theta_1.$$

Since x_1 is independent of θ_1 and θ_2, which have zero mean, this problem is equivalent to the following deterministic optimization problem:[83]

$$\min_{v \in \mathrm{R}}\{[(\frac{2}{3})(x_1 - v) - \alpha(v - \frac{5}{14}x_1)]^2 + v^2\}$$

where we have written v for u^2, to simplify the notation. Now, a simple optimization shows that for the value of $\alpha = 8/27$, this optimization problem admits the unique solution $v = (5/14)x_1 \equiv \gamma^{2^t}(x_1)$, and hence the policy pair

$$\gamma^1(x_2, x_1, v) = \frac{1}{3}x_2 + \frac{8}{27}(v - \frac{5}{14}x_1), \; \gamma^2(x_1) = \frac{5}{14}x_1$$

provides a global Stackelberg solution. This is in fact the unique such solution in the linear class. □

Stochastic decision problems of the type above, where the leader is allowed to have access to past actions of the follower are known as *stochastic incentive problems*, which are the stochastic counterparts of those briefly discussed in subsection 7.4.4. In stochastic incentive problems, the information structure may not always be *nested* (for the leader), as in the example above, where the leader has access to all the information that the follower has access to (plus more). If, for instance, in Example 7.6 the leader has only access to x_2 and u^2, then we have a problem with a *nonnested* information structure, to which the methodology presented above does not apply, since the dynamic information for the leader no longer exhibits redundancy. Discussion of such problems, where the follower possesses *private information* not known to the leader, is beyond the scope of our coverage here; the interested reader can consult with Ho, Luh and Olsder (1982) and Başar (1984, 1989a)). For stochastic incentive problems with nested information, however, the methodology used in Example 7.6 can be

[83]This equivalence holds as far as its optimum solution goes (which is what we seek), but not for the corresponding minimum values.

developed into a general procedure as briefly discussed below for a special class of such problems.

Consider a two-person stochastic incentive problem with the cost functions $L^1(u^1, u^2; \theta)$ and $L^2(u^1, u^2; \theta)$, for **P1** (leader) and **P2** (follower), respectively, where θ is some random vector with a known distribution function. Let $y^1 = h^1(\theta)$ be the measurement of **P1** on θ, and $y^2 = h^2(\theta)$ be **P2**'s measurement, with the property that what **P2** knows is also known by **P1** (but not necessarily *vice versa*).[84] Let Γ^i be the set of all measurable policies of the form $u^i = \gamma^i(y^i)$, $i = 1, 2$, and $\hat{\Gamma}^1$ be the set of all measurable policies of the form $u^1 = \gamma^1(y^1, u^2)$. Introduce the pair of policies

$$(\gamma^{1^t}, \gamma^{2^t}) \triangleq \arg \min_{\gamma^1 \in \Gamma^1, \gamma^2 \in \Gamma^2} E_\theta\{L^1(\gamma^1(h^1(\theta)), \gamma^2(h^2(\theta)), \theta)\}$$

assuming that the underlying team problem admits a minimizing solution. Then, a *representation* of the leader's strategy γ^{1^t} under the pair above is an element of $\hat{\Gamma}^1$, say $\hat{\gamma}^1$, with the property

$$\hat{\gamma}^1(h^1(\theta), \gamma^{2^t}(h^2(\theta))) = \gamma^{1^t}(h^1(\theta)), \quad \text{a.s.}^{85} \tag{7.57}$$

The following result now readily follows:

Proposition 7.6 *For the stochastic incentive decision problem with nested information as formulated above, the pair $(\hat{\gamma}^1, \gamma^{2^t})$ constitutes a global Stackelberg solution, leading to the best possible outcome for the leader. Equivalently, if a strategy $\hat{\gamma}^1 \in \hat{\Gamma}^1$ exists satisfying (7.57), the stochastic decision problem is* incentive controllable.[86]

Remark 7.13 For the special class of linear-quadratic-Gaussian (LQG) problems, where the decision variables (u^1, u^2) belong to finite dimensional Euclidean spaces, θ is a Gaussian random vector, h^1 and h^2 are linear, and L^1 is jointly quadratic in the triple (u^1, u^2, θ) and strictly convex in (u^1, u^2) for each θ, the team-optimal policies γ^{1^t} and γ^{2^t} exist, are unique and linear in y^1 and y^2, respectively (see any standard book on stochastic control, such as (Bertsekas, 1987)). If, furthermore, L^2 is also a quadratic function, then except for some isolated cases one can restrict the search to linear representations of γ^{1^t}:

$$\hat{\gamma}^1(y^1, u^2) = \gamma^{1^t}(y^1) + P[u^2 - \gamma^{2^t}(y^2)], \tag{7.58}$$

where P is a matrix of appropriate dimensions. Now, invoking the condition (7.57) one can obtain an equation for P, whose solution (when used in (7.58)) leads to a linear incentive policy. This then makes the decision problem linear incentive controllable; for details see (Başar, 1979d). □

[84] In mathematical terms, this requirement can be stated as the sigma-field generated by y^1 including the sigma-field generated by y^2.

[85] The equality should hold for almost all values of θ, under its assumed distribution function.

[86] The terminology we have used here is the natural counterpart (in the stochastic case) of the one introduced in Def. 7.1 for deterministic incentive problems.

The development above does not cover (even in the LQG framework) the most general class of dynamic nested stochastic incentive problems, because the measurements of the decision makers have been taken to be *static*—not depending on the past actions. If the leader's measurement at stage k depends on the past actions of the follower $(u_\ell^2, \ell < k)$, then the approach discussed above can easily be adjusted to apply to such multi-stage problems too. If, however, the follower also has access to the leader's past control actions, then because of the nestedness of the information structure for the leader (which does not allow for the follower to have access to all measurements of the leader) the associated dynamic team problem becomes what is called a *nonclassical stochastic control problem*, for which no general theory exists. Issues such as learning, inference, and filtering become of relevance then, whose treatment requires background in stochastic processes, information theory and control, much beyond the level of our coverage here.

For static stochastic incentive problems where there is one leader and several followers, with the followers playing according to the Nash concept, a natural counterpart of (7.45) exists, which is:

$$\arg\min_{\gamma^i\in\Gamma^i} J^i(\gamma^1(y^1;u^2,\ldots,u^M),\gamma^2,\ldots,\gamma^M) = \gamma^{i^t},$$
$$\text{with}\quad \gamma^j = \gamma^{j^t}\ \ \forall j\neq i,\ \ i,j=2,\ldots,M.$$

Since the Nash equilibrium is defined here over strategy spaces, it admits natural (conceptual) extensions to the multi-stage case, where the informations of the players (leader as well as the followers) might be intertwined in more complicated ways.

7.6 Stackelberg Solution of Differential Games

In this section we discuss the continuous-time counterparts of some of the results presented in the previous sections, and in particular the open-loop and feedback Stackelberg solutions.

7.6.1 The open-loop information structure

As counterparts of the results of Section 7.2 in the continuous time, we present here the open-loop Stackelberg solution of deterministic two-person differential games of prescribed fixed duration (cf. Def. 5.5), when **P1** acts as a leader. The class of games under consideration are described by the state equation

$$\dot{x}(t) = f(t,x(t),u^1(t),u^2(t)); \qquad x(0)=x_0, \tag{7.59a}$$

and cost functionals

$$L^i(u^1,u^2) = \int_0^T g^i(t,x(t),u^1(t),u^2(t))\,\mathrm{d}t + q^i(x(T));\ i=1,2, \tag{7.59b}$$

where $[0,T]$ denotes the fixed prescribed duration of the game, x_0 is the initial state known by both players, $x(t) \in \mathbf{R}^n$ and $u^i(t) \in S^i \subseteq \mathbf{R}^{m_i}$ $(i = 1,2)$, $\forall t \in [0,T]$. The underlying information structure is open-loop for both players, so that the controls $u^1(\cdot)$ and $u^2(\cdot)$ depend only on the time variable t and the initial state x_0. We allow only for controls (or synonymously, in this case, strategies) that are continuous in the time variable; and this determines the strategy sets (Γ^1, Γ^2) which are in this case equivalent to the control function sets (U^1, U^2). We further assume that f satisfies the conditions of Thm. 5.1 so that a unique continuously differentiable trajectory exists as a solution to (7.59a) for every permissible control pair (u^1, u^2).

To determine the set of relations to be satisfied by an open-loop Stackelberg solution, we first obtain the optimal reaction of the follower (**P2**) to every announced control u^1 of the leader by minimizing $L^2(u^1, u^2)$ over $u^2 \in u^2$.

Lemma 7.3 *In addition to the conditions of Thm. 5.1, let*

i) $f(t,\cdot,u^1,u^2)$ *be continuously differentiable on* $\mathbf{R}^n$ $\forall t \in [0,T]$,

ii) $g^2(t,\cdot,u^1,u^2)$ *and* $q^2(\cdot)$ *be continuously differentiable on* $\mathbf{R}^n$ $\forall t \in [0,T]$.

Then, if $u^1 \in U^1$ *is a fixed control of* **P1**, *and* $u^{2^\circ} \in U^2$ *denotes a corresponding optimal response of* **P2**, *there exists a function* $p(\cdot) : [0,T] \to \mathbf{R}^n$ *such that the following relations are satisfied:*

$$\begin{aligned} \dot{x} &= f(t, x(t), u^1(t), u^{2^\circ}(t)); \; x(0) = x_0, \\ u^{2\circ}(t) &= \arg\min_{u^2 \in S^2} \; H^2(t, p(t), x, u^1(t), u^2), \\ \dot{p}'(t) &= -\frac{\partial}{\partial x} H^2(t, p(t), x, u^1(t), u^{2^\circ}(t)); \; p'(T) = \frac{\partial}{\partial x} q^2(x(T)), \end{aligned} \tag{7.60}$$

where

$$H^2(t, p, u^1, u^2) \;=\; g^2(t, x, u^1, u^2) + p' f(t, x, u^1, u^2), \;\; t \in [0,T].$$

If, furthermore,

iii) $H^2(t,p,x,u^1,\cdot)$ *is continuously differentiable and strictly convex on* S^2 *which is taken as an open set, the second relation above is replaced by*

$$\nabla_{u^2} H^2(t, p, x, u^1, u^{2^\circ}) \;=\; 0.$$

Proof. This result follows directly from Thm. 5.4. □

To proceed further, we now assume that there exists a unique $u^{2^\circ} \in U^2$ under which the set of relations of Lemma 7.3 is satisfied for a given $u^1 \in U^1$, a sufficient condition for which is strict convexity of $L^2(u^1, \cdot)$ on U^2, with $x(\cdot)$ substituted from (7.59a). Then, to determine his Stackelberg strategy, the leader will be faced with the optimal control problem:

$$\min_{u^1, u^2} L^1(u^1, u^2) \tag{7.61a}$$

such that

$$\dot{x} = f(t, x, u^1, u^2); x(0) = x_0, \tag{7.61b}$$

$$\dot{p}' = -\frac{\partial}{\partial x} H^2(t, p, x, u^1, u^2); \;\; p'(T) = \frac{\partial}{\partial x} q^2(x(t)), \tag{7.61c}$$

$$\nabla_{u^2} H^2(t, p, x, u^1, u^2) = 0. \tag{7.61d}$$

This optimization problem is not of the type covered by Thm. 5.4 since its differential constraints involve specified boundary conditions at both ends. The problem, however, is still tractable, and can be solved by utilizing some other standard results of optimal control theory, particularly form (Bryson and Ho, 1975, p. 65). Toward this end, we introduce the "Hamiltonian"

$$H^1 = g^1(t, x, u^1, u^2) + \lambda_1' f(t, x, u^1, u^2) - \lambda_2' \left(\frac{\partial H^2}{\partial x} \right)' + \nabla_{u^2} H^2 \lambda_3, \tag{7.62}$$

where $\lambda_1(\cdot) : [0, T] \to \mathbf{R}^n$ is the costate function corresponding to (7.61b), $\lambda_2(\cdot) : [0, T] \to \mathbf{R}^n$ is the costate function corresponding to (7.61c) and $\lambda_3(\cdot) : [0, T] \to \mathbf{R}^{m_2}$ is the Lagrange multiplier function associated with the equality constraint (7.61d). Under suitable differentiability conditions (to be made precise in Thm. 7.5 to follow), $\lambda_1(\cdot)$ and $\lambda_2(\cdot)$ satisfy the set of differential equations

$$\begin{aligned}
\dot{\lambda}_1' &= -\frac{\partial}{\partial x} H^1(t, p, \lambda_1, \lambda_2, \lambda_3, u^1, u^2); \\
\lambda_1'(T) &= \frac{\partial}{\partial x} q^1(x(T)) - \frac{\partial^2}{(\partial x)^2} q^2(x(T)) \lambda_2(T), \\
\dot{\lambda}_2' &= -\frac{\partial}{\partial p} H^1(t, p, \lambda_1, \lambda_2, \lambda_3, u^1, u^2); \;\; \lambda_2(0) = 0;
\end{aligned}$$

and with S^1 taken as an open set, the Stackelberg open-loop control of **P1** satisfies the relations

$$\nabla_{u^1} H^1 = 0, \qquad \nabla_{u^2} H^2 = 0.$$

These results are now summarized in Thm. 7.5 below.

Theorem 7.5 *For the class of two-person differential games under consideration in this subsection, assume, in addition to i), ii) and iii) of Lemma 7.3, that*

iv) $f(t, \cdot, u^1, u^2)$ *is twice continuously differentiable on* $\mathbf{R}^n$ $\forall t \in [0, T]$,

v) $g^2(t, \cdot, u^1, u^2)$ *and* $q^2(\cdot)$ *are twice continuously differentiable on* $\mathbf{R}^n$ $\forall t \in [0, T]$,

vi) $g^1(t, \cdot, u^1, u^2)$ *and* $q^1(\cdot)$ *are twice continuously differentiable on* $\mathbf{R}^n$ $\forall t \in [0, T]$, *vii)* $f(t, x, \cdot, u^2)$, $g^i(t, x, \cdot, u^2)$, $i = 1, 2$, *are continuously differentiable on* $\mathbf{R}^{m_1}$ $\forall t \in [0, T]$,

viii) S^1 *is an open set.*

Then, if $\{\gamma^{i*}(t, x_0) = u^{i*}(t); i = 1, 2\}$ *provides an* open-loop Stackelberg solution with **P1** as the leader, *and* $\{x^*(t), 0 \le t \le T\}$ *denotes the corresponding state trajectory, there exist continuously differentiable functions* $p(\cdot)$, $\lambda_1(\cdot)$, $\lambda_2(\cdot)$: $[0, T] \to \mathbf{R}^n$, *and a continuous function* $\lambda_3(\cdot)$: $[0, T] \to \mathbf{R}^{m_2}$, *such that the following relations are satisfied:*

$$\begin{aligned}
&\dot{x}^*(t) = f(t, x^*(t), u^{1*}(t), u^{2^*}(t)); \quad x^*(0) = x,\\
&\dot{p}'(t) = -\frac{\partial}{\partial x} H^2(t, p, x^*, u^{1^*}, u^{2^*}); \quad p'(T) = \frac{\partial}{\partial x} q^2(x^*(T)),\\
&\dot{\lambda}_1'(t) = -\frac{\partial}{\partial x} H^1(t, p, \lambda_1, \lambda_2, \lambda_3, x^*, u^{1^*}, u^{2^*}),\\
&\lambda_1'(T) = \frac{\partial}{\partial x} q^1(x^*(T)) - \frac{\partial^2}{(\partial x)^2} q^2(x^*(T))\lambda_2(T), \qquad (7.63)\\
&\dot{\lambda}_2'(t) = -\frac{\partial}{\partial p} H^1(t, p, \lambda_1, \lambda_2, \lambda_3, x^*, u^{1^*}, u^{2^*}); \quad \lambda_2(0) = 0,\\
&\nabla_{u^1} H^1(t, p, \lambda_1, \lambda_2, \lambda_3, x^*, u^{1^*}, u^{2^*}) = 0,\\
&\nabla_{u^2} H^1(t, p, \lambda_1, \lambda_2, \lambda_3, x^*, u^{1^*}, u^{2^*}) = \nabla_{u^2} H^2(t, p, x^*, u^{1^*}, u^{2^*}) = 0,
\end{aligned}$$

where H^2 *is defined in Lemma 7.3, and* H^1 *is defined by (7.62).*

Proof. Since the optimization problem faced by the leader is (7.61a)-(7.61d), the theorem follows, in view of Lemma 7.3, from a standard optimal control result that can, for instance, be found on p. 65 of the text by Bryson and Ho (1975). □

As a specific application of this theorem, we now consider the special class of linear-quadratic differential games (cf. Def. 6.5), wherein we also assume $R^{12}(\cdot) > 0$, $Q^i(\cdot) \ge 0$, $i = 1, 2$. The Hamiltonians H^1 and H^2 can be written as

$$\begin{aligned}
H^2 &= \frac{1}{2}(x'Q^2x + u^{2'}R^{22}u^2 + u^{1'}R^{21}u^1) + p'(Ax + B^1u^1 + B^2u^2),\\
H^1 &= \frac{1}{2}(x'Q^1x + u^{1'}R^{11}u^1 + u^{2'}R^{12}u^2) + \lambda_1'(Ax + B^1u^1 + B^2u^2)\\
&\quad -\lambda_2'(Q^2x + A'p) + \lambda_3'(R^{22}u^2 + B^{2'}p)
\end{aligned}$$

and thus the set of relations (7.62) reduce to (with the arguments suppressed)

$$\begin{aligned}
\dot{x}^* &= Ax^* + B^1u^{1*} + B^2u^{2*}; \quad x^*(0) = x_0, &(7.64a)\\
\dot{p} &= -Q^2x^* - A'p; \quad p(T) = Q_f^2x^*(T), &(7.64b)\\
\dot{\lambda}_1 &= -Q^1x^* - A'\lambda_1 + Q^2\lambda_2; &(7.64c)\\
\lambda_1(T) &= Q_f^1x^*(T) - Q_f^2\lambda_2(T), &\\
\dot{\lambda}_2 &= A\lambda_2 - B^2\lambda_3; \quad \lambda_2(0) = 0, &(7.64d)\\
u^{1*} &= -R^{11^{-1}}B^{1'}\lambda_1, &(7.64e)
\end{aligned}$$

$$u^{2*} = -R^{22^{-1}}B^{2'}p, \tag{7.64f}$$
$$\lambda_3 = -R^{22^{-1}}(B^{2'}\lambda_1 + R^{12}R^{22^{-1}}B^{2'}p). \tag{7.64g}$$

Hence, Thm. 7.5 says, for this special case, that every open-loop Stackelberg solution of the linear-quadratic differential game is determined by the set (7.64a)-(7.64g). Because of the specific structure of the problem, we can actually obtain a stronger result, which is that the linear quadratic differential game admits a *unique* solution, and this, in turn, implies that (7.64a)-(7.64g) admits a unique solution set. To establish this result, let us first observe that[87] there exist, for the linear-quadratic problem, bounded linear operators $L_1^1 : U^1 \to U^1$, $L_1^2 : U^1 \to U^1$, $L_2^1 : U^2 \to U^2$, $L_2^2 : U^2 \to U^2$, $L_{12}^1 : U^2 \to U^1$, $L_{12}^2 : U^2 \to U^1$, with $L_2^2 > 0$, $\begin{pmatrix} L_1^1 & L_{12}^1 \\ L_{12}^{1*} & L_2^1 \end{pmatrix} > 0$ and elements $r_1^i \in U^1$, $r_2^i \in U^2$, $r_3^i \in \mathbf{R}$, $i = 1, 2$, so that

$$L^i(u^1, u^2) = \frac{1}{2}\langle u^1, L_1^i u^1\rangle_1 + \langle u^1, L_{12}^i u^2\rangle_1 + \frac{1}{2}\langle u^2, L_2^i u^2\rangle_2$$
$$+\langle u^1, r_1^i\rangle_1 + \langle u^2, r_2^i\rangle_2 + r_3^i, i = 1, 2,$$

where

$$\langle u, v\rangle_i = \int_0^T u'(t)v(t)\,dt$$

denotes the inner product of two vectors in U^i which is assumed to be structured as a complete vector space. Now, by hypothesis, $L^2(u^1, \cdot)$ is quadratic and strictly convex on U^2, and $L^1(\cdot, \cdot)$ is quadratic and strictly convex on $U^1 \times U^2$, and therefore it follows from Prop. 4.7, together with Remark 4.8 (interpreted in infinite dimensional spaces), that the linear-quadratic Stackelberg game under consideration admits a unique solution given by

$$u^{1*} = -L^{-1}(L_{12}^2 L_2^{2^{-1}} L_2^1 L_2^{2^{-1}} r_2^2 - L_{12}^{1^*} L_2^{2^{-1}} r_2^2 + r_1^1 - L_{12}^2 L_2^{2^{-1}} r_2^1),$$
$$u^{2*} = -(L_2^2)^{-1}(L_{12}^{2*} u^{1*} + r_2^2),$$

where

$$L \triangleq L_1^1 + L_{12}^2 L_2^{2^{-1}} L_2^1 L_2^{2^{-1}} L_{12}^{2^*} - L_{12}^{1*} L_2^{2^{-1}} L_{12}^{2*} - L_{12}^2 L_2^{2^{-1}} L_{12}^1 > 0.$$

This solution, written in operator form, should clearly correspond to the one given by (7.64a)-(7.64g), and this can also be verified directly (see Simaan and Cruz (1973a)).

Finally, let us note that, since the solutions of the coupled set of differential equations (7.64a)-(7.64d) depend linearly on x_0, and since $x^*(t)$ can be recovered

[87]The analysis to follow is in the same spirit as the proof of Thm. 6.14, and requires some prior knowledge of functional analysis. Our treatment here is rather informal and without details; for a more complete version the reader is referred to Simaan and Cruz (1973a). Furthermore, for the terminology and notation used here, the reader is referred to the proof of Thm. 6.14.

from $x^*(s)$ by linear invertible transformation for all t, $s \geq 0$, the set (7.64a)-(7.64d) can be replaced by a set of matrix differential equations independent of x_0, which are obtained by letting $\lambda_1(t) = \Lambda_1(t)x^*(t)$, $\lambda_2(t) = \Lambda_2(t)x^*(t)$, $p(t) = P(t)x^*(t)$, where Λ_1, Λ_2 and P are the corresponding matrices. This set of coupled matrix differential equations is

$$\begin{aligned}
&\dot{P} + P(A - B^1 R^{11^{-1}} B^{1'} \Lambda_1 - B^2 R^{22^{-1}} B^{2'} P) + A'P + Q^2 = 0; \\
&\qquad\qquad\qquad\qquad\qquad\qquad\qquad\qquad P(T) = Q_f^2, \\
&\dot{\Lambda}_1 + \Lambda_1(A - B^1 R^{11^{-1}} B^{1'} \Lambda_1 - B^2 R^{22^{-1}} B^{2'} P) + A'\Lambda_1 \\
&\qquad\qquad + Q^1 - Q^2\Lambda_2 = 0; \;\; \Lambda_1(T) = Q_f^1 - Q_f^2 \Lambda_2(T), \\
&\dot{\Lambda}_2 + \Lambda_2(A - B^1 R^{11^{-1}} B^{1'} \Lambda_1 - B^2 R^{22^{-1}} B^{2'} P) - A\Lambda_2 \\
&\qquad + B^2 R^{22^{-1}} R^{12} B^{2'} P - B^2 R^{22^{-1}} B^{2'} \Lambda_1 = 0; \;\; \Lambda_2(0) = 0.
\end{aligned} \tag{7.65}$$

The following theorem now summarizes the result.

Theorem 7.6 *The two-person linear-quadratic differential game (cf. Def. 6.5) characterized by the additional parametric restrictions $R^{12}(\cdot) > 0$, $Q^i \geq 0$, $Q_f^i \geq 0$, $i = 1, 2$, admits a unique open-loop Stackelberg solution with* **P1** *acting as the leader, which is given by*

$$\begin{aligned}
\gamma^{1*}(t, x_0) &= -R^{11^{-1}}(t) B^{1'}(t) \Lambda_1(t) x^*(t), \\
\gamma^{2*}(t, x_0) &= -R^{22^{-1}}(t) B^{2'}(t) P(t) x^*(t),
\end{aligned}$$

where P and Λ_1 are uniquely determined from (7.65), and $x^*(\cdot)$ satisfies the differential equation

$$\dot{x}^*(t) = (A - B^1 R^{11^{-1}} B^{1'} \Lambda_1 - B^2 R^{22^{-1}} B^{2'} P) x^*(t); \; x^*(0) = x_0.$$

Proof. It follows from the preceding discussion. □

Remark 7.14 All results of this subsection (with the exception of uniqueness of follower response strategy) are valid if the follower is allowed to have access to CLPS information, and the discussion of Remark 7.1 has a natural counterpart here. □

7.6.2 The CLPS information pattern

In discrete-time dynamic games (see Sections 7.3, 7.4), the CLPS information (for the leader) has led to two types of equilibria—the global Stackelberg and feedback Stackelberg solutions. In the continuous time, we have direct counterparts of these, which we briefly discuss below.

Global Stackelberg solution

For the global Stackelberg solution, a direct approach is again quite unwieldy, since the optimum response of the follower to an arbitrary CLPS strategy of the leader cannot be obtained using standard methods of optimal control—the major difficulty stemming from the fact that a general control strategy for the leader will also incorporate memory, and may even be discontinuous. To circumvent this difficulty two alternatives exist.

One alternative is to make some structural assumption on the permissible strategies of the leader (such as, taking γ^1 to be a general CLNM policy, or one that is parameterized in terms of its dependence on the current and past values of the state), which then would make it possible to write down a set of necessary (and/or sufficient) conditions for the follower's reaction function to satisfy. Then, the leader will have to solve a dynamic optimization problem over an infinite dimensional strategy space subject to these infinite dimensional dynamic constraints. A theory for such nonclassical control problems can be found in Papavassilopoulos and Cruz (1979a), which is also applied there to Stackelberg problems. If, on the other hand, the strategy set of the leader is parameterized (say, by a finite dimensional vector), then the leader will have to solve an optimization problem on the corresponding parameter space, which might be more feasible[88] but definitely sub-optimal.

The second alternative approach to the global Stackelberg solution is the counterpart of the methodology discussed in Section 7.4, which is to find an equivalent team problem the global solution of which forms a tight lower bound on the leader's cost, and then to obtain a particular representation of the team solution (for the leader) on the optimum team trajectory that achieves this bound. This methodology has been applied in (Başar and Olsder, 1980b) to linear-quadratic differential games, and conditions have been obtained for a finite dimensional linear representation (i.e., a linear dynamic compensator) to provide a global Stackelberg solution. Counterparts of these results when the leader has access to sampled state information has been presented in (Başar, 1981b); here representation of the leader's team strategy will have to meet the additional side condition that it can depend only on the current and past sampled values of the state. Details of these derivations can be found in the references cited.

Feedback Stackelberg solution

For the feedback Stackelberg solution, we have to extend the definition from discrete to continuous time. Let us recall that in discrete time a feedback Stackelberg solution is one that retains the Stackelberg property at every stage—with the leader having only *stagewise* advantage over the follower.

The continuous-time problem can be viewed as the limit of the discrete-time game as the number of stages becomes unbounded in a finite interval, which means that two consecutive decision points get arbitrarily close to each other. Hence, in a continuous-time dynamic game stagewise advantage of the leader (on the follower) turns into instantaneous advantage. Formally, the feedback

[88]This has been demonstrated by Medanić (1977) for linear-quadratic games by taking the initial state uniformly distributed on the unit sphere.

Stackelberg solution in a differential game can be obtained as the limit of the feedback Stackelberg solutions of a sequence of discrete-time dynamic games, each one obtained by time-discretization of the original differential game, with the $k+1$st game in the sequence corresponding to a finer discretization (sampling) than the kth one. One possible construction would be to divide the finite interval $[0,T]$ into k uniform subintervals, and to assume that over the ℓth subinterval $\left[\frac{T}{k}(\ell-1), \frac{T}{k}\ell\right)$, where $0<\ell\leq k$, the players' policies depend only on the values of state prior to and including $t=\frac{T}{k}(\ell-1)$, and that the leader can enforce his policy on the follower over each such interval—thus defining the kth game in the sequence. If necessary, one can take a subsequence of these games, corresponding to $k=1,2,4,8,\ldots$, so that the information set of each time-discretized game is strictly included in the information set of the next such game. An appropriate terminology to use for these games is "sampled-state", as the only time-discretization is in the information set, which comprises the state vector.

The feedback Stackelberg solution of each state-sampled game can be obtained by again using a dynamic programming-type argument, by solving a sequence of open-loop Stackelberg games of the type covered in Section 7.2, with the one on the time interval $\left[\frac{T}{k}\ell, \frac{T}{k}(\ell+1)\right)$ having the initial state $x\left(\frac{T}{k}\ell\right)$. The Stackelberg costs associated with this open-loop game (which will be expressed in terms of $x\left(\frac{T}{k}\ell\right)$, will constitute the terminal state cost part of the cost functions of the respective players in the open-loop game defined on the next subinterval (in retrograde time) $\left[\frac{T}{k}(\ell-1), \frac{T}{k}\ell\right)$.

A complete analysis of the convergence of these solutions as $k\to\infty$ is beyond the scope of our coverage here; but, if a limit exits, then the limiting solution should involve solutions of a sequence of open-loop Stackelberg games, each one defined on an infinitesimally small subinterval, which means that we now have to obtain Stackelberg solutions based on incremental costs at each time t. If $V^i(t,x)$ denotes the feedback Stackelberg cost-to-go of Pi at time t, at state x, then the counterpart of the pair of Nash equilibrium PDEs (6.52) are in this case (for the game described by (7.59a)-(7.59b))

$$\begin{Bmatrix} -\dfrac{\partial V^1(t,x)}{\partial t} \\[2ex] -\dfrac{\partial V^2(t,x)}{\partial t} \end{Bmatrix} = \mathsf{sol} \begin{Bmatrix} \dfrac{\partial V^1(t,x)}{\partial x} f(t,x,u^1,u^2) + g^1(t,x,u^1,u^2) \\[2ex] \dfrac{\partial V^2(t,x)}{\partial x} f(t,x,u^1,u^2) + g^2(t,x,u^1,u^2) \end{Bmatrix}$$
$$V^i(T,x) = q^i(x), \qquad i=1,2, \tag{7.66}$$

where "sol" stands for the static Stackelberg solution with **P1** as the leader. By analogy with the discrete-time case, we will call any set of policies obtained from (7.66) the *continuous-time feedback Stackelberg solution*, which is clearly *strongly time consistent* (by definition).

To bring (7.66) to a more explicit form, let us introduce the instantaneous reaction function of **P2** by

$$T^2(t,x;u^1,\frac{\partial V^2}{\partial x}) = \arg\min_{u^2\in S^2}\left[\frac{\partial V^2}{\partial x}f(t,x,u^1,u^2)+g^2(t,x,u^1,u^2)\right]. \quad (7.67)$$

Substituting this into the RHS of the first line of (7.66) and minimizing the resulting expression over $u^1 \in S^1$ yields the instantaneous Stackelberg solution for the leader:

$$\begin{aligned}\gamma^{1^*}(t,x) \;=\; \arg\min_{u^1\in S^1} & \left[\frac{\partial V^1}{\partial x}f(t,x,u^1,T^2(t,x;u^1,\frac{\partial V^2}{\partial x}))\right.\\ & \left. +g^2(t,x,u^1,T^2(t,x;u^1,\frac{\partial V^2}{\partial x}))\right]\end{aligned} \quad (7.68a)$$

with the corresponding one for the follower being

$$\gamma^{2^*}(t,x) \;=\; T^2(t,x;\gamma^{1^*}(t,x),\frac{\partial V^2}{\partial x}). \quad (7.68b)$$

Using these, (7.66) can equivalently be written as

$$\begin{aligned}\frac{\partial V^i(t,x)}{\partial t} \;&=\; \frac{\partial V^i(t,x)}{\partial x}f(t,x,\gamma^{1^*}(t,x),\gamma^{2^*}(t,x))\\ &\quad +g^i(t,x,\gamma^{1^*}(t,x),\gamma^{2^*}(t,x))\\ V^i(T,x) \;&=\; q^i(x), \qquad i=1,2,\end{aligned} \quad (7.69)$$

which is a pair of coupled PDEs, whose solution yields the feedback Stackelberg cost-to-go pair. Hence, we have the following.

Proposition 7.7 *For the differential game described by (7.59a)-(7.59b), and under the CLPS information pattern, the pair of policies (7.68a)-(7.68b) constitutes a feedback Stackelberg solution, where $V^i(t,x)$, satisfying (7.69), is the corresponding cost-to-go function for* **P***i* $(i = 1,2)$.

Since the asymmetry in the roles of the players in a continuous-time feedback Stackelberg solution is only incremental, one may be led to the conclusion that the feedback Stackelberg solution should coincide with (or be very close to) the feedback Nash solution (cf. Thm. 6.16). This, however, is not necessarily the case as the feedback Nash solution corresponds to choosing the static sol operator in (7.66) as Nash equilibrium, whereas in the present case it is the Stackelberg equilibrium, and the two are not generally the same. To illustrate this point, as well as the possibility that they may sometimes be equivalent, we consider in the sequel a class of linear-quadratic games with a coupling term between the controls of the players:

$$f \;=\; Ax+B^1u^1+B^2u^2, \quad (7.70a)$$

$$g^i \;=\; \frac{1}{2}(x'Q^ix+u^{i'}u^i-2u^{i'}R_{ij}u^j),\; j\neq i, i,j=1,2; \quad (7.70b)$$

$$q^i \;=\; \frac{1}{2}x'Q^i_fx, \quad S^i=\mathbf{R}^{m_i},\; i=1,2. \quad (7.70c)$$

Then, the instantaneous optimum response function of the follower is

$$T^2 = R_{21}u^1 + B^{2'}\left(\frac{\partial V^2}{\partial x}\right)'$$

leading to the following structure for (7.68a)-(7.68b):

$$\begin{aligned}\gamma^{1^*}(t,x) &= -(I - R_{12}R_{21} - R_{21}'R_{12}')^{-1}, \\ &\quad \left[R_{12}B^{2'}\left(\frac{\partial V^2}{\partial x}\right)' + \left(B^{1'} + R_{21}'B^{2'}\right)\left(\frac{\partial V^2}{\partial x}\right)'\right] \\ &\triangleq -K_1(\partial V^1/\partial x)' - K_2(\partial V^2/\partial x)', \end{aligned} \tag{7.71a}$$

$$\begin{aligned}\gamma^{2^*}(t,x) &= -R_{21}K_1\left(\partial V^1/\partial x\right)' - (B^{2'} + R_{21}K_2)(\partial V^2/\partial x)' \\ &\triangleq -L_1(\partial V^1/\partial x)' - L_2(\partial V^2/\partial x)'. \end{aligned} \tag{7.71b}$$

The PDEs (7.69) then become

$$\begin{aligned}\frac{\partial V^1}{\partial t} &= -\frac{\partial V^1}{\partial x}\left[Ax + \bar{B}_1\left(\frac{\partial V^1}{\partial x}\right)' + \bar{B}^2\left(\frac{\partial V^2}{\partial x}\right)'\right] - \frac{1}{2}x'Q^1x \\ &-\frac{1}{2}\left[K_1\left(\frac{\partial V^1}{\partial x}\right)' + K_2\left(\frac{\partial V^2}{\partial x}\right)'\right]'\left[K_1\left(\frac{\partial V^1}{\partial x}\right)' + K_2\left(\frac{\partial V^2}{\partial x}\right)'\right] \\ &+\left[K_1\left(\frac{\partial V^1}{\partial x}\right)' + K_2\left(\frac{\partial V^2}{\partial x}\right)'\right]' R_{12}\left[L_1\left(\frac{\partial V^1}{\partial x}\right)' - L_2\left(\frac{\partial V^2}{\partial x}\right)'\right], \end{aligned} \tag{7.72a}$$

$$\begin{aligned}\frac{\partial V^2}{\partial t} &= -\frac{\partial V^2}{\partial x}\left[Ax + \bar{B}_1\left(\frac{\partial V^1}{\partial x}\right)' + \bar{B}^2\left(\frac{\partial V^2}{\partial x}\right)'\right] - \frac{1}{2}x'Q^2x \\ &-\frac{1}{2}\left[L_1\left(\frac{\partial V^1}{\partial x}\right)' + L_2\left(\frac{\partial V^2}{\partial x}\right)'\right]'\left[L_1\left(\frac{\partial V^1}{\partial x}\right)' + L_2\left(\frac{\partial V^2}{\partial x}\right)'\right] \\ &+\left[L_1\left(\frac{\partial V^1}{\partial x}\right)' + L_2\left(\frac{\partial V^2}{\partial x}\right)'\right]' R_{21}\left[K_1\left(\frac{\partial V^1}{\partial x}\right)' - K_2\left(\frac{\partial V^2}{\partial x}\right)'\right], \end{aligned} \tag{7.72b}$$

$$V^i(T,x) = (1/2)x'Q_f^i x, \qquad i = 1,2, \tag{7.72c}$$

where

$$\bar{B}_1 \triangleq -(B_1K_1 + B_2L_1) \qquad \bar{B}_2 \triangleq -(B_1K_2 + B_2L_2). \tag{7.73}$$

For the feedback Nash equilibrium solution, however, the relevant set of PDEs is in the same form as (7.72a)-(7.72c) but with K_1, K_2, L_1, L_2, $\bar{B}_1$ and $\bar{B}_2$, respectively, replaced by the "hat'ted" quantities

$$\begin{aligned} &\hat{K}_1 = (I - R_{12}R_{21})^{-1}B^{1'}, && \hat{K}_2 = (I - R_{12}R_{21})^{-1}R_{12}B^{2'}, \\ &\hat{L}_1 = (I - R_{21}R_{12})^{-1}R_{21}B_1', && \hat{L}_2 = (I - R_{21}R_{12})^{-1}B^{2'}, \\ &\hat{\bar{B}}_1 = B^1\hat{K}_1 + B^2\hat{L}_1, && \hat{\bar{B}}_2 = B^1\hat{K}_2 + B^2\hat{L}_2. \end{aligned}$$

Note that if the cross terms in the cost functions are absent (i.e., $R_{12} = 0$, $R_{21} = 0$), we have the simple relations

$$K_1 = \hat{K}_1 = B^{1'}, \quad K_2 = \hat{K}_2 = 0,$$
$$L_1 = \hat{L}_1 = 0, \qquad L_2 = \hat{L}_2 = B^{2'}$$

which imply that the two sets of PDEs become identical, thus admitting the same set of solutions. If the cross terms are not absent, however, the two sets of PDEs are intrinsically different and admit different sets of solutions. Hence, even in linear-quadratic games with generalized quadratic cost functionals, the feedback Stackelberg and Nash solutions may be different.

We now conclude this subsection by reporting a result on the existence and structure of the feedback Stackelberg solution of the linear-quadratic differential game described by (7.70a)-(7.70c).

Proposition 7.8 *If T is sufficiently small and the matrix inverse in (7.71a) exists, the linear-quadratic differential game described by (7.70a)-(7.70c), and with* **P1** *as the leader, admits a feedback Stackelberg solution given by*

$$\gamma^{1^*}(t,x) = -(K_1P_1 + K_2P_2)x, \tag{7.74a}$$
$$\gamma^{2^*}(t,x) = -(L_1P_1 + L_2P_2)x, \tag{7.74b}$$

where $\{P_1(t), P_2(t)\}$ are symmetric solutions of the set of coupled matrix Riccati differential equations

$$\begin{aligned}\dot{P}_1 = {} & -P_1L - L'P_1 - Q^1 - (K_1P_1 + K_2P_2)'(K_1P_1 + K_2P_2) \\ & +(L_1P_1 + L_2P_2)'R_{12}'(K_1P_1 + K_2P_2) \\ & +(K_1P_1 + K_2P_2)'R_{12}(L_1P_1 + L_2P_2); \qquad P_1(T) = Q_f^1,\end{aligned} \tag{7.75a}$$

$$\begin{aligned}\dot{P}_2 = {} & -P_2L - L'P_2 - Q^2 - (L_1P_1 + L_2P_2)'(L_1P_1 + L_2P_2) \\ & +(L_1P_1 + L_2P_2)'R_{21}(K_1P_1 + K_2P_2) \\ & +(K_1P_1 + K_2P_2)'R_{21}'(L_1P_1 + L_2P_2); \qquad P_2(T) = Q_f^2,\end{aligned} \tag{7.75b}$$

$$L \triangleq A + \bar{B}_1P_1 + \bar{B}_2P_2. \tag{7.75c}$$

Proof. This result follows by substituting $V^i = (1/2)x'P_ix$ into the pair of PDEs (7.72a)-(7.72b) and observing that (7.75a)-(7.75b) imply satisfaction of (7.72a)-(7.72b) by such quadratic cost-to-go functions. Existence of a (unique) solution to (7.75a)-(7.75b) when T is sufficiently small follows from a standard property of ordinary differential equations with continuous right-hand sides. □

Remark 7.15 Proposition 5.1 has a natural counterpart in the context of feedback Nash equilibria, simply with K_i, L_i and $\bar{B}_i$ replaced by their corresponding "hat'ted" versions introduced earlier. The solution to the resulting set of Riccati equations will *not* be the same as the solution to (7.75a)-(7.75b), unless $R_{21} = 0$, $R_{12} = 0$. For the latter case, these equations are equivalent to the two-player version of (6.17a) with $R^{12} = R^{21} = 0$. □

7.7 Problems

1. Show that, for the special case of two-person zero-sum dynamic games (i.e., with $L^1 \equiv -L^2$), Thm. 7.1 coincides with Thm. 6.3, and Corollary 7.1 coincides with Thm. 6.4.

2. Obtain the open-loop Stackelberg solution of the two-stage dynamic game described by (7.23), (7.24a) and (7.24b) when (i) **P1** is the leader, (ii) **P2** is the leader. For what values of the parameter β is the open-loop Stackelberg solution concurrent? (See Section 4.5 for terminology.)

3. Obtain the feedback Stackelberg solution of the dynamic game of the previous problem under the CLPS information pattern when (i) **P1** is the leader, (ii) **P2** is the leader.

4. Consider the scalar K-stage dynamic game described by the state equation

$$x_{k+1} = 0.5x_k + u_k^1 - 2u_k^2, \quad x_1 = 1,$$

and cost functionals

$$\begin{aligned} L^1 &= \sum_{k=1}^{K} [(x_k)^2 + (u_k^1)^2 + 2(u_k^2)^2], \\ L^2 &= \sum_{k=1}^{K} [(x_k)^2 + 0.5(u_k^2)^2 + 1.5(u_k^1)^2]. \end{aligned}$$

Obtain the global and feedback Stackelberg solutions of this dynamic game under CLPS information pattern when (i) **P1** is the leader, (ii) **P2** is the leader. Compare the realized values of the cost functions of the players under these different Stackelberg solutions for the cases when $K = 5$ and $K \to \infty$.

5. Consider the two-stage dynamic game characterized by the two-dimensional state equation

$$\begin{aligned} x_2 &= x_1 - u^2, \quad x_1 = (1,1)', \\ x_3 &= x_1 - (1,1)'u^1 \end{aligned}$$

and cost functionals

$$\begin{aligned} L^1 &= x_3'x_3 + (u^1)^2 + u^{2'} \begin{pmatrix} 2 & 0 \\ 0 & 1 \end{pmatrix} u^2, \\ L^2 &= x_3' \begin{pmatrix} 1 & 0 \\ 0 & 2 \end{pmatrix} x_3 + u^{2'} \begin{pmatrix} 1 & 0 \\ 0 & 2 \end{pmatrix} u^2. \end{aligned}$$

Here, u^1 is the scalar control variable of **P1** (the leader) who has access to

$$y = (1,1)x_2,$$

and u^2 is the two-dimensional control vector of **P2** (the follower) who has open-loop information. Show that the global minimum of L^1 cannot constitute a tight lower bound on the leader's Stackelberg cost function and a tight lower bound can be obtained by first minimizing L^2 over $u^2 \in \mathbf{R}^2$ subject to the constraint

$$2 - (1,1)u^2 = y$$

and then minimizing L^1 over $u^1 \in \mathbf{R}$ and $y \in \mathbf{R}$, with u^2 determined as above. (For more details on this problem and the method of solution, see Başar (1980a, 1982).)

6. Consider the scalar dynamic game of subsection 7.4.1 with $x_1 = 1$, $\beta = 2$, and under the additional restriction $-1 \le u^i \le 0$, $i = 1,2$. Show (by graphical display) that the Stackelberg cost of the leader is higher than the global minimum value of his cost function, and a strategy that leads to this cost value (which is $J_1^* = 8/9$) is $y_1^* = 1/3$.

7. Prove Prop. 7.4 by carrying out the two steps preceding it in subsection 7.5.1.

8. Consider the scalar stochastic two-stage dynamic game described by the state equation

$$x_{k+1} = 0.5x_k + u_k^1 - 2u_k^2 + \theta_k, \quad x_1 = 1$$

and cost functionals

$$\begin{aligned} L^1 &= (x_3)^2 + \sum_{k=1}^{2}[(u_k^1)^2 + 2(u_k^2)^2], \\ L^2 &= (x_3)^2 + \sum_{k=1}^{2}[1.5(u_k^1)^2 + 0.5(u_k^2)^2], \end{aligned}$$

where θ_1 and θ_2 are statistically independent Gaussian variables with mean zero and equal variance 1. The underlying information structure is CLPS for both players.

Determine the feedback Stackelberg solution and the linear (global) Stackelberg solution with **P1** as the leader. Can the leader improve his performance in the latter case by employing nonlinear strategies?

9. Obtain the linear (global) Stackelberg solution of the previous problem when the stochastic terms are correlated so that $E[\theta_1\theta_2] = 1$ (in other words, there exists a single Gaussian variable θ with mean zero and variance 1, so that $\theta_1 = \theta_2 = \theta$).

10. Consider the three-player incentive problem where **P1** is the leader and **P**i, $i = 2,3$, are the followers. The controls are $u^1 = (u_1^1, u_2^1)$ for the leader

and u^i (scalar quantities) for **P**i, $i = 2, 3$. The cost functions are

$$\begin{aligned} L^1 &= (u_1^1)^2 + (u_2^1)^2 + (u^2)^2 + (u^3)^2; \\ L^2 &= u_1^1 - 3u_2^1 + (u^2 - 1)^2 + (u^3 - 1)^2; \\ L^3 &= u_1^1 + u_2^1 + (u^2 + 1)^2 + (u^3 + 1)^2. \end{aligned}$$

Determine an incentive mechanism $u_i^1 = \gamma_i^1(u^2, u^3)$, $i = 1, 2$, such that the two followers face a team problem; i.e., after substitution of the incentive policy into the cost functions of the followers, these cost functions become identical. Moreover, the solution of this resulting "team" problem of the two followers must lead to the team solution of the leader, i.e., $u_1^1 = u_2^1 = u^2 = u^3 = 0$. Next determine another incentive mechanism such that the followers face a zero-sum game and such that the saddle-point solution of this game leads to the team solution of the leader.

11. Consider the two-person incentive problem with **P1** the leader and **P2** the follower. The decision variables are scalars and the cost functions are:

$$\begin{aligned} L^1 &= (u^1)^2 + (u^2)^2; \\ L^2 &= (u^1 - \theta)^2 + (u^2 + 1)^2, \end{aligned}$$

where θ is a zero-mean Gaussian random variable with unit covariance. The value of θ is known to **P2** only, i.e., the information is nonnested. Show that any continuous and deterministic incentive mechanism γ^1 will not lead to $L^1 = 0$ (depict the contours for the possible L^2-functions). If, however, the leader applies the following incentive mechanism in mixed strategies:

$$\mu^1 = \begin{cases} 0 & \text{w.p.} \quad 1 \quad \text{for} \quad u^2 \geq 0; \\ -Nu^2 & \text{w.p.} \quad 0.5 \quad \text{for} \quad u^2 < 0; \\ +Nu^2 & \text{w.p.} \quad 0.5 \quad \text{for} \quad u^2 < 0\,, \end{cases}$$

where N is a constant larger than 1, show that it indeed leads to achievement of his optimum team cost.

12. Extend the result of Thm. 7.5 to generalized linear-quadratic differential games, where

$$\begin{aligned} f &= Ax + B^1u^1 + B^2u^2 + c, \\ g^i &= \frac{1}{2}(x'Q^ix + \sum_{j=1}^{2} u^{j'}[R_i^{ji}u^i + Y_i^jx]),\ i = 1, 2; \\ q^i &= \frac{1}{2}x'Q_f^ix, \quad S^i = \mathbf{R}^{m_i},\ i = 1, 2. \end{aligned}$$

Here $c(\cdot)$ is a continuous vector-valued function of dimension n, and the entries of all matrices are continuous in t.

What conditions on the weighting matrices in the cost functions will ensure the existence and uniqueness of open-loop Stackelberg solution?

13. Obtain the counterpart of Prop. 7.7 for a three-player differential game with CLPS information pattern (with general state dynamics, and general cost functions), where **P1** is the leader, and **P2** and **P3** are followers who play according to the feedback Nash equilibrium for every announced strategy of the leader. Furthermore, the equilibrium solution concept adopted between the two levels of the hierarchy is the feedback Stackelberg.

    Under what conditions on the state dynamics and/or the cost functions would the equilibrium strategies of the players (obtained under this two-level hierarchy) also constitute a Nash equilibrium solution for the three-player differential game (without any decision hierarchy)?

7.8 Notes

Sections 7.2 and 7.3. The open-loop Stackelberg solution concept in infinite dynamic games was first treated in the continuous time in the works of Chen and Cruz (1972) and Simaan and Cruz (1973a,b), where the latter has also introduced the feedback Stackelberg concept in the context of finite and discrete-time infinite dynamic games. Some other references that discuss the open-loop and feedback Stackelberg solutions in discrete-time infinite dynamic games are Kydland (1975), Hämäläinen (1976), Başar (1979b) and Cruz (1978), where the latter also discusses derivation of feedback Stackelberg solution of continuous-time dynamic games under sampled-data state information. For applications in microeconomics, see Okuguchi (1976).

Section 7.4. Derivation of the global Stackelberg solution in infinite dynamic games with CLPS information has remained a challenge for a long time, because of the difficulties explained in this section prior to subsection 7.4.1 and illustrated in subsection 7.4.1. The indirect approach presented in this section was first introduced in Başar and Selbuz (1979a,b) and later in Tolwinski (1981b) and Başar (1980a); the specific examples included in subsections 7.4.1 and 7.4.2 are taken from Başar (1980a, 1982). This approach was extended later to more general (infinite dimensional Hilbert and Banach) spaces in (Zheng and Başar, 1982). The analysis of subsection 7.4.3 follows closely the one of Başar and Selbuz (1979a) which also includes an extension to $N(>2)$-person games with one leader and $N-l$ followers, and with the followers playing according to the Nash equilibrium solution concept among themselves. Extensions of this approach to other types of $N(>2)$-person dynamic games (with different types of hierarchy in decision making) can be found in Başar (1981a). Other representative articles devoted to this topic are Başar (1981b), Olsder (1977a), Ho, Luh and Olsder (1982) and Tolwinski (1980, 1981a). Subsection 7.4.4 on incentives contains material from Ho, Luh and Olsder (1982); see this reference, as well as Ho and Olsder (1981) and Zheng, Başar and Cruz (1984) for results on multi-stage decision problems, the nonnested case, and problems with multiple hierarchies. The method presented in Example 7.4 as to what the leader can achieve if his team solution is beyond reach is due to *Tolwinski*. The fact that the optimum incentive mechanism is generally nonunique has opened the possibilities for further refinement, by choosing the one that is least sensitive to changes in the values of some parameters defining the game. Some results along this direction have been presented in (Cansever and Başar, 1983).

Section 7.5. This section follows the lines of Başar (1979b) and extends some of the results given there. The difficulties illustrated by Example 7.5 were first pointed out in Castanon (1976). For a discussion of the derivation of the Stackelberg solution when players have access to noisy (but redundant) state information, see Başar (1979a); this is related to the topic of stochastic incentives as discussed in subsection 7.5.3. Furthermore, Başar (1979c) discusses the feedback Stackelberg solution in linear-quadratic stochastic dynamic games with noisy observation. A general existence theory for stochastic incentive problems, with nested information, can be found in Başar (1984), which was later extended to problems with multiple levels of hierarchy in Başar (1989a). Cansever and Başar (1985a,b) discuss a further refinement of the notion of optimum schemes in stochastic incentive problems based on insensitivity to system parameters.

Section 7.6. The open-loop Stackelberg solution to two-player differential games was first studied by Chen and Cruz (1972), and subsequently refined by Simaan and Cruz (1973a). Wishart and Olsder (1979) have shown (based on necessary conditions) that the open-loop Stackelberg solution of some simple investment problems could be discontinuous. Details of the indirect approach discussed in subsection 7.6.2 for the global Stackelberg solution of differential games with CLPS information pattern can be found in Başar and Olsder (1980b) and Papavassilopoulos and Cruz (1980). The material on the continuous-time feedback Stackelberg solution in this subsection is from Başar and Haurie (1984). This reference, besides developing a theory for the feedback Stackelberg solution in differential games, has also introduced and analyzed a general framework for differential games where the role of leadership, and the nature of the solution concept to be adopted by the players change as the game evolves, with the change being prompted by the past policy choices of the players, as well as the outcome of a chance mechanism (modeled as a jump process). An application of the continuous-time feedback Stackelberg solution in economics can be found in (Başar, Haurie and Ricci, 1985).

Chapter 8

Pursuit-Evasion Games

8.1 Introduction

This chapter deals with two-person deterministic zero-sum differential games with variable terminal time. We have already discussed zero-sum differential games in Chapter 6, but as a special case of nonzero-sum games and with fixed terminal time. The class of problems to be treated in this chapter, however, are of the pursuit-evasion type for which the duration of the game is not fixed. Actually, it was through this type of problems (i.e., through the study of pursuit and evasion between two objects moving according to simple kinematic laws) that the theory of differential games was started in the early 1950s. Extensions to nonzero-sum dynamic games, as treated in Chapters 6 and 7, were subsequently considered in the late 1960s.

Section 8.2 discusses the necessary and sufficient conditions for existence of saddle-point equilibrium strategies. Sufficiency conditions are provided by a natural two-person extension of the Hamilton–Jacobi–Bellman equation, which is called the "Isaacs equation", after Isaacs—the acknowledged father of pursuit-evasion games. A geometric derivation of the Isaacs equation is given, which utilizes the principle of dynamic programming, and by means of which the concept of semipermeable surfaces is introduced; such surfaces play an important role in the remaining sections of the chapter. Subsequently, necessary conditions are derived, which form the two-person extension of the Pontryagin minimum principle (cf. Section 5.5). These conditions are valid not for feedback strategies but for their open-loop representations, and in order to obtain the feedback strategies these open-loop solutions have to be synthesised. We also briefly discuss upper and lower value functions (which are important if the so-called Isaacs condition does not hold), and their relation with viscosity solutions.

In Section 8.3, we treat "capturability", which addresses the question of whether the pursuer can "catch" the evader or not. The answer to this is completely determined by the kinematics, initial conditions and the target set; the cost function does not play any role here. It is in this section that the homicidal chauffeur game and the two-cars model are introduced.

In Section 8.4, we return to the derivation of saddle-point strategies for quantitative games. What makes the application of necessary and sufficient conditions intellectually interesting and challenging is the presence of a large number of singular surfaces. These surfaces are manifolds in the state space across which the backward solution process of dynamic programming fails because of discontinuities in the value function or in its derivatives.

The complexity that arises in the derivation of the solution of a differential game due to the presence of singular surfaces is best illustrated by the game of the "lady in the lake," which is the main topic of Section 8.5. In Section 8.6, a problem in maritime collision avoidance is treated: under what circumstances can one ship avoid collision with another? The approach taken is one of worst-case analysis, which quite naturally leads to a differential game formulation. The solution method, however, is also applicable to other collision avoidance problems which do not necessarily feature noncooperative decision making.

In Section 8.7 we address the problem of role determination in a differential game wherein the roles of the players (viz. pursuing or evading) are not specified at the outset, but rather are determined as functions of the initial conditions. Role determination is applied to an aeronautical problem, which involves a dogfight between two airplanes.

8.2 Necessary and Sufficient Conditions for Saddle-Point Equilibria

The systems considered in this chapter can all be described by

$$\dot{x}(t) = f(t, x(t), u^1(t), u^2(t)), \quad x(0) = x_0, \tag{8.1}$$

where $x(t) \in S^\circ \subset \mathrm{R}^n$, $u^i(t) \in S^i \subset \mathrm{R}^{m_i}$, $i = 1, 2$. The function f is continuous in t, u^1 and u^2, and is continuously differentiable in x. The first player, **P1**, who chooses u^1, is called the *pursuer*, which we shall abbreviate as **P**. The second player is called the *evader* and we shall refer to him as **E** instead of **P2**. The final time T of the game is defined by

$$T = \inf\{t \in \mathrm{R}^+ : (x(t), t) \in \Lambda\}, \tag{8.2}$$

where Λ is a closed subset, called the target set, in the product space $S^\circ \times \mathrm{R}^+$. The boundary $\partial\Lambda$ of Λ is assumed to be an n-dimensional manifold, i.e., a hypersurface, in the product space $\mathrm{R}^+ \times \mathrm{R}^n$, characterized by a scalar function $\ell(t, x) = 0$. This function is assumed to be continuous in t and continuously differentiable in x, unless stated differently.

Since all differential games in this chapter will be of the zero-sum type, we have a single objective function

$$L(u^1, u^2) = \int_0^T g(t, x(t), u^1(t), u^2(t))\,\mathrm{d}t + q(T, x(T)), \tag{8.3}$$

where g is continuous in t, u^1 and u^2, and continuously differentiable in x; q is continuous in T and continuously differentiable in $x(T)$. We define J as the cost function of the game in normal form, which is determined from L in terms of the strategies $\gamma^i(t, x(t)) = u^i(t)$, $i = 1, 2$. Only feedback strategies will be considered, and we assume $\gamma^i(t, x)$ to be piecewise continuous in t and x. With the permissible strategy spaces denoted by Γ^1 and Γ^2, let us recall that a strategy pair $(\gamma^{1*}, \gamma^{2*})$ is in (saddle-point) equilibrium if

$$J(\gamma^{1*}, \gamma^2) \leq J(\gamma^{1*}, \gamma^{2*}) \leq J(\gamma^1, \gamma^{2*}) \tag{8.4}$$

for all $\gamma^i \in \Gamma^i$. The reason for considering feedback strategies, and not strategies of a more general class (such as strategies with memory), lies in the continuous-time counterpart of Thm. 6.9: as long as a saddle point in feedback strategies exists, it is not necessary to consider saddle points with respect to other classes of strategies. Actually, as it will be shown in the sections to follow, "solutions" to various problems are, in general, first obtained in open-loop strategies, which may then be synthesized to feedback strategies, provided that they both exist.

8.2.1 The Isaacs equation

In Corollary 6.6 a sufficiency condition (in the form of the solution of a partial differential equation) was obtained for the feedback saddle-point equilibrium of zero-sum differential games with a fixed final time. It will now be shown that this result essentially holds true also if the final time, instead of being fixed, is determined by the terminal constraint (8.2). The underlying idea here, too, is the principle of optimality; wherever the system is at any given arbitrary time, from then onwards the pursuer (respectively, evader), minimizes (respectively, maximizes) the remaining portion of the cost function under the feedback information structure. The function describing the minimax (upper) value of the cost function, when started from the position (t, x), is

$$\begin{aligned}\bar{V}(t, x) = \min_{\gamma^1} \max_{\gamma^2} \Big\{ \int_t^T & g(s, x(s), \gamma^1(s, x(s)), \gamma^2(s, x(s)))\, \mathrm{d}s \\ & + q(T, x(T)) \Big\},\end{aligned} \tag{8.5}$$

which is called the *value function*. Under the assumption that such a function exists and is continuously differentiable in x and t, it satisfies the partial differential equation

$$-\frac{\partial \bar{V}}{\partial t} = \min_{u^1 \in S^1} \max_{u^2 \in S^2} \left[\frac{\partial \bar{V}}{\partial x} f(t, x, u^1, u^2) + g(t, x, u^1, u^2) \right], \tag{8.6}$$

which is known as the Isaacs equation. The lower (maximin) value function is defined analogously, with (8.5) and (8.6) replaced, respectively, by

$$\begin{aligned}\underline{V}(t, x) = \max_{\gamma^1} \min_{\gamma^2} \Big\{ \int_t^T & g(s, x(s)), \gamma^1(s, x(s)), \gamma^2(s, x(s))\, \mathrm{d}s \\ & + q(T, x(T)) \Big\}\end{aligned} \tag{8.7}$$

$$-\frac{\partial \underline{V}}{\partial t} = \max_{u^2\in S^2} \min_{u^1\in S^1} \left[\frac{\partial \underline{V}}{\partial x} f(t,x,u^1,u^2) + g(t,x,u^1,u^2)\right]. \tag{8.8}$$

If the differential game has equal upper and lower values, then we denote this common value by V, which satisfies (8.6) or equivalently (8.8), with the ordering of the *min* and *max* operations being irrelevant here.[89] For future reference, let us rewrite this PDE explicitly:

$$\begin{aligned} -\frac{\partial V}{\partial t} &= \min_{u^1\in S^1} \max_{u^2\in S^2} \left[\frac{\partial V}{\partial x} f(t,x,u^1,u^2) + g(t,x,u^1,u^2)\right] \\ &= \max_{u^2\in S^2} \min_{u^1\in S^1} \left[\frac{\partial V}{\partial x} f(t,x,u^1,u^2) + g(t,x,u^1,u^2)\right]. \end{aligned} \tag{8.9}$$

We now give a geometrical derivation of (8.9) in the case of a variable terminal time (as determined by (8.2)). It is also possible to extend the derivation given in Thm. 6.16 by allowing the terminal time to be variable, but we prefer the geometric derivation here, since it provides more insight into the problem, in particular if there is only a terminal pay-off. Accordingly, we assume now that $g \equiv 0$, which in fact leads to no loss of generality because of the argument of Remark 5.1 (suitably modified to apply to continuous-time systems).

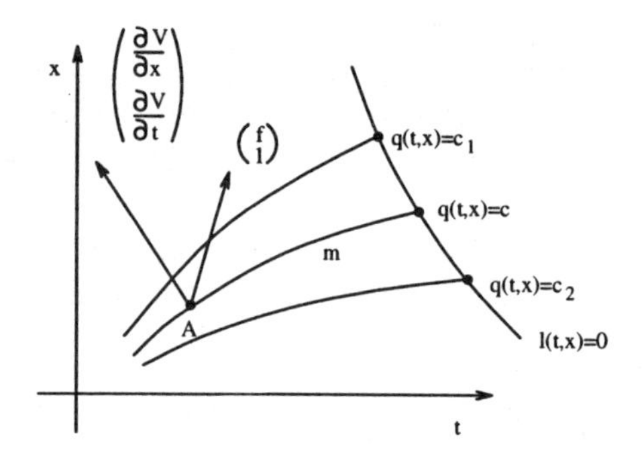

Figure 8.1: Geometrical derivation of the Isaacs equation, where $c_1 > c > c_2$.

The value function V, if it exists, is then defined by

$$V(t,x) = \min_{\gamma^1} \max_{\gamma^2} q(T,x(T)) = \max_{\gamma^2} \min_{\gamma^1} q(T,x(T)), \tag{8.10}$$

when the initial position is (t,x). One particular saddle-point trajectory is sketched in Fig. 8.1; it is indicated by curve m and the value corresponding to any point of m is c. We now make the assumption that for initial points in a certain neighborhood below m, $V(t,x) < c$, and similarly for initial points

[89]If not explicitly stated otherwise, it will be assumed henceforth that the upper and lower values are equal, that is, the *max* and *min* operations in (8.5) (or (8.7)) commute. Only in the next subsection, where we deal with viscosity solutions, will these operations not necessarily commute.

above m, $V(t,x) > c$.[90] At point A, on curve m, **P** will choose $u^1 \in S^1$ so as to make the inner product between $(\partial V/\partial x, \partial V/\partial t)'$ and $(f',1)'$ as small as possible, i.e., he tries to make the vector $(f',1)'$ point downward towards points corresponding to lowest possible costs. On the other hand, **E** wants to choose $u^2 \in S^2$ in such a way so as to maximize this inner product, because then the system will move in **E**'s favorite direction, i.e., towards points corresponding to the highest possible costs. Hence u^{1*} and u^{2*} are chosen as the arguments of

$$\min_{u^1} \max_{u^2} \left(\frac{\partial V}{\partial x} f + \frac{\partial V}{\partial t} \right) = \max_{u^2} \min_{u^1} \left(\frac{\partial V}{\partial x} f + \frac{\partial V}{\partial t} \right). \tag{8.11}$$

Since m is an equilibrium trajectory, the system will move along m if u^{1*} and u^{2*} are applied, and that results in

$$\min_{u^1 \in S^1} \max_{u^2 \in S^2} \left(\frac{\partial V}{\partial x} f + \frac{\partial V}{\partial t} \right) = 0, \tag{8.12}$$

which is equivalent to (8.6) if $g \equiv 0$.

If **E** plays optimally, i.e., he chooses u^{2*}, then

$$\frac{\partial V}{\partial x} f(t,x,u^1,u^{2*}) + \frac{\partial V}{\partial t} \geq 0 \tag{8.13}$$

for all $u^1 \in S^1$, which implies that the outcome of the game will be $\geq c$, since the system will move into the area where $V \geq c$. However, **P** can keep the outcome at c by playing u^{1*}, since then the equality sign will apply in (8.13). Thus, without **E**'s cooperation (i.e., without **E** playing non-optimally), **P** cannot make the system move from A towards the area where $V < c$. Analogously, by interchanging the roles of the players in this discussion, **E** cannot make the system move from A in the "upward" direction if **P** plays optimally. Only if **P** plays non-optimally, **E** will be able to obtain an outcome higher than c. Because of these features, the n-dimensional manifold in $\mathrm{R}^n \times \mathrm{R}^+$, comprising all initial positions (t,x) with the same value, is called *semipermeable*, provided that $g = 0$.

In conclusion, if the value function $V(t,x)$ is continuously differentiable, then the saddle-point strategies are determined from (8.9). This equation also provides a sufficiency condition for saddle-point strategies as stated in the following theorem, the proof of which is a straightforward two-player extension of Thm. 5.3; compare it also with Corollary 6.6.

Theorem 8.1 *If (i) a continuously differentiable function $V(t,x)$ exists that satisfies the Isaacs equation (8.9), (ii) $V(T,x) = q(T,x)$ on the boundary of the target set, defined by $\ell(t,x) = 0$, and (iii) either $u^{1*}(t) = \gamma^{1*}(t,x)$, or $u^{2*}(t) = \gamma^{2*}(t,x)$, as derived from (8.9), generates trajectories that terminate in finite time (whatever γ^2, respectively γ^1, is), then $V(t,x)$ is the value function and the pair $(\gamma^{1*}, \gamma^{2*})$ constitutes a saddle point.*

[90] If this assumption does not hold (which is the case of V being constant in an $(n+1)$-dimensional subset of $\mathrm{R}^n \times \mathrm{R}^+$) the derivation to follow can easily be adjusted.

Remark 8.1 The underlying assumption of interchangeability of the *min* and *max* operations in the Isaacs equation is often referred to as the *Isaacs condition.* The slightly more general condition

$$\min_{u^1} \max_{u^2} [p'f + g] = \max_{u^2} \min_{u^1} [p'f + g],$$

for all n-vectors p, is sometimes also referred to as the Isaacs condition. Regardless of which definition is adopted, the Isaacs condition will hold if both f and g are separable in u^1 and u^2, i.e., they can be written as

$$\begin{aligned} f(t,x,u^1,u^2) &= f_1(t,x,u^1) + f_2(t,x,u^2), \\ g(t,x,u^1,u^2) &= g_1(t,x,u^1) + g_2(t,x,u^2). \end{aligned}$$

If the Isaacs condition does not hold, and if only feedback strategies are allowed, then one has to seek for equilibria in the general class of mixed strategies defined on feedback strategy spaces. In this chapter, we will not extend our investigation to mixed strategies, and deal only with the class of differential games for which the Isaacs condition holds (almost) everywhere.

Note, however, that, even if separability holds, this does not necessarily mean that the order of the actions of the players is irrelevant, since the underlying assumption of V being continuously differentiable may not always be satisfied. In regions where V is not continuously differentiable, the order in which the players act may be crucial. □

In the derivation of saddle-point strategies, equation (8.9) cannot directly be used, since V is not known at the outset. An alternative is to use Thm. 8.2, given below, which provides a set of necessary conditions for an *open-loop representation (or realization) of the feedback saddle-point solution.* In spite of the fact that it does not deal with feedback solutions directly, the theorem is extremely useful in the computation (synthesis) of feedback saddle-point solutions. It can be viewed as the variable terminal-time extension of Thm. 6.13 and the two-player extension of Thm. 5.4. The proof follows by direct application of Thm. 5.4, which is also valid under the weaker assumption of f and g being measurable in t (instead of being continuous in t) (see Berkovitz (1974), p. 52).

Theorem 8.2 *Given a two-person zero-sum differential game, described by (8.1)-(8.4), suppose that the pair $\{\gamma^{1*}, \gamma^{2*}\}$ provides a saddle-point solution in feedback strategies, with $x^*(t)$ denoting the corresponding state trajectory. Furthermore, let its open-loop representation $\{u^i(t) = \gamma^i(t, x^*(t)), i = 1,2\}$ also provide a saddle-point solution (in open-loop policies). Then there exists a costate func-*

tion $p(\cdot):[0,T]\to \mathbf{R}^n$ such that the following relations are satisfied:

$$\left.\begin{array}{rcl} \dot{x}^*(t) & = & f(t,x^*(t),u^{1*}(t),u^{2*}(t)),\ x^*(0)=x_0, \\ & & H(t,p,x^*,u^{1*},u^2)\le H(t,p,x^*,u^{1*},u^{2*})\le H(t,p,x^*,u^1,u^{2*}), \\ & & \qquad \forall u^1\in S^1, \qquad \forall u^2\in S^2, \\ \dot{p}'(t) & = & -\dfrac{\partial}{\partial x}H(t,p(t),x^*(t),u^{1*}(t),u^{2*}(t)), \\ p'(T) & = & \dfrac{\partial}{\partial x}q(T,x^*(T))\ \ along\ \ell(T,x)=0, \end{array}\right\} \quad (8.14)$$

where

$$H(t,p,x,u^1,u^2)\triangleq g(t,x,u^1,u^2)+p'f(t,x,u^1,u^2).$$

Remark 8.2 Provided that both open-loop and feedback equilibria exist, conditions of Thm. 8.2 will lead to the desired feedback saddle-point solution. Even if an open-loop saddle-point solution does not exist, this method can still be utilized to obtain the feedback saddle-point solution, and, by and large, this has been the method of approach adopted in the literature for solving pursuit-evasion games. For an example of a differential game in which this method is utilized, but an open-loop (pure) saddle point does not exist, see the "lady in the lake" game treated in Section 8.5.

Also, for certain classes of problems, the value function $V(t,x)$ is not continuously differentiable, and therefore the sufficiency conditions of Thm. 8.1 cannot be employed. Sufficiency conditions for certain types of nonsmooth $V(t,x)$ have been discussed in Bernhard (1977), but they are beyond the scope of our treatment here. □

In the literature on pursuit-evasion differential games, it is common practice to write V_x instead of p' (for the costate vector), and this will also be adopted here. The reader should note, however, that in this context, V_x is only a function of time (but not of x), just the way p' is. After synthesis, the partial derivative with respect to x, of the value function corresponding to the feedback equilibrium solution, coincides with this function of time, thus justifying the notation.

We now illustrate the procedure outlined above by means of a simple example. As with most other problems to be considered in this chapter, this example is time-invariant, and consequently the value function and the saddle-point strategies do not explicitly depend on t (cf. Remark 5.5). Therefore the t-dependence has been dropped from the notation.

Example 8.1 Consider the two-dimensional system described by

$$\left.\begin{array}{rcl} \dot{x}_1 & = & (2+\sqrt{2})u^2-\sin u^1, \\ \dot{x}_2 & = & -2-\cos u^1, \end{array}\right\} \quad (i)$$

where the initial value of the state lies in the half plane $x_2>0$, u^2 satisfies the constraint $|u^2|\le 1$ and no restrictions are imposed on u^1. The target set is

$x_2 \leq 0$ and the objective functional is $L = x_1(T)$. Note that the game terminates when the x_1-axis is reached, and since $\dot{x}_2 \leq -1$, it will always terminate. In order to apply Thm. 8.2, we first write the Hamiltonian H for this pursuit-evasion game as

$$H = V_{x_1}\left\{\left(2+\sqrt{2}\right)u^2 - \sin u^1\right\} + V_{x_2}\{-2-\cos u^1\},$$

where $V_{x_1}(\equiv p_1)$ and $V_{x_2}(\equiv p_2)$ satisfy, according to (8.14),

$$\dot{V}_{x_1} = 0, \quad \dot{V}_{x_2} = 0. \qquad (ii)$$

Since the value function does not explicitly depend on time, we have $V(x_1, x_2 = 0) = x_1$, and hence $V_{x_1} = 1$ along the x_1-axis. The final condition for V_{x_2} along the x_1-axis cannot be obtained directly; it will instead be obtained through the relation $\min_{u^1} \max_{u^2} H = 0$ at $t = T$. Carrying out these min-max operations, we get $u^2 = \text{ sgn } (V_{x_1}) = 1$ and hence $\gamma^{2*}(x) = 1$, and the vector $(\sin u^1, \cos u^1)$ is parallel to (V_{x_1}, V_{x_2}). Substitution of these equilibrium strategies into H leads to

$$2 + \sqrt{2} - \sqrt{(1 + V_{x_2}^2)} - 2V_{x_2} = 0,$$

which yields $V_{x_2} = 1$ at $t = T$. From (ii) it now follows that $V_{x_1} = 1$ and $V_{x_2} = 1$ for all t. In this example it is even possible to obtain V explicitly: $V(x_1, x_2) = x_1 + x_2$. Hence, the feedback saddle-point strategies are constants: $\gamma^{1*}(x) = \pi/4$, $\gamma^{2*}(x) = 1$. The corresponding state trajectories are straight lines, making an angle of $\pi/4$ radians with the negative x_2-axis (see Fig. 8.2).

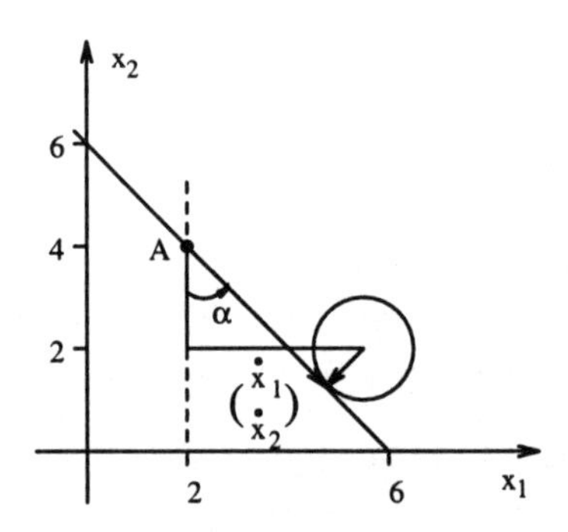

Figure 8.2: Vectogram for Example 8.1.

Figure 8.2 also helps to visualize the geometric derivation of the saddle-point strategies. Suppose that the system is at point A corresponding to the value 6, say. Emanating from point A, the velocity vector $(\dot{x}_1, \dot{x}_2)'$ has been drawn in terms of the vectors $(0, -2)'$, $((2+\sqrt{2})\gamma^{2*}, 0)'$ and $(-\sin\gamma^{1*}, -\cos\gamma^{1*})'$, which add up to the RHS of (i). **E**, the maximizer, would like to make the angle α (see Fig. 8.2) as large as possible. The best he can do, irrespective of **P**'s decision, is to choose $\gamma^{2*} = +1$. If **E** would play $\gamma^2 \neq 1$, i.e., $\gamma^2 < 1$, the angle α will become smaller, and consequently the outcome will become smaller than 6.

Similarly, **P**, the minimizer, would like to have a vector $(\dot{x}_1, \dot{x}_2)'$ with angle α as small as possible, which leads to his optimal strategy $\gamma^{1*} = \pi/4$; for any other γ^1, the vector $(\dot{x}_1, \dot{x}_2)'$ will point more to the right, and this is exactly what **P** wants to avoid.

All conditions of Thm. 8.1 are satisfied and therefore the strategies obtained indeed constitute a saddle-point solution in feedback strategies. □

In addition to the necessary conditions of Thm. 8.2, we now exemplify some of the intricacies which might occur with respect to termination.

Example 8.2 Consider the one-dimensional system

$$\dot{x} = u^1 + u^2, \quad x(0) = x_0 > 0,$$

where the controls u^1 and u^2 are constrained to $-1 \leq u^i \leq 0$, $i = 1, 2$. The target set is the half line $x \leq 0$. The cost functional is

$$L = \int_0^T (1 - x)\, dt.$$

The costate equation for this problem is

$$\dot{p} = -\frac{\partial H}{\partial x} = 1,$$

and the optimal $\{u^{1*}, u^{2*}\}$ satisfy

$$u^{1*} = \begin{cases} -1 \text{ if } p > 0, \\ \text{undetermined if } p = 0, \\ 0 \text{ if } p < 0, \end{cases} \quad u^{2*} = \begin{cases} -1 \text{ if } p < 0, \\ \text{undetermined if } p = 0, \\ 0 \text{ if } p > 0. \end{cases} \tag{i}$$

How do we determine $p(T)$? The appropriate condition in (8.14) cannot be used since that condition, which gives $p(T)$ along the boundary of the target set, requires variations in x within $\ell(T, x) = 0$; but this latter relation only yields $x = 0$. In its stead, we make use of the definition of the derivative for $p(T)$ and also make use of the functional form $V(\Delta x) = \Delta x - \frac{1}{2}(\Delta x)^2$ for sufficiently small (positive) Δx (which is obtained through some analysis) to arrive at the value $p(T) = 1$. (See Problem 1 in Section 8.8 for another derivation.)

Integration of the costate equation and substitution into (i) leads to

$$u^{1*} = \gamma^{1*}(x) = -1, u^{2*} = \gamma^{2*}(x) = 0, \text{ for } 0 \leq x < 1. \tag{ii}$$

For $x = 1$ we have $p = 0$ and (i) does not determine the equilibrium strategies. For $\gamma^{i*} = 0$ $(i = 1, 2)$, the system would remain at $x = 1$ and the game would not terminate in finite time. To exclude the occurrence of such singular cases, we have to include an additional restriction in the problem statement, which forces **P** to terminate the game in finite time; this can be achieved by defining $L = \infty$ at $T = \infty$. Then, at state $x = 1$, the preceding restriction forces **P** to choose $\gamma^{1*}(x) = -1$. Now we can integrate further in retrograde time. For

$t \le T-1$, $p(t) = V_x(x^*(t))$ changes sign, and therefore, for $x > 1$, we obtain from (i):

$$u^{1*} = \gamma^{1*}(x) = 0, \quad u^{2*} = \gamma^{2*}(x) = -1, \quad x > 1. \qquad (iii)$$

Now the question is whether the strategy pair $\{\gamma^{1*}, \gamma^{2*}\}$, defined by (ii) and (iii), constitutes a saddle point. To this end, Thm. 8.1 will be consulted. The value function is easily determined to be $V(x) = x(2-x)/2$, and all conditions of Thm. 8.1 hold, except condition (iii). The reason why condition (iii) does not hold is because, for initial points $x > 1$, the strategy pair $\{\gamma^{1*}, \gamma^{2*}(\cdot) \equiv 0\}$ does not lead to termination in finite time, and hence, by playing $\gamma^2 \equiv 0$, **E** would be much better off. However, in such a situation, **P** can do much better if he is allowed to have an informational advantage in the form

$$u^1 = \tilde{\gamma}^1(x(t), u^2(t)), \qquad (iv)$$

i.e., at each instant of time, **P** is informed of **E**'s current decision (open-loop value).[91] If **P** chooses his new strategy as

$$\tilde{\gamma}^{1*} = \begin{cases} -1 \text{ if } x \le 1 \text{ or } \left(x > 1 \text{ and } u^2 > -\frac{1}{2}\right), \\ 0 \text{ otherwise}, \end{cases}$$

the saddle-point equilibrium will be restored again by means of the pair $(\tilde{\gamma}^{1*}, \gamma^{2*})$, which follows from direct comparison.

Another way of getting out of the "termination dilemma" is to restrict the class of permissible strategies to "playable pairs", i.e., to pairs for which termination occurs in finite time (cf. Example 5.1 and Def. 5.7). Such an assumption excludes for instance the pair $\{\gamma^1 = 0, \gamma^2 = 0\}$ from consideration. □

8.2.2 Upper and lower values, and viscosity solutions

In the previous subsection it was tacitly assumed (with the exception of (8.6) and (8.8)) that the Isaacs condition holds. In situations where it does not, one common approach is to endow one player with an instantaneous informational advantage over the other player, which we now briefly discuss.[92] An assumption throughout this subsection (only) will be that the final time T is fixed. With Γ^i taken as the set of all open-loop controls for **P**i, let us introduce a mapping μ^2 from Γ^1 to Γ^2 with the property that for each $t \in [0,T]$ and $u^1, \overline{u}^1 \in \Gamma^1$ the following holds: if $u^1(\tau) = \overline{u}^1(\tau)$ for almost all $\tau \in [0,t]$, then $\mu^2[u^1](\tau) = \mu^2[\overline{u}^1](\tau)$ almost everywhere. In more popular terms this says that $u^2(t)$ will depend on the past and current values of u^1, but not on its future values. It should be emphasized that the current value of u^1 is part of the information for **P2**. The set of all such strategies μ^2 for **P2** will be denoted by Δ^2. Similarly, μ^1

[91] This is, of course, possible if **E** has access to only open-loop information, and can therefore only pick open-loop policies. If **E** has access to feedback information, however, (iv) is not physically realizable and a time delay has to be incorporated.

[92] At the outset, this is different from the case of minimax ((8.5)) or maximin ((8.7)) values where one of the players is assumed to know not only the current but also the future values of the other player's controls.

and Δ^1 are defined with the roles of the players reversed. Formally, a mapping μ^1 from Γ^2 to Γ^1 will be called a strategy for **P1** provided that for each $t \in [0,T]$ and $u^2, \overline{u}^2 \in \Gamma^2$ the following holds: If $u^2(\tau) = \overline{u}^2(\tau)$ for almost all $\tau \in [0,t]$, then $\mu^1[u^2](\tau) = \mu^1[\overline{u}^2](\tau)$ almost everywhere.

Definition 8.1 *The value V^+ of the zero-sum differential game described by (8.1) and (8.3), if the maximizer has instantaneous informational advantage over the minimizer, is defined as*

$$V^+(t,x) \triangleq \inf_{u^1\in\Gamma^1} \sup_{\mu^2\in\Delta^2} \left(\int_t^T g(t,x(t),u^1(t),\mu^2[u^1](t))\, dt + q(t,x(T)) \right),$$

and the value V^- of the same game, but now with the minimizer having instantaneous informational advantage over the maximizer, is defined as

$$V^-(t,x) \triangleq \sup_{u^2\in\Gamma^2} \inf_{\mu^1\in\Delta^1} \left(\int_t^T g(t,x(t),\mu^1[u^2](t),u^2(t))\, dt + q(t,x(T)) \right).$$

Remark 8.3 Clearly, $V^+(t,x) \geq V^-(t,x)$. □

Remark 8.4 In the study of the existence of strategies that achieve the values V^+ and V^-, the starting point is to partition the time interval of interest, $[0,T]$, into n subintervals each of length $\delta = T/n$. The informational advantage of one player with respect to the other then concerns the entire subinterval to which the current time belongs. Thus one can define the so-called upper δ-value and lower δ-value. Under some regularity conditions, it can be shown that in the limit as $n \to \infty$ and hence $\delta \to 0$, these values become equal to V^+ and V^-, respectively. □

Definition 8.2 *The* upper Hamiltonian *and the* lower Hamiltonian *are, respectively,*

$$H^+(x,t,p) = \min_{u^1\in S^1} \left[\max_{u^2\in S^2} (p'f(t,x,u^1,u^2) + g(t,x,u^1,u^2)) \right]$$

and

$$H^-(x,t,p) = \max_{u^2\in S^2} \left[\min_{u^1\in S^1} (p'f(t,x,u^1,u^2) + g(t,x,u^1,u^2)) \right].$$

Remark 8.5 To be very explicit, the maximizing u^2-value in the definition of H^+ will in general depend on the minimizing u^1-value, i.e., $u^2(u^1)$, whereas in the definition of H^- it will be exactly the other way around. Note also that $H^+ \geq H^-$ for all values of their arguments. □

The following theorem now makes the connection between Defs. 8.1 and 8.2, and also relates them to (8.5) and (8.7). The precise regularity conditions under which it is valid, as well as its proof can be found in Barron, Evans and Jensen (1984).

Theorem 8.3 *Subject to appropriate regularity conditions on the functions f, g and q, the functions V^+ and V^- are uniformly Lipschitz continuous, and hence differentiable almost everywhere, and they are solutions, respectively, of*

$$\frac{\partial V^+}{\partial t} + H^+\left(x, t, \frac{\partial V^+}{\partial x}\right) = 0, \qquad V^+(T, x) = q(T, x(T)),$$

$$\frac{\partial V^-}{\partial t} + H^-\left(x, t, \frac{\partial V^-}{\partial x}\right) = 0, \qquad V^-(T, x) = q(T, x(T)),$$

and hence are equal to $\bar{V}$ and $\underline{V}$, respectively.

One can also define viscosity solutions for zero-sum differential games. In Chapter 5 the starting point to define viscosity solutions for optimal control problems was (5.65) (or (5.66)). In zero-sum differential games the equivalent starting point is (8.6) and (8.8), or their time-invariant equivalents if V does not directly depend on t. Since (5.65) and (8.6) are similar PDEs, the theory of viscosity solutions as presented in Section 5.8 remains equally valid for differential games, simply with $\bar{H}$ replaced by $\bar{H}^+$ or $\bar{H}^-$. The following theorem, proved in Barron, Evans and Jensen (1984), and in Evans and Souganidis (1984) for problems with fixed terminal time, relates upper and lower value functions to viscosity solutions.

Theorem 8.4 *The upper value function V^+ is the viscosity solution of (8.6) and the lower value function V^- is the viscosity solution of (8.8).*

8.3 Capturability

Consider the class of pursuit evasion games formulated in Section 8.2, under the additional assumption of time-invariance, i.e., f, g and ℓ are independent of t, and q does not depend on T. The target set Λ is therefore a tube in $\mathrm{R}^n \times \mathrm{R}^+$, with rays parallel to the time axis. The projection of Λ onto the R^n-space is denoted by $\bar{\Lambda}$.

Before studying the derivation of a saddle-point solution, we must first address the more fundamental question of whether the target set can be reached at all. If it cannot be reached, we simply say that the pursuit-evasion game is not well defined. Accordingly, we deal, in this section, with the following qualitative pursuit-evasion game. The evader tries to prevent the state from reaching $\bar{\Lambda}$, whereas the pursuer seeks the opposite. Once it is known, for sure, that the target set can be reached, one can return to the original (quantitative) pursuit-evasion game and investigate existence and derivation of the saddle-point solution.

For the differential game of *kind* introduced above, we define an auxiliary cost functional as

$$J = \begin{cases} -1 \text{ if } (x(t), t) \in \Lambda \text{ for some } t, 0 \le t < \infty, \\ +1 \text{ otherwise,} \end{cases}$$

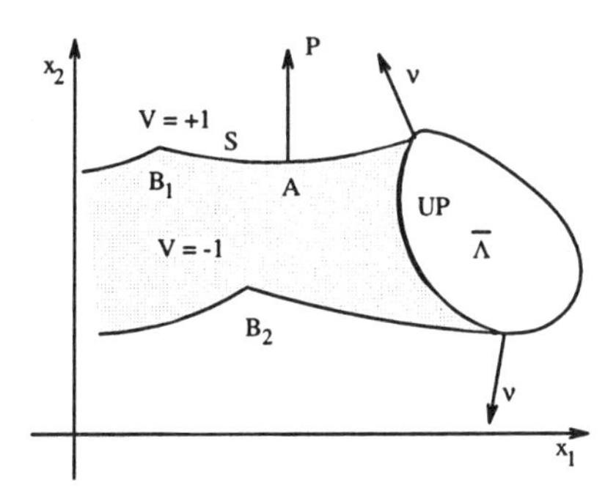

Figure 8.3: The construction of the barrier.

thus converting it into a differential game of *degree*. The value function for this differential game can take only two values, $V(x) = -1$ and $V(x) = +1$, where the former corresponds to initial states that lead to capture, while the latter corresponds to initial states from where capture cannot be assured. Consequently, the value function is discontinuous, and therefore the theory of subsection 8.2.1 is not directly applicable. It can, however, be modified suitably to fit the present framework, which we now discuss below.

Let us first note that only those points x of $\partial\bar{\Lambda}$ for which

$$\min_{u^1} \max_{u^2} \nu' f(x, u^1, u^2) = \max_{u^2} \min_{u^1} \nu' f(x, u^1, u^2) \leq 0 \tag{8.15}$$

are candidates for a terminal position of the game. Here, the n-dimensional vector ν is the outward normal of $\bar{\Lambda}$ at x (see Fig. 8.3).

If strict inequality holds in (8.15), then the state will penetrate $\bar{\Lambda}$; furthermore, the points x for which the equality sign holds may only be grazing or touching points. The set of all points x of $\partial\bar{\Lambda}$ that satisfy (8.15) is called the *usable part* (UP) of $\partial\bar{\Lambda}$. Let us suppose that the initial points situated in the shaded region of Fig. 8.3, and only those, can be steered to the target set by **P**, regardless of how **E** acts, and investigate whether we can determine the surface **S** that separates these initial states from the others which do not necessarily lead to termination. This surface **S** is sometimes referred to as a *barrier*. The points that are common to **S** and $\bar{\Lambda}$ are called the boundary of the usable part (BUP), it consists of those points of the UP for which the equality sign holds in (8.15).

Consider a point **A** on **S**, at which a tangent hyperplane relative to **S** exists. The outward normal at **A**, indicated by p, is unique, apart from its magnitude. The surface **S** is then determined by

$$\min_{u^1} \max_{u^2} p' f(x, u^1, u^2) = 0, \tag{8.16}$$

which is only a necessary condition and therefore does not characterize **S** completely. This semipermeable surface should in fact have the property that without **P**'s cooperation, **E** cannot make the state cross **S** (from the shaded area

$V = -1$ to the non-shaded area $V = +1$) and conversely, without **E**'s cooperation, **P** cannot make the system cross **S** (from $V = +1$ to $V = -1$). Hence if a "tangential" penetration of **S** is impossible, (8.16) is also sufficient.

Points of **S** at which no tangent hyperplane exists (such as $\mathbf{B}_1$ and $\mathbf{B}_2$ in Fig. 8.3) must be considered separately. Such points will be considered later, in Sections 8.6 and 8.7, while discussing intersections of several smooth semipermeable surfaces.

To construct **S**, we now substitute the saddle-point solution of (8.16), to be denoted by $u^{i*} = \gamma^{i*}(x)$, $i = 1, 2$, into (8.16) to obtain the identity

$$p'(x)f(x, \gamma^{1*}(x), \gamma^{2*}(x)) \equiv 0.$$

Differentiation with respect to x leads to (also in view of the relation $(\partial/\partial u^1)(p'f)\cdot(\partial\gamma^i/\partial x) = 0$, which follows from the discussion included in subsection 5.5.3 right after (5.42))

$$\frac{\mathrm{d}p}{\mathrm{d}t} = -\left(\frac{\partial f}{\partial x}\right)' p, \tag{8.17}$$

which is evaluated along **S**. The boundary condition for (8.17) is

$$p(T) = \nu, \tag{8.18}$$

where ν is the outward normal of $\bar{\Lambda}$ at **S**.

We now introduce various features of the barrier **S** by means of the following example, which is known as the "homicidal chauffeur game".

Example 8.3 Consider a pursuer and an evader, both moving in a two-dimensional plane (with coordinates x_1, x_2) according to

$$\left.\begin{array}{rclc} \dot{x}_{1_p} &=& v_1 \sin\theta_p, & \dot{x}_{1_e} = v_2 \sin\theta_e, \\ \dot{x}_{2_p} &=& v_1 \cos\theta_p, & \dot{x}_{2_e} = v_2 \cos\theta_e, \\ \dot{\theta}_p &=& \omega_1 u^1, & \dot{\theta}_e = \omega_2 u^2, \end{array}\right\} \tag{8.19}$$

where the subscripts p and e stand for pursuer and evader, respectively. The controls u^1 and u^2 satisfy the constraints $|u^i| \le 1$, $i = 1, 2$. The scalars v_1 and v_2 are the constant speeds, and ω_1 and ω_2 are the maximum angular velocities of **P** and **E**, respectively. Since the angular velocities are bounded, three coordinates (x_1, x_2 and θ) are needed for each player to describe his position. For $u^1 = 0$ (respectively, $u^2 = 0$), **P** (respectively, **E**) will move in a straight line in the (x_1, x_2) plane. The action $u = +1$ stands for the sharpest possible turn to the right; $u = -1$ analogously stands for a left turn. The minimum turn radius for **P** is given by $R_1 \triangleq \omega_1/v_1$ and for **E** by $R_2 \triangleq \omega_2/v_2$.

For this differential game, yet to be formulated, it is the relative positions of **P** and **E** that are important (and relevant), and not the absolute positions by means of which the model (8.19) has been described. Model (8.19) is six-dimensional; however, by means of the relative coordinates, defined as

$$\left.\begin{array}{rcl} x_1 &=& (x_{1_e} - x_{1_p})\cos\theta_p - (x_{2_e} - x_{2_p})\sin\theta_p, \\ x_2 &=& (x_{1_e} - x_{1_p})\sin\theta_p + (x_{2_e} - x_{2_p})\cos\theta_p, \\ \theta &=& \theta_e - \theta_p, \end{array}\right\} \tag{8.20}$$

it can be converted into a three-dimensional model, as to be shown in the sequel. In this relative coordinate system, (x_1, x_2) denotes **E**'s position with respect to **P** and the origin coincides with **P**'s position. The x_2-axis is aligned with **P**'s velocity vector, the x_1-axis is perpendicular to the x_2-axis (see Fig. 8.4), and the angle θ is the relative direction (heading) in which the players move.

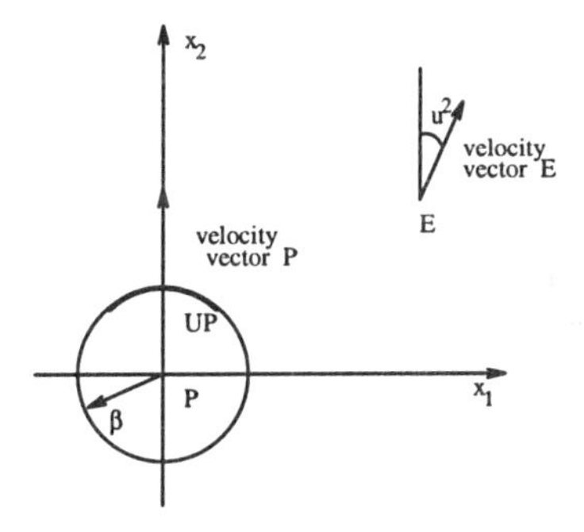

Figure 8.4: The relative coordinate system.

With respect to the relative coordinates (8.20), model (8.19) can be described as

$$\left.\begin{aligned} \dot{x}_1 &= -\omega_1 u^1 x_2 + v_2 \sin\theta, \\ \dot{x}_2 &= -v_1 + \omega_1 u^1 x_1 + v_2 \cos\theta, \\ \dot{\theta} &= \omega_2 u^2 - \omega_1 u^2, \end{aligned}\right\} \tag{8.21}$$

which is sometimes referred to as the *two-cars model.* We shall consider a special version of (8.21), namely the one in which **E** is allowed to change his direction instantaneously (i.e., $\omega_2 = \infty$). Then **E** will have complete control over θ, which we therefore take as **E**'s new control u^2. Also, normalizing both **P**'s speed and minimum turn radius to one, we obtain

$$\left.\begin{aligned} \dot{x}_1 &= -u^1 x_2 + v_2 \sin u^2, \\ \dot{x}_2 &= -1 + u^1 x_1 + v_2 \cos u^2, \end{aligned}\right\} \tag{8.22}$$

with no bounds imposed on u^2, while u^1 still satisfies the constraint $|u^1| \leq 1$. The two normalizations introduced above determine the distance scale and time scale, respectively. Figure 8.4 depicts model (8.22); note that the angle u^2 is measured clockwise with respect to the positive x_2-axis. This convention of measuring u^2 has been initiated by Isaacs (1975), and since then it has been adopted in the literature on pursuit-evasion games. Note that the coordinate system is affixed to **P** and moves along with **P** through the "real" space.

The target set $\bar{\Lambda}$ is defined through the inequality

$$x_1^2 + x_2^2 \leq \beta^2,$$

that is, **P** tries to get within a distance β of the evader, whereas the evader tries to avoid this. The game thus formulated is referred to as the *homicidal chauffeur game*, in which a chauffeur (**P**) tries to overrun a pedestrian (**E**). The homicidal chauffeur game is trivial for $v_2 > 1 = v_1$, since capture will then never be possible, provided that **E** plays the appropriate strategy. Therefore we shall henceforth assume $0 \leq v_2 < 1$.

The UP (part of the circle described by $x_1^2+x_2^2 = \beta^2$) is determined by those (x_1, x_2) for which the following relation holds (note that at the point (x_1, x_2), the vector $(\nu_1, \nu_2)' = (x_1, x_2)'$ is an outward normal to $\bar{\Lambda}$):

$$\begin{aligned}&\min_{u^1} \max_{u^2} \{x_1(-u^1 x_2 + v_2 \sin u^2) + x_2(-1 + u^1 x_1 + v_2 \cos u^2)\}\\&= -x_2 + v_2\beta \leq 0.\end{aligned}$$

Hence only those x_2 for which $x_2 > v_2\beta$ on the circle $x_1^2 + x_2^2 = \beta^2$ constitute the UP. Intuitively this is clear: directly in front of **P**, **E** cannot escape; if **E** is close to **P**'s side, he can side-step and avoid capture, at least momentarily; if **E** is behind **P**, i.e., $x_2 < 0$, there is no possibility for immediate capture.

In order to construct the surface **S**, we shall start from the end point $(x_1 = \beta\sqrt{(1 - v_2^2)}, x_2 = v_2\beta)$, on the capture set. Another branch of **S** can be obtained if we start from the other end of the UP, which is $(x_1 = -\beta\sqrt{(1 - v_2^2)}, x_2 = v_2\beta)$, but that one is the mirror image of the first one; i.e., if the first branch is given by $(\bar{x}_1(t), \bar{x}_2(t))$, then the second one will be given by $(-\bar{x}_1(t), \bar{x}_2(t))$. Therefore we shall carry out the analysis only for the "first" branch. Equation (8.16) determines **S** and reads in this case

$$\min_{u^1} \max_{u^2} \{p_1(-u^1 x_2 + v_2 \sin u^2) + p_2(-1 + u^1 x_1 + v_2 \cos u^2)\} \equiv 0,$$

whereby

$$\left.\begin{aligned}\sin u^{2*} &= \frac{p_1}{\sqrt{(p_1^2 + p_2^2)}}, \qquad \cos u^{2*} = \frac{p_2}{\sqrt{(p_1^2 + p_2^2)}},\\ u^{1*} &= \operatorname{sgn} s, \text{with}\\ s &\triangleq p_1 x_2 - x_1 p_2.\end{aligned}\right\} \tag{8.23}$$

The differential equations for p_1 and p_2 are, from (8.17),

$$\dot{p}_1 = -u^1 p_2, \quad \dot{p}_2 = u^1 p_1. \tag{8.24}$$

The final condition for p, as expressed by (8.18), will be normalized to unit magnitude, since only the direction of the vector p plays a role, i.e.,

$$p_1(T) = \sqrt{(1 - v_2^2)} \triangleq \cos\alpha; \quad p_2(T) = v_2 \triangleq \sin\alpha \tag{8.25}$$

for some appropriate angle α. In order to solve for $p(t)$, we need to know u^{1*}. At $t = T$, $s = 0$ and therefore $u^1(T)$ cannot be determined directly. For an indirect derivation we first perform

$$\frac{\mathrm{d}s}{\mathrm{d}t} = \frac{\mathrm{d}}{\mathrm{d}t}(p_1 x_2 - x_1 p_2) = -p_1. \tag{8.26}$$

Since $p_1 > 0$ at $t = T$, we have $\mathrm{d}s/\mathrm{d}t < 0$ at $t = T$, which leads to $s(t) > 0$ for t sufficiently close to T, and therefore $u^1 = +1$ just before termination. The solution to (8.24), together with (8.25) and $u^1 = +1$, now becomes

$$p_1(t) = \cos(t - T + \alpha), \quad p_2(t) = \sin(t - T + \alpha), \tag{8.27}$$

and the equations determining the branch of **S** emanating from $x_1 = \beta\sqrt{(1 - v_2^2)}$, $x_2 = \beta v_2$ become

$$\left.\begin{array}{rclrcl} \dot{x}_1 & = & -x_2 + v_2\cos(t - T + \alpha), & x_1(T) & = & \beta\sqrt{(1 - v_2^2)} \\ \dot{x}_2 & = & -1 + x_1 + v_2\sin(t - T + \alpha), & x_2(T) & = & \beta v_2, \end{array}\right\}$$

which are valid as long as $s(t) > 0$. The solution is

$$\left.\begin{array}{rcl} x_1(t) & = & (\beta + v_2(t - T))\cos(t - T + \alpha) + 1 - \cos(t - T), \\ x_2(t) & = & (\beta + v_2(t - T))\sin(t - T + \alpha) - \sin(t - T), \end{array}\right\} \tag{8.28}$$

which leaves the circle $x_1^2 + x_2^2 = \beta^2$ tangentially. What we have to ensure is that the trajectory (8.28) lies outside Λ. This is what we tacitly assumed in Fig. 8.3, but which unfortunately does not always hold true as we shall shortly see. If we define $r \triangleq \sqrt{(x_1^2 + x_2^2)}$, we require $\ddot{r} > 0$ at $t = T$ (note that $\dot{r} = 0$ at $t = T$), and some analysis then leads to the conclusion that this holds if, and only if,

$$\beta^2 + v_2^2 < 1. \tag{8.29}$$

For parameter values (β, v_2) for which $\beta^2 + v_2^2 > 1$, the semipermeable surface **S** moves inside $\bar{\Lambda}$ and thereby loses its significance.[93] In the latter case the proposed construction (which was to split up the state space into two parts, $V = +1$ and $V = -1$) obviously fails. In the present example we henceforth assume that (8.29) holds.

Let us now explore the trajectories described by (8.28). The following two possibilities may be distinguished: (i) $x_1(t_2) = 0$ for some $t_2 < T$ and (ii) $x_1(t) > 0$ for all $t < T$, i.e., as long as $s(t) > 0$. These two cases have been sketched in Fig. 8.5.

From (8.28) it follows that the two branches of **S** intersect (if they do, the point of intersection will be on the positive x_2-axis) if

$$\beta \leq v_2 \arcsin(v_2) + \sqrt{(1 - v_2^2)} - 1, \tag{8.30}$$

whose verification is left as an exercise to the reader.

Case 1 (Fig. 8.5a)

For the situation depicted in Fig. 8.5a, let us first assume that **E** is initially somewhere within the shaded area. Then he cannot escape from the area, except

[93] In fact, it is still a semipermeable surface, but for a different ("dual") game in which **E** is originally within the set $x_1^2 + x_2^2 < \beta^2$ and tries to remain there, whereas **P** tries to make the state cross the circle $x_1^2 + x_2^2 = \beta^2$.

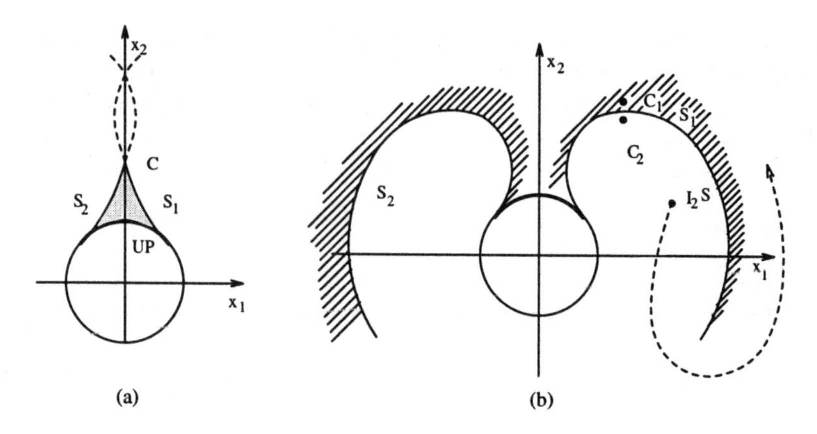

Figure 8.5: Possible configurations for the barriers.

for moving through the UP (= capture), provided that **P** plays optimally, which is that **P** will never let **E** pass through the two semipermeable surfaces indicated by $\mathbf{S}_1$ and $\mathbf{S}_2$. If **E** is initially outside the shaded area, however, he will never be captured (assuming that he plays optimally when in the neighborhood of $\mathbf{S}_1$ or $\mathbf{S}_2$).

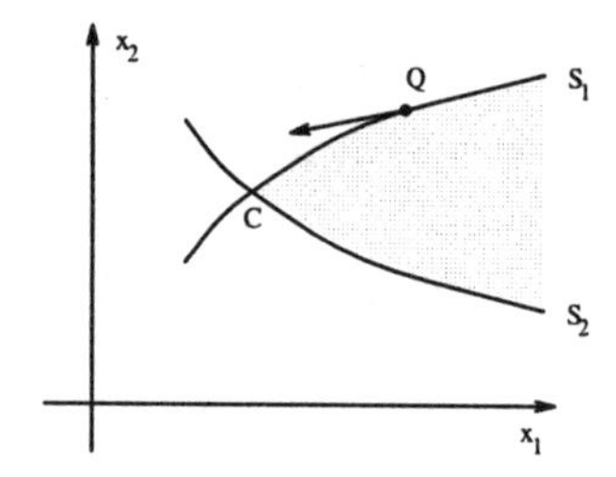

Figure 8.6: A leaking corner.

If **E** is initially in the shaded area, the intersection point C requires closer attention. Toward this end let us assume that the state is initially on $\mathbf{S}_1$; then **P** should try to force it to move in the direction towards the UP (which fortunately can be done in this example), and not towards C and pass C eventually. This latter possibility has been depicted in Fig. 8.6, where **P** tries to keep **E** in the shaded region. Suppose that, in Fig. 8.6, the state is initially at point Q. In order for **P** to keep the state within or on the boundary of the shaded area, and for **E** to try to escape from it, both players should employ their strategies according to (8.16). If these strategies cause the state to move in the direction of C, and eventually pass C, then **P** cannot keep **E** within the shaded area, though he can prevent **E** from passing either $\mathbf{S}_1$ or $\mathbf{S}_2$. The intersection point of the two semipermeable surfaces "leaks" and it is therefore called a *leaking corner*. As already noted, point C in Fig. 8.5a does not leak.

The next question is: once **E** is inside the shaded area, can **P** terminate the game in finite time? In other words, would it be possible for **E** to maneuver within the shaded area in such a way so that he can never be forced to pass the UP? Even though such a stalemate situation may in general be possible, in the present example **E** can always be captured in finite time, which, for the time being, is left to the reader to verify; Example 8.5 in Section 8.4 will in fact provide the solution.

Case 2 (Fig. 8.5b)

We now turn our attention to the situation depicted in Fig. 8.5b in which case

$$\beta > v_2 \arcsin(v_2) + \sqrt{(1 - v_2^2)} - 1. \tag{8.31}$$

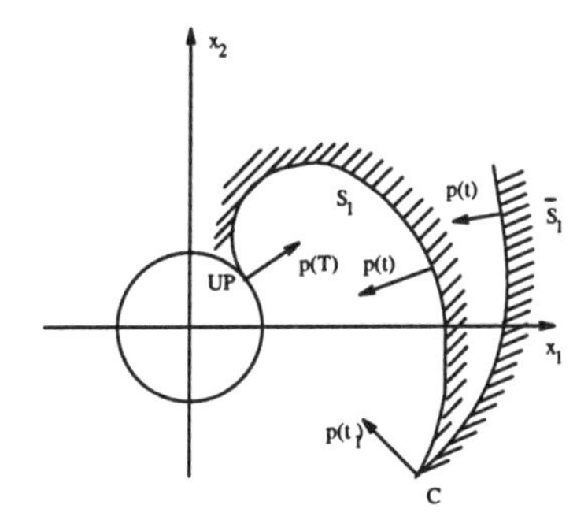

Figure 8.7: The failure of continuation of the semipermeable surface.

In the figure, the semipermeable surfaces have been drawn up to $t_1 = T - \pi - 2\alpha$ which corresponds to the first switching time of u^1 in retrograde time from $t = T$. (From (8.26) and (8.27) it follows that $s(t) = -\sin(t - T + \alpha) + \sin\alpha$.) For $t < t_1$ we have to continue with $u^{1*} = -1$. Carrying out the corresponding analysis, it follows that the resulting continuation of $\mathbf{S}_1$, designated by $\overline{\mathbf{S}}_1$, would be as sketched in Fig. 8.7.

The trajectory leaves point C in exactly the opposite direction as it arrived, and the semipermeable surface has a sharp corner (which is also known as a *cusp*). For the semipermeable surface $\mathbf{S}_1$, **P** can keep the state on the shaded side of $\mathbf{S}_1$, and the same is true for $\overline{\mathbf{S}}_1$ (the reader should note the directions of $p(t)$ in Fig. 8.7). Clearly, $\mathbf{S}_1$ and $\overline{\mathbf{S}}_1$ together do not constitute a semipermeable surface. Therefore, $\overline{\mathbf{S}}_1$ will be disregarded altogether, and we shall consider only the semipermeable surface of Fig. 8.5b.

It is now claimed that termination is possible from any initial point. If the state is originally at I, say, (see Fig. 8.5b), **E** can prevent the state from passing $\mathbf{S}_1$; **P**, on the other hand, can force the system to move along the dotted line, and therefore termination will occur. We have to ensure that no little "island" exists anywhere in the (x_1, x_2) plane, surrounded by semipermeable surfaces, and not connected to $\bar{\Lambda}$, and wherein **E** is safe. Such a possibility is easily ruled out in this example, on account of the initial hypothesis that **P**'s velocity is greater than **E**'s, since then **P** can always force the state to leave such an island, and therefore such an island cannot exist to begin with. □

In conclusion, the foregoing example has displayed various features of the barriers which separate states from where capture is possible from those from which it is not possible. If in particular two semipermeable surfaces intersect, then we have to ensure that the composite surface is also semipermeable, i.e., there should not be a leaking corner. Furthermore, in the construction discussed above, within the context of Example 8.3, we have treated only the case where the barrier does not initially move inside the target set.

8.4 Singular Surfaces

An assumption almost always made at the outset of every pursuit-evasion game is that the state space can be split up into a number of mutually disjoint regions, the value function being continuously differentiable in each of them. The behavior and methods of construction of $V(t,x)$ are well understood in such regions. The boundaries of these regions are called *singular surfaces*, or *singular lines* if they involve one-dimensional manifolds, and the value function is not continuously differentiable across them. In this book we in fact adopt a more general definition of a singular surface. A singular surface is a manifold on which (i) the equilibrium strategies are not uniquely determined by the necessary conditions (8.14), or (ii) the value function is not continuously differentiable, or (iii) the value function is discontinuous.

In general, singular surfaces cannot be obtained by routinely integrating the state and costate equations backward in time. A special singular surface that manifests itself by backward integration is the *switching surface*, or equivalently, a *transition surface*. In Example 5.2, we have already encountered a switching line within the context of an optimal control problem, and its appearance in pursuit-evasion games is quite analogous. The procedure for locating a switching or transition line is fairly straightforward, since it readily follows from the state and costate equations.

A singular surface which does not manifest itself by backward integration of state and costate equations is called a *dispersal surface*. The following example now serves to demonstrate what such a surface is and how it can be detected.

Example 8.4 The system is the same as that of Example 8.1:

$$\begin{aligned}\dot{x}_1 &= \left(2+\sqrt{2}\right)u^2 - \sin u^1, \\ \dot{x}_2 &= -2 - \cos u^1,\end{aligned}$$

with $|u^2| \leq 1$ and with no bounds on u^1. The initial state lies in the positive x_2 half plane and the game terminates when $x_2(t) = 0$. Instead of taking $L = x_1(T)$ as in Example 8.1, we now take $L = x_1^2(T)$. In order to obtain the solution, we initially proceed in the standard manner. The Hamiltonian is

$$H = V_{x_1}\left\{\left(2+\sqrt{2}\right)u^2 - \sin u^1\right\} + V_{x_2}\left\{-2-\cos u^1\right\}, \qquad (i)$$

where $\dot{V}_{x_1} = 0$, $\dot{V}_{x_2} = 0$. Since V is the value function, we obviously have $V_{x_1} = 2x_1$ along the x_1-axis. Then, as candidate equilibrium strategies, we obtain

$$\left.\begin{array}{l} u^2 = \text{ sgn } (V_{x_1}) = \text{ sgn } (x_1(T)), \\ (\sin u^1, \cos u^1) \| (V_{x_1}, V_{x_2}), \end{array}\right\} \qquad (ii)$$

where $\|$ stands for "parallel to". Note that the u^1 and u^2 in (ii) are candidates for open-loop strategies and therefore the expression $u^2(t) =$ sgn $(x_1(T))$ does not contradict with causality. Substituting (ii) and $V_{x_1} = 2x_1$ into (i) with $t = T$ and setting the resulting expression equal to zero, we obtain $V_{x_2}(x(T)) = 2|x_1(T)|$. The costate variables V_{x_1} and V_{x_2} are completely determined now.

For terminal state conditions of $(x_1 > 0, x_2 = 0)$ we find, as candidate equilibrium strategies,

$$u^1 = \frac{\pi}{4}, \quad u^2 = 1,$$

and in the case of terminal state conditions $(x_1 < 0, x_2 = 0)$ their counterparts are

$$u^1 = -\frac{\pi}{4}, \quad u^2 = -1.$$

The corresponding state trajectories are sketched in Fig. 8.8, wherefrom we observe that they intersect. Starting from an initial position I_1, there are two possibilities, but only one of them—the trajectory going in the "south-east" direction—is optimal. Starting from I_2, the trajectory in the "south-west" direction would be optimal. The separation line between the south-east (SE) and south-west (SW) directions is obviously the x_2-axis, since it is on this line where the two values (one corresponding to a trajectory going SW, the other one to SE) of the cost function match. While integrating the optimal paths backward in time, we should stop once the x_2-axis is reached, which is called a *dispersal line*. More precisely, it is a dispersal line for the evader, since, if the initial condition is on this line, it is **E** who chooses one of the two optimal directions along which to proceed; along both of these directions, the system trajectory leaves the dispersal line and both lead to the same outcome. □

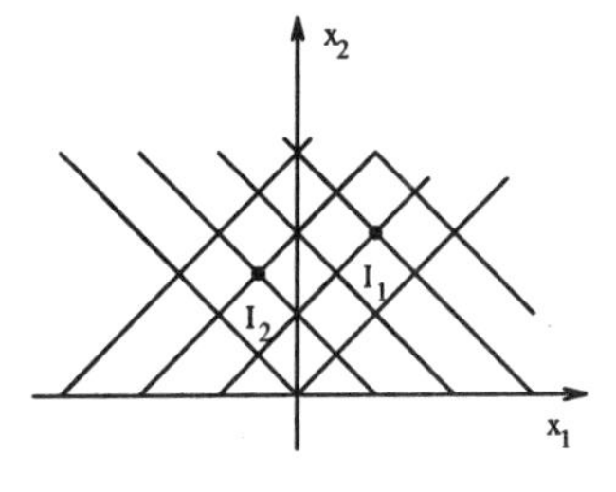

Figure 8.8: Intersection of candidate optimal trajectories.

In the foregoing example we have discovered a dispersal line for the evader. Of course, dispersal lines for the pursuer also exist. Toward that end, consider the homicidal chauffeur game introduced in Example 8.3. Suppose that the parameters of the game (β, v_2) are such that termination is always possible and that the cost functional stands for the time elapsed before capture. If **E** is far behind **P** (i.e., if, in Fig. 8.4, the state is $x_1 = 0, x_2 \ll -\beta$), then **P** will turn either to the right or to the left as quickly as possible, so as to have **E** directly in front of him. Therefore, the half line $(x_1 = 0, x_2 \leq \alpha)$ for some negative α, is a dispersal line for the pursuer.

By means of the same game we now introduce another type of a singular surface, viz. the so-called *universal line* or *universal surface.* Assume that (8.29) and (8.31) hold, which means that **E** can be captured by **P**, whatever the initial position is. Suppose that **E** is situated far in front of **P**, i.e., $x_2 \gg \beta$, relatively close to the x_2-axis. Then, the best **P** can do is to turn towards **E**, such that, in the relative (x_1, x_2) space, **E**'s position moves towards the x_2-axis. Once **E** is on the x_2-axis, the remaining phase of the game is trivial, since it is basically a chase along a straight line. A line, or more generally a surface, to which optimal trajectories enter from both sides and then stay on, is called a *universal line* (*surface*). It turns out that, in the case when both (8.29) and (8.30) hold, the x_2-axis is also a universal line, as shown in the following example.

Example 8.5 Consider the homicidal chauffeur game as introduced in Example 8.3, in which the cost for **P** is the "time to capture", and where the parameters β and v_2 are such that (8.29) and (8.30) hold. In order to construct candidates for optimal paths by backward integration from the UP, we now follow the analysis of Example 8.3 rather closely, along with the formulas (8.23) and (8.24).

For the present example, equation (8.6) becomes

$$\min_{u^1} \max_{u^2} \left\{ V_{x_1}(-u^1 x_2 + v_2 \sin u^2) + V_{x_2}(-1 + u^1 x_1 + v_2 \cos u^2) + 1 \right\} = 0,$$

whereby

$$u^1 = \operatorname{sgn}\,(V_{x_1} x_2 - V_{x_2} x_1); \quad (\sin u^2, \cos u^2) \| (V_{x_1}, V_{x_2}), \tag{8.32}$$

where V_{x_1} and V_{x_2} satisfy

$$\frac{\mathrm{d}}{\mathrm{d}t} V_{x_1} = -u^1 V_{x_2}; \ \frac{\mathrm{d}}{\mathrm{d}t} V_{x_2} = u^1 V_{x_1},$$

along optimal trajectories. The final conditions for the costate equations are

$$V_{x_1} = \frac{\sin\theta(T)}{\cos\theta(T) - v_2}, \quad V_{x_2} = \frac{\cos\theta(T)}{\cos\theta(T) - v_2},$$

the derivation of which is left as an exercise to the reader. (Rewrite the problem in polar coordinates (r, θ), and utilize the boundary condition $V_\theta = 0$ at termination.)

Exactly as in Example 8.3, it can be shown that $u^{1*} = +1$ near the end. For the right half plane, the terminal conditions are

$$x_1(T) = \beta \sin\theta_0, \quad x_2(T) = \beta\cos\theta_0,$$

where θ_0 is the parameter determining the final state, and the solution of the state equation close to termination can be written as

$$\left.\begin{array}{rcl} x_1(\tau) & = & 1-\cos\tau + (\beta - v_2\tau)\sin(\theta_0+\tau), \\ x_2(\tau) & = & \sin\tau + (\beta - v_2\tau)\cos(\theta_0+\tau), \end{array}\right\} \tag{8.33}$$

where τ is the retrograde time $T-t$. Equation (8.33) displays two facts:

(i) $x_2(\tau) > 0$ for $\tau > 0$, which means that part of the shaded region in Fig. 8.5a is not yet filled with (candidate) optimal trajectories,

(ii) for $\tau = \beta/v_2$, the state $(x_1(\tau), x_2(\tau))$ is independent of the arrival angle θ_0.

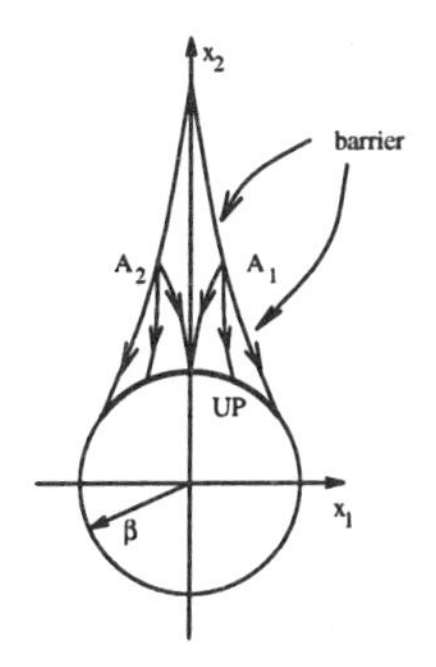

Figure 8.9: The decision points in the homicidal chauffeur game.

Figure 8.9 depicts these trajectories. Point A_1 in this figure has the coordinates $x_1(\beta/v_2), x_2(\beta/v_2)$; it is a so-called *decision point*; **E** has various options, all leading to the same pay-off. Point A_2 is the mirror image of A_1 with respect to the x_2-axis. In order to fill the gap with optimal paths, we try a singular arc for which the switching function is identically zero, i.e.,

$$V_{x_1}x_2 - V_{x_2}x_1 = 0. \tag{8.34}$$

For this to be true, its time derivative must also be identically zero, which yields

$$V_{x_1}(-1 + v_2\cos u^2) - V_{x_2}v_2\sin u^2 = 0. \tag{8.35}$$

For (8.34) and (8.35) to be true when not both V_{x_1} and V_{x_2} are zero, the determinant of the coefficients must be zero, which results in

$$x_1 - v_2(x_1\cos u^2 - x_2\sin u^2) = 0.$$

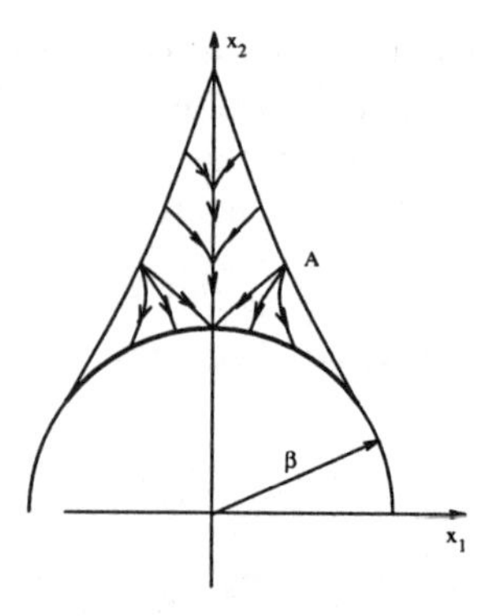

Figure 8.10: The x_2-axis is a universal line.

Because of (8.32) and (8.34), $x_1 \cos u^2 - x_2 \sin u^2 \equiv 0$, which ultimately leads to $x_1 = 0$ as the only candidate for a singular surface. Along this line, the switching function is $V_{x_1} x_2$ and therefore we must have $V_{x_1} \equiv 0$, leading to $\{u^{1*} \equiv 0, u^{2*} \equiv 0\}$. Along the line $x_1 = 0$, both **P** and **E** follow the same straight path in real space. From the singular arc, paths can leave in retrograde time at any instant, leading to Fig. 8.10, thus filling the whole shaded area in Fig. 8.5b with optimal trajectories. The line $x_1 = 0$ is a universal line. It should be noted that both V and $\partial V/\partial x$ are continuous across the universal line (see Problem 2, Section 8.8). □

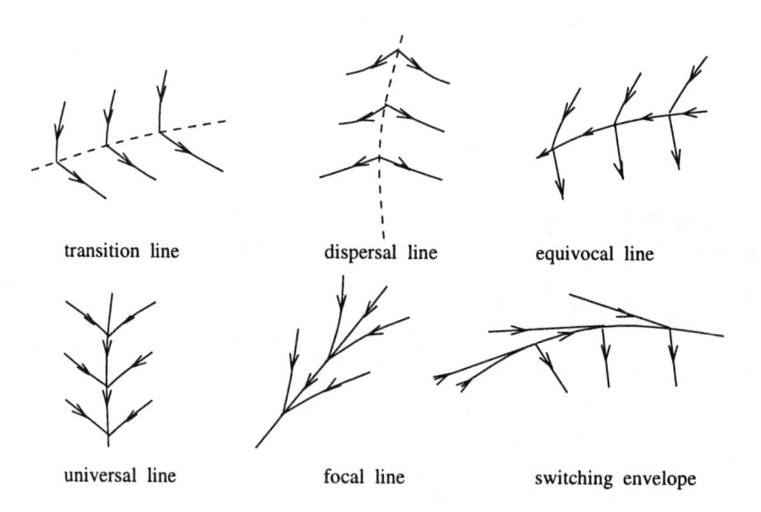

Figure 8.11: Classification of singular surfaces.

There are also other types of singular surfaces. In Fig. 8.11 we have sketched all hitherto known singular surfaces of co-dimension one (i.e., these surfaces are manifolds of dimension one less than that of the state space) and across which $V(x)$ is continuous. It is believed that no other such singular surface exists, but a verification of this claim is not yet available. The singular surfaces not treated heretofore in this chapter are the *equivocal line*, *focal line* and *switching envelope*. They will not be treated in this book; for examples of games where equivocal lines arise, the reader is referred to Isaacs (1975), Lewin and Olsder (1979) and Lewin (1994); the focal line appears for instance in the "obstacle tag" example (Breakwell, 1971) and a game with a switching envelope is treated in Breakwell (1973). Problem 7 in Section 8.8 is also believed to feature a switching envelope in its solution.

In the cases of the equivocal line and the switching envelope, one of the players has the option of either staying on the singular surface or leaving it; both possibilities lead to the same outcome. The difference between these two surfaces is, that in the switching envelope case the trajectories enter (in forward time) tangentially, whereas in the equivocal line case they do not. An analogous difference exists between the universal line and the focal line.

Some of the singular surfaces can only be *approximated* by strategies of the form $u^1 = \gamma^1(x), u^2 = \gamma^2(x)$. In the case of the focal line, for instance, one player, say **P**, tries to get the state vector off the focal line, but **E** brings it back to the focal line. Strategies of the form $u^1 = \gamma^1(x), u^2 = \gamma^2(x, u^1)$, with an instantaneous informational advantage to the evader, such as those described in Section 8.2 with respect to upper and lower values, would be able to keep the state precisely on the focal line.

Singular surfaces across which $V(x)$ is discontinuous are the *barrier* and the *safe contact*. We have already encountered barriers in Example 5.2 in the context of optimal control, and in Section 8.3 where games of kind were treated. Their appearance in games of degree will now be delineated within the context of the homicidal chauffeur game. An example of a safe contact can be found in Chigir (1976).

Example 8.6 Consider the homicidal chauffeur game with the cost function defined as the time-to-go. In Example 8.3 we have described how semipermeable surfaces, starting from the end points of the UP of the target, could be constructed. We now treat the case depicted in Fig. 8.5b, i.e., the case when the parameters v_2 and β are such that (8.31) holds. If **E** is initially situated at point C_2 (see Fig. 8.5b), **P** cannot force him across $\mathbf{S}_1$, but instead he can force him to go around it. Therefore it is plausible—and this can be proven rigorously (see Merz, 1971), that capture from C_2 will take considerably more time than capture from C_1. Hence, the value function is discontinuous across $\mathbf{S}_1$ which, therefore, is a barrier. □

This concludes the present section on singular surfaces. In conclusion, the crucial problem in the construction of the value function is to locate the singular surfaces, but hitherto this problem has not been solved in a systematic way. On the other hand, once a particular V has been constructed, which is continuously differentiable in each of a finite number of mutually disjoint regions of the state space, some conditions (known as "junction conditions") exist to check whether the $V(x)$ obtained is really the value function or not. Since these conditions are not yet very well understood, we do not treat them here; but for some discussion on this topic the reader is referred to Bernhard (1977).

8.5 Solution of a Pursuit-Evasion Game: The Lady in the Lake

In this section we obtain the complete solution of a pursuit-evasion game called "the lady in the lake" by utilizing the techniques developed in the previous sections. This game features a dispersal surface for the evader; there is also a decision point.

A lady (**E**) is swimming in a circular pond with a maximum constant speed v_2. She can change the direction in which she swims instantaneously. A man (**P**), who has not mastered swimming, and who wishes to intercept the lady when she reaches the shore, is on the side of the pond and can run along the perimeter with maximum speed 1. He, also, can change his direction instantaneously. Furthermore, it is assumed that both **E** and **P** never get tired. **E** does not want to stay in the lake forever, though; she wishes eventually to come out without being caught by the man. (On land, **E** can run faster than **P**.) **E**'s goal is to maximize the pay-off, which is the angular distance **PE** viewed from the center of the pond, at the time **E** reaches the shore (see Fig. 8.12). **P** obviously wants to minimize this pay-off. To make the game nontrivial, it is assumed that $v_2 < 1$.

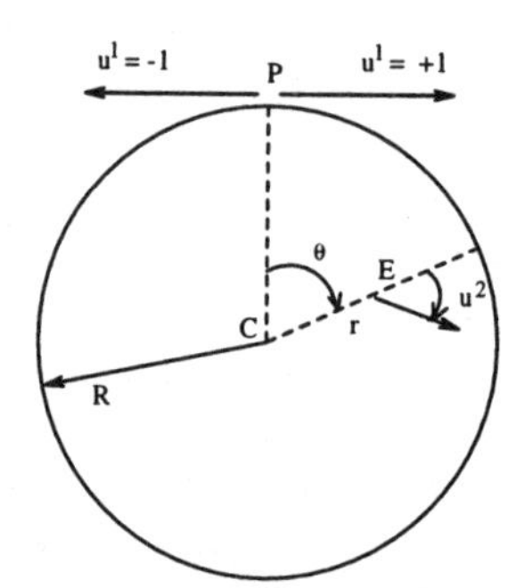

Figure 8.12: The "lady in the lake".

Even though the problem could be cast in a rectangular coordinate system fixed in space, a description in the *relative* coordinates θ and r turns out to be simpler, where θ is the angle between **P** and **E** (see Fig. 8.12 for the direction) and r is **E**'s distance from the center of the lake. This coordinate system is called *relative* since it is attached to **P**; it is not fixed in space, and yet it describes the game completely. The kinematics of the game are

$$
\begin{aligned}
\dot{\theta} &= \frac{v_2 \sin u^2}{r} - \frac{u^1}{R}, \\
\dot{r} &= v_2 \cos u^2,
\end{aligned}
$$

where R is the radius of the pond. **P**'s control u^1 is restricted by $|u^1(t)| \leq 1$, while **E**'s control u^2, which stands for the angle of **E**'s velocity vector with respect to the radius vector CE (see Fig. 8.12), is not restricted in any way. We now seek for the equilibrium strategies in feedback form. The cost function is $|\theta(T)|$, where T is defined as $T = \min\{t : r(t) = R\}$, and $-\pi \leq \theta(t) \leq +\pi$.

The Isaacs equation for this problem is

$$\min_{u^1} \max_{u^2} \left\{ \frac{\partial V(\theta, r)}{\partial r} v_2 \cos u^2 + \frac{\partial V(\theta, r)}{\partial \theta} \left(\frac{v_2 \sin u^2}{r} - \frac{u^1}{R} \right) \right\} = 0, \tag{8.36}$$

whence

$$u^{1*} = \operatorname{sgn} \left(\frac{\partial V}{\partial \theta} \right), \ (\cos u^{2*}, \sin u^{2*}) \| \left(\frac{\partial V}{\partial r}, \frac{1}{r} \frac{\partial V}{\partial \theta} \right). \tag{8.37}$$

The differential equations for the costate variables along the optimal trajectories are

$$\frac{\mathrm{d}}{\mathrm{d}t} \left(\frac{\partial V(\theta, r)}{\partial \theta} \right) = 0, \quad \frac{\mathrm{d}}{\mathrm{d}t} \left(\frac{\partial V(\theta, r)}{\partial r} \right) = \left(\frac{\partial V}{\partial \theta} \right) \cdot \frac{v_2 \sin u^2}{r^2},$$

and the value function at $t = T$ is given by

$$V(\theta(T), r(T)) = |\theta(T)|.$$

The optimal control u^{1*}, taken as the open-loop representation of the feedback equilibrium strategy,[94] can now be written as $u^{1*}(t) = \operatorname{sgn}(\partial V / \partial \theta) = \operatorname{sgn}(\theta(T))$. Since we assumed $-\pi \leq \theta(t) \leq +\pi$, this implies that **P** will move in **E**'s direction along the smallest angle. Substituting u^{1*} and u^{2*} into (8.36), we obtain

$$\sin u^{2*}(t) = \frac{R v_2}{r(t)} \operatorname{sgn} \theta(T). \tag{8.38}$$

Two observations can be made at this stage. First, (8.38) is valid as long as $r(t) > R v_2$; for $r(t) \leq R v_2$ the outlined derivation does not hold. Second, **E** swims along a straight line in real space, tangent to a circle of radius $R v_2$ (see Fig. 8.13a).

This result could also have been obtained geometrically (see Fig. 8.13b). In the relative coordinate system, **E**'s velocity $\vec{v}_R$ is made up of two components: (i) **E**'s own velocity vector in real space, $\vec{v}_2$, and (ii) a vector $\vec{v}_1$, obtained by rotation around the center C, opposite to **P**'s direction of motion (recall that **P** tries to reduce $|\theta(T)|$), and with magnitude $r(t)/R$. The best policy for **E** is to make $\vec{v}_2$ and the resulting velocity vector $\vec{v}_R$ perpendicular to each other. For any other choice of u^2 (i.e., direction of $\vec{v}_2$ in Fig. 8.13b), the angle between C**E** and $\vec{v}_R$ will become smaller, or, in other words, **E** will move faster into **P**'s direction, which is precisely what **E** tries to avoid. Therefore, **E**'s equilibrium strategy is to keep the angle between $\vec{v}_2$ and $\vec{v}_R$ at $\pi/2$ radians, or equivalently, $\sin u^{2*} = R v_2 / r(t)$.

[94]Note that, although the open-loop representation of the feedback saddle-point solution (if it exists) may exist, an open-loop saddle-point solution surely does not exist for this pursuit evasion game. Whether a mixed open-loop saddle-point equilibrium exists is an altogether different question which we do not address here.

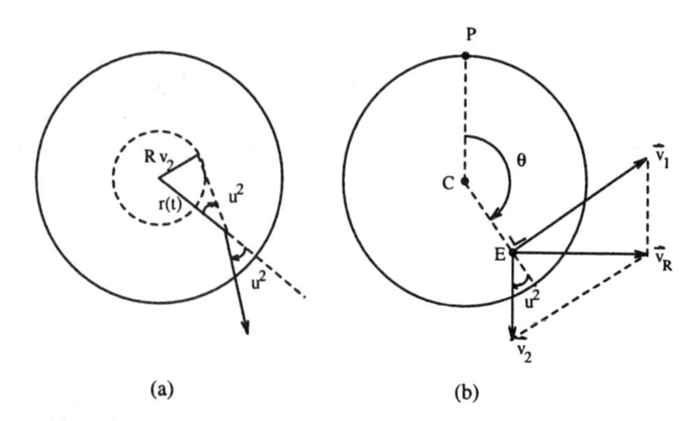

Figure 8.13: **E**'s equilibrium strategy for the "lady in the lake".

This way of constructing **E**'s optimal strategy fails for $r(t) < Rv_2$ in which case **E** can achieve a larger angular velocity (with respect to the center C) than **P**, and therefore she can always maneuver herself into the position $\theta(t) = \pi$, i.e., to a position diametrically opposite from **P**. Note that, in order to maintain $\theta(t) \equiv \pi$, **E** would have to know **P**'s current action. If she does not have access to this (which is a more realistic situation), she can stay arbitrarily close to $\theta(t) \equiv \pi$ as she moves outward towards the circle with radius Rv_2. In case **E** would know **P**'s current action, she can reach the point $(\theta = \pi, r = Rv_2)$ in finite time (see Problem 3 in Section 8.8); if she does not know **P**'s current action, she can get arbitrarily close to this point, also in finite time.[95] From the position $(\theta = \pi, r = Rv_2)$, **E** moves right or left along the tangent. More precisely, **E** will first swim outward to a position $r > Rv_2$ which is sufficiently close to $r = Rv_2$. Then **P** has to make a decision as to whether to run clockwise or anti-clockwise around the pond (**P** cannot wait since, as **E** moves outward, his pay-off will then become worse). Once **E** knows which direction has been chosen by **P**, she will swim "away" from **P**, along the tangent line just described. In the region $r(t) > Rv_2$, **P**'s angular velocity is larger, and therefore he will continue to run in the direction he had chosen before.

The outcome of the game is readily calculated to be

$$|\theta(T)| = \pi + \text{ arc } \cos v_2 - \frac{1}{v_2}\sqrt{(1 - v_2^2)}, \tag{8.39}$$

which holds true for all initial positions inside the circle of radius Rv_2. The lady can escape from the man if $|\theta(T)| > 0$, which places a lower bound on **E**'s speed: $v_2 > 0.21723\ldots$. From all initial positions inside the pond, **E** can always first swim to the center and then abide by the strategy just described, resulting in the outcome (8.39). From some initial positions she can do better, namely, the positions in the shaded area in Fig. 8.14 which is bounded by the pond and

[95]This would inevitably involve some delay in the current state information of **E** and thereby some memory in her information. Restriction to only feedback strategies would lead to some subtle measurability questions.

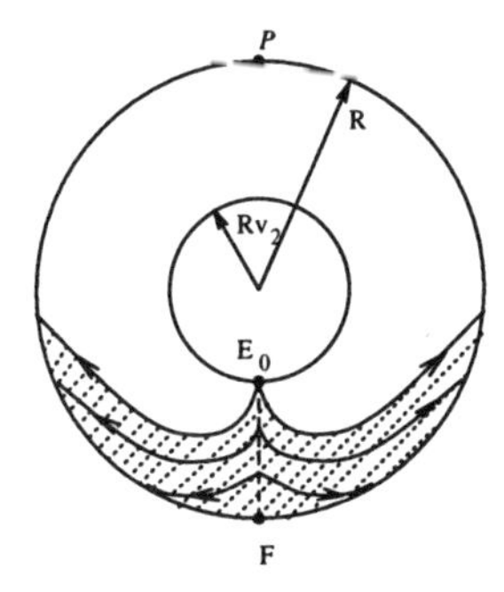

Figure 8.14: Optimal trajectories in the relative space.

the two equilibrium trajectories which, constructed in retrograde time, end at $(r = Rv_2, \theta = \pi)$. Within the shaded area the optimal strategies are determined by (8.37). The backward construction is only valid until $|\theta(t)| = \pi$; the line segment $(\theta = \pi, Rv_2 \leq r < R)$ obviously forms a dispersal line on which it is **P** who decides whether to go "to the left" or "to the right". In the non-shaded area of the pond in Fig. 8.14, the value function is constant and equals $|\theta(T)|$ given by (8.39). Properly speaking, this value can only be obtained if there is an informational advantage to **E** as explained above; see also Problem 9 in Section 8.8. If this is not the case and only feedback strategies are allowed, only an approximation to this value function can be obtained. Therefore, in such a case, a saddle point will not exist, but an ϵ-saddle-point equilibrium will (cf. Section 4.2, specifically Def. 4.2).

The solution obtained for the "lady in the lake" has a very special feature. If **E** would accidentally choose a wrong maneuver, she can always return to the center of the lake and start all over again. The point E_0 in the relative coordinate system (see Fig. 8.14), which plays a special role, is sometimes referred to as a *decision point.*

8.6 An Application in Maritime Collision Avoidance

In this section, collision avoidance during a two-ship encounter in the open sea will be treated as a problem in the theory of differential games. Given two ships close to each other in the open sea, a critical question to be addressed is whether a collision can be avoided. If, for instance, we assume that the helmsman of ship 1 (**P**1) has complete information on the state of ship 2, whereas the helmsman of ship 2 (**P**2) is not aware of the presence of the first ship, this lack of information on the part of **P**2 may lead to a hazardous outcome. It is quite possible that **P**2 may perform a maneuver leading to a collision which would not have occurred if **P**2 had had full knowledge of **P**1's position. Hence, in such a situation, it would be reasonable to undertake a worst-case analysis, by assuming that one of the ships (**P1**) tries to avoid and the other one (**P2**) tries to cause a collision.

One can envisage other situations, such as the case in which both ships are aware of each other's position and they both try to prevent a collision, i.e., they cooperate. This case, although not a differential game in the proper sense, can be solved by similar techniques. Roughly speaking, the *minmax* operation in the differential game case must be replaced by the *maxmax* operation in the cooperative case. Still another situation arises when one ship holds course and speed (is "standing on") because of the "rules of the road", whereas the other ship must evade.

The dynamic system comprises two ships, **P1** and **P2**, maneuvering on a sea surface which is assumed to be homogeneous, isotopic, unbounded and undisturbed. The kinematic equations to be used are identical to those of the two-cars model (see Example 8.3):

$$\left.\begin{aligned} \dot{x}_1 &= -\omega_1 u^1 x_2 + v_2 \sin\theta, \\ \dot{x}_2 &= -v_1 + \omega_1 u^1 x_1 + v_2 \cos\theta, \\ \dot{\theta} &= \omega_2 u^2 - \omega_1 u^1. \end{aligned}\right\} \qquad (8.40)$$

This simple model may not be (too) unreasonable if only short duration maneuvers are considered, during which the ships cannot change their speeds markedly. Each ship is characterized by two parameters: the maximum angular velocity ω_i and the constant forward speed v_i. The control of **P***i* is u^i, which is the rudder angle, and which is bounded by $|u^i(t)| \leq 1$. Extensions of this model are possible, such as, for instance, the case of variable forward speed (the engine setting will then become a control variable as well), but these would make the analysis to follow much more complicated. Some other effects (hydrodynamic and inertial) have been ignored in the model (8.40), such as the u^i-dependence of v_i.

The cost function is described in terms of the closest point of approach (CPA), the distance corresponding to it is called the *miss distance*. The terminal condition is given by the CPA at first pass, characterized by

$$\frac{\mathrm{d}r}{\mathrm{d}t} = 0, \frac{\mathrm{d}^2 r}{\mathrm{d}t^2} > 0,$$

where $r = \sqrt{(x_1^2 + x_2^2)}$, and the terminal range $r(T)$ is the cost function. This constitutes a game of degree. In this section we shall consider, instead, the directly related game of kind, characterized by a given number r_m, the minimum range or minimum miss distance. If $r(T) > r_m$, no collision takes place, and for $r(T) \leq r_m$, a collision will take place.

In the three-dimensional state space, the target set is described by the cylinder $x_1^2 + x_2^2 = r_m^2$. The UP, i.e., the set of points on this cylinder at which a collision can take place, is determined by

$$v_2(x_1 \sin\theta + x_2 \cos\theta) - v_1 x_2 \leq 0.$$

The terminal condition wherefrom a barrier can be constructed is determined from this relation with the inequality sign replaced by an equality sign. Substituting $x_1 = r_m \sin\alpha$ and $x_2 = r_m \cos\alpha$, where α is the bearing, into this equality, we obtain, at the final time,

$$\left.\begin{aligned} r(T) &= r_m, \\ \sin\alpha(T) &= \epsilon(v_1 - v_2\cos\theta(T))/w, \\ \cos\alpha(T) &= \epsilon v_2(\sin\theta(T))/w, \end{aligned}\right\} \tag{8.41}$$

where $w \triangleq \sqrt{(v_1^2 + v_2^2 - 2v_1v_2\cos\theta(T))}$, and $\epsilon = \pm 1$. For each $\theta(T)$ we obtain two values for $\alpha(T)$, one corresponding to $\epsilon = +1$ and the other one to $\epsilon = -1$. These two values correspond to a right and left barrier, respectively; **P2** can just miss **P1**, either on the right or on the left side.

From the final conditions thus obtained, we must construct equilibrium trajectories backward in time in order to obtain the barrier which separates the points from where avoidance is possible from the points from where it is not. The barrier—actually composed of two parts, each corresponding to a different value of ϵ in (8.41)—is depicted in Fig. 8.15. In this figure the two parts of the barrier intersect, which indicates that the enclosed points are guaranteed collision points for **P2** (provided that the intersection of the two parts does not leak).

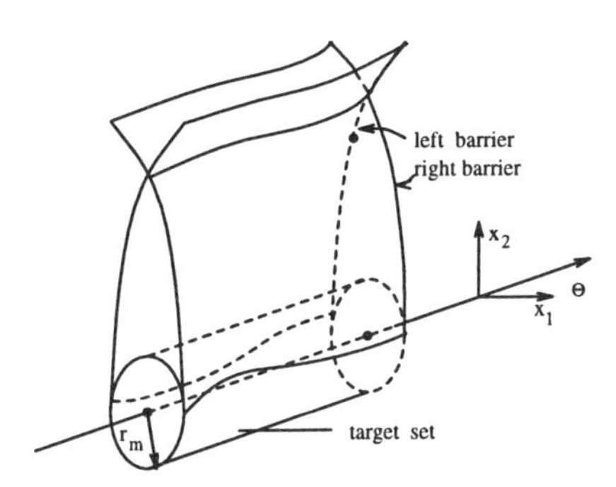

Figure 8.15: A schematic picture of the barrier.

We now turn to the actual construction of the barriers. The optimal u^1 and u^2 are the arguments of the corresponding Isaacs equation, thus leading to

$$u^{1*} = \text{sgn}\,(x_1V_{x_2} - x_2V_{x_1} - V_\theta),\ u^{2*} = \text{sgn}\,(-V_\theta). \tag{8.42}$$

Furthermore, the costate variables satisfy

$$\left.\begin{aligned} \dot V_{x_1} &= -\omega_1 u^1 V_{x_2}, & V_{x_1}(T) &= \sin\alpha(T); \\ \dot V_{x_2} &= \omega_1 u^1 V_{x_1}, & V_{x_2}(T) &= \cos\alpha(T); \\ \dot V_\theta &= v_2(V_{x_2}\sin\theta - V_{x_1}\cos\theta), & V_\theta(T) &= 0. \end{aligned}\right\} \tag{8.43}$$

As long as the arguments of the sgn-relations in (8.42) are nonzero, it is simple to obtain the solution of (8.40) and (8.43); but, substitution of the final values for V_{x_1}, V_{x_2}, V_θ into (8.43) shows that the arguments of the sgn-relations are zero at $t = T$. The situation can, however, be saved by replacing the arguments with their retrograde time derivatives, which leads to

$$\begin{aligned} u^1(T) &= \epsilon \ \text{sgn}\left(\frac{v_1}{v_2} - \cos\theta(T)\right), \\ u^2(T) &= -\epsilon \ \text{sgn}\left(\frac{v_2}{v_1} - \cos\theta(T)\right), \end{aligned}$$

where $\epsilon = \pm 1$. Hence, the equilibrium strategies at the terminal situation are determined for almost all values of $\theta(T)$. For $v_2/v_1 < 1$, we can expect a singular control for **P2** if $\theta(T) = \pm\arccos(v_2/v_1)$ for the last part of the maneuver. Similarly, if $v_1/v_2 < 1$, **P1** may have a singular control.

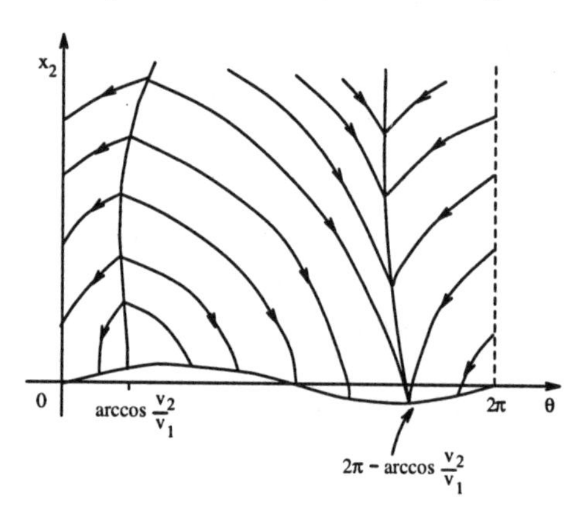

Figure 8.16: The trajectories which make up the right barrier, projected in the (x_2, θ) plane.

Let us proceed with the case $v_2 < v_1 < 1$. Closer scrutiny reveals that there are no equilibrium paths leading to the singular point $\theta(T) = \epsilon\arccos(v_2/v_1)$, but, in its stead, a dispersal line ends at this point. The other point, $\theta(T) = -\epsilon\arccos(v_2/v_1)$, however, is the end point of a singular line (with $u^2 = 0$), which is a universal line. Integrating backward in time from the final condition, we first start with the singular control $u^2 = 0$ which may be terminated at any time (switching time) we desire; and going further backward in time we obtain a nonsingular u^2. In Fig. 8.16, it is shown how the right barrier ($\epsilon = +1$) can be constructed by determining the optimal trajectories backward in time. In the figure, the projections of the trajectories on the (x_2, θ) plane have been drawn. In order to construct these trajectories, $\theta(T)$ can be taken as a parameter, and for $\theta(T) = -\epsilon\arccos(v_2/v_1)$ we have another parameter, viz. the time instant at which the singular control switches to a nonsingular one (there are two options: $u^2 = +1$ and $u^2 = -1$) in retrograde time. The dispersal line is found as the intersection of two sets of trajectories obtained through backward integration, namely that set of trajectories which have $\theta(T)$ smaller than $\arccos(v_2/v_1)$, modulo 2π, and the set of trajectories with $\theta(T)$ greater than $\arccos(v_2/v_1)$. The left barrier can be constructed analogously. If the two barriers intersect over the whole range $0 \leq \theta \leq 2\pi$, and if the line of intersection does not leak, then the enclosed points are guaranteed collision points for **P2**. Also a tacit assumption which makes this whole construction work is that the two parts which define the dispersal line (for both the right and left barriers) do indeed intersect. If they do not intersect, then there would not be a dispersal line; instead, there would be a hole in the surface that encloses the points we described above. Consequently, this set of points would not be enclosed anymore, and **P1** might be able to force escape by steering the state through this hole. As yet, no sufficiency conditions are known which ensure enclosure; hence, for each problem one has to verify numerically that no holes exist. For this reason, the method can be used conveniently only for systems with state space dimension not greater than three.

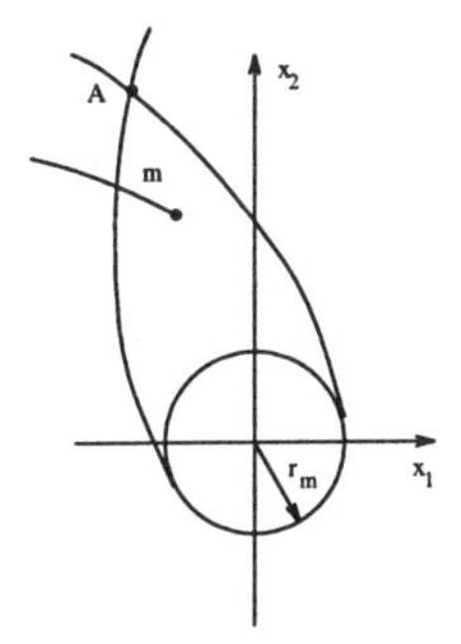

Figure 8.17: Intersection of the semipermeable surfaces with the plane $\theta = \theta_0$.

A possible and appealing way of constructing the barriers which enclose the points wherefrom collision is unavoidable, is to consider cuts in the (x_1, x_2, θ) space for which θ is a constant, say $\theta = \theta_0$. For each parameter value $\theta(T)$ and also for the retrograde time parameter when $\theta(T) = -\epsilon \arccos(v_2/v_1)$ one calculates the equilibrium trajectories until they hit the plane $\theta = \theta_0$—this being performed for both the right and left barriers. For different values of the parameters, one obtains, in general, different points in the plane $\theta = \theta_0$, with the ultimate picture looking like the one sketched in Fig. 8.17. In this approach existence of a dispersal line cannot be detected; one simply "by-integrates" it. The semipermeable surface (line) m in Fig. 8.17 is the result of such a procedure. However, this is not a serious problem, since what one seeks in the plane $\theta = \theta_0$ is a region completely surrounded by semipermeable surfaces; the connecting corners must not leak. Hence, in a picture like Fig. 8.17, one has to check whether point A leaks or not. If it does not, then one can disregard the semipermeable line m. If it leaks, however, then this is an indication that either the problem is ill-posed (i.e., no enclosed region exists from where collision can be guaranteed), or there must be other semipermeable surfaces which will define, together with (parts of) the already existing semipermeable surfaces, a

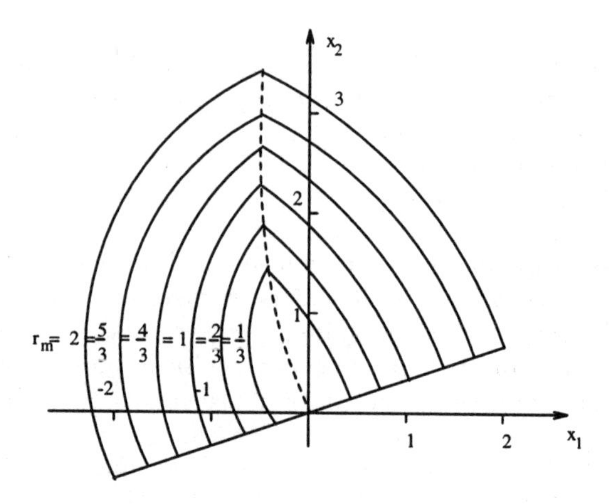

Figure 8.18: Initial points (x_1, x_2), with $\theta = \pi/2$, for which distance of CPA is r_m.

new enclosed region.

The next figure, Fig. 8.18, taken from Miloh and Sharma (1976), depicts several enclosed areas for different r_m-values. The figure shows the barrier cross-sections at $\theta = \pi/2$. The numerical data are as follows: $v_1 = 1$, $v_2 = \frac{1}{2}$, $\omega_1 = \omega_2 = 1$. This means that the minimum turn radii of **P1** and **P2** are 1 and $\frac{1}{2}$, respectively. By constructing similar pictures for different sets of parameter values (different speed ratios and different maneuvering capabilities), it is possible to gain insight into collision sensitivity with respect to these parameters.

As a final remark, collision was defined here as "two ships (considered as points in the plane) approaching each other closer than a given distance". If ships would actually collide this way, then they must have the shape of round disks, as viewed from above. The construction presented can, however, be applied to more realistic shapes, which is though slightly more complicated; such an example will be presented in Section 8.7. Problem 5 (Section 8.8) presents yet another slightly different problem in which the target set is still a round disk, but it is not centered at the origin.

8.7 Role Determination and an Application in Aeronautics

Heretofore we have considered the class of pursuit-evasion games wherein the roles of the players are prescribed at the outset of the problem. These problems are sometimes called one-target games which model situations such as a missile chasing an airplane. However, in a "dogfight" that takes place between two planes or ships which are both armed and capable of destroying their opponents, it may not be apparent at the outset who pursues whom. Accordingly, we now introduce *two-target games*, where either player may be the pursuer or the evader, depending on their current configuration, and each target set is determined by the shooting range of the respective player. Each player's task is to destroy his opponent (i.e., maneuver the opponent into his target set) and to avoid destruction by his opponent.

The solution method to two-target games comprises essentially two stages:

(i) Given the initial conditions, determine the relative roles of the players; who pursues whom? It is clear that in a deterministic differential game the roles of the players are determined completely by the initial state and will not change during the course of the game—provided that only pure strategies are permissible and that the players act rationally.

(ii) With the roles of the players determined, what are the optimal, e.g., time-optimal, strategies for the players? Note that the question of who pursues whom has nothing to do with the cost function.

Here we shall be concerned with the first stage. The state space consists of three mutually disjoint regions: two regions corresponding to victory by one or the other player, and the third region corresponding to a draw, i.e., neither player is capable of destroying his opponent. The states within the draw region may correspond to different situations. A particular possibility is a stalemate: both players play certain strategies such that neither player will be destroyed and, moreover, a deviation from his equilibrium strategy by one of the players will lead to his own destruction. Another possibility is that the faster player can escape—tacitly assuming that this player cannot win the game in finite time.

The region R_1 of the state space, which corresponds to victory by $\mathbf{P1}$, may be expected to be bounded by a surface comprising a part of the surface of Λ_1, the target set for $\mathbf{P1}$, together with a semipermeable surface Σ_1 which prevents the state from leaving R_1 provided that $\mathbf{P1}$ plays optimally in the neighborhood of Σ_1. The interior of R_1 does not contain any points of $\mathbf{P2}$'s target set Λ_2, and thus, assuming that the game terminates in finite time (i.e., a stalemate cannot arise), victory for $\mathbf{P1}$ is guaranteed for initial states inside R_1 (see Fig. 8.19a). That portion of Λ_1, which forms part of the boundary of R_1, is permeable only

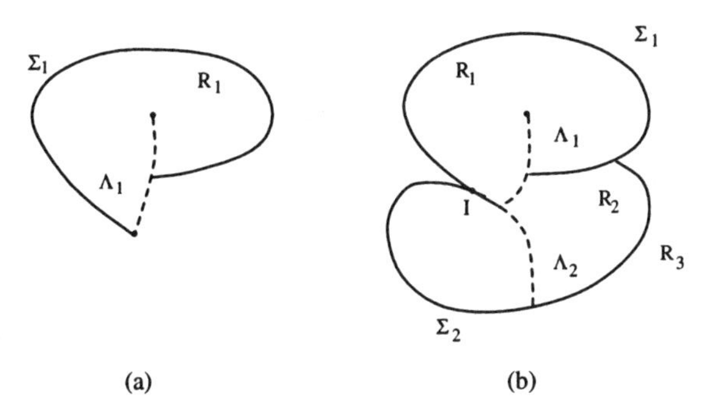

Figure 8.19: Regions R_1 and R_2 corresponding to victory by $\mathbf{P1}$ and $\mathbf{P2}$, respectively.

in the direction away from R_1. In a similar way we can construct R_2—the region corresponding to victory by **P2**. If R_1 and R_2 do not fill up the whole state space, then points belonging to neither R_1 nor R_2 belong to R_3, the draw region. Such a situation is illustrated schematically in Fig. 8.19b. Victory by **P2** in region R_2 is guaranteed if we assume that no stalemate is possible, so that the target set Λ_2 is reached in finite time.

States belonging to both Σ_1 and Σ_2 (such as I in Fig. 8.19b) will terminate on both Λ_1 and Λ_2, thereby giving rise to "victory" by both players, or, what is essentially the same, to a *simultaneous confrontation*. States belonging to Σ_2 and not to Σ_1, or belonging to Σ_1 and not to Σ_2, give rise to *near miss* situations.

We now present an example of a two-target game which has the additional feature that the two target sets are nonsmooth.

Consider an aerial duel or dogfight, wherein each combatant wishes to destroy his opponent without himself being destroyed. In order to keep the analysis at a reasonable level, we take the dimension of the state space to be three, and assume that the two players move in a single horizontal plane. Equations of motion within this plane are the same as those in the ship collision avoidance problem, and they are given by (8.40). The constraints on the controls are also the same. We assume that $v_1 > v_2 > 0$ and $0 < \omega_1 < \omega_2$, i.e., **P1** has the advantage of having a higher speed, whereas **P2** is more maneuverable. **P*i*** has a confrontation range (which is his target set Λ_i) consisting of a line segment of length ℓ_i in the direction of his velocity vector. If **P1** crosses the confrontation range of **P2**, or vice versa, the game ends. For the analysis to follow, we make the additional assumption $0 < \ell_1 < \ell_2$.

We start with the construction of the barrier Σ which is the composition of Σ_1 and Σ_2 introduced earlier. Each portion of Σ, whether it leads to simultaneous confrontation or to near miss, is obtainable by solving a particular "local" differential game, with a terminal cost function defined only in the immediate neighborhood of the simultaneous confrontation or near miss. If we adopt the convention that the local game has a positive (respectively, negative) value when it ends in favor of **P2** (respectively, **P1**), the equilibrium controls are determined from

$$\min_{u^1} \max_{u^2} (V_{x_1}\dot{x}_1 + V_{x_2}\dot{x}_2 + V_\theta\dot{\theta}) = 0,$$

which leads to

$$\left.\begin{aligned} u^{1*} &= \operatorname{sgn}\,(x_2 V_{x_1} - x_1 V_{x_2} + V_\theta), \\ u^{2*} &= \operatorname{sgn}\, V_\theta. \end{aligned}\right\} \tag{8.44}$$

The costate equations are the same as in (8.43), but in this case the terminal conditions are different as will be seen shortly. In fact, the terminal conditions will be different for each local game.

From (8.44) it follows that, if u^1 remains constant for a duration of at least τ units of time before the final time T is reached, then, apart from an insignificant constant multiplier, assumed henceforth to be unity, the costates at time $\tau =$

$T - t$ satisfy

$$\begin{aligned} V_{x_1}(\tau) &= \cos(\phi - \omega_1 u^1 \tau), \\ V_{x_2}(\tau) &= \sin(\phi - \omega_1 u^1 \tau), \end{aligned} \tag{8.45}$$

where ϕ is determined by the final conditions. From (8.44) and (8.43) it further follows that

$$\dot{V}_\theta(\tau) = -v_2 \cos(\theta + \phi - \omega_1 u^1 \tau) = -v_2 \cos(\theta(T) + \phi - \omega_2 u^2 \tau), \tag{8.46}$$

so that, if $u^2 \neq 0$,

$$V_\theta(\tau) = V_\theta(T) - \frac{v_2}{\omega_2 u^2} \left\{ \sin(\theta(T) + \phi - \omega_2 u^2 \tau) - \sin(\theta(T) + \phi) \right\}.$$

By introducing $A = x_2 V_{x_1} - x_1 V_{x_2} + V_\theta$, we also find $\dot{A} = -v_1 V_{x_1}$, so that, if $u^1 \neq 0$,

$$A(\tau) = A(T) + \frac{v_1}{\omega_1 u^1}(V_{x_2}(T) - V_{x_2}(\tau)). \tag{8.47}$$

Since many of the local games will include singular controls, we now investigate their properties. From (8.44) it follows that a singular control for **P**2 is only possible for $V_\theta \equiv 0$, which results in $\dot{V}_\theta \equiv 0$. Equation (8.46) now leads to the conclusion that the only possible singular control for **P**2 is $u^2 \equiv 0$. Similarly, it can be shown that the only possible singular control for **P**1 is $u^1 \equiv 0$.

Returning to the actual construction of Σ, we shall first deal with simultaneous confrontation and thereafter with the near miss situations. For each situation, several types may be distinguished, since simultaneous confrontation and near miss occur in different ways.

Simultaneous confrontation

Consider the situation depicted in Fig. 8.20, where the trajectories of **P**1 and **P**2 in real space have been sketched, together with the relative coordinate system at terminal time T. In the figure, **P**1 turns to the right as fast as possible, and

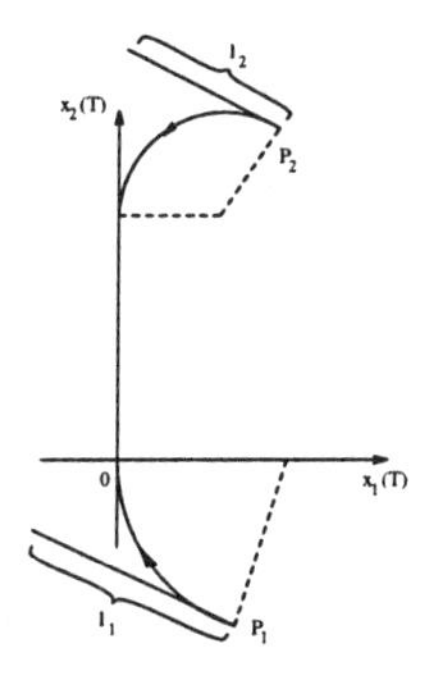

Figure 8.20: A situation of simultaneous confrontation.

P2 does likewise, but to the left. The final state is given as follows: $x_1(T) = 0$, $\theta(T) = \pi$ and $x_2(T)$ is a variable that must satisfy

$$0 \leq x_2(T) \leq \min(\ell_1, \ell_2). \tag{8.48}$$

The equilibrium strategies are $\gamma^{1*}(\cdot) = +1$, $\gamma^{2*}(\cdot) = -1$. Initial conditions associated with these strategies and terminal conditions can easily be found by backward integration of the state equations. The value of the game corresponding to these initial conditions is neither positive (victory for **P2**) nor negative (victory for **P1**) and is therefore determined as $V = 0$. It should be clear from Fig. 8.20 that γ^{1*} and γ^{2*} are indeed in equilibrium since a deviation from either strategy will lead to a destruction of the deviating player. This is also confirmed by application of the necessary conditions. Let us assume for the moment that the terminal values of the costate variables satisfy the requirements

$$V_{x_1}(T) > 0, \quad V_{x_2}(T) = 0, \quad V_\theta(T) < 0. \tag{8.49}$$

Then, it follows that $\phi = 0$ in (8.45), which, together with (8.44) and (8.47), indeed leads to $u^{1*} = +1$, $u^{2*} = -1$. But why should (8.49) be a requirement? Let us consider the case of $V_{x_1}(T)$: a small deviation of **P2**'s position in the positive x_1-direction in the final situation (keeping $x_2(T) = 0$, $\theta(T) = \pi$) will be to the advantage of **P2**, since he will be able to win the game from the new relative position. Hence, after this deviation, we have $V > 0$ instead of $V = 0$, and therefore $V_{x_1}(T) > 0$. A similar argument verifies the sign of $V_\theta(T)$. Furthermore, since a change in $x_2(T)$ does not change the outcome of the game (it remains $V = 0$), we have $V_{x_2}(T) = 0$.

Let us now investigate how this simultaneous confrontation affects the barrier Σ in the (x_1, x_2, θ) space, which separates the initial points where from **P1** can win from those corresponding to **P2**'s victory. Since $x_2(T)$ is a parameter, subject to the constraint (8.48), a whole family of equilibrium trajectories exists, all leading to a simultaneous confrontation of the type described. This family forms a semipermeable surface in the (x_1, x_2, θ) space. For the actual construction we shall consider the intersection of this surface with the planes $\theta =$ constant, say $\theta = \theta_0$. Equations (8.40), (8.43), (8.44) and (8.49) are integrated backwards in time until $\theta = \theta_0$ is reached and for varying $x_2(T)$ we obtain different points in the $\theta = \theta_0$ plane, which make up a line segment.

In Fig. 8.21, the solution of a specific example has been sketched, for which the characteristics of the airplanes are

$$v_1 = \omega_1 = 1, \quad v_2 = \frac{3}{4}, \quad \omega_2 = 3, \quad \ell_1 = 2 \text{ and } \ell_2 = \infty. \tag{8.50}$$

In the figure, $\theta_0 = \pi/3$. The line segment which separates the states corresponding to victory by **P1** from those corresponding to **P2**'s victory, according to the simultaneous confrontation described above, has number 1 attached to it. From initial positions slightly above this line, **P1** will win, and from initial positions slightly below, **P2** will win.

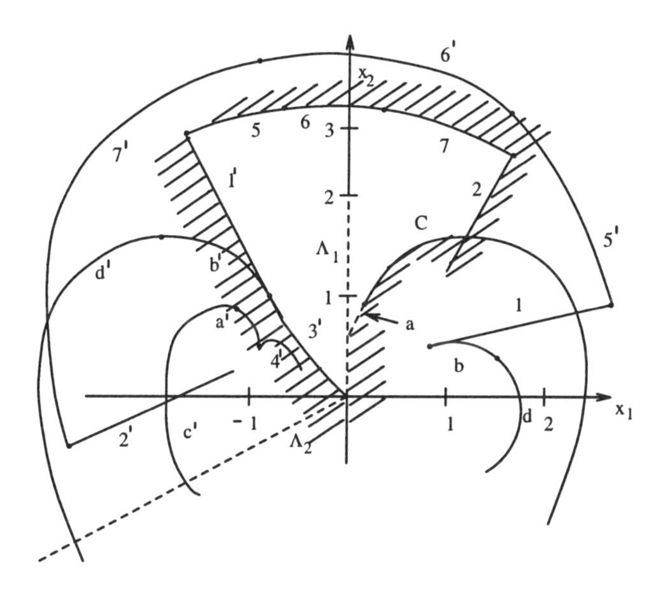

Figure 8.21: Intersection of semipermeable surfaces with the plane $\theta = \pi/3$.

Different types of simultaneous confrontation exist, each one corresponding to a different maneuver. These different types are shown in Fig. 8.22. Type 1 has already been discussed extensively. Type 2 is similar, but now **P2** turns right as fast as possible; $x_2(T)$ is again the parameter. Types 3 and 4 are different in the sense that destruction is not caused by the confrontation range, but instead by collision; the angle $\theta(T)$ is the parameter. Type 3 is a collision with **P1** turning right and **P2** turning left. In type 4, both turn right. Type 5 calls for an S-maneuver by **P1** to get within the range of **P2**, who has the longer confrontation range, before **P2**, who is turning right, faces directly towards **P1**, at a distance of $x_2(T) = l_1$; the turn angle α of the final bend of **P1** is the parameter. Type 6 is an extension of the S-maneuver of **P1** in type 5, in which the final bend with a fixed angle α_1 is preceded by a straight-line segment (singular arc) whose length functions as a parameter. This singular arc itself is preceded by a left turn. Type 7 is similar to type 6, the only difference being that now **P1** starts with a right turn. In addition to these 7 types, there are 7 other types, indicated by $1', 2', \ldots, 7'$, obtained from the first 7 types by left-right reflections. For instance, in case $1'$, **P1** turns left and **P2** turns right.

In all these types, the parameter is subject to certain constraints. We will not treat detailed analyses of all these types here; they are somewhat technical and not enlightening for the results to be obtained. The details can be found in Olsder and Breakwell (1974). Those types which give rise to curves lying in the $\theta = \pi/3$ plane have been indicated in Fig. 8.21.

A question that comes into mind now is whether these fourteen different types of simultaneous confrontation constitute the whole picture. In fact, we do not know for sure whether there are other types or not. The ones that have

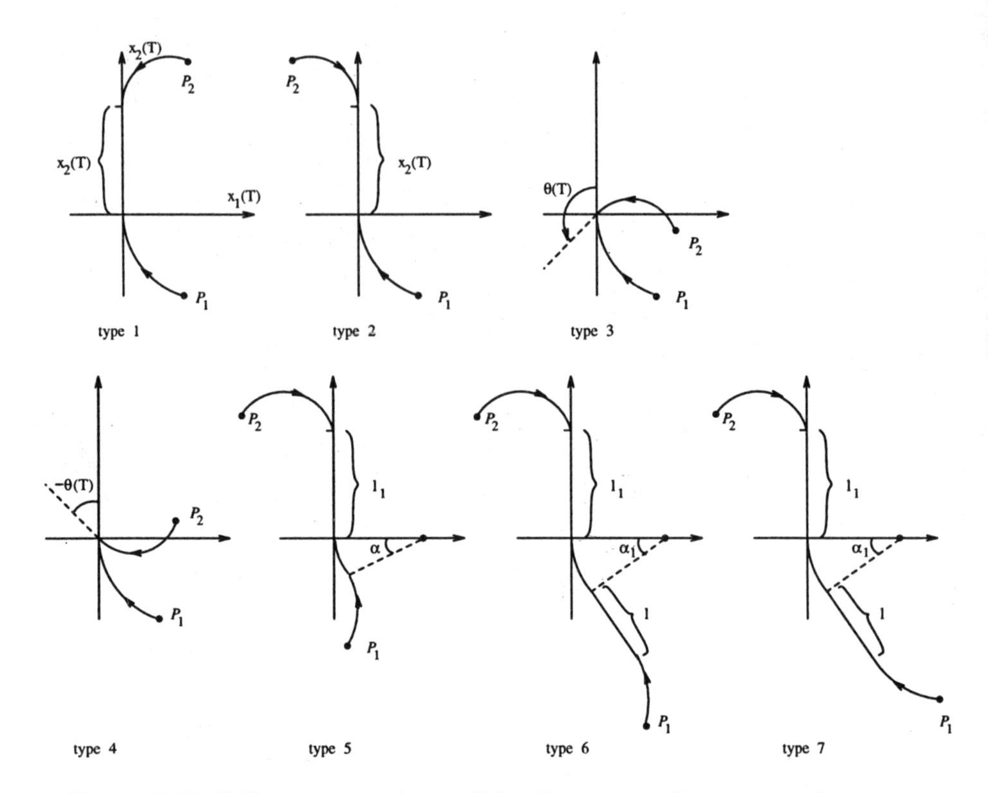

Figure 8.22: Different maneuvers, all leading to simultaneous confrontation.

been depicted in Fig. 8.21 have been obtained using intuitive reasoning, and the analysis indeed showed that they give rise to semipermeable surfaces. However, together with the semipermeable surfaces corresponding to near miss situations, for which the discussions will follow shortly, they separate (and enclose) the states where from **P1** can guarantee victory from the other states, i.e., there is no hole in Σ (cf. Section 8.6). If, however, another simultaneous confrontation maneuver (or another yet-unknown near miss maneuver) is discovered, it may lead to either a larger or a smaller $\mathbf{R}_1$.

Near miss

Essentially ten different types of near miss situations have been discovered so far, which are labeled as $a, b, \ldots, j$ in Fig. 8.23, which depicts the corresponding trajectories in real space. Another set of ten types exists, obtained by interchanging left and right, as in the case of simultaneous confrontation. We now give a brief rundown of the first ten types.

In type a both players move to the right as fast as possible; **P2**'s trajectory in relative space looks as sketched in Fig. 8.24, and its characteristic feature is that it touches the boundary of **P1**'s confrontation range, but does not cross it—hence the name "near miss". The parameter can be taken as $x_2(T)$, which

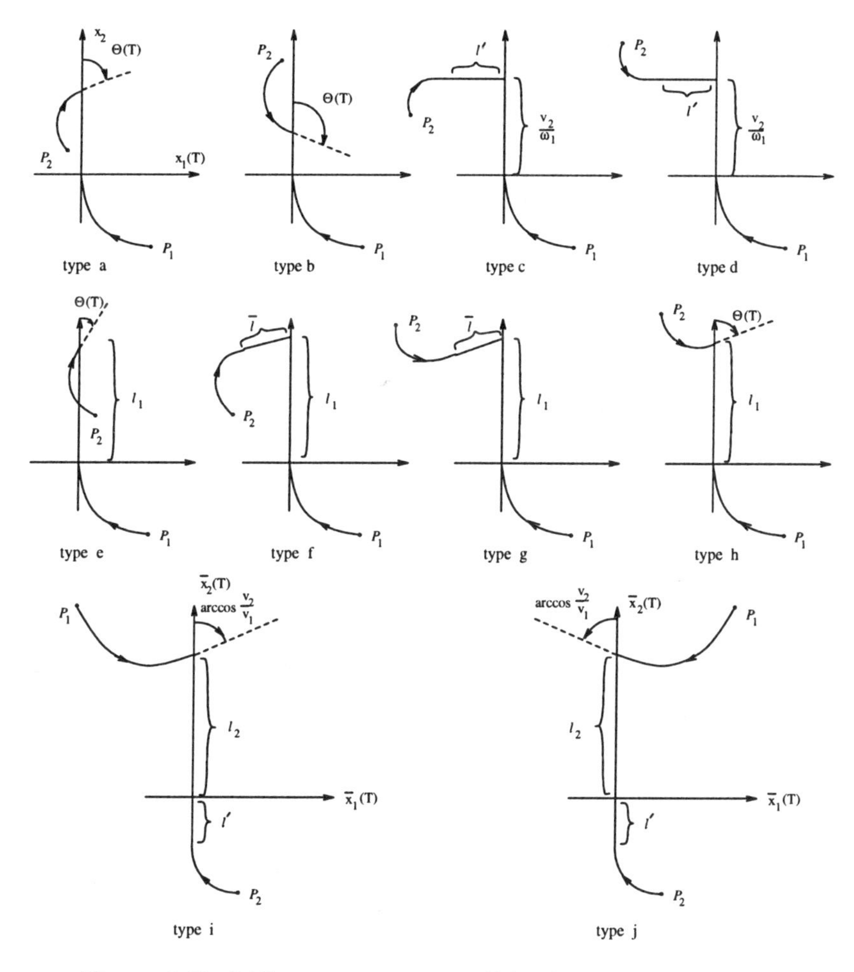

Figure 8.23: Different maneuvers, all leading to near misses.

satisfies the constraint $0 \leq x_2(T) \leq \min(\ell_1, v_2/\omega_1)$ and determines $\theta(T)$ through $\sin\theta(T) = \omega_1 x_2(T)/v_2$. The terminal conditions for the costate equations are $V_{x_1}(T) > 0$, $V_{x_2}(T) = V_\theta(T) = 0$. Type b resembles type a, the difference being that $u^{2*} = -1$. Types c and d have $\theta(T) = \pi/2$ and end with a singular arc by **P2**. The length of this arc is the parameter. These two types only occur if $\omega_1\ell_1 \geq v_2$. For $\omega_1\ell_1 < v_2$, the types c and d do not arise, but instead we have the types e, f, g and h, for which $x_2(T) = \ell_1$. For types e and h the parameter is $\theta(T)$, for f and g it is the length of the singular arc of **P2**'s trajectory. Types i and j resemble e and h, respectively, but now **P1** is almost destroyed at the far end of **P2**'s confrontation range. In e and h it is **P2** who can barely escape the far end of **P1**'s confrontation range. Note that in the figures corresponding to i and j, the relative coordinate system at $t = T$ has been drawn with respect to **P2** (instead of **P1**, as in all other types). In these two latter types, **P2** has a

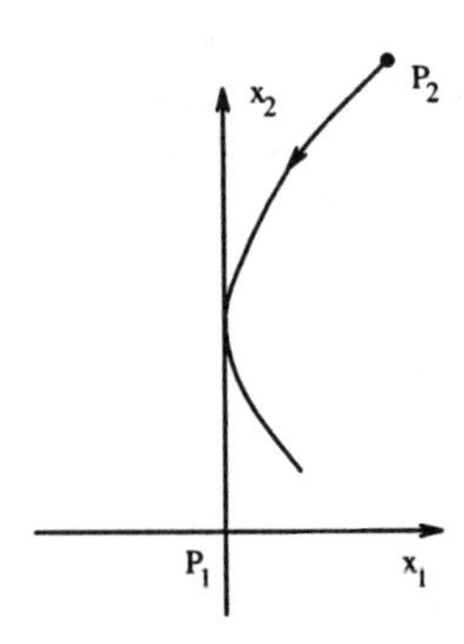

Figure 8.24: The trajectory corresponding to type a in Fig. 8.23, now in relative space.

singular arc, the length of which is the parameter.

In Fig. 8.21, we have drawn the line segments corresponding to the applicable near miss cases for the example with characteristics (8.50), which are the types a, b, c and d and their mirror images a', b', c', d'. The crucial question now is whether all the line segments (both simultaneous confrontation and near miss) enclose a region, together with Λ_1. It should be clear from the figure that two such regions exist: an "inner one" and an "outer one". Closer scrutiny reveals that some corners of the outer region leak, and the semipermeable surfaces which enclose the inner region constitute the barrier which separates $\mathbf{R}_1$ from $\mathbf{R}_2$.

To summarize, we have given a complete characterization of the regions leading to victory for **P1** and **P2**, and have arrived at the qualitative result that if **P1** plays optimally along the boundary of his region of victory ($\mathbf{R}_1$), then **P2** is captured within $\mathbf{R}_1$ (and likewise with the roles of **P1** and **P2** interchanged).[96] If he cannot stay within $\mathbf{R}_1$ forever, then he has to leave this region via Λ_1 which leads to termination of the game (with **P1**'s victory).

8.8 Problems

1. Instead of the pursuit-evasion game of Example 8.2, consider the equivalent problem described by

$$\begin{aligned} \dot{x}_1 &= u^1 + u^2, \quad x_1(0) > 0, \\ \dot{x}_2 &= 1 - x_1, \quad x_2(0) = 0, \\ L &= x_2(T), \end{aligned}$$

where $T = \min\{t : x_1(t) = 0\}$, $-1 \leq u^i \leq 0$, $i = 1, 2$. Obtain the final condition for the costate equation by making use of (8.14), and show that the result is in complete agreement with the corresponding condition derived in Example 8.2.

[96] Note that since $\ell_1 = \infty$, $\mathbf{R}_3$ is empty.

2. Calculate the value function $V(x)$ for Example 8.5, and show that both V and $\partial V/\partial x$ are continuous along the universal surface.

3. Consider the "lady in the lake" with an instantaneous informational advantage to **E**, i.e., the strategies are of the form $u^1(t) = \gamma^1(x(t)), u^2(t) = \gamma^2(x(t), u^1(t))$. **E** starts at the center of the pond and spirals outward as fast as possible, while maintaining $\theta(t) \equiv \pi$. Show that the circle $r = Rv_2$ will be reached in finite time.

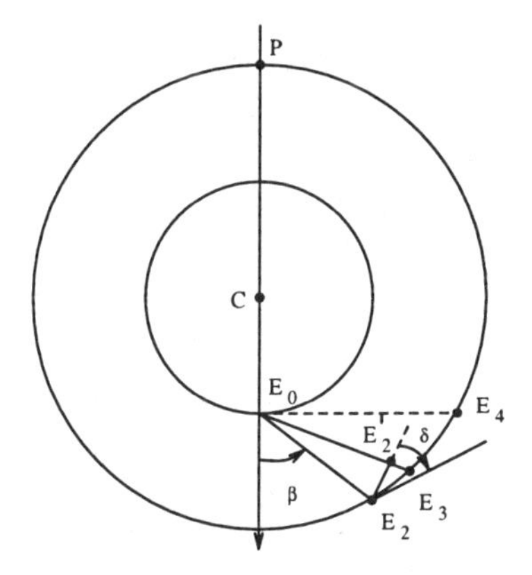

Figure 8.25: Geometrical derivation of the saddle-point solution.

4. This problem addresses the derivation of the equilibrium solution of the "lady in the lake" geometrically, without utilizing Thm. 8.2. Start from the decision point E_0 in Fig. 8.25, and suppose that **P** chooses $u^1 = -1$ (i.e., he moves such that the pond is to his left in real space); the case $u^1 = +1$ can be dealt with analogously. **E** will then swim in a straight line to the shore. The question is: in which direction will she swim? Suppose that she swims in the direction β (see Fig. 8.25).

   (i) Why can β be restricted to $|\beta| \leq \pi/2$?

   (ii) Show geometrically that swimming towards E_3 is better for **E** then to swim to E_2, where E_3 is only an ϵ-distance away from E_2. (Show that the time needed for **P** to cover the distance E_2E_3 is greater than the time needed for **E** to cover the distance $E_2'E_3$, where E_2' is the projection of E_2 on E_0E_3.)

   (iii) Show that swimming towards E_4 is the best **E** can do. (Make use of the fact that the circle through the points E_4, E_0 and C lies inside the pond.)

5. Consider the following two-cars model:

$$\dot{x}_1 = \left(\sqrt{2}/2\right) \sin\theta - x_2 u^1,$$

$$\begin{aligned}\dot{x}_2 &= \left(\sqrt{2}/2\right)\cos\theta + x_1 u^1 - 1,\\ \dot{\theta} &= -u^1 + 2\sqrt{2}u^2,\end{aligned}$$

together with the constraints $|u^i(t)| \leq 1$, $i = 1, 2$. The target set $\bar{\Lambda}$ is defined by $x_1^2 + (x_2 + R)^2 - R^2 = 0$, where $R = \sqrt{2}/8$. **P1** tries to avoid $\bar{\Lambda}$, whereas **P2** would like to have the state enter $\bar{\Lambda}$.

(i) Determine the barrier(s) which separate(s) points from where avoidance is possible from those from where it is not, and show that each one lies partly inside $\bar{\Lambda}$. Therefore, the semipermeable surface determined cannot be the barrier(s) sought.

(ii) Determine the actual barrier, in a way similar to that employed in the dolichobrachistochrone problem (see Chigir (1976) or the first edition of Başar and Olsder (1982)).

6. Consider the system described by

$$\begin{aligned}\dot{x}_1 &= x_2 + 1 + 2\sin u^2,\\ \dot{x}_2 &= -3u^1 + 2\cos u^2,\end{aligned}$$

with $|u^1| \leq 1$ and no constraints imposed on u^2. The target set is the half line $(x_1 > 0, x_2 = 0)$. **P1** wants to steer the system to the target set as soon as possible, whereas **P2** wants to do the opposite. The game is played in the half plane $x_2 \geq 0$.

(i) Show that a barrier starts at the origin $(x_1 = 0, x_2 = 0)$ and ends at $x_2 = 1$.

(ii) Show that the continuation of the barrier, for $x_2 > 1$, is an equivocal line.

7. The differential game addressed in this problem is the same as that of Problem 6, except that the dynamics are replaced by

$$\begin{aligned}\dot{x}_1 &= x_2 + 1 + u_1^1 + 2\sin u^2,\\ \dot{x}_2 &= -3u_2^1 + 2\cos u^2,\end{aligned}$$

where **P1** chooses u_1^1 and u_2^1, subject to the constraint $(u_2^1)^2 + (u_1^1)^2/\epsilon \leq 1$, where ϵ is a small positive number. **P2** chooses, as before, u^2, without any restrictions. For $\epsilon = 0$ this problem reduces to Problem 6. It is conjectured that the present more general version features a switching envelope. Prove or disprove this conjecture, and also investigate the limiting behavior of the solution when $\epsilon \downarrow 0$.

8. The kinematics of this problem are described by (8.22) with $v_2 = \frac{1}{2}$. The only restriction on the controls is $|u^1(t)| \leq 1$. The target set of **P1** is given by

$$\Lambda_1 = \left\{(x_1, x_2) : x_1^2 + x_2^2 \leq 4 \text{ and } \arctan\frac{x_2}{x_1} \leq \frac{2\pi}{3}\right\}$$

and the target set of **P2** is given by

$$\Lambda_2 = \left\{(x_1, x_2) : x_1^2 + x_2^2 \leq 1\right\}.$$

P1's objective is to force (x_1, x_2) into Λ_1 without passing through Λ_2, whereas **P2** wants to have the state enter Λ_2 without passing through Λ_1. Show that the region of victory $\mathbf{R}_i$ for $\mathbf{P}i$ is as indicated in the following figure:

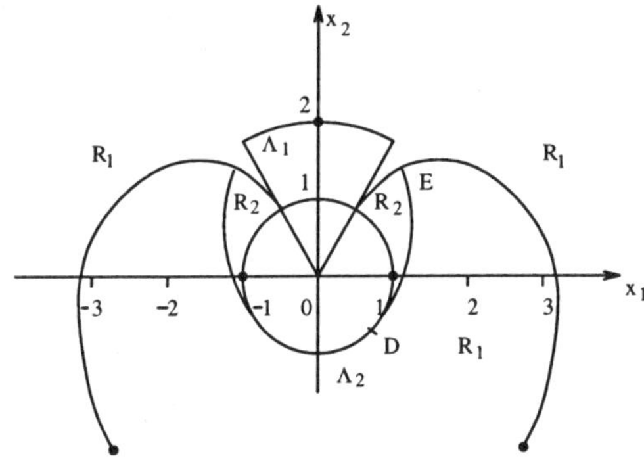

Check also that the semipermeable curve DE lies completely outside Λ_2 and that the corner E does not leak.

9. We are given a zero-sum differential game for which the Isaacs condition does not hold everywhere, but instead the Isaacs equation admits two different solutions depending on the order of the *min* and *max* operations. Prove that the associated differential game does not admit a saddle-point solution in feedback strategies, but it admits a saddle point when one or the other player is given an informational advantage on the current action of his opponent. Make use of this result in providing a possible verification for the existence of a saddle point in the differential game "lady in the lake."

8.9 Notes

Section 8.2. The theory of deterministic pursuit-evasion differential games was single-handedly created by Isaacs in the early 1950s, which culminated in his book (Isaacs, 1975; first edn. 1965). Blaquière and Leitmann, independently of Isaacs, also obtained Thms. 8.1 and 8.2 (see Blaquière et al., 1969). Their geometric approach is essentially the same as that of Isaacs, but it is stated in a mathematically more precise language. In this chapter, we follow essentially Isaacs' approach. Starting in the 1960s, many researchers have worked on a rigorous establishment of the validity of the Isaacs equation. The starting point is often a time-discretization and a carefully defined information structure. Depending on this information structure, lower and upper values of the game are defined, and then appropriate limiting arguments are incorporated, under which these upper and lower values approach each other; the resulting limiting value, in each case, is declared to be the value of the game. References in this connection are Fleming (1961, 1964), Varaiya and Lin (1969), Roxin (1969),

Friedman (1971), Elliott and Kalton (1972), Elliott (1977) and Elliott et al. (1973). The relationship with viscosity solutions is given in Evans and Souganidis (1984) and in Barron, Evans and Jensen (1984). These two references contain results for fixed terminal time problems only; see also Elliott (1987). For results on viscosity solutions for problems with variable terminal time the reader is referred to Bardi and Soravia (1991a, b). For extended solution concepts of ordinary differential equations, in which, for instance, the f-function in the state equation is not Lipschitz-continuous, the reader is referred to Krasovskii and Subbotin (1988), which also includes a rigorous definition of strategy, and also to Hajek (1979).

The notions of playability and nontermination aided in the understanding of the so-called "bang-bang-bang surface" introduced in Isaacs (1969); see also Ciletti (1970) and Lewin (1976).

Section 8.3. The concept of capturability also originates in Isaacs' works. A mathematically more rigorous approach to capturability is given in Krasovskii and Subbotin (1988), leading to the so-called "Theorem of the Alternative". A different approach to capturability has been developed by Pontryagin (1967) and subsequently by Hajek (1975). In this set-up the pursuer knows the current action of the evader, which he "neutralizes" by means of his own control. The left-over power (if any) is used by the pursuer to steer the system to the target set.

The famous "homicidal chauffeur game" is very versatile in that its solution features many singular surfaces. We do not provide here the complete solution to this game; it was partly solved in Isaacs (1975). The complete solution is spread out over Merz (1971) and Breakwell (1973). The notion of a leaking corner was introduced in Bernhard (1971).

Section 8.4. The transition, dispersal, universal and equivocal lines were first introduced in Isaacs (1975). The focal line and the switching envelope appeared for the first time in Merz (1971) and Breakwell (1973).

Introduction of noise in the systems equations tends to "smoothen" the singular surfaces which will therefore no longer exist as such. The addition of noise to the system dynamics is the starting point for viscosity solutions; see Chapter 5. For a "noisification" of the homicidal chauffeur game, see Pachter and Yavin (1979).

Section 8.5. The "lady in the lake" appeared in the Russian translation of Isaacs' book; see also Breakwell (1977). Another well-known differential game, which has a *safe contact* in its solution, is the dolichobrachistochrone problem, stated and partly solved in Isaacs (1975). The complete solution was given later in Breakwell (1971) and Chigir (1976). In the survey article (Ho and Olsder, 1983), a rather exhaustive list of other known solved or partly solved zero-sum differential games is given, including such games as "lion and man", "obstacle tag", "conic surveillance evasion" and the "suicidal pedestrian". References to extensions with "two pursuers and one evader" or "one pursuer and two evaders" are also given there. A recent book in the vein of Isaacs is Lewin (1994); it contains new developments and examples.

Section 8.6. The application in maritime ship collision avoidance follows the works of Sharma (1976), and Vincent and Peng (1973). See also Merz (1973) and Olsder and Walter (1978).

Section 8.7. Role determination and its application in aeronautics were first discussed in Olsder and Breakwell (1974). Similar problems, but with different target sets, have been considered in Merz (1976). For other aeronautical applications see Peng and Vincent (1975) and Vincent et al. (1976).

Section 8.8. Problem 5 is from Vincent and Peng (1973). Problem 6 is a special version of a problem considered in Isaacs (1975), Problem 8 is from Getz and Pachter (1980).

Appendix A

Mathematical Review

This appendix provides some background material on those aspects of real analysis and optimization which are frequently used in the text; it also serves to introduce the reader to our notation and terminology. For a more detailed exposition on the topics covered here, a standard reference is Luenberger (1969).

A.1 Sets

A set S is a collection of elements. If s is a member (element) of S, we write $s \in S$; if s does not belong to S, we write $s \notin S$. If S contains a finite number of elements, it is called a *finite set*; otherwise it is called an *infinite set*. If the number of elements of an infinite set is countable (i.e., if there is a one-to-one correspondence between its elements and positive integers), we say that it is a denumerable (countable) set, otherwise it is a *nondenumerable* set.

A set S with some specific structure attached to it is called a *space*, and it is called a linear *(vector) space* if this specific structure is of algebraic nature with certain well-known properties which we assume that the reader is familiar with. If S is a vector space, a subset of S which is also a vector space is called a *subspace*. An example of a vector space is the *n-dimensional Euclidean space* (denoted by R^n) each element of which is determined by n real numbers. An $x \in \mathrm{R}^n$ can either be written as a *row vector* $x = (x_1, \ldots, x_n)$, where $x_1, \ldots, x_n$ are real numbers and denote the components of x, or as a *column vector* which is the "transpose" of $(x_1, \ldots, x_n)$ (written as $x = (x_1, \ldots, x_n)'$). We shall adopt the latter convention in this text, unless indicated otherwise.

Linear independence

Given a finite set of vectors $s_1, \ldots, s_n$ in a vector space S, we say that this set of vectors is *linearly independent* if the equation $\sum_{i=1}^{n} \alpha_i s_i = 0$ implies $\alpha_i = 0$, $i = 1, \ldots, n$. Furthermore, if every element of S can be written as a linear combination of these vectors, we say that this set of vectors *generates* S. Now, if S is generated by such a linearly independent finite set (say, X), it is said to be *finite dimensional* with its unique "dimension" being equal to the number of elements of X; otherwise, S is *infinite dimensional.*

A.2 Normed Linear (Vector) Spaces

A *normed linear vector space* is a linear (vector) space S which has some additional structure of topological nature. This structure is induced on S by a real-valued function which maps each element $u \in S$ into a real number $\|u\|$ called the *norm* of u. The norm satisfies the following three axioms:

(1) $\|u\| \geq 0 \; \forall u \in S$; $\|u\| = 0$ if, and only if, $u = 0$.

(2) $\|u + v\| \leq \|u\| + \|v\|$ for each u, $v \in S$.

(3) $\|\alpha u\| = |\alpha| \cdot \|u\| \; \forall \alpha \in \mathrm{R}$ and for each $u \in S$.

Convergent sequences, Cauchy sequence

An infinite sequence of vectors $\{s_1, s_2, \ldots, s_i \ldots\}$ in a normed vector space S is said to *converge* to a vector s if, given an arbitrary $\epsilon > 0$, there exists an N, which may depend on ϵ, such that $\|s - s_i\| < \epsilon$ for all $i \geq N$. In this case, we write $s_i \to s$, or $\lim_{i\to\infty} s_i = s$, and call s the *limit point* of the sequence $\{s_i\}$. More generally, a point s is said to be a *limit point* of an infinite sequence $\{s_i\}$ if the latter has an infinite subsequence $\{s_{i_k}\}$ that converges to s.

An infinite sequence $\{s_i\}$ in a normed vector space is said to be a *Cauchy sequence* if, given an $\epsilon > 0$, there exists an N such that $\|s_n - s_m\| < \epsilon$ for all n, $m \geq N$. A normed vector space S is said to be *complete*, or a *Banach space*, if every Cauchy sequence in S is convergent to an element of S.

Open, closed and compact sets

Let S be a normed vector space. Given an $s \in S$ and an $\epsilon > 0$, the set $N_\epsilon(s) = \{x \in S : \|x - s\| < \epsilon\}$ is said to be an *ϵ-neighborhood* of s. A subset X of S is *open* if, for every $x \in X$, there exists an $\epsilon > 0$ such that $N_\epsilon(x) \subset X$. A subset X of S is *closed* if its complement in S is open; equivalently, X is closed if every convergent sequence in X has its limit point in X. Given a set $X \subset S$, the largest subset of X which is open is called the *interior* of X and denoted as $\mathring{X}$.

A subset X of a normed vector space S is said to be *compact* if every infinite sequence in X has a convergent subsequence whose limit point is in X. If X is finite dimensional, compactness is equivalent to being closed and bounded.

Transformations and continuity

A mapping f of a vector space S into a vector space T is called a *transformation* or a *function*, and is written symbolically $f : S \to T$ or $y = f(x)$, for $x \in S$, $y \in T$; f is said to be a *functional* if $T = \mathrm{R}$.

Let $f : S \to T$, where S and T are normed linear spaces. The mapping f is said to be *continuous* at $x_0 \in S$ if, for every $\epsilon > 0$, there exists a $\delta > 0$ such that $f(x) \in N_\epsilon(f(x_0))$ for every $x \in N_\delta(x_0)$. If f is continuous at every point of S it is said to be *continuous everywhere* or, simply, *continuous*.

A.3 Matrices

An $(m \times n)$ matrix A is a rectangular array of numbers, called *elements* or *entries*, arranged in m rows and n columns. The element in the ith row and jth column of A is denoted by a subscript ij, such as a_{ij} or $[A]_{ij}$, in which case we write $A = \{a_{ij}\}$. A matrix is said to be *square* if it has the same number of rows and columns; an $(n \times n)$ square matrix A is said to be an *identity matrix* if $a_{ii} = 1$, $i = 1, \ldots, n$, and $a_{ij} = 0$, $i \neq j$, $i, j = 1, \ldots, n$. An $(n \times n)$ identity matrix will be denoted by I_n or, simply, by I whenever its dimension is clear from the context.

The *transpose* of an $(m \times n)$ matrix A is the $(n \times m)$ matrix A' with elements $a'_{ij} = a_{ji}$. A square matrix A is symmetric if $A = A'$; it is nonsingular if there is an $(n \times n)$ matrix, called the inverse of A, denoted by A^{-1}, such that $A^{-1}A = I = AA^{-1}$.

Eigenvalues and quadratic forms

If A is a square matrix, a scalar λ and a nonzero vector x satisfying the equation $Ax = \lambda x$ are said to be, respectively, an *eigenvalue* and an *eigenvector* of A.

A square symmetric matrix A whose eigenvalues are all positive (respectively, nonnegative) is said to be *positive definite* (respectively, *nonnegative definite* or *positive semidefinite*). An equivalent definition is as follows. A symmetric $(n \times n)$ matrix A is said to be positive definite (respectively, nonnegative definite) if $x'Ax > 0$ (respectively, $x'Ax \geq 0$) for all nonzero $x \in \mathrm{R}^n$. The matrix A is said to be *negative definite* (respectively, *nonpositive definite*) if the matrix $(-A)$ is positive (respectively, nonnegative) definite. We symbolically write $A \geq 0$ (respectively, $A \geq 0$) to denote that A is positive (respectively, nonnegative) definite.

A.4 Convex Sets and Functionals

A subset C of a vector space S is said to be *convex* if for every u, $v \in C$ and every $\alpha \in [0, 1]$, we have $\alpha u + (1 - \alpha)v \in C$. A functional $f : C \to \mathrm{R}$ defined over a convex subset C of a vector space S is said to be *convex* if, for every u, $v \in C$ and every scalar $\alpha \in [0, 1]$, we have $f(\alpha u + (1 - \alpha)v) \leq \alpha f(u) + (1 - \alpha)f(v)$. If this is a strict inequality for every $\alpha \in (0, 1)$, then f is said to be *strictly convex*. The functional f is said to be *concave* if $(-f)$ is convex, and *strictly concave* if $(-f)$ is strictly convex.

A functional $f : \mathrm{R}^n \to \mathrm{R}$ is said to be differentiable if, with $x = (x_1, \ldots, x_n)' \in \mathrm{R}^n$, the partial derivatives of f with respect to the components of x exist, in which case we write

$$\nabla f(x) = [\partial f(x)/\partial x_1, \ldots, \partial f(x)/\partial x_n].$$

$\nabla f(x)$ is called the *gradient* of f at x and is a row vector. We shall also use the notation $f_x(x)$ or $\mathrm{d}f(x)/\mathrm{d}x$ to denote the same quantity. If we partition x into two vectors y and z of dimensions n_1 and $n-n_1$, respectively, and are interested only in the partial derivatives of f with respect to the components of y, then we use the notation $\nabla_y f(y,z)$ or $\partial f(y,z)/\partial y$ to denote this partial gradient.

Let $g : \mathrm{R}^n \to \mathrm{R}^m$ be a vector-valued function whose components are differentiable with respect to the components of $x \in \mathrm{R}^n$. Then, we say that $g(x)$ is differentiable, with the derivative $\mathrm{d}g(x)/\mathrm{d}x$ being an $(m \times n)$ matrix whose ijth element is $\partial g_i(x)/\partial x_j$. (Here g_i denotes the ith component of g.) The gradient $\nabla f(x)$ being a vector, its derivative (which is the second derivative of $f : \mathrm{R}^n \to \mathrm{R}$) will thus be an $(n \times n)$ matrix, assuming that $f(x)$ is twice continuously differentiable in terms of the components of x. This matrix, denoted by $\nabla^2 f(x)$, is symmetric, and is called the *Hessian matrix* of f at x. This Hessian matrix is nonnegative definite for all $x \in \mathrm{R}^n$ if, and only if, f is convex.

Separating hyperplane theorem

Given a convex set S, one can always find a (hyper)plane such that S lies on one side of it. This result, which should be intuitive in finite-dimensional spaces, also holds in infinite-dimensional linear vector spaces, and is referred to as *support theorem* or *separating hyperplane theorem*. Toward giving a precise statement for the most general case, let us first define what a hyperplane is: A *hyperplane* H in a linear vector space S is a *linear variety* H (that is, translation of a subspace of S), with the property that $H \neq S$, and if V is any linear variety containing H, then either $V = S$ or $V = H$. Then the *separating hyperplane theorem* says that if C is a convex set (as a subset of S) with nonempty interior, and x a point in S and not an interior point of C, there is a *closed* hyperplane H containing x such that H lies on one side of C. The point x can of course be chosen as a boundary point of C.

A.5 Optimization of Functionals

Given a functional $f : S \to \mathrm{R}$, where S is a vector space, and a subset $X \subseteq S$, by the optimization problem

$$\text{minimize } f(x) \text{ subject to } x \in X$$

we mean the problem of finding an element $x^* \in X$ (called a *minimizing element* or an *optimal solution*) such that

$$f(x^*) \leq f(x) \qquad \forall x \in X.$$

If such an $x^* \in X$ exists, then we use the notation $x^* = \arg\min_{x \in X} f(x)$. This is sometimes also referred to as a *globally minimizing solution*, in order to differentiate it from the other alternative—a *locally minimizing solution.* An element $x^\circ \in X$ is called a locally minimizing solution if we can find an $\epsilon > 0$ such that

$$f(x^\circ) \leq f(x) \qquad \forall x \in N_\epsilon(x^\circ) \cap X,$$

i.e., we compare $f(x^\circ)$ with values of $f(x)$ in that part of a certain ϵ-neighborhood of x°, which lies in X.

For a given optimization problem, it is not necessary that an optimal solution exists; an optimal solution will exist if the set of real numbers $\{f(x) : x \in X\}$ is bounded below and there exists an $x^* \in X$ such that $\inf\{f(x) : x \in X\} = f(x^*)$, in which case we write

$$f(x^*) = \inf_{x \in X} f(x) = \min_{x \in X} f(x).$$

If such an x^* cannot be found, even though $\inf\{f(x) : x \in X\}$ is finite, we simply say that an optimal solution does not exist; but we declare the quantity

$$\inf\{f(x) : x \in X\} \text{ or } \inf_{x \in X} f(x)$$

as the *optimal value* of the optimization problem. If $\{f(x) : x \in X\}$ is not bounded below, i.e., $\inf_{x \in X} f(x) = -\infty$, then neither an optimal solution nor an optimal value exists.

An optimization problem which involves maximization instead of minimization may be converted into a minimization problem by simply replacing f by $-f$. Any optimal solution of this minimization problem is also an optimal solution for the initial maximization problem, and the optimal value of the latter, denoted $\sup_{x \in X} f(x)$, is equal to minus the optimal value of the former. If a *maximizing element* $x^* \in X$ exists, then $\sup_{x \in X} f(x) = \max_{x \in X} f(x) = f(x^*)$.

Existence of optimal solutions

In the minimization problem formulated above, an optimal solution exists if X is a finite set, since then there is only a finite number of comparisons to make. If X is not finite, however, existence of an optimal solution is not always guaranteed; it is guaranteed if f is continuous and X is compact—a result known as the *Weierstrass theorem.* For the special case when X is finite dimensional, we should recall that compactness is equivalent to being closed and bounded.

Necessary and sufficient conditions for optimality

Let $S = \mathbf{R}^n$, and $f : \mathbf{R}^n \to \mathbf{R}$ be a differentiable function. If X is an open set, a first-order necessary condition for an optimal solution to satisfy is

$$\nabla f(x^*) = 0.$$

If, in addition, f is twice continuously differentiable on $\mathbf{R}^n$, a second-order necessary condition is

$$\nabla^2 f(x^*) \geq 0.$$

The pair of conditions $\{\nabla f(x^*) = 0, \nabla^2 f(x^*) > 0\}$ is sufficient for $x^* \in X$ to be a locally minimizing solution. These conditions are also sufficient for global optimality if, in addition, X is a convex set and f is a convex functional on X.

These results, by and large, hold also for the case when S is infinite dimensional, but then one has to replace the gradient vector and the Hessian matrix by first and second Gateaux (or Fréchet) derivatives, and the positive-definiteness requirement by "strong positiveness" of an operator. See Luenberger (1969) for these extensions.

Appendix B

Some Notions of Probability Theory

This appendix briefly presents some notions of probability theory which are used in the text. For more complete exposition the reader should consult standard texts on probability theory, such as Feller (1971), Papoulis (1965), Ash (1972) and Loève (1963). In two of the sections of this book (viz. Sections 5.3 and 6.7) we have used some material on stochastic processes and stochastic differential equations which is, however, not covered in this appendix; for this, the reader should consult Wong and Hajek (1985), Gikhman and Skorohod (1972), Fleming and Rishel (1975), Fleming and Soner (1993) and the references cited in Section 6.7.

B.1 Ingredients of Probability Theory

Let Ω denote a set whose elements are the outcomes of a random experiment. This random experiment might be the toss of a coin (in which case Ω has only two elements) or selection of an integer from the set $[0, \infty)$ (in which case Ω is countably infinite) or the continuous roulette wheel which corresponds to a nondenumerable Ω. Any subset of Ω on which a probability measure can be defined is known as an *event*. Specifically, if $\mathbf{F}$ denotes the class of all such events (i.e., subsets of Ω), then it has the following properties:

(1) $\Omega \in \mathbf{F}$.

(2) If $A \in \mathbf{F}$, then its complement $A^c = \{\omega \in \Omega : \omega \notin A\}$ also belongs to $\mathbf{F}$. (The empty set, ϕ, being the complement of Ω, also belongs to $\mathbf{F}$.)

(3) If $A_1, A_2 \in \mathbf{F}$, then $A_1 \cap A_2$ and $A_1 \cup A_2$ also belong to $\mathbf{F}$.

(4) If $A_1, A_2, \ldots, A_i, \ldots$ denote a countable number of events, the countable intersection $\cap_{i=1}^{\infty} A_i$ and the countable union $\cup_{i=1}^{\infty} A_i$ are also events (i.e., they belong to $\mathbf{F}$).

The class $\mathbf{F}$, thus defined, is called a *sigma algebra* (*σ-algebra*) and a *probability measure* $\mathcal{P}$ is a nonnegative functional defined on the elements of this σ-algebra. The probability measure $\mathcal{P}$ also satisfies the following axioms:

(1) For every event $A \in \mathbf{F}$, $0 \leq \mathcal{P}(A) \leq 1$, and $\mathcal{P}(\Omega) = 1$.

(2) If $A_1, A_2 \in \mathbf{F}$ and $A_1 \cap A_2 = \phi$ (i.e., A_1 and A_2 are disjoint events), $\mathcal{P}(A_1 \cup A_2) = \mathcal{P}(A_1) + \mathcal{P}(A_2)$.

(3) Let $\{A_i\}$ denote a (countably) infinite sequence in F, with the properties $A_{i+1} \subset A_i$ and $\cap_{i=1}^{\infty} A_i = \phi$. Then, the limit of the sequence of real numbers $\{\mathcal{P}(A_i)\}$ is zero (i.e., $\lim_{i\to\infty} \mathcal{P}(A_i) = 0$).

The triple $(\Omega, \mathbf{F}, \mathcal{P})$ defined above is known as a *probability space*, while the pair $(\Omega, \mathbf{F})$ is called a *measurable space.* If $\Omega = \mathbf{R}^n$, then its subsets of interest are the n-dimensional rectangles, and the smallest σ-algebra generated by these rectangles is called the n-dimensional *Borel σ-algebra* and is denoted $\mathbf{B}^n$. Elements of $\mathbf{B}^n$ are *Borel sets*, and the pair $(\mathbf{R}^n, \mathbf{B}^n)$ is a *Borel (measurable) space.* A probability measure defined on this space is known as a *Borel probability measure*.

Finite and countable probability spaces

If Ω is a finite set (say, $\Omega = \{\omega_1, \omega_2, \ldots, \omega_n\}$), we can assign probability weights on individual elements of Ω, instead of on subsets of Ω, in which case we write p_i to denote the probability of the single event ω_i. We call the n-tuple $(p_1, p_2, \ldots, p_n)$ a *probability distribution* over Ω. Clearly, we have the restriction $0 \leq p_i \leq 1 \; \forall i = 1, \ldots, n$, and furthermore as the elements of Ω are all disjoint events, we have the property $\sum_{i=1}^{n} p_i = 1$. The same convention applies when Ω is a countable set (i.e., $\Omega = \{\omega_1, \omega_2, \ldots, \omega_i, \ldots\}$), in which case we simply replace n by ∞.

B.2 Random Vectors

Let $(\Omega_1, \mathbf{F}_1)$ and $(\Omega_2, \mathbf{F}_2)$ be two measurable spaces and f be a function defined from the domain set Ω_1 into the range set Ω_2. If for every $A \in \mathbf{F}_2$ we have $f^{-1}(A) \triangleq \{\omega \in \Omega_1 : f(\omega) \in A\} \in \mathbf{F}_1$, then f is said to be a *measurable function*, or a *measurable transformation* from $(\Omega_1, \mathbf{F}_1)$ into $(\Omega_2, \mathbf{F}_2)$. If the latter measurable space is a Borel space, then f is said to be a *Borel function*, in which case we denote it by x. For the special case when the Borel space is $(\Omega_2, F_2) = (\mathbf{R}, \mathbf{B})$, the Borel function x is called a *random variable.* For the case when $(\Omega_2, F_2) = (\mathbf{R}^n, \mathbf{B}^n)$, x is known as an *n-dimensional random vector.*

If there is a probability measure $\mathcal{P}$ defined on $(\Omega_1, \mathbf{F}_1)$—which we henceforth write simply as $(\Omega, \mathbf{F})$—then the random vector x will induce a probability measure P_x on the Borel space $(\mathbf{R}^n, \mathbf{B}^n)$, so that for every $B \in \mathbf{B}^n$ we have $\mathcal{P}_x(B) = \mathcal{P}(x^{-1}(B))$. Since every element of $\mathbf{B}^n$ is an n-dimensional rectangle, the arguments of P_x are in general infinite sets; however, considering the collection of sets $\{\xi \in \mathbf{R}^n : \xi_i < a_i, i = 1, \ldots, n\}$ in $\mathbf{B}^n$ where a_i $(i = 1, \ldots, n)$ are real numbers, restriction of P_x to this class is also a probability measure whose argument is now a finite set. We denote this probability measure by $P_x = P_x(a_1, a_2, \ldots, a_n)$ and call it a *probability distribution function* of the random vector x. Note that

$$P_x(a_1, a_2, \ldots, a_n) = \mathcal{P}(\{\omega \in \Omega : x_1(\omega) < a_1, x_2(\omega) < a_2, \ldots, x_n(\omega) < a_n\}),$$

where x_i is a random variable denoting the ith component of x. Whenever $n > 1$, P_x is sometimes also called the *cumulative (joint) probability* distribution function. It is a well-established fact that there is a one-to-one correspondence between P_x and $\mathcal{P}_x$ and the subspace on which P_x is defined can generate the whole $\mathbf{B}^n$ (cf. Loève, 1963).

Independence

Given the probability distribution function of a random vector $x = (x_1, \ldots, x_n)'$ the (*marginal*) distribution function of each random variable x_i can be obtained from

$$P_{x_i}(a_i) = \lim_{a_j \to \infty, j \neq i} P_x(a_1, \ldots, a_n).$$

The random variables $x_1, \ldots, x_n$ are said to be (statistically) *independent* if

$$P_x(a_1, \ldots, a_n) = P_{x_1}(a_1) P_{x_2}(a_2) \cdots P_{x_n}(a_n),$$

for all scalars $a_1, \ldots, a_n$.

Probability density function

A measure defined on subintervals of the real line and which equals the length of the corresponding subinterval(s) is called a *Lebesgue measure.* It assigns zero weight to countable subsets of the real line, and its definition can readily be extended to n-dimensional rectangles in $\mathbf{R}^n$.

Let $\mathcal{P}$ be a Borel probability measure on $(\mathbf{R}^n, \mathbf{B}^n)$ such that any element of $\mathbf{B}^n$ which has a Lebesgue measure of zero has also a $\mathcal{P}$-measure of zero; then we say that $\mathcal{P}$ is *absolutely continuous* with respect to the Lebesgue measure. Now, a well-established result of probability theory says that (cf. Loève, 1963), if $x : (\Omega, \mathbf{F}, P) \to (\mathbf{R}^n, \mathbf{B}^n, \mathcal{P}_x)$ is a random vector and if $\mathcal{P}_x$ is absolutely continuous with respect to the Lebesgue measure, there exists a nonnegative Borel function $p_x(\cdot)$ such that, for every $A \in \mathbf{B}^n$,

$$\mathcal{P}_x(A) = \int_A p_x(\xi)\, \mathrm{d}\xi.$$

Such a function $p_x(\cdot)$ is called the *probability density function* of the random vector x. In terms of the distribution function P_x, the preceding relation can be written as

$$P_x(a_1, \ldots, a_n) = \int_{-\infty}^{a_1} \cdots \int_{-\infty}^{a_n} p_x(\xi_1, \ldots, \xi_n)\, \mathrm{d}\xi_1 \cdots \mathrm{d}\xi_n$$

for every scalar $a_1, \ldots, a_n$.

B.3 Integrals and Expectation

Let $x : (\Omega, \mathbf{F}, \mathcal{P}) \to (\mathbf{R}^n, \mathbf{B}^n, \mathcal{P}_x)$ be a random vector and $f : (\mathbf{R}^n, \mathbf{B}^n) \to (\mathbf{R}^m, \mathbf{B}^m)$ be a nonnegative Borel function. Then, f can also be considered as a random vector from $(\Omega, \mathbf{F})$ into $(\mathbf{R}^m, \mathbf{B}^m)$, and its *average value (expected value)* is defined either by $\int_\Omega f(x(\omega))\, \mathcal{P}(\mathrm{d}\omega)$ or by $\int_{\mathbf{R}^n} f(\xi)\, \mathcal{P}_x(\mathrm{d}\xi)$ depending on which interpretation we adopt. Both of these integrals are well defined and are uniquely equal in value. If f changes signs, then we take $f = f^+ - f^-$ where both f^+ and f^- are nonnegative, and write the expected value of f as

$$E[f(x)] = \int_{\mathbf{R}^n} f^+(\xi)\, \mathcal{P}_x(\mathrm{d}\xi) - \int_{\mathbf{R}^n} f^-(\xi)\, \mathcal{P}_x(\mathrm{d}\xi) \triangleq \int_{\mathbf{R}^n} f(\xi)\, \mathcal{P}_x(\mathrm{d}\xi),$$

provided that at least one of the pair $E[f^+(x)]$ and $E[f^-(x)]$ is finite. Since, by definition, $\mathcal{P}_x(d\xi) = \mathcal{P}_x(\xi + d\xi) - \mathcal{P}_x(\xi)$, this integral can further be written as

$$E[f(x)] = \int_{\mathbf{R}^n} f(\xi)\, \mathrm{d}\mathcal{P}_x(\xi)$$

which is a Lebesgue–Stieltjes integral and which is the convention that we shall adopt. For the special case when $f(x) = x$ we have

$$E[x] \triangleq \int_{\mathbf{R}^n} \xi\, \mathrm{d}\mathcal{P}_x(\xi) \triangleq \bar{x}$$

which is known as the *mean (expected) value* of x. The covariance of the n-dimensional random vector x is defined as

$$E[(x - \bar{x})(x - \bar{x})'] = \int_{\mathbf{R}^n} (\xi - \bar{x})(\xi - \bar{x})'\, \mathrm{d}\mathcal{P}_x(\xi) \triangleq \operatorname{cov}(x)$$

which is a nonnegative definite matrix of dimension $(n \times n)$. Now, if $\mathcal{P}_x$ is absolutely continuous with respect to the Lebesgue measure, $E[f]$ can equivalently be written, in terms of the corresponding density function p_x, as

$$E[f(x)] = \int_{\mathbf{R}^n} f(\xi) p_x(\xi)\, \mathrm{d}\xi.$$

If Ω consists only of a finite number of disjoint events $\omega_1, \omega_2, \ldots, \omega_n$, then the integrals are all replaced by the single summation

$$E[f(x(\omega))] = \sum_{i=1}^{n} f(x(\omega_i)) p_i,$$

where p_i denotes the probability of occurrence of event ω_i. For a countable set Ω, we have the counterpart:

$$E[f(x(\omega))] = \lim_{n \to \infty} \sum_{i=1}^{n} f(x(\omega_i)) p_i.$$

B.4 Norms and the Cauchy–Schwarz Inequality

Given a random vector $x : (\Omega, \mathbf{F}, \mathcal{P}) \to (\mathrm{R}^n, \mathbf{B}^n, \mathcal{P}_x)$, the quantity defined by

$$\|x\| = \{E[x'x]\}^{1/2},$$

provided that it is finite, is called the ($\mathcal{L}^2$) norm of x, and indeed satisfies all the properties of a *norm* introduced in the previous appendix. A random vector with a finite $\mathcal{L}^2$-norm is called a *second-order random vector*. Note that a second-order random vector has a well-defined covariance.

If x and y are two n-dimensional second-order random vectors defined on the same probability space, then we have the following useful inequality, known as the *Cauchy–Schwarz inequality*:

$$|E[x'y]| \leq \|x\| \, \|y\|.$$

The inequality here is an equality if, and only if, $x = \lambda y$, w.p. 1 for some scalar λ, or $y = 0$ w.p. 1.

Appendix C

Fixed Point Theorems

In this appendix we give, without proof, three theorems (on fixed points) which have been used in Chapters 3 and 4 of the text. Proofs of these theorems are rather lengthy and can be found in the references cited. Some general references for these as well as other results on fixed points are (Istrătescu, 1981), (Joshi and Bose, 1985), and (Smart, 1974).

Fixed point theorems

Theorem C.1 *(Brouwer fixed point theorem) If S is a compact and convex subset of R^n and f is a continuous function mapping S into itself, then there exists at least one $x \in S$ such that $f(x) = x$.*

Several proofs exist for this fixed point theorem, one of the most elementary ones being given by Kuga (1974). Its original version has appeared in 1910 (Brouwer, 1910). A generalization of this theorem is the Kakutani fixed point theorem (Kakutani, 1941) given below.

Definition C.1 *(Upper semicontinuity) Let f be a function defined on a normed linear space X, and associating with each $x \in X$ a subset $f(x)$ of some (other) normed linear space Y. Then, f is said to be* upper semicontinuous *(usc) at a point $x_0 \in X$ if, for any sequence $\{x_i\}$ converging to x_0 and any sequence $\{y_i \in f(x_i)\}$ converging to y_0, we have $y_0 \in f(x_0)$. The function f is upper semicontinuous if it is usc at each point of X.*

Theorem C.2 *(Kakutani) Let S be a compact and convex subset of R^n, and let f be an upper semicontinuous function which assigns to each $x \in S$ a closed and convex subset of S. Then there exists some $x \in S$ such that $x \in f(x)$.*

Another generalization of the Brouwer fixed point theorem is due to Schauder (1930), and involves the space of real-valued bounded continuous functions on a subset S of R^n, to be denoted $C(S)$. Here, we will first need the definition of equicontinuity:

Definition C.2 *(Equicontinuity) Let the space of all real-valued bounded continuous functions on S, denoted $C(S)$, be endowed with the* sup *norm. A subset F of $C(S)$ is* equicontinuous *if for every $\epsilon > 0$ there exists a $\delta > 0$ such that*

$$\|x - y\| < \delta \ \ \textit{implies} \ \ \|f(x) - f(y)\| < \epsilon, \quad \forall f \in F.$$

Theorem C.3 *(Schauder) Let S be a bounded subset of R^n, and let $C(S)$ be the space of real-valued bounded continuous functions on S, endowed with the* sup *norm. Let $F \subset C(S)$ be nonempty, closed, bounded and convex. Then if the mapping $T : F \to F$ is continuous and the family $T(F)$ is equicontinuous, T has a fixed point in F.*

A proof of this theorem can be found in (Istrătescu, 1981) and (Hutson and Pym, 1980). This is a special case of a more general theorem (also due to Schauder) which says that "every real-valued continuous function mapping a convex compact subspace of a Banach space into itself has a fixed point." The space $C(S)$ defined above is indeed one such space, and equicontinuity assures that convex compact subsets of $C(S)$ are mapped into themselves.

For interesting discussions on the Brouwer fixed point theorem and its various extensions, see Franklin (1980).

Bibliography

S. ALPERN, *Games with repeated decisions*, SIAM Journal on Control and Optimization, 26 (1988), pp. 468–477.

S. ALPERN, *Cycles in extensive form perfect information games*, Journal of Mathematical Analysis and Applications, 159 (1991), pp. 1–17.

B. ANDERSON AND J. B. MOORE, *Optimal Control*, Prentice-Hall, Englewood Cliffs, NJ, 1989.

R. B. ASH, *Real Analysis and Probability*, Academic Press, New York, NY, 1972.

J. AUBIN, *Mathematical Methods of Game and Economic Theory*, North- Holland, Amsterdam, The Netherlands, 1980.

R. J. AUMANN, *Borel structures for function spaces*, Illinois Journal of Mathematics, 5 (1961), pp. 614–630.

———, *Mixed and behavior strategies in infinite extensive games*, in Advances in Game Theory, M. Dresher, L. S. Shapley, and A. W. Tucker, eds., Princeton University Press, Princeton, NJ, 1964, pp. 627–650.

R. J. AUMANN AND M. MASCHLER, *Some thoughts on the minimax principle*, Management Science, 18 (1972), pp. 54–63.

A. BAGCHI AND G. J. OLSDER, *Linear stochastic pursuit evasion games*, Journal of Applied Mathematics and Optimization, 7 (1981), pp. 95–123.

A. V. BALAKRISHNAN, *Applied Functional Analysis*, Springer-Verlag, New York, NY, 1976.

R. B. BAPAT AND T. RAGHAVAN, *Nonnegative Matrices and Applications*, Cambridge University Press, Cambridge, England, 1997.

M. BARDI, *A boundary value problem for the minimum-time function*, SIAM J. Control and Optimization, 27 (1989), pp. 776–785.

M. BARDI AND C. SARTORI, *Convergence results for Hamilton-Jacobi equations in variable domains*, Differential and Integral Equations, 5 (1992), pp. 805–816.

M. BARDI AND P. SORAVIA, *Hamilton-Jacobi equations with singular boundary conditions on a free boundary and applications to differential games*,

Transactions of the American Mathematical Society, 325 (1991), pp. 205–229.

M. BARDI AND V. STAICU, *The Bellman equation for time-optimal control of non-controllable nonlinear systems*, Tech. Report SISSA 132/91/M, SISSA-ISAS, Strada Costiera 11, 34014 Trieste, Italy, 1991.

E. N. BARRON, L. C. EVANS, AND R. JENSEN, *Viscosity solutions of Isaacs' equations and differential games with Lipschitz controls*, Journal of Differential Equations, 53 (1984), pp. 213–233.

T. BAŞAR, *On the relative leadership property of Stackelberg strategies*, Journal of Optimization Theory and Applications, 11 (1973), pp. 655–661.

——, *A counter example in linear-quadratic games: Existence of non-linear Nash solutions*, Journal of Optimization Theory and Applications, 14 (1974), pp. 425–430.

——, *Nash strategies for M-person differential games with mixed information structures*, Automatica, 11 (1975), pp. 547–551.

——, *On the uniqueness of the Nash solution in linear-quadratic differential games*, International Journal of Game Theory, 5 (1976a), pp. 65–90.

——, *Some thoughts on saddle-point conditions and information structures in zero-sum differential games*, Journal of Optimization Theory and Applications, 18 (1976b), pp. 165–170.

——, *Existence of unique equilibrium solutions in nonzero-sum stochastic differential games*, in Differential Games and Control Theory II, E. O. Roxin, P. T. Liu, and R. Sternberg, eds., Marcel Dekker, Inc., 1977a, pp. 201–228.

——, *Informationally nonunique equilibrium solutions in differential games*, SIAM Journal on Control and Optimization, 15 (1977b), pp. 636–660.

——, *Multicriteria optimization of linear deterministic and stochastic systems: A survey and some new results.* Marmara Research Institute Publication, Applied Mathematics Division, no. 37,Gebze Kocaeli, Turkey, 1977c.

——, *Two general properties of the saddle-point solutions of dynamic games*, IEEE Transactions on Automatic Control, AC-22 (1977d), pp. 124–126.

——, *Decentralized multicriteria optimization of linear stochastic systems*, IEEE Transactions on Automatic Control, AC-23 (1978), pp. 233–243.

——, *Hierarchical decision making under uncertainty*, in Dynamic Optimization and Mathematical Economics, P. T. Liu, ed., Plenum Press, New York, NY, 1979a, pp. 205–221.

——, *Information structures and equilibria in dynamic games*, in New Trends in Dynamic System Theory and Economics, M. Aoki and A. Marzollo, eds., Academic Press, New York and London, 1979b, pp. 3–55.

——, *Stochastic stagewise Stackelberg strategies for linear-quadratic systems*, in Stochastic Control Theory and Stochastic Differential Systems, M. Kohlmann and W. Vogel, eds., Lecture Notes in Control and Information Sciences, Springer-Verlag, New York, 1979c, ch. 16, pp. 264–276.

——, *Memory strategies and a general theory for Stackelberg games with partial dynamic information*, in Proceedings of the 4th International Conference on the Analysis and Optimization of Systems, Versailles, France, December 1980a, Springer-Verlag, New York, pp. 397–415.

——, *On the existence and uniqueness of closed-loop sampled-data Nash controls in linear-quadratic stochastic differential games*, in Optimization Techniques, K. Iracki et al., eds., Lecture Notes in Control and Information Sciences, Springer-Verlag, New York, 1980b, ch. 22, pp. 193–203.

——, *Equilibrium strategies in dynamic games with multi-levels of hierarchy*, Automatica, 17 (1981a), pp. 749–754.

——, *A new method for the Stackelberg solution of differential games with sampled-data state information*, in Proceedings of the 8th IFAC World Congress, Kyoto, Japan, Pegamon Press, New York, August 1981b, pp. 139–144.

——, *On the saddle-point solution of a class of stochastic differential games*, Journal of Optimization Theory and Applications, 33 (1981c), pp. 539–556.

——, *A general theory for Stackelberg games with partial state information*, Large Scale Systems, 3 (1982), pp. 47–56.

——, *Affine incentive schemes for stochastic systems with dynamic information*, SIAM Journal on Scientific and Statistical Computing, 22 (1984), pp. 199–210.

——, ed., *Dynamic Games and Applications in Economics*, vol. 265 of Lecture Notes in Economics and Mathematical Systems, Springer-Verlag, Berlin, 1986a.

——, *A tutorial on dynamic and differential games*, in Dynamic Games and Applications in Economics, T. Başar, ed., vol. 265 of Lecture Notes in Economics and Mathematical Systems, Springer-Verlag, New York, 1986b, pp. 1–25.

——, *Relaxation techniques and asynchronous algorithms for on-line computation of noncooperative equilibria*, Journal of Economic Dynamics and Control, 71 (1987), pp. 531–549.

——, *Stochastic incentive problems with partial dynamic information and multiple levels of hierarchy*, European J. Political Economy, 5 (1989a), pp. 203–217.

——, *Time consistency and robustness of equilibria in noncooperative dynamic games*, in Dynamic Policy Games in Economics, F. Van der Ploeg and A. de Zeeuw, eds., North–Holland, Amsterdam, 1989b, pp. 9–54.

——, *A dynamic games approach to controller design: Disturbance rejection in discrete time*, IEEE Transactions on Automatic Control, AC-36 (1991a), pp. 936–952.

——, *Generalized Riccati equations in dynamic games*, in The Riccati Equation, S. Bittanti, A. Laub, and J. C. Willems, eds., Springer-Verlag, Berlin, 1991b, pp. 293–333.

——, *On the application of differential game theory in robust control design for economic systems*, in Dynamic Economic Models and Optimal Control, G. Feichtinger, ed., North–Holland, Amsterdam, 1992, pp. 171–186.

T. Başar and P. Bernhard, eds., *Differential Games and Applications*, Lecture Notes in Control and Information Sciences, vol. 119, Springer-Verlag, Berlin, 1989.

——, *H^∞-Optimal Control and Related Minimax Design Problems: A Dynamic Game Approach*, Birkhäuser, Boston, MA, 2nd ed., 1995.

T. Başar and A. Haurie, *Feedback equilibria in differential games with structural and modal uncertainties*, vol. 1 of Advances in Large Scale Systems, J. B. Cruz, Jr., ed. JAI Press Inc., Connecticut, 1984, pp. 163–201.

T. Başar and A. Haurie, eds., *Advances in Dynamic Games and Applications*, Birkhäuser, Boston, MA, 1994.

T. Başar, A. Haurie, and G. Ricci, *On the dominance of capitalists leadership in a feedback Stackelberg solution of a differential game model of capitalism*, Journal of Economic Dynamics and Control, 9 (1985), pp. 101–125.

T. Başar and Y. C. Ho, *Informational properties of the Nash solutions of two stochastic nonzero-sum games*, Journal of Economic Theory, 7 (1974), pp. 370–387.

T. Başar and S. Li, *Distributed algorithms for the computation of Nash equilibria in linear stochastic differential games*, SIAM Journal on Control and Optimization, 27 (1989), pp. 563–578.

T. Başar and M. Mintz, *On the existence of linear saddle-point strategies for a two-person zero-sum stochastic game*, in Proceedings of the IEEE 11th Conference on Decision and Control, New Orleans, LA, 1972, IEEE Computer Society Press, Los Alamitos, CA, pp. 188–192.

——, *A multistage pursuit-evasion game that admits a Gaussian random process as a maximum control policy*, Stochastics, 1 (1973), pp. 25–69.

T. Başar and G. J. Olsder, *Mixed Stackelberg strategies in continuous-kernel games*, IEEE Transactions on Automatic Control, AC-25 (1980a), pp. 307–309.

——, *Team-optimal closed-loop Stackelberg strategies in hierarchical control problems*, Automatica, 16 (1980b), pp. 409–414.

——, *Dynamic Noncooperative Game Theory*, Academic Press, London/New York, 1982. (Second printing 1989.)

T. Başar and H. Selbuz, *Properties of Nash solutions of a 2-stage nonzero-sum game*, IEEE Transactions on Automatic Control, AC-21 (1976), pp. 48–54.

——, *Closed-loop Stackelberg strategies with applications in the optimal control of multilevel systems*, IEEE Transactions on Automatic Control, AC-24 (1979a), pp. 166–179.

——, *A new approach for derivation of closed-loop Stackelberg strategies*, in Proceedings of the IEEE 17th Conference on Decision and Control, San Diego, CA, January 1979b, IEEE Computer Society Press, Los Alamitos, CA, pp. 1113–1118.

T. BAŞAR, S. J. TURNOVSKY, AND V. D'OREY, *Optimal macroeconomic policies in interdependent economies: A strategic approach*, in Dynamic Games and Applications in Economics, T. Başar, ed., Lecture Notes in Economics and Mathematical Systems, volume 265, Springer-Verlag, Berlin, 1986, pp. 134–178.

R. BELLMAN, *Dynamic Programming*, Princeton University Press, Princeton, NJ, 1957.

——, *Introduction to Matrix Analysis*, McGraw-Hill, New York, NY, 2nd ed., 1970.

A. BENSOUSSAN, *Saddle points of convex concave functionals*, in Differential Games and Related Topics, H. W. Kuhn and G. P. Szegö, eds., North-Holland, Amsterdam, The Netherlands, 1971, pp. 177–200.

——, *Points de Nash dans le cas de fonctionelles quadratiques et jeux differentiels lineaires a N personnes*, SIAM Journal on Control and Optimization, 12 (1974), pp. 237–243.

——, *Perturbation Methods in Optimal Control*, John Wiley, Gauthier-Villars, Chichester, England, 1988.

L. D. BERKOVITZ, *A variational approach to differential games*, in Advances in Game Theory, M. Dresher, L. S. Shapley, and A. W. Tucker, eds., Princeton University Press, Princeton, NJ, 1964, pp. 127–174.

——, *Optimal Control Theory*, Springer-Verlag, Berlin, 1974.

P. BERNHARD, *Condition de coin pour les jeux differentiels.* Seminaire des 21–25 Juin, Les Jeux Differentiels, Centre d'Automatique de l'Ecole Nationale Supérieure de Mines de Paris, 1971.

——, *Singular surfaces in differential games: an introduction*, in Differential Games and Applications, P. Hagedorn, H. W. Knobloch, and G. J. Olsder, eds., Springer-Verlag, Berlin, 1977, pp. 1–33. Lecture Notes in Control and Information Sciences, vol. 3.

——, *Linear-quadratic two-person zero-sum differential games: Necessary and sufficient conditions*, Journal of Optimization Theory and Applications, 27 (1979), pp. 51–69.

D. P. BERTSEKAS, *Dynamic Programming: Deterministic and Stochastic Models*, Prentice-Hall, Englewood Cliffs, NJ, 1987.

D. P. BERTSEKAS AND J. N. TSITSIKLIS, *Parallel and Distributed Computation: Numerical Methods*, Prentice-Hall, Englewood Cliffs, NJ, 1989.

D. P. Bertsekas and J. N. Tsitsiklis, *Convergence rate and termination of asynchronous interative algorithms*, in Proceedings of the 1989 International Conference on Supercomputing, Irakleion, Greece, 1989, pp. 561–470.

——, *Some aspects of parallel and distributed iterative algorithms – a survey*, Automatica, 27 (1991), pp. 3–21.

K. Binmore, *Fun and Games*, D. C. Heath and Company, Lexington, MA, 1992.

D. Blackwell and M. A. Girshick, *Theory of Games and Statistical Decisions*, John Wiley and Sons, New York, NY, 1954.

A. Blaquière, ed., *Topics in Differential Games*, North–Holland, Amsterdam, The Netherlands, 1973.

——, *Une géneralisation du concept d'optimalité et des certaines notions geometriques qui's rattachent.* Institute des Hautes Etudes de Belgique, Cahiers du centre d'études de recherche operationelle, vol. 18, no. 1–2, Bruxelles, pp. 49–61, 1976.

——, *Differential games with piece-wise continuous trajectories*, in Differential Games and Applications, P. Hagedorn, H. W. Knobloch, and G. J. Olsder, eds., Springer-Verlag, Berlin, 1977, ch. 3, pp. 34–69.

A. Blaquière, F. Gerard, and G. Leitmann, *Quantitative and Qualitative Games*, Academic Press, New York and London, 1969.

V. G. Boltyanski, *Optimal Control of Discrete Systems*, Hallsted Press, John Wiley, New York, NY, 1978.

E. Borel, *The theory of play and integral equations with skew symmetrical kernels: On games that involve chance and skill of the players*, On systems of linear forms of skew symmetric determinants and the general theory of play, trans. by L. J. Savage, Econometrica, 21 (1953), pp. 97–117.

D. Braess, *Uber eine Paradoxen aus der Verkehersplanung*, Unternehmensforschung, 12 (1968), pp. 258–268.

J. V. Breakwell, *Examples élementaires.* Séminaire des 21–25 Juin, Les Jeux Differentiels, Centre d'Automatique de l'Ecole Nationale Superieure des Mines de Paris, 1971.

——, *Some differential games with interesting discontinuities.* Internal Report, Stanford University, Stanford, CA, 1973.

——, *Zero-sum differential games with terminal payoff*, in Differential Games and Applications, P. Hagedorn, H. W. Knobloch, and G. J. Olsder, eds., Springer-Verlag, Berlin, 1977, ch. 3, pp. 70–95.

T. F. Bresnahan, *Duopoly models with consistent conjectures*, American Economic Review, 71 (1981), pp. 934–945.

R. W. Brockett, *Finite Dimensional Linear Systems*, John Wiley and Sons, New York, NY, 1970.

L. E. J. BROUWER, *Uber abbildung von mannigfaltigkeiten*, Math. Annalen., 71 (1910), pp. 97–115.

G. W. BROWN, *Iterative solutions of games by fictitious play*, in Activity Analysis of Production and Allocation, T. C. Koopmans, ed., John Wiley and Sons, New York, NY, 1951, ch. 13, pp. 374–376. Cowles Commission Monograph.

G. W. BROWN AND J. VON NEUMANN, *Solutions of games by differential equations*, in Contributions to the Theory of Games, H. W. Kuhn and A. W. Tucker, eds., Princeton University Press, Princeton, NJ, 1950, pp. 73–79. Vol. I, no. 24.

A. E. BRYSON, JR. AND Y. C. HO, *Applied Optimal Control*, Hemisphere, Washington, DC, 1975.

E. BURGER, *Einfuhrung in die Theorie der Spiele, (2 Auflage)*, Walter de Gruyter and Company, Berlin, 1966.

M. D. CANON, C. D. CULLUM, JR., AND E. POLAK, *Theory of Optimal Control and Programming*, McGraw-Hill, New York, NY, 1970.

D. H. CANSEVER AND T. BAŞAR, *A minimum sensitivity approach to incentive design problems*, Large Scale Systems, 5 (1983), pp. 233–244.

——, *On stochastic incentive control problems with partial dynamic information*, Systems & Control Letters, 6 (1985a), pp. 69–75.

——, *Optimum/near optimum incentive policies for stochastic decision problems involving parametric uncertainty*, Automatica, 21 (1985b), pp. 575–584.

J. H. CASE, *Equilibrium Points of N-person differential games*, PhD thesis, University of Michigan, Ann Arbor, MI, 1967. Department of Industrial Engineering, Tech. report no. 1967–1.

——, *Toward a theory of many player differential games*, SIAM Journal on Control and Optimization, 7 (1969), pp. 179–197.

——, *Applications of the theory of differential games to economic problems*, in Differential Games and Related Topics, H. W. Kuhn and G. P. Szegö, eds., North-Holland, Amsterdam, The Netherlands, 1971, pp. 345–371.

——, *Economics and the Competitive Process*, New York University Press, New York, NY, 1979.

D. A. CASTANON, *Equilibria in Stochastic Dynamic Games of Stackelberg Type*, PhD thesis, M.I.T. Electronics Systems Laboratory, Cambridge, MA, 1976.

D. CHAZAN AND W. MIRANKER, *Chaotic relaxation*, Linear Algebra and Applications, 2 (1969), pp. 199–222.

C. I. CHEN AND J. B. CRUZ JR., *Stackelberg solution for two-person games with biased information patterns*, IEEE Transactions on Automatic Control, AC-17 (1972), pp. 791–798.

S. A. CHIGIR, *The game problem on the dolichobrachistochrone*, PMM, 30 (1976), pp. 1003–1013.

M. D. CILETTI, *On the contradiction of the bang-bang-bang surfaces in differential games*, Journal of Optimization Theory and Applications, 5 (1970), pp. 163–169.

S. CLEMHOUT, G. LEITMANN, AND H. Y. WAN, JR., *A differential game model of oligopoly*, Journal of Cybernetics, 3 (1973), pp. 24–39.

E. A. CODDINGTON AND N. LEVINSON, *Theory of Ordinary Differential Equations*, McGraw-Hill, New York, NY, 1955.

J. E. COHEN, *The counterintuitive in conflict and cooperation*, American Scientist, (1988).

J. E. COHEN AND P. HOROWITZ, *Paradoxal behaviour of mechanical and electrical networks*, Nature, 352 (1991), pp. 699–701.

A. COURNOT, *Recherches sur les principes mathematiques de la theorie des richesses.* Hachette, Paris. (English edition published in 1960 by Kelley, New York, under the title 'Researches into the Mathematical Principles of the Theory of Wealth', trans. by N. T. Bacon, 1838.

M. CRANDALL AND P. LIONS, *Viscosity solutions of Hamilton-Jacobi equations*, Transactions of the American Mathematical Society, 277 (1983), pp. 1–42.

M. G. CRANDALL, L. C. EVANS, AND P. L. LIONS, *Some properties of viscosity solutions of Hamilton-Jacobi equations*, Transactions of the American Mathematical Society, 282 (1984), pp. 487–502.

J. B. CRUZ, JR., *Leader-follower strategies for multilevel systems*, IEEE Transactions on Automatic Control, AC-23 (1978), pp. 244–255.

G. B. DANTZIG, *Linear Programming and Extensions*, Princeton University Press, Princeton, NJ, 1963.

M. DAVIS, *On the existence of optimal policies in stochastic control*, SIAM Journal on Control and Optimization, 11 (1973), pp. 587–594.

M. DAVIS AND P. P. VARAIYA, *Dynamic programming conditions for partially observable stochastic systems*, SIAM Journal on Control and Optimization, 11 (1973), pp. 226–261.

M. A. DAVIS, *Linear Estimation and Stochastic Control*, Chapman and Hall, London, 1977.

E. DOCKNER AND G. FEICHTINGER, *Dynamic advertising and pricing in an oligopoly: A Nash equilibrium approach*, in Dynamic Games and Applications in Economics, T. Başar, ed., Lecture Notes in Economics and Mathematical Systems, volume 265, Springer-Verlag, Berlin, 1986, pp. 238–251.

J. DOYLE, K. GLOVER, P. KHARGONEKAR, AND B. FRANCIS, *State-space solutions to standard H_2 and H_∞ control problems*, IEEE Transactions on Automatic Control, AC-34 (1989), pp. 831–847.

P. Dubey and M. Shubik, *Information conditions, communications and general equilibrium*, Mathematics of Operations Research, 6 (1981), pp. 186–189.

F. Y. Edgeworth, *Paper Relating to Political Economy*, vol. 1, Macmillan, London, 1925.

A. Ehrenfeucht and J. Mycielski, *Positional strategies for mean payoff games*, International Journal of Game Theory, 8 (1979), pp. 109–113.

T. Eisele, *Nonexistence and nonuniqueness of open-loop equilibria in linear-quadratic differential games*, Journal of Optimization Theory and Applications, 37 (1982), pp. 443–468.

R. J. Elliott, *The existence of value in stochastic differential games*, SIAM Journal on Control and Optimization, 14 (1976), pp. 85–94.

——, *Feedback strategies in deterministic differential games*, in Differential Games and Applications, P. Hagedorn, H. W. Knobloch, and G. J. Olsder, eds., Springer-Verlag, Berlin, 1977, pp. 136–142. Lecture Notes in Control and Information Sciences, vol. 3.

——, *Viscosity Solutions and Optimal Control*, Longman Scientific and Technical, Wiley, New York, NY, 1987.

R. J. Elliott and N. J. Kalton, *The existence of value in differential games of pursuit and evasion*, Journal of Differential Equations, 12 (1972), pp. 504–523.

R. J. Elliott, N. J. Kalton, and L. Markus, *Saddle-points for linear differential games*, SIAM Journal on Control and Optimization, 11 (1973), pp. 100–112.

J. C. Engwerda, *On the open-loop Nash equilibrium in LQ-games*, Journal of Economic Dynamics and Control, 22 (1998), pp. 729–762.

R. A. Epstein, *The Theory of Gambling and Statistical Logic*, Academic Press, New York and London, 1967.

L. Evans and P. Souganidis, *Differential games and representation formulas for solutions of Hamilton-Jacobi-Isaacs equations*, Indiana Univ. Math. J., 33 (1984), pp. 773–797.

K. Fan, *Minimax theorems*, Proceedings of the Academy of Sciences, 39 (1953), pp. 42–47.

W. Feller, *An Introduction to Probability Theory and Its Applications*, vol. 2, John Wiley and Sons, New York, NY, 1971.

W. H. Fleming, *The convergence problem for differential games*, Journal of Mathematical Analysis and Applications, 3 (1961), pp. 102–116.

——, *The convergence problem for differential games*, in Advances in Game Theory, M. Dresher, L. S. Shapley, and A. W. Tucker, eds., Princeton University Press, Princeton, NJ, 1964, pp. 195–210. Annals of Mathematics Studies, no. 52.

——, *Optimal continuous-parameter stochastic control*, SIAM Review, 11 (1969), pp. 470–509.

W. H. Fleming and R. W. Rishel, *Deterministic and Stochastic Optimal Control*, Springer-Verlag, Berlin, 1975.

W. H. Fleming and H. M. Soner, *Controlled Markov Processes and Viscosity Solutions*, Applications of Mathematics, vol. 25, Springer-Verlag, 1993.

J. G. Foreman, *The princess and the monster on the circle*, SIAM Journal on Control and Optimization, 15 (1977), pp. 841–856.

M. Fox and G. S. Kimeldorf, *Noisy duels*, SIAM Journal of Applied Mathematics, 17 (1969), pp. 353–361.

B. A. Francis, *A Course in H_∞ Control Theory*, vol. 88 of Lecture Notes in Control and Information Sciences, Springer-Verlag, New York, NY, 1987.

J. Franklin, *Methods of Mathematical Economics*, Springer-Verlag, Berlin, 1980.

A. Friedman, *Differential Games*, John Wiley and Sons, New York, NY, 1971.

——, *Stochastic differential games*, Journal of Differential Equations, 11 (1972), pp. 79–108.

J. W. Friedman, *Oligopoly and the Theory of Games*, North–Holland, Amsterdam, The Netherlands, 1977.

D. Fudenberg and J. Tirole, *Game theory*, MIT Press, Cambridge, MA, 1991.

Z. Gajic, D. Petkovski, and X. Shen, *Singularly Perturbed and Weakly Coupled Linear Control Systems: A Recursive Approach*, Springer-Verlag, New York, NY, 1990.

S. Gal, *Search Games*, Academic Press, New York, 1980.

W. M. Getz and M. Pachter, *Two-target pursuit-evasion differential games in the plane*, Journal of Optimization Theory and Applications, 34 (1981), pp. 383–403.

I. I. Gikhman and A. V. Skorohod, *Stochastic Differential Equations*, Springer-Verlag, Berlin, 1972.

I. L. Glicksberg, *Minimax theorem for upper and lower semicontinuous payoffs.* Rand Corporation Research Memorandum RM-478, Santa Monica, CA, 1950.

C. Gonzaga, *Path-following methods for linear programming*, SIAM Review, 34 (1992), pp. 167–224.

M. Green and D. Limebeer, *Linear Robust Control*, Prentice-Hall, Englewood Cliffs, NJ, 1995.

P. Hagedorn, H. W. Knobloch, and G. J. Olsder, eds., *Differential Games and Applications*, vol. 3 of Lecture Notes in Control and Information Sciences, Springer-Verlag, Berlin, 1977.

O. HAJEK, *Pursuit Games*, Academic Press, New York and London, 1975.

——, *Discontinuous differential equations, I and II*, Journal of Differential Equations, 32 (1979), pp. 149–185.

A. HALANAY, *Differential games with delay*, SIAM Journal on Control and Optimization, 6 (1968), pp. 579–593.

J. HALE, *Theory of Functional Differential Equations*, Springer-Verlag, Berlin, 1977.

R. P. HÄMÄLÄINEN, *Nash and Stackelberg solutions to general linear-quadratic two-player difference games*, Systems Theory Laboratory B-29, Helsinki University of Technology publication, Espoo, Finland, 1976.

R. P. HÄMÄLÄINEN AND H. K. EHTAMO, eds., *Differential Games: Developments in Modelling and Computation*, vol. 156, Springer-Verlag, New York, August 1991.

H. HERMES AND J. LASALLE, *Functional analysis and time optimal control*, Academic Press, New York, 1969.

Y. C. HO, *Differential games, dynamic optimization and generalized control theory, survey paper*, Journal of Optimization Theory and Applications, 6 (1970), pp. 179–209.

Y. C. HO, A. E. BRYSON, JR., AND S. BARON, *Differential games and optimal pursuit-evasion strategies*, IEEE Transactions on Automatic Control, AC-10 (1965), pp. 385–389.

Y. C. HO, P. B. LUH, AND G. J. OLSDER, *A control-theoretic view on incentives*, in Proceedings of the Fourth International Conference on Analysis and Optimization of Systems, A. Bensoussan and J. L. Lions, eds., Springer-Verlag, Berlin, 1980, pp. 359–383. Lecture Notes in Control and Information Sciences, vol. 28.

——, *A control-theoretic view on incentives*, Automatica, 18 (1982), pp. 167–180.

Y. C. HO AND S. K. MITTER, *Directions in Large Scale Systems*, Plenum Press, New York, NY, 1976.

Y.-C. HO AND G. OLSDER, *Aspects of the Stackelberg game problem,— incentive, bluff and hierarchy*, in Proceedings of the 8th IFAC world congress, Kyoto, Japan, 1981, IFAC, Laxenburg, 1982, pp. 1359–1363.

Y.-C. HO AND G. J. OLSDER, *Differential games: Concepts and applications*, in Mathematics of Conflict, M. Shubik, ed., North–Holland, Amsterdam, The Netherlands, 1983, pp. 127–186.

N. HOWARD, *Paradoxes of Rationality: Theory of Metagames and Political Behavior*, The MIT Press, Cambridge, MA, 1971.

V. HUTSON AND J. S. PYM, *Applications of Functional Analysis and Operator Theory*, Operator Theory, Academic Press, London, 1980.

M. D. Intriligator, *Mathematical Optimization and Economic Theory*, Prentice-Hall, Englewood Cliffs, NJ, 1971.

R. Isaacs, *Differential games I, II, III, IV.* Rand Cooperation Research Memorandum RM-1391, 1399, 1411, 1468, Santa Monica, CA, 1954-1956.

——, *Differential games: Their scope, nature and future*, Journal of Optimization Theory and Applications, 3 (1969), pp. 283–295.

——, *Differential Games*, Kruger Publishing Company, Huntington, NY, 2nd ed., 1975. (First edition: Wiley, NY, 1965.)

J. Isbell, *Finitary games*, in Annals of Mathematical Studies, vol. 39, Princeton University Press, Princeton, NJ, 1957, pp. 79–95.

H. Ishii, *Perron's method for Hamilton-Jacobi equations*, Duke Math. J, 55 (1987), pp. 369–384.

——, *A boundary value problem of the dirichlet type for Hamilton–Jacobi equations*, Ann. Sc. Norm. Sup. Pisa, XVI (1989), pp. 105–135.

V. I. Istrăţescu, *Fixed Point Theory*, D. Reidel Publishing Company, Dordrecht, Holland, 1981.

D. H. Jacobson, *On values and strategies for infinite-time linear quadratic games*, IEEE Transactions on Automatic Control, AC-22 (1977), pp. 490–491.

S. Jorgensen, *Optimal dynamic pricing in an oligopolistic market: A survey*, in Dynamic Games and Applications in Economics, T. Başar, ed., Lecture Notes in Economics and Mathematical Systems, volume 265, Springer-Verlag, Berlin, 1986, pp. 179–237.

M. C. Joshi and R. K. Bose, *Some Topics in Nonlinear Functional Analysis*, John Wiley and Sons, New York, NY, 1985.

T. Kailath, *Linear Systems*, Prentice-Hall, Englewood Cliffs, NJ, 1980.

S. Kakutani, *Generalization of Brouwer's fixed point theorem*, Duke Journal of Mathematics, 8 (1941), pp. 457–459.

M. I. Kamien and N. L. Schwartz, *Conjectural variations*, Canadian J. Economics, 16 (1983), pp. 191–211.

S. Karlin, *Matrix Games, Programming and Mathematical Economics*, vols. I and II, Addison-Wesley, Reading, MA, 1959.

N. Karmarkar, *A new polynomial-time algorithm for linear programming*, Combinatorica, 4 (1984), pp. 373–395.

P. P. Khargonekar, *State-space H_∞ control theory and the LQG control problem*, in Mathematical System Theory: The Influence of R. E. Kalman, A. C. Antoulas, ed., Springer-Verlag, Berlin, 1991, pp. 159–176.

P. P. Khargonekar, K. N. Nagpal, and K. R. Poolla, *H^∞ control with transients*, SIAM Journal on Control and Optimization, 29 (1991), pp. 1373–1393.

E. Kohlberg and J.-F. Mertens, *On the strategic stability of equilibria*, Econometrica, 54 (1986), pp. 1003–1037.

N. Krasovskii and A. Subbotin, *Game-theoretical Control Problems*, Springer-Verlag, New York, 1988.

D. Kreps, *Game Theory and Economic Modelling*, Oxford University Press, Oxford, England, 1990.

D. M. Kreps and R. Wilson, *Sequential equilibria*, Econometrica, 50 (1982), pp. 863–894.

K. Kuga, *Brouwer's fixed point theorem: An alternative proof*, SIAM Journal of Mathematical Analysis, 5 (1974), pp. 893–897.

H. W. Kuhn, *Extensive games and the problems of information*, in vol. 29 of Contributions to the Theory of Games, Princeton University Press, Princeton, NJ, 1953, pp. 193–216. H. W. Kuhn and A. W. Tucker, eds.

H. W. Kuhn and G. P. Szegö, eds., *Differential Games and Related Topics*, North–Holland, Amsterdam, The Netherlands, 1971.

H. W. Kuhn and A. W. Tucker, eds., *Contributions to the Theory of Games*, vol. I, no. 24, Annals of Mathematical Studies, Princeton University Press, Princeton, NJ, 1950.

——, eds., *Contributions to the Theory of Games*, vol. I, no. 28, Annals of Mathematical Studies, Princeton University Press, Princeton, NJ, 1953.

P. R. Kumar and T. H. Shiau, *Existence of value and randomized strategies in zero-sum discrete-time stochastic dynamic games*, SIAM Journal on Control and Optimization, 19 (1981), pp. 617–634.

F. Kydland, *Noncooperative and dominant player solutions in discrete dynamic games*, International Journal of Economic Review, 16 (1975), pp. 321–335.

F. Kydland and E. Prescott, *Rules rather than discretion: The inconsistency of optimal plans*, Journal of Political Economy, 85 (1977), pp. 473–491.

S. Lakshmivarahan and K. S. Narendra, *Learning algorithms for two-person zero-sum stochastic games with incomplete information*, Mathematics of Operations Research, 6 (1981), pp. 379–286.

G. Leitmann, *Cooperative and Noncooperative Many Player Differential Games*, vol. 190 of CISM Monograph, Springer-Verlag, Vienna, 1974.

——, *Many player differential games*, in Differential Games and Applications, P. Hagedorn, H. W. Knobloch, and G. J. Olsder, eds., Springer-Verlag, Berlin, 1977, pp. 153–171. Lecture Notes in Control and Information Sciences, vol. 3.

——, *On generalized Stackelberg strategies*, Journal of Optimization Theory and Applications, 26 (1978).

G. LEITMANN AND H. Y. WAN, JR., *Macro-economic stabilization policy for an uncertain dynamic economy*, in New Trends in Dynamic System Theory and Economics, M. Aoki and A. Marzollo, eds., Academic Press, London and New York, 1979, pp. 105–136.

C. E. LEMKE AND J. F. HOWSON, *Equilibrium points of bi-matrix games*, SIAM Journal on Applied Mathematics, 12 (1964), pp. 413–423.

J. LEVINE, *Two-person zero-sum differential games with incomplete information–a Bayesian model*, in Differential Games and Control Theory III, P. T. Liu and E. O. Roxin, eds., Plenum Press, New York, NY, 1979, pp. 119–151. Proceedings of the 3rd Kingston Conference 1978, Part A. Lecture Notes in Pure and Applied Mathematics, vol. 44.

R. E. LEVITAN AND M. SHUBIK, *Noncooperative equilibrium and strategy spaces in an oligopolistic market*, in Differential Games and Related Topics, H. W. Kuhn and G. P. Szegö, eds., North–Holland, Amsterdam, The Netherlands, 1971, pp. 429–448.

J. LEWIN, *The bang-bang-bang problem revisited*, Journal of Optimization Theory and Applications, 18 (1976), pp. 429–432.

——, *Differential Games: Theory and Methods for Solving Game Problems with Singular Surfaces.* Springer-Verlag, New York, 1994.

J. LEWIN AND G. J. OLSDER, *Conic surveillance evasion*, Journal of Optimization Theory and Applications, 19 (1979), pp. 107–125.

S. LI AND T. BAŞAR, *Distributed algorithms for the computation of noncooperative equilibria*, Automatica, 23 (1987), pp. 523–533.

D. LIMEBEER, B. ANDERSON, P. KHARGONEKAR, AND M. GREEN, *A game theoretic approach to H_∞ control for time-varying systems*, SIAM J. Control Optim., 30 (1992), pp. 262–283.

P. LIONS, *Generalized Solutions of Hamilton-Jacobi Equations*, Pitman Advanced Publishing Program, 1982.

M. LOÈVE, *Probability Theory*, Van Nostrand, Princeton, NJ, 3rd ed., 1963.

R. D. LUCE AND H. RAIFFA, *Games and Decisions*, John Wiley and Sons, New York, NY, 1957.

D. G. LUENBERGER, *Optimization by Vector Space Methods*, John Wiley and Sons, New York, NY, 1969.

——, *Introduction to Linear and Nonlinear Programming*, Addison-Wesley, Reading, MA, 1973.

D. L. LUKES, *Equilibrium feedback control in linear games with quadratic costs*, SIAM Journal on Control and Optimization, 9 (1971), pp. 234–252.

D. L. LUKES AND D. L. RUSSELL, *A global theory for linear quadratic differential games*, Journal of Mathematical Analysis and Applications, 33 (1971), pp. 96–123.

E. F. MAGEIROU, *Values and strategies for infinite duration linear quadratic games*, IEEE Transactions on Automatic Control, AC-21 (1976), pp. 547–550.

A. MAITRA AND T. PARTHASARATHY, *On stochastic games*, Journal of Optimization Theory and Applications, 5 (1970), pp. 289–300.

M. MARCUS AND H. MINC, *A Survey of Matrix Theory and Matrix Inequalities*, Allyn and Bacon, Inc., Boston, MA, 1964.

J. MCKINSEY, *Introduction to the Theory of Games*, McGraw-Hill, New York, NY, 1952.

J. MEDANIĆ, *Closed-loop Stackelberg strategies in linear-quadratic problems*, in Proceedings of the 1977 JACC, San Francisco, CA, 1977, pp. 1324–1329.

J. F. MERTENS, *Formulation of Bayesian analysis for games with incomplete information*, International Journal of Game Theory, 14 (1985), pp. 1–29.

——, *Stable equilibria – a reformulation. Part II Discussion of the definition and further results*, Mathematics of Operations Research, 16 (1991), pp. 694–753.

A. W. MERZ, *The homicidal chauffer–a differential game*, Tech report, Guidance and Control Laboratory 418, Stanford University, Stanford, CA, 1971.

——, *Optimal evasive maneuvers in maritime collision avoidance*, Navigation, 20 (1973), pp. 144–152.

——, *A differential game solution to the coplanar tail-chase aerial combat problem.* Tech report NASA-Cr-137809, NASA Langley Research Center, Hampton, VA, 1976.

T. MILOH AND S. D. SHARMA, *Maritime collision avoidance as a differential game*, Berich 329, Institut für Schiffbau der Universität Hamburg, 1976.

R. B. MYERSON, *Refinement of the Nash equilibrium concept*, International Journal of Game Theory, 7 (1978), pp. 73–80.

——, *Game Theory: Analysis of Conflict*, Harvard University Press, Cambridge, MA, 1991.

J. NASH, *Noncooperative games*, Annals of Mathematics, 54 (1951), pp. 286–295.

H. NIKAIDO AND K. ISODA, *Note on non-cooperative convex games*, Pacific Journal of Mathematics, 5 (1955), pp. 807–815.

K. OKUGUCHI, *Expectations and Stability in Oligopoly Models*, Lecture Notes in Economics and Mathematical Systems, Springer-Verlag, Berlin, 1976, ch. 138.

G. OLSDER, *A critical analysis of a new equilibrium concept*, in Axiomatics and Pragmatics of Conflict Analysis, J. Paeling and P. H. Vossen, eds., Gower Pub. Co., Aldershot, England, 1987, pp. 80–100.

G. J. Olsder, *Some thoughts about simple advertising models as differential games and the structure of coalitions*, in Directions in Large Scale Systems, Y. C. Ho and S. K. Mitter, eds., Plenum Press, New York, 1976, pp. 187–206.

——, *Information structures in differential games*, in Differential Games and Control Theory II, E. O. Roxin, P. T. Liu, and R. L. Sternberg, eds., Marcel Dekker, New York, NY, 1977a, pp. 99–136.

——, *On observation costs and information structures in stochastic differential games*, in Differential Games and Applications, P. Hagedorn, H. W. Knobloch, and G. J. Olsder, eds., Springer-Verlag, Berlin, 1977b, ch. 3, pp. 172–185.

G. J. Olsder and J. V. Breakwell, *Role determination in an aerial dogfight*, International Journal of Game Theory, 3 (1974), pp. 47–66.

G. J. Olsder and G. P. Papavassilopoulos, *About when to use the searchlight*, Journal of Mathematical Analysis and Applications, 136 (1988a), pp. 466–478.

——, *A Markov chain with dynamic information*, Journal of Optimization Theory and Applications, 59 (1988b), pp. 467–486.

G. J. Olsder and J. Walter, *Collision avoidance of ships*, in Proceedings 8th IFIP Conference on Optimization Techniques, J. Stoer, ed., Springer-Verlag, Berlin, 1978, ch. 6, pp. 264–271.

G. Owen, *Game Theory*, Saunders, Philadelphia, PA, 1968.

——, *Existence of equilibrium pairs in continuous games*, International Journal of Game Theory, 5 (1974), pp. 97–105.

——, *Game Theory*, Academic Press, New York, NY, 1982.

M. Pachter and Y. Yavin, *A stochastic homicidal chauffeur pursuit-evasion differential game*, Technical Report TWISK 95, WNNR, CSIR, Pretoria, South Africa, 1979.

Z. Pan and T. Başar, *H^∞-optimal control for singularly perturbed systems. Part I: Perfect state measurements*, Automatica, 29 (1993), pp. 401–423.

——, *H^∞-optimal control for singularly perturbed systems. Part II: Imperfect state measurements*, IEEE Transactions on Automatic Control, AC-39 (1994a), pp. 280–299.

——, *H^∞-optimal control of singularly perturbed systems with sampled-state measurements*, in Advances in Dynamic Games and Applications, T. Başar and A. Haurie, eds., Birkhäuser, Boston, MA, 1994b, pp. 23–55.

G. P. Papavassilopoulos and J. B. Cruz, Jr., *Nonclassical control problems and Stackelberg games*, IEEE Transactions on Automatic Control, AC-24 (1979a), pp. 155–166.

——, *On the existence of solutions to coupled matrix Riccati differential equations in linear quadratic Nash games*, IEEE Transactions on Automatic Control, AC-24 (1979b), pp. 127–129.

——, *Sufficient conditions for Stackelberg and Nash strategies with memory*, Journal of Optimization Theory and Applications, 31 (1980), pp. 233–260.

G. P. PAPAVASSILOPOULOS, J. V. MEDANIĆ, AND J. B. CRUZ, JR., *On the existence of Nash strategies and solutions to coupled Riccati equations in linear-quadratic games*, Journal of Optimization Theory and Applications, 28 (1979), pp. 49–76.

G. P. PAPAVASSILOPOULOS AND G. J. OLSDER, *On the linear-quadratic, closed-loop, no-memory Nash game*, Journal of Optimization Theory and Applications, 42 (1984), pp. 551–560.

A. PAPOULIS, *Probability, Random Variables and Stochastic Processes*, McGraw-Hill, New York, NY, 1965.

T. PARTHASARATHY AND T. RAGHAVAN, *Some Topics in Two-person Games*, Elsevier, New York, NY, 1971.

T. PARTHASARATHY AND M. STERN, *Markov games – a survey*, in Differential Games and Control Theory II, E. O. Roxin, P. T. Liu, and R. L. Sternberg, eds., Marcel Dekker, New York, NY, 1977, pp. 1–46.

W. Y. PENG AND T. L. VINCENT, *Some aspects of aerial combat*, AIAA Journal, 13 (1975), pp. 7–11.

R. S. PINDYCK, *Optimal economic stabilization policies under decentralized control and conflicting objectives*, IEEE Transactions on Automatic Control, AC-22 (1977), pp. 517–530.

M. POHJOLA, *Applications of dynamic game theory to macroeconomics*, in Dynamic Games and Applications in Economics, T. Başar, ed., Lecture Notes in Economics and Mathematical Systems, volume 265, Springer-Verlag, Berlin, 1986, pp. 103–133.

L. S. PONTRYAGIN, *Linear differential games, I*, Soviet Math. Doklady, 8 (1967), pp. 769–771.

L. S. PONTRYAGIN, V. G. BOLTYANSKII, R. V. GAMKRELIDZE, AND E. F. MISHCHENKO, *The Mathematical Theory of Optimal Processes*, Interscience Publishers, New York, NY, 1962.

W. POUNDSTONE, *Prisoner's Dilemma*, Doubleday, New York, NY, 1992.

O. POURTALLIER AND B. TOLWINSKI, *Discretization of Isaacs' equation: A convergence result*, Tech. report, INRIA, Sophia-Antipolis, France, 1992.

R. RADNER, *Monitoring cooperative agreements in a repeated principal-agent relationship*, Econometrica, 49 (1981), pp. 1127–1148.

——, *Repeated principal-agent games with discounting*, Econometrica, 53 (1985), pp. 1173–1198.

T. Raghavan and J. Filar, *Algorithms for Stochastic Games – A Survey*, ZOR – Methods and Models of Operations Research, 35 (1991), pp. 437–472.

W. T. Reid, *Riccati Differential Equations*, Academic Press, New York and London, 1972.

R. Restrepo, *Tactical problems involving several actions*, in Contributions to the Theory of Games, M. Dresher, A. W. Tucker, and P. Wolfe, eds., Princeton University Press, Princeton, NJ, 1957, ch. III, pp. 313–335.

J. Robinson, *An iterative method of solving a game*, Annals of Mathematics, 54 (1951), pp. 296–301.

J. B. Rosen, *Existence and uniqueness of equilibrium points for concave n-person games*, Econometrica, 33 (1965), pp. 520–534.

E. O. Roxin, *The axiomatic approach in differential games*, Journal of Optimization Theory and Applications, 3 (1969), pp. 153–163.

——, *Differential games with partial differential equations*, in Differential Games and Applications, P. Hagedorn, K. W. Knobloch, and G. J. Olsder, eds., Springer-Verlag, Berlin, 1977, vol. 3, pp. 186–204. Lecture Notes in Control and Information Sciences.

W. Rupp, *ϵ-Gleichgewichtspunkte in n-Personenspielen*, in Mathematical Economics and Game Theory; Essays in Honor of Oskar Morgenstern, R. Henn and O. Moeschlin, eds., Springer-Verlag, Berlin, 1977, pp. 128–138.

H. Sagan, *Introduction to the Calculus of Variations*, McGraw–Hill, New York, NY, 1969.

R. C. Scalzo, *n-person linear quadratic differential games with constraints*, SIAM Journal on Control, 12 (1974), pp. 419–425.

H. Scarf, *The approximation of fixed points of a continuous mapping*, SIAM Journal on Applied Mathematics, 15 (1967), pp. 1328–1343.

J. Schauder, *Der Fixpunktsatz in Funktionalraumen*, Studia Mathematica, 2 (1930), pp. 171–180.

W. E. Schmitendorf, *Existence of optimal open-loop strategies for a class of differential games*, Journal of Optimization Theory and Applications, 5 (1970), pp. 363–375.

R. Selten, *Reexamination of the perfectness concept for equilibrium points in extensive games*, International Journal of Game Theory, 4 (1975), pp. 25–55.

L. S. Shapley, *Stochastic games*, in Proceedings of the National Academy of Sciences, vol. 39, 1953, pp. 1095–1100.

——, *A note on the Lemke-Howson algorithm. Pivoting and extensions*, Mathematical Programming Study, 1 (1974), pp. 175–189.

L. D. Sharma, *On ship maneuverability and collision avoidance*. International report, Institut für Schiffabu der Universität Hamburg, 1976.

M. SHUBIK, *Game Theory in the Social Sciences*, MIT Press, Cambridge, MA, 1983.

M. SIMAAN AND J. B. CRUZ, JR., *Additional aspects of the Stackelberg strategy in nonzero sum games*, Journal of Optimization Theory and Applications, 11 (1973a), pp. 613–626.

——, *On the Stackelberg strategy in nonzero sum games*, Journal of Optimization Theory and Applications, 11 (1973b), pp. 533–555.

R. R. SINGLETON AND W. F. TYNDAL, *Games and Programs*, Freeman, San Francisco, CA, 1974.

M. SION, *On general minimax theorems*, Pacific Journal of Mathematics, 8 (1958), pp. 171–176.

M. SION AND P. WOLFE, *On a game without a value*, in Contributions to the Theory of Games, M. Dresher, A. W. Tucker, and P. Wolfe, eds., Princeton University Press, Princeton, NJ, 1957, ch. III, pp. 209–306.

D. R. SMART, *Fixed Point Theorems*, Cambridge University Press, London, 1974.

M. J. SOBEL, *Noncooperative stochastic games*, Ann. of Math. Statist., 42 (1971), pp. 1930–1935.

R. SRIKANT AND T. BAŞAR, *Iterative computation of noncooperative equilibria in nonzero-sum differential games with weakly coupled players*, Journal of Optimization Theory and Applications, 71 (1991), pp. 137–168.

——, *Sequential decomposition and policy iteration schemes for M-player games with partial weak coupling*, Automatica, 28 (1992), pp. 95–106.

V. STAICU, *Minimal time function and viscosity solutions*, Journal of Optimization Theory and Applications, 60 (1989), pp. 81–91.

A. W. STARR AND Y. C. HO, *Further properties of nonzero-sum differential games*, Journal of Optimization Theory and Applications, 3 (1969a), pp. 207–219.

——, *Nonzero-sum differential games*, Journal of Optimization Theory and Applications, 3 (1969b), pp. 184–206.

A. A. STOORVOGEL, *The H_∞ Control Problem: A State Space Approach*, Prentice-Hall, New York, 1992.

J. SZEP AND F. FORGO, *Introduction to the Theory of Games*, D. Reidel Publishing Co, Boston, MA, 1985.

S. H. TIJS, *Semi-infinite and infinite matrix games and bimatrix games*, PhD thesis, University of Nijmegen, The Netherlands, 1975.

B. TOLWINSKI, *Numerical_solution of N-person non-zero-sum differential games*, Control and Cybernetics, 7 (1978a), pp. 37–50.

——, *On the existence of Nash equilibrium points for differential games with linear and nonlinear dynamics*, Control and Cybernetics, 7 (1978b), pp. 57–69.

——, *Stackelberg solution of dynamic game with constraints*, in Proceedings of the IEEE 19th Conference on Decision and Control, Albuquerque, NM, 1980, IEEE Computer Society Press, Los Alamitos, CA.

——, *Closed-loop Stackelberg solution to multi-stage linear-quadratic games*, Journal of Optimization Theory and Applications, 35 (1981a), pp. 485–502.

——, *Equilibrium solutions for a class of hierarchical games*, in Applications of Systems Theory to Economics, Management and Technology, J. Outenbaum and M. Niergodka, eds., PWN, Warsaw, Poland, 1981b.

J. N. TSITSIKLIS, *On the stability of asynchronous iterative processes*, Math. Systems Theory, 20 (1987), pp. 137–153.

——, *A comparison of Jacobi and Gauss-Seidel parallel iterations*, Applied Mathematical Letters, 2 (1989), pp. 167–170.

S. J. TURNOVSKY, T. BAŞAR, AND V. D'OREY, *Dynamic strategic monetary policies and coordination in interdependent economics*, The American Economic Review, 78 (1988), pp. 341–361.

K. UCHIDA, *On the existence of Nash equilibrium point in n-person nonzero-sum stochastic differential games*, SIAM Journal on Control and Optimization, 16 (1978), pp. 142–149.

——, *A note on the existence of a Nash equilibrium point in stochastic differential games*, SIAM Journal on Control and Optimization, 17 (1979), pp. 1–4.

K. UCHIDA AND M. FUJITA, *On the central controller: Characterizations via differential games and LEQG control problems*, Systems & Control Letters, 13 (1989), pp. 9–13.

E. VAN DAMME, *A relation between perfect equilibria in extensive games and proper equilibria in normal form*, International Journal of Game Theory, 13 (1984), pp. 1–13.

——, *Stability and Perfection of Equilibria*, Springer-Verlag, New York, 1987.

——, *Stable equilibria and forward induction*, Journal of Economic Theory, 48 (1989), pp. 476–496.

P. P. VARAIYA, *The existence of solutions to a differential game*, SIAM Journal on Control, 5 (1967), pp. 153–162.

——, *N-person nonzero-sum differential games with linear dynamics*, SIAM Journal on Control, 8 (1970), pp. 441–449.

——, *N-player stochastic differential games*, SIAM Journal on Control and Optimization, 14 (1976), pp. 538–545.

P. P. VARAIYA AND J. G. LIN, *Existence of saddle-point in differential games*, SIAM Journal on Control, 7 (1969), pp. 141–157.

M. VIDYASAGAR, *A new approach to N-person, nonzero-sum, linear differential games*, Journal of Optimization Theory and Applications, 18 (1976), pp. 171–175.

J. A. VILLE, *Sur la théorie générale des jeux ou intervient l'habilitédes joueurs.* In E. Borel, Traite du calcul des probabilites et de ses applications IV, 1938, pp. 105–113.

T. L. VINCENT AND W. Y. PENG, *Ship collision avoidance. Workshop on differential games*, Naval Academy, Annapolis, MO, 1973.

T. L. VINCENT, D. L. STRICHT, AND W. Y. PENG, *Aircraft missile avoidance*, Journal of Operations Research, 24 (1976).

J. VON NEUMANN, *Zur theorie der gesellschaftspiele*, Mathematische Annalen, 100 (1928), pp. 295–320.

——, *Uber ein okonomisches gleichungssytem und eine verallgemeinerung des Brouwerschen fixpunktsatzes*, Ergebnisse eines Mathematik Kolloquiums, 8 (1937), pp. 73–83.

——, *A numerical method to determine optimum strategy*, Naval Research Logistics Quarterly, 1 (1954), pp. 109–115.

J. VON NEUMANN AND O. MORGENSTERN, *Theory of Games and Economic Behavior*, Princeton University Press, Princeton, NJ, 2nd ed., 1947.

H. VON STACKELBERG, *Marktform und Gleichgewicht.* Springer-Verlag, Vienna. (An English translation appeared in 1952 entitled *The Theory of the Market Economy*, published by Oxford University Press, Oxford, England, 1934.

N. H. VOROB'EV, *Game Theory*, Springer-Verlag, Berlin, 1977.

A. WALD, *Generalization of a theorem by von Neumann concerning zero-sum two-person games*, Annals of Mathematics, 46 (1945), pp. 281–286.

J. WARGA, *Optimal Control of Differential and Functional Equations*, Academic Press, New York and London, 1972.

A. WASHBURN, *Deterministic graphical games*, Journal of Mathematical Analysis and Applications, 153 (1990), pp. 84–96.

R. L. WEIL, *Game theory and eigensystems*, SIAM Review, 10 (1968), pp. 360–367.

J. C. WILLEMS, *Least squares stationary optimal control and the algebraic Riccati equation*, IEEE Transactions on Automatic Control, AC-16 (1971), pp. 621–634.

J. D. WILLIAMS, *The Compleat Strategyst*, McGraw-Hill, New York, NY, 1954.

J. L. WILLMAN, *Formal solutions for a class of stochastic pursuit-evasion games*, IEEE Transactions on Automatic Control, AC-14 (1969), pp. 504–509.

D. J. WILSON, *Differential games with no information*, SIAM Journal on Control and Optimization, 15 (1977), pp. 233–246.

D. WISHART AND G. J. OLSDER, *Discontinuous Stackelberg solutions*, International Journal of Systems Science, 10 (1979), pp. 1359–1368.

H. J. WITSENHAUSEN, *On the relations between the values of a games and its information structure*, Information and Control, 19 (1971).

H. S. WITSENHAUSEN, *On information structures, feedback and causality*, SIAM Journal on Control, 9 (1971), pp. 149–160.

——, *Alternatives to the tree model for extensive games*, in The Theory and Applications of Differential Games, J. D. Grote, ed., Reidel Publishing Company, The Netherlands, 1975, pp. 77–84.

E. WONG AND B. E. HAJEK, *Stochastic Processes in Engineering Systems*, Springer-Verlag, New York, NY, 1985.

L. A. ZADEH, *The concepts of system, aggregate and state in system theory*, in System Theory, L. A. Zadeh and E. Polak, eds., McGraw-Hill, New York, NY, 1969, pp. 3–42.

G. ZAMES, *Feedback and optimal sensitivity: Model reference transformation, multiplicative seminorms and approximate inverses*, IEEE Transactions on Automatic Control, AC-26 (1981), pp. 301–320.

Y. P. ZHENG AND T. BAŞAR, *Existence and derivation of optimal affine incentive schemes for Stackelberg games with partial information: A geometric approach*, International Journal of Control, 35 (1982), pp. 997–1012.

Y. P. ZHENG, T. BAŞAR, AND J. B. CRUZ, JR., *Incentive Stackelberg strategies for deterministic multi-stage decision processes*, IEEE Transactions on Systems, Man and Cybernetics, SMC-14 (1984), pp. 10–24.

Corollaries, Definitions, Examples, Lemmas, Propositions, Remarks and Theorems

Chapter 1

Chapter 2

Propositions

Remarks

Theorems

Chapter 3

Corollaries

Definitions

Chapter 4

Chapter 5

Chapter 6

Chapter 7

Chapter 8

Appendix C

Index